中国植物病理学会
2016年学术年会论文集

◎ 彭友良 王源超 主编

Proceedings of the Annual Meeting of Chinese Society for Plant Pathology (2016)

中国农业科学技术出版社

图书在版编目（CIP）数据

中国植物病理学会 2016 年学术年会论文集 / 彭友良，王源超主编 .—北京：中国农业科学技术出版社，2016. 7

ISBN 978 – 7 – 5116 – 2654 – 7

Ⅰ. ①中…　Ⅱ. ①彭… ②王…　Ⅲ. ①植物病理学 – 学术会议 – 文集　Ⅳ. ①S432. 1 – 53

中国版本图书馆 CIP 数据核字（2016）第 154249 号

责任编辑　姚　欢　邹菊华
责任校对　马广洋

出 版 者　中国农业科学技术出版社
北京市中关村南大街 12 号　邮编：100081
电　　话　(010)82106636(发行部)　(010)82106631(编辑室)
(010)82109703(读者服务部)
传　　真　(010) 82106631
网　　址　http://www.castp.cn
经 销 者　各地新华书店
印 刷 者　北京富泰印刷有限责任公司
开　　本　889 mm × 1 194 mm　1/16
印　　张　35
字　　数　1000 千字
版　　次　2016 年 7 月第 1 版　2016 年 7 月第 1 次印刷
定　　价　100. 00 元

《中国植物病理学会 2016 年学术年会论文集》

编辑委员会

前　言

由中国植物病理学会主办、南京农业大学植物保护学院等单位承办的中国植物病理学会2016年学术年会将于2016年8月5—9日在江苏省南京市举行。会议主题为“植物病理学与创新发展”，交流我国植物病理学科理论与实践方面的研究进展。

自2016年2月发布第一轮通知，国内外同仁踊跃报名与投稿，截至6月15日，共收到论文和摘要500余篇。大会论文编辑组对这些论文和摘要进行了初步整理和编辑，从论文和摘要的题目看，涵盖了植物病理学科的各个方面，尤其是真菌与真菌病害、抗病性与分子互作、病毒及病害防治方面成果颇多，基本反映了我国植物病理学科的发展现状和研究水平。

本论文集收录的论文和摘要数量多，多数论文和摘要集中在6月5日前后提交，编辑工作量大，时间紧迫，我们抽调了6位骨干教师参与了论文的分类和初步编辑工作，本着尊重作者原意和文责自负的原则，对论文内容未做改动，只是按照会议论文及摘要的格式要求进行了统一整理与规范，保留了作者原有的写作风貌。尽管力求严谨，由于时间仓促和经验不足，对于论文集可能的分类与格式的不妥之处，恳请作者和读者谅解。

大会筹办过程中，得到了中国植物病理学会、南京农业大学、江苏省植物病理学会、江苏省农业科学院植物保护研究所、江苏省科学技术协会等单位及众多专家的支持，在此我们表示由衷感谢！

最后，谨以此论文集祝贺中国植物病理学会2016年学术年会召开，预祝大会圆满成功！

编　者

2016年7月

目　录

第一部分　真　菌

第二部分　卵　菌

第三部分 病 毒

第四部分　细　菌

第五部分　线　虫

第六部分　抗病性

第七部分　病害防治

第八部分　其　他

第一部分　真　菌

草坪草币斑病菌群体遗传结构研究*

刘清源**，杨静雅，李　婕，马子元，胡　健***
（南京农业大学草业学院，南京　210095）

摘　要：币斑病菌（*Sclerotinia homoeocarpa* F. T. Bennett）属于核盘菌属（*Sclerotinia*），其寄主范围广泛，可以侵染几乎所有冷季型和暖季型草坪草，引起草坪币斑病，是造成草坪上经济损失最严重的植物病原菌之一。目前，已收集2008年以来我国8个地区23个采样点6种寄主上的币斑病菌205株。ITS测序结果显示币斑病菌在我国存在3种类型（Ⅰ、Ⅱ、Ⅲ型），分别占24.4%、16.1%、59.5%，其中，Ⅰ、Ⅱ型币斑病菌在国外其他地区已有报道，分别主要侵染C_3和C_4草坪草，而Ⅲ型则尚未有中国以外地区的报道。进一步分析显示Ⅰ型主要分布于我国的北方地区，Ⅱ型主要分布于我国南方的地区，Ⅲ型则主要分布于我国的海南地区。上述结果表明，我国币斑病菌群体结构呈现出复杂和独特的特点，寄主选择和环境因素可能造成币斑病群体遗传结构的分化，但需要通过进一步研究来明确。

关键词：币斑病菌；草坪草；群体遗传；核盘菌

* 基金项目：中央高校基本科研业务费自主创新重点项目（KYZ201554）
** 第一作者：刘清源，硕士，主要从事草地微生物相关研究；E-mail：15001813068@163.com
*** 通迅作者：胡健，讲师，主要从事草地病害研究；E-mail：jaffyhu@njau.edu.cn

水稻纹枯病菌氮胁迫条件下转录组差异表达分析*

曾　泉，史国英，农泽梅，岑贞陆，胡春锦**
（广西农业科学院微生物研究所，南宁　530007）

摘　要：水稻纹枯病是水稻的三大病害之一，分布在世界各稻区，严重影响水稻的产量和品质。目前公认水稻纹枯病的病原菌为立枯丝核菌融合群 1 的 IA 亚群（*Rhizoctonia solani* AG1 - IA）。由于 *R. solani* AG1 - IA 具有宽广的寄主范围和较强的环境适应性，导致其致病机制的相关研究进展比较缓慢。笔者前期研究结果显示，低氮胁迫条件下培养的水稻纹枯病菌株对水稻的初侵染力明显强于在完全营养条件下培养的菌株；同时还证明了氮素浓度相同条件下，以甘氨酸作为氮源培养的纹枯菌致力比较弱。为进一步了解氮素营养对水稻纹枯病菌生长及其致病性能的影响，本文以水稻纹枯病菌一强致病力菌株 GX - 2 为研究对象，将该菌株接种在以硝酸钠为氮源且氮素终浓度分别为 0g/L（无氮培养）、0.082g/L（低氮胁迫）、0.494g/L（正常培养使用氮浓度）和 1.235g/L（高氮胁迫）以及以甘氨酸为氮源且氮素含量为 0.494g/L 的改良查氏培养基上培养，3d 后收集菌丝体，将上述 5 种不同培养处理得到的菌丝体样本依次编号为 RA1 ~ RA5，交由上海泉脉生物科技有限公司进行转录组测序。测序采用 Illumina HiSeq™2500 测序平台进行，结果显示，RA1 ~ RA5 样品分别获得 51M、66M、52M、51M 和 68M clean reads，Q30 均高于 90%，GC 含量在 51% ~52%。通过使用 Trinity 软件进行序列拼接，共获得 10 414 条 Unigene，平均长度 1 636bp，N50 为 3058，测序质量符合后期分析要求。以 RA3 为对照进行基因表达差异分析，分别在 RA1、RA2、RA4 和 RA5 中找到 220 个、224 个、260 个、250 个表达差异显著基因。其中，RA1 相对 RA3 上调 60 个基因，RA2 上调 130 个基因，RA4 上调 62 个基因，RA5 上调 78 个基因，RA2 样品基因上调个数明显高于其他样品。由于目前尚无立枯丝核菌 AG1 - IA 亚群全基因组测序的完整数据，所以，该菌的转录测序结果在与各数据库的比对过程中，能够获得注释的转录组数据较少，如 Unigene 在不同数据库中注释的结果分别为：NR（1 754，16.84%），SWISSPROT（2 705，25.97%），KOG（2 387，22.92%），KEGG（1 177，11.30%），GO（789，7.58%）。鉴于该菌基因注释的缺乏，下一步计划先对 RA2 上调基因中已得到注释的进行分析，明确其功能与注释是否相符，拟获得与该病原菌致病性相关的因子，为进一步研究该菌的致病机制提供科学依据。

关键词：水稻纹枯病；立枯丝核菌；转录组；差异表达

---

* 基金项目：国家自然科学基金项目（30860159）；广西自然科学基金项目（2014GXNSFAA118095）

** 通讯作者：胡春锦；E-mail：chunjin-hu@126.com

First Detection of *Cercospora canescens* (Ellis & Martin) Based on Cytochrome *b* Gene Using PCR Technique*

CUI Jia**, LIU Zheng, ZUO Yu-hu, LIU Tong***

(*Institute of Plant Pathology and Applied Microbiology, Heilongjiang Bayi Agricultural University, Daqing, Heilongjiang* 163319, *P. R. China*)

Abstract: [*Cercospora canescens* (Ellis & Martin)], an important pathogen of mungbean, which lead to serious yield losses. But, the molecular detection of this pathogen is still undeveloped. In this study, we designed one specific PCR primer pairs based on the cytochrome *b* (*cytb*) gene, and detected *C. canescens* and other 15 pathogenic fungi, including *Fusarium graminearum*, *Pyricularia grisea*, *Helminthosporiumturcicum*, *Pass Curvularia lunata Boedijn*, *Alternaria solani. Sorauer*, *Phytophthora megasperma f*, *Bipolaris sorokiniana*, *Rhizoctonia solani* et al. The gels electrophoresis results showed that the *C. canescens* were only positive, and other fungi were negative using the specific primer pair, indicating the primer pair could be used to detect *C. canescens*. Furthermore, the detection limit was 100pg/μL of genomic DNA. To our knowledge, this is the first report successfully detecting *C. canescens* based on cytb gene.

Key words: *Cercospora canescens*; Cytochrome b; PCR; Detection

* 基金项目：国家科技支撑项目“杂豆病虫草安全高效防控技术”（2014BAD07B05－06）

** 第一作者：崔佳，硕士研究生，从事植物病理学研究；E-mail：jiaxin19@outlook.com

*** 通讯作者：刘铜，副教授，从事植物病理学研究；E-mail：liutongamy@sina.com

大葱炭疽病在甘肃省首次发生报道

陈爱昌，魏周全，刘小娟
（甘肃省定西市植保植检站，定西　743000）

摘　要：在甘肃省定西市陇西县大葱种植田，自 2014 年以来，移栽的大葱每年 8—9 月都陆续出现长势弱、枯死的现象。通过调查，严重发生田块大葱提早枯死病株率达 60%。对提早枯死大葱植株进行病原菌分离培养，经形态学观察、致病性测定和分子生物学 ITS 序列分析，巢式 PCR 中用特异性引物 Cc1NF1/Cc2NR1 扩增 *C. circinans* 得到了 500bp 特异性条带。致死大葱提早枯死的病原菌为洋葱炭疽刺盘孢菌［*Colletotrichum circinans*（Berk.）Vogl］。这是大葱炭疽病在甘肃省的首次报道。

关键词：大葱；田间；早死；炭疽病；病原鉴定

莪术炭疽病病原鉴定*

宁 平[1**]，郭堂勋[2,3]，莫贱友[2,3]，黄穗萍[2,3]，唐利华[2,3]，李其利[2,3***]
（1. 广西农业职业技术学院生物技术系，南宁 530007；2. 广西农业科学院植物保护研究所，南宁 530007；3. 广西作物病虫害生物学重点实验室，南宁 530007）

摘 要：莪术具有行气破血、消积止痛之功效，在我国主产于四川、福建、浙江、广西壮族自治区（全书简称广西）、贵州等地。2012 年 12 月，在广西隆安县发现了一种为害严重的莪术叶斑病。该病开始出现圆形或椭圆形小斑点，伴有黄色晕圈，随后病斑逐渐扩大，严重时病斑融合，导致叶片枯死。在病害发生后期，病斑处可见很多分生孢子盘散生于叶面，黑色，有明显的刚毛（图）。从广西隆安县采集具有典型症状的叶斑病病叶，采用常规组织分离法分离获得一种真菌。形态学观察结果表明，该菌在 PDA 上菌落为圆形，菌落由白色逐渐变为灰黑色至黑色。菌落无明显气生菌丝，后期可产生粉红色的分生孢子团。分生孢子单孢，无色透明，镰刀形，大小为（16～22）μm ×（4～5.5）μm（平均值 19.7μm × 4.7μm）。采用核糖体转录间隔区（ITS）、肌动蛋白（ACT）、几丁质合成酶 A（CHS-I）、3－磷酸甘油醛脱氢酶（GAPDH）等基因的通用引物对该菌基因组 DNA 进行 PCR 扩增、测序（登录号 KT004507，KT004508，KT004509，KT004510）。结果表明，该菌的 ITS、ACT、CHS-I 和 GAPDH 序列与 GenBank 数据库中的 *Colletotrichum curcumae* 的同源性均达 100%。结合形态学观察和分子鉴定结果，将该菌鉴定为 *Colletotrichum curcumae*。印度在 1988 年报道过 *Colletotrichum curcumae* 可引起姜黄炭疽病，但在中国这是首次报道该病原可引起莪术炭疽病。

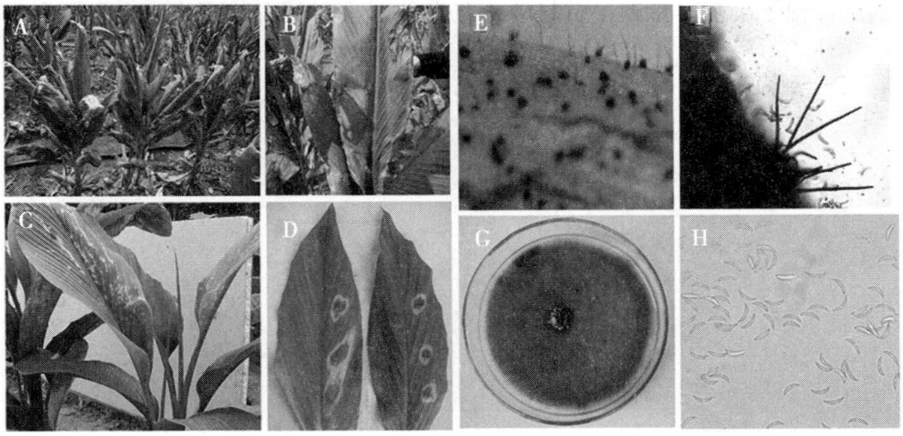

图 莪术炭疽病症状及病原形态

A，B：莪术炭疽病田间症状；C：分生孢子液活体接种症状；D：菌丝块离体接种症状；E，F：病斑上分生孢子盘及分生孢子；G：菌落形态；H：分生孢子形态。

---

* 基金项目：国家自然科学基金（31560526）；广西农科院科技发展基金（桂农科 2016JZ13）；广西作物病虫害生物学重点实验室基金（15－140－45－ST－2）

** 第一作者：宁平，讲师，主要研究方向为植物真菌病害及其防治；E-mail：382279601@ qq. com
*** 通讯作者：李其利，副研究员，主要研究方向为植物真菌病害及其防治；E-mail：liqili@ gxaas. net

玉米大斑病菌 *MAT* 基因在两性菌株有性生殖中的作用*

戴冬青**，许苗苗，杨　阳，马双新，刘　俊，曹志艳***，董金皋***
（河北农业大学，真菌毒素与植物分子病理学实验室，保定　071001）

摘　要：玉米大斑病是近年来玉米生产上的主要病害，病原为玉米大斑病菌［*Setosphaeria turcica*（Tuttrell）Leonard *et* Suggs.］，属异宗配合真菌。自然界玉米大斑病菌存在 3 种交配型菌株，其中 A 交配型菌株中含 *MAT*1 基因；a 交配型菌株中含 *MAT*2 基因；Aa 交配型菌株中含有 *MAT*1 和 *MAT*2 两个基因，并且 *MAT*1 基因编码 α1 转录因子，*MAT*2 基因编码高迁移率蛋白（high mobility group，HMG）家族。根据玉米大斑病菌菌株中交配基因类型不同以及其杂交和自交的育性不同，推测 *MAT* 基因在有性生殖过程中均发挥着重要作用，本论文在前期研究的基础上，利用 qRT-PCR 的方法分析了 *StMAT*1 和 *StMAT*2 基因在有性态诱导过程中的表达规律，利用原生质体转化的方法创制 *MAT*2 基因的缺失突变体，经对突变体的无性繁殖和有性生殖能力进行分析，初步明确了 *StMAT*2 基因在玉米大斑病菌两性交配型菌株有性生殖中的作用，主要结果如下。

（1）有性态诱导前、诱导 15d、30d 和 60d 时 *StMAT*1 和 *StMAT*2 基因均有表达，随着诱导时间的延长，*StMAT*2 基因在不同诱导组合中的表达量都呈显著增高的趋势，说明 *StMAT*2 在玉米大斑病菌有性生殖过程中起着重要作用。

（2）通过原生质体转化的方法获得了 3 株 *StMAT*2 基因缺失转化子，利用 PCR、qRT-PCR 和 Southern blot 的方法确定了 3 株转化子均为 *StMAT*2 基因缺失突变体。

（3）Δ*StMAT*2 不能产生分生孢子，且 Δ*StMAT*2 菌株自交以及与 A 交配型菌株或 a 交配型菌株杂交均不能产生子囊壳、子囊和子囊孢子。

关键词：玉米大斑病菌；交配型；*MAT* 基因；qRT-PCR

---

* 基金项目：河北省高等学校科学技术研究项目（ZD2014053）；现代农业产业技术体系（CARS－02）

** 第一作者：戴冬青，硕士研究生，研究方向为玉米大斑病育性的研究；E-mail：m15933562795@163. com
*** 通讯作者：曹志艳，E-mail：caoyan208@126. com；董金皋，E-mail：dongjingao@126. com

An Argonaute gene of *Valsa mali* Plays Distinct Roles in the Stress Responses and Pathogenicity

FENG Hao, XU Ming, LIU Yang-yang, GAO Xiao-ning, YIN Zhi-yuan, HUANG Li-li *

(*College of Plant Protection and State Key Laboratory of Crop Stress Biology for Arid Areas, Northwest A&F University, Yangling, Shaanxi* 712100, *China*)

Abstract: *Valsa mali* (*V. mali*), the causative agent of apple tree Valsa canker, causes heavy damage to apple production. Exploration of the pathogenesis of *V. mali* will contribute to the generation of effective disease control strategies. The Argonaute proteins (AGOs), as the core components of the RNA-induced silencing complex (RISC), play key roles in RNA interference, which is important for various cellular processes. In this study, one AGO gene of *V. mali* was found to be dramatically up-regulated during infection of the pathogen. Functions of this gene in growth, stresses response and pathogenicity of *V. mali* was analyzed by constructing gene knock-out mutant. Compared to the wild-type strain 03 - 8 (WT), the colonial morphology and growth rate of mutant showed no significant difference on PDA plates, but the hypha morphology of mutant showed increased branches. Moreover, no obvious altered colonial morphology was observed for mutant under treatment of NaCl (0. 1mol/L, 0. 5mol/L and 1. 0mol/L), KCl (0. 5mol/L, 1. 0mol/L and 1. 5mol/L) and different pH value (2. 0 ~ 11. 0). However, application of 0. 05% H_2O_2 led to the growth stop of mutant. In addition, the mutant exhibited decreased pathogenicity. Our results revealed the important roles of an AGO gene in stress tolerance and pathogenicity of *V. mali*, which may lay foundation for exploring the mechanisms of RNA interference of *V. mali*.

Key words: Argonaute protein; Pathogenicity; Small RNA; Stress responses; Apple *Valsa* canker

* Corresponding author: HUANG Li-li; E-mail: huanglili@ nwsuaf. edu. cn

江苏省稻瘟病标样中分离真菌的鉴定及在稻瘟病发生过程中的作用*

杜　艳**，齐中强，俞咪娜，余振仙，刘永锋***

（江苏省农业科学院植物保护研究所稻病与生防研究室，南京　210014）

摘　要：梨孢属真菌（*Pyricularia*）是一类重要的植物病原菌，除了感染水稻引起稻瘟病以外，还能感染多种杂草包括禾本科和莎草科等。由于江苏地处沿海温带亚热带区域，稻病区常见优势杂草的种类有稗、水苋菜、异型莎草和千金子等植物。田间杂草植物易受梨孢属真菌的侵染形成病斑，病斑上的孢子会散落在稻株上，影响稻瘟病的发生。因此，研究其他寄主的梨孢菌与稻瘟病菌在水稻植株上的互作关系，对于水稻抗病品种的选育及稻瘟病的防治具有重要意义。

本研究对江苏省连续 6 年分离的稻瘟病菌标样进行回接，发现 10% 以上的分离菌株不能引起稻瘟病菌普感的水稻品种丽江新团的黑谷发病。以稻瘟病菌 Guy11 为对照，从中随机选取 2 个致病的分离菌株 2011 －9 －1、2011 －214 和 4 个不致病的分离菌株 2012 －18 －1、2012 －18 －2、2011 －122 －1 和 2011 －132 －2，观察它们在分生孢子形态和大小、分生孢子梗形态等特性，发现 6 个分离菌株与 Guy11 均较为相近。另外，4 个不致病菌株与野生型在生长生长速率方面没有明显差别，但是在菌落形态、气生菌丝密度及色泽方面与野生型差别较大。进一步对分离菌株进行 PCR 鉴定，根据 actin、β-tubulin 和 calmodulin 等基因序列设计引物并进行 PCR 扩增，将比对结果绘制相应的系统发育树，鉴定分离的 2 个致病菌株 2011 －9 －1、2011—214 属于 *Magnaporthe oryzae*，而 4 个不致病菌株 2012 －18 －1、2012 －18 －2、2011 －122 －1 和 2011 －132 －2 为梨孢属真菌 *Pyricularia*。附着胞形成试验发现，4 个不致病的分离菌株在疏水表面均能形成附着胞，但膨压显著低于野生型。通过水稻叶鞘侵染显微观察发现，单独接种 4 个不致病菌株均不能侵染水稻叶鞘细胞，而混合接种野生型和不致病菌株均有明显的侵入和细胞内扩展。另外，根部接种试验表明，4 个不致病菌株均能引起丽江新团黑谷根部发病。

本研究结果表明，田间分离的不致病菌株与稻瘟病的发生存在一定的关系。江苏省各稻区杂草分布广泛，且田间杂草种类繁多，推测这些寄主在稻瘟病菌的积累和传播方面具有重要的作用，研究结果对研究其他梨孢属真菌在稻瘟病的发生过程的作用以及稻瘟病的防治具有重要意义。

关键词：稻瘟病菌；梨孢属真菌；鉴定；防治

---

* 基金项目：江苏省农业科技自主创新资金子项目（CX（15）1054）

** 第一作者：杜艳，助理研究员，分子植物病理学；E-mail：dy411246508@126.com

*** 通讯作者：刘永锋，研究员；E-mail：liuyf@jaas.ac.cn

橡胶白粉病菌 HOG-MAPK 级联途径中 *PBS2* 基因的结构分析及功能研究*

冯 霞**，刘文波，林春花，何其光，梁 鹏，缪卫国***，郑服丛***
（海南省热带生物资源可持续利用重点实验室/海南大学环境与植物保护学院，海口 570228）

摘 要：本文旨在研究橡胶白粉菌 *pbs*2 基因在橡胶炭疽菌中的表达情况，为橡胶树白粉病的防治提供科学理论依据。制备无 *pbs*2 基因的橡胶炭疽菌菌株（突变体菌株）的原生质体：把培养好的突变体菌株菌丝用 0.8mol/L NaCl 溶液洗涤后，加入 10mg/mL 溶壁酶溶液 20mL，28℃下恒温振荡（80r/min）酶解 3h，离心去上清液得到纯的突变体菌株原生质体，用 1.2 × STC 溶液把原生质体的浓度调整到 $1\times10^7 \sim 3\times10^7$ 个/mL 备用；外源基因（*pbs*2）的转化：在含有突变体菌株原生质体的缓冲液各加入橡胶白粉菌 *pbs*2 基因片段 1.0μg，冰浴 30min；加入 2 mL 60% PEG3350，在室温下放置 20min；加入适量 PDA + 1.5mol/L 山梨醇再生培养液，使终体积为 5mL，轻轻混匀，水平放置离心管 25℃、培养 12 ~ 24h；摇匀后，加入 PDA + 1.5mol/L 山梨醇培养基，混匀，倒平板；28℃黑暗培养观察；转化子的鉴定：以突变体菌株和转化子的菌丝基因组 DNA 为模板，进行 PCR 扩增然后测序，鉴定 *pbs*2 基因在橡胶炭疽菌中的表达情况；表型测定：把转化子，野生橡胶树炭疽菌和突变体菌株分别培养在含有不同浓度 NaCl 的 MM 培养基上，观察菌丝生长情况。经 PDA + 1.5mol/L 山梨醇培养基筛选后，在 PDA + 1.5mol/L 山梨醇培养基平板上长出白色的菌落，突变体菌株作为阴性对照的平板上没有长出菌落，说明 *pbs*2 基因被成功转入了突变体菌株并得到了初步的表达；PCR 鉴定结果表明转化子整合了 *pbs*2 基因；表型鉴定结果表明转化子在含 NaCl 的 MM 培养基上菌丝生长较快，野生型次之，突变体菌株最慢，且随着 NaCl 浓度的提高 3 种菌菌丝生长都减慢。橡胶白粉菌作为专性寄生真菌，证明了 *pbs*2 基因在橡胶树炭疽菌中的功能，其为以后研究橡胶白粉菌中其他基因提供一个潜在的可能，对今后开展橡胶白粉病的防治研究具有重要意义。

关键词：橡胶树白粉病菌；*pbs*2 基因；遗传转化

---

* 基金项目：海南省重点研发计划项目（ZDYF2016208）；海南自然科学基金创新研究团队项目（2016CXTD002）；973 计划前期研究专项（2011CB111612）

** 第一作者：冯霞，在读硕士，研究方向为热带植物病理学；E-mail：952372934@ qq. com

*** 通讯作者：缪卫国，博士，教授，研究方向为分子植物病理等；E-mail：weiguomiao1105@ 126. com
郑服丛，教授，研究方向为植物病理学；E-mail：zhengfucong@ 126. com

玉米大斑病菌分生孢子产生条件的分析及产孢机制初探*

冯胜泽**，赵　洁，刘星晨，赵立卿，
郑亚男，巩校东，韩建民，谷守芹***，董金皋***
（河北农业大学真菌毒素与植物分子病理学实验室，保定　071001）

摘　要：玉米大斑病（Northern Leaf Blight of Corn）是由大斑刚毛座腔菌（*Setosphaeria turcica*）引起的一种威胁玉米生产的重要叶部病害。分生孢子是子囊菌、半知菌及担子菌的无性孢子，具有抵抗高温、低温、干燥和营养缺乏等不良环境的能力，同时也是真菌病害的重要传播媒介。在基因功能研究中也常作为转化起始材料用于突变体的创制。因此，分生孢子产生的数量、活力对遗传转化的效率至关重要。

本研究以玉米大斑病菌野生型菌株 01－23 为试材，探索了不同的培养基、温度、培养基 pH、碳源、氮源、光照等条件对分生孢子产生的影响。试验结果表明，在培养基中添加玉米茎秆、玉米叶片时对分生孢子的产生有促进作用，玉米叶片葡萄糖琼脂培养基中分生孢子产量最高；在不同温度条件下，发现25℃时产孢量最多，4℃、10℃、15℃时不产孢，30℃时产孢量极少，温度对分生孢子产生的影响较为显著；不同 pH 值条件下，分生孢子在 pH 值为 8 时产孢量最高，pH 值为 4 时分生孢子产量明显低于其他研究条件，说明弱碱性环境更有利于分生孢子的产生；分别利用不同碳源等量替换 LCA 培养基中乳糖时，发现分生孢子的产量均有不同程度的减少，而乳糖为碳源时产孢量最高；在 LCA 培养基中添加 0.5% 的不同氮源，研究发现添加 KNO_3时更有利于分生孢子的产生；不同光照条件下，12h 光照 12h 黑暗交替培养时产孢量高于 24h 全黑暗和 24h 全光照两种培养条件，且其最佳光照强度为 6 000lx。

进一步利用 Real-time PCR 技术分析菌丝生长阶段及分生孢子产生阶段 GATA 家族 5 个基因的相对表达量情况，结果表明，与菌丝时期相比分生孢子时期 *StGATA2* 基因表达明显上调，*StGATA3*、*StGATA4*、*StGATA5* 三个基因相对表达量均明显下调，其中，*StGATA4* 下调最为显著，因此，初步推测 *StGATA2* 对分生孢子的产生具有正调节作用，*StGATA3*、*StGATA4*、*StGATA5* 对分生孢子的产生具有负调控作用。

本研究不仅明确了玉米大斑病菌分生孢子产生的因素、确定了 *GATA* 家族参与调控分生孢子产生，为进一步确定病菌分生孢子产生最佳条件及产孢机制奠定基础。

关键词：玉米大斑病菌；分生孢子产生；产孢机制

---

* 基金项目：国家自然科学基项目（31271997；31371897）
** 第一作者：冯胜泽，硕士研究生，研究方向为植物分子与生理病理学；E-mail：fengshengze1126@163.com
*** 通讯作者：谷守芹，教授，博士生导师；E-mail：gushouqin@126.com
董金皋，教授，博士生导师；E-mail：dongjingao@126.com

甘薯爪哇黑腐病病原鉴定及生物学特性*

黄立飞**，叶芍君，陈景益，房伯平***，黄实辉
（广东省农业科学院作物研究所，广州　510640）

摘　要：明确引起甘薯储藏性黑腐病—爪哇黑腐病的病原菌，了解其生物学特性，为甘薯爪哇黑腐病的防治及抗病育种提供依据。对取自广东省广州市和海南省儋州市的病薯块进行病菌分离纯化。采用甘薯薯块接种法进行致病性测定；观察病原菌在马铃薯葡萄糖琼脂培养基（potato dextrose agar，PDA）上的菌落形态；采用光学显微镜对病原菌在甘薯薯块上产生的分生孢子形态进行观察。对菌株 CRI-LP1 和 CRI-LP8－1 的内部转录间隔区（ITS）和延伸因子基因（EF－1α）序列进行扩增和测序。采用 MEGA6. 06 软件和邻接法，将 2 个菌株与 *Lasiodiplodia* 属内 24 个种的代表性菌株基于 ITS 和 EF－1α 序列进化分析。此外，探索了培养基、温度、光照、pH 值等条件对菌株 CRI-LP1 菌丝生长的影响，并测定了菌丝的致死温度。从病样上共分离纯化到 11 个致病菌株，其中：来源广州市 7 个菌株，儋州市 4 个菌株。不同菌株在 PDA 培养基平板上的菌落都是由灰白变黑色，并没有显著差异。接种到甘薯薯块上都可以产生大量的分生孢子，不成熟时呈透明状单胞，成熟后成为棕色至黑色的具有纵纹的椭圆形厚壁双胞状。基于 ITS 和 EF－1α 基因序列进化分析表明分离自海南儋州的菌株 CRI-LP8－1 与假可可毛色二孢（*L. pseudotheobromae*）的模式菌株 CBS 116459 聚为一簇，分离自广州的菌株 CRI-LP1 与可可毛色二孢（*L. theobromae*）的模式菌株 CBS 164. 96 聚为一簇。随后对所有致病菌株的 ITS 和 EF－1α 序列分析表明：除菌株 CRI-LP8－1 外，其他菌株都与菌株 CRI-LP1 聚为一簇。菌株 CRI-LP1 在温度为 30℃，pH 值为 4，PDA 的条件下利于病原菌营养生长；病原菌菌丝的致死温度为 54℃。根据分离病菌的形态特征、柯赫氏法则证病及两个基因系统进化分析结果，确定了引起甘薯储藏性黑腐病——爪哇黑腐病的主要病原菌为 *L. theobromae*，并首次发现 *L. pseudotheobromae* 也是甘薯爪哇黑腐病的病原菌。此外，明确了病原菌的生物学特征，为该病的防治提供参考。

* 基金项目：国家甘薯产业技术体系专项资金（CARS－11－B－5，CARS－11－C－17）
** 第一作者：黄立飞，助理研究员；E-mail：hlf157@163. com
*** 通讯作者：房伯平，研究员；E-mail：bpfang01@163. com

广西油茶 1 种新的炭疽病病原菌鉴定*

廖旺姣**，邹东霞，黄乃秀，邓　艳，吴耀军
（广西壮族自治区林业科学研究院，国家林业局中南速生材繁育实验室，
广西优良用材林资源培育重点实验室，南宁　530002）

摘　要：油茶是我国南方特有木本食用油料植物。其产品具有良好的经济价值和市场前景，被赋予重要地位，我国现有油茶栽培面积约 4 400hm^2，主要分布在长江流域及以南的 18 个省（区），其中又以湖南、江西、广西三省（区）为集中栽培区。与任何植物一样，油茶在生长过程中会遭遇有害生物的为害，其中炭疽病就是为害油茶的主要病害。该病主要为害叶片、嫩芽及果实，导致落叶、芽枯及落果，严重影响油茶的生长和产量。2014—2015 年，笔者对广西林科院油茶苗圃中的岑软 2 ~3 号，陆川油茶、香花油茶、博白大果有茶和普通软枝油茶苗进行调查，发现苗圃炭疽病发生普遍，感病率 10% ~70%，因不同品种感病率略有不同，其中以陆川油茶感病最重，感病率高达 70%，香花油茶感病率最轻，感病率为 10%。对病叶样品进行病原菌分离、纯化并依柯赫氏法则确定分离物的致病性，选取代表性菌株 YC1 ~ YC6 进行病原菌形态和分子鉴定，以期为油茶炭疽病的防控提供理论依据。菌株 YC1 ~ YC6 均为同一类型真菌，在 PDA 培养基菌落浅灰色至灰黑色，边缘色稍浅，中间色深，绒毛状，边缘整齐，气生菌丝发达。菌落生长速率为 10. 78mm/d。培养后期产生黑色分生孢子器，其上可见橘红色黏液，分生孢子为无色单细胞，圆柱状，两端钝圆或一端稍尖，光滑，具有 1 ~2 个油球，大小为（14. 56 ±0. 87）μm ×（4. 99 ±0. 33）μm。分生孢子附着胞近椭圆形，浅褐色，边缘完整，大小为（6. 72 ±0. 77）μm ×（5. 45 ±0. 68）μm，形态特征与已报道的核果炭疽菌特征相符，结合菌株 ITS、TUB2、ACT、GPDH 和 CHSI 多基因序列分析，确认分离获得的油茶炭疽病病原菌为核果炭疽菌（*Colletotrichum fructicola* Prihastuti，L. Cai&K. D. Hyde.）。与已报道的广西油茶炭疽病菌胶孢炭疽菌（*Colletotrichum gloeosporioides* Penz.）不相同，因此，确定本次分离获得油茶炭疽病菌核果炭疽菌是 1 种新的油茶炭疽病原菌。

关键词：油茶；炭疽病；核果炭疽菌

---

* 基金项目：广西林业科技项目（桂林科字〔2012〕9 号）

** 第一作者：廖旺姣，硕士，工程师，从事林木病害研究；E-mail：liaowangjiao@ 126. com

效应因子 AVR1 - CO39 与其受体结合结构域复合物的结构解析

郭力维，刘　强，彭友良，刘俊峰
（中国农业大学植物保护学院植物病理学系，北京　100193）

摘　要：水稻是世界三大粮食作物之一，稻瘟菌引起的稻瘟病严重影响水稻的产量和品质。水稻与稻瘟病菌已成为研究植物与病原物互作的模式系统之一，二者互作符合“基因对基因”假说。因此，研究稻瘟病菌效应因子与其水稻受体的相互作用机理具有理论和实践意义。

前人的研究结果表明，RGA5 - A 是水稻抗病基因 *Pia* 的一个转录本翻译产物，在 RGA4 存在的条件下，RGA5 - A 能够识别效应因子 AVR1 - CO39，激发寄主产生抗病反应。与典型的 R 蛋白不同，RGA5 - A 通过自身 C 末端 LRR 区域下游的 RATX1 结构域参与效应因子的识别，其识别机理尚不明确。本研究利用原核表达系统将 RGA5 - A _ S（RATX1）和 AVR1 - CO39 进行重组表达，通过筛选不同的重组表达菌株和载体组合，借助层析技术获得高质量的重组表达蛋白。在酵母双杂实验体内验证二者相互识别的同时，利用凝胶排阻层析和等温量热滴定等技术体外验证重组表达的 RGA5 - A_ S（RATX1）和 AVR1 - CO39 存在直接的相互作用，二者间结合的解离常数约为 2. 13μM。采用座滴气相扩散法获得了复合物的晶体，收集了衍射分辨率为 2. 2Å 的数据，解析了该复合物的晶体结构，正在开展基于结构的突变体功能分析。本研究利用结构生物学等手段解析了水稻抗病蛋白 Pia RATX1 结构域识别相应因子的结构基础，这些结果将为深入分析植物抗病基因识别病原物的分子机理和水稻抗病基因的改良等方面奠定基础。

胶胞炭疽病菌小柱孢酮脱水酶基因的克隆与敲除载体构建*

刘娅楠**，蒲金基，张 贺，周芳雪，杨永利，张恺莉，刘晓妹***

（1. 海南大学环境与植物保护学院，海南大学热带作物种质资源保护与开发利用教育部重点实验室，海口 570228；2. 中国热带农业科学院环境与植物保护研究所·农业部热带作物有害生物综合治理重点实验室，海口 571101）

摘 要：芒果炭疽病（*Colletotrichum gloeosporioides*）是为害芒果的重要病害之一，主要由半知菌类炭疽菌属的胶孢炭疽菌（*Colletotrichum gloeosporides* Penz.）侵染引起。为了明确小柱孢酮脱水酶基因与芒果炭疽病菌致病力之间的相互关系，本实验以芒果炭疽病菌 DNA 为模板，利用同源克隆技术扩增该基因，分析其序列特征、推测了其蛋白的保守结构域，并借助 In-Fusion © HD Cloning Kit 技术进行敲除载体构建。结果表明，该基因 DNA 和 cDNA 全长分别为 796bp、564bp，编码区有两个内含子（大小为 52bp、180bp），推测编码 187 个氨基酸，其分子量约为 21. 52kDa，等电点 PI 为 5. 90，与 NCBI 网站中已公布的基因进行 Blastp 比对，发现该序列与香蕉炭疽菌（*C. musae*）、无花果炭疽菌（*C. caricae*）（登录号 GQ266389. 1、GQ266386. 1）的小柱孢酮脱水酶基因相似性分别为 98% 和 97%。该基因的敲除载体已构建成功，为下一步获得基因敲除突变体，研究该基因功能打下了材料基础。

关键词：芒果；胶孢炭疽菌；小柱孢酮脱水酶基因；克隆；敲除载体

---

* 基金项目：国家自然科学基金（31460455）；公益性行业（农业）科研专项项目：芒果产业技术研究与示范(201203092 - 2)

** 第一作者：刘娅楠，在读硕士研究生，研究方向为分子植物病理学；E-mail：625325769@ qq. com

*** 通讯作者：刘晓妹，博士，教授，研究方向为分子植物病理等；E-mail：lxmpll@ 126. com

河南周口、安徽阜阳小麦叶锈菌 EST-SSR 遗传多样分析*

张 林**，孟庆芳，张梦雅，闫红飞***，刘大群***
（河北农业大学植物保护学院/国家北方山区农业工程技术研究中心/
河北省农作物病虫害生物防治工程技术研究中心，保定 071000）

摘 要：小麦叶锈菌（*Puccinia triticinia*）引起的叶锈病是小麦上的一种重要真菌病害，在世界各麦区均有发生，对我国小麦也危害严重。近年来，小麦叶锈病在我国发生呈逐年加重的趋势。2013 年，山东、河南及新疆局部地区发生严重，2015 年在黄淮海麦区发生大流行。因此，加强小麦叶锈菌小种及群体的研究，可为有效防控小麦叶锈病提供指导。

河南周口和安徽阜阳两地相邻，位于河南省与安徽省交界处，其地理环境相似，气候条件差异较小。为明确小麦叶锈菌在这两个地区的小种组成及群体结构的差异，本研究以 2015 年采集于河南周口和安徽阜阳的 110 株小麦叶锈菌菌株为材料，并利用 21 对 EST-SSR 引物对其进行分子多态性分析。聚类结果显示，110 株小麦叶锈菌株间的遗传相似系数为 0.68～1.00，表明这些菌株间的遗传相似性较高。两个地区的群体内遗传多样性 Hs 为 0.2381，群体间遗传多样性 Dst 为 0.017 1，表明这两个地区小麦叶锈菌群体内存在一定的遗传多样性，且群体内的多样性较群体间多样性更丰富；基因流强度 N_m 为 6.931 9，表明在两个群体间基因交流较强。另外，群体遗传分化分析结果显示，两个地区小麦叶锈菌在群体间和群体内都存在一定的遗传分化，群体内遗传变异占总变异的 88.93%，群体间遗传变异占总变异的 11.07%，表明群体内遗传变异是小麦叶锈菌遗传变异的主要来源。

关键词：小麦叶锈菌；EST-SSR；遗传多样性

* 基金项目：国家重点基础研究发展计划（2013CB127700）；河北省自然基金项目（C2015204105）
** 第一作者：张林，在读硕士生研究生，研究方向为分子植物病理学；E-mail：zhanglin42@163.com
*** 通讯作者：闫红飞，副教授，主要从事植物病害生物防治与分子植物病理学研究；E-mail：hongfeiyan2006@163.com
刘大群，教授，主要从事植物病害生物防治与分子植物病理学研究

橡胶树白粉菌效应蛋白预测及功能初步研究*

何其光**，梁　鹏，刘文波，林春花，缪卫国***
（海南省热带生物资源可持续利用重点实验室/海南大学环境与植物保护学院，海口　570228）

摘　要：白粉病是橡胶树重要叶部病害之一，由专性寄生菌橡胶树白粉菌（*Oidium heveae* B. A. Steinmann）引起，其分子致病机理尚不明确。本实验根据实验室前期完成橡胶树白粉菌HO－73基因组及互作转录组数据，通过生物信息学分析，我们初步筛选到橡胶树白粉菌候选效应蛋白233个，具有以下3个特征：存在预测信号肽序列；无跨膜域结构（移除前20个氨基酸）；除白粉菌外，在NCBI nr数据库中无其他同源序列（BLASTP，$e<10^{-5}$）。多序列比对分析发现其中7个橡胶树白粉菌预测效应蛋白为白粉菌共保守效应蛋白。进化聚类分析证实橡胶树白粉菌候选效应蛋白之间变异较大。同时，通过保守结构域分析（SMART：http：//smart.embl-heidelberg.de/）发现部分效应蛋白含有公共保守模序，其中Y/F/WxC效应蛋白共106个，RxLR效应蛋白14个。由于橡胶树白粉菌目前无法进行遗传操作，因此，我们通过构建植物瞬时表达载体PVX：：effectors，利用农杆菌介导的烟草瞬时表达技术，在模式植株烟草上瞬时表达橡胶树白粉菌预测效应蛋白，对橡胶树白粉菌效应蛋白的毒性或无毒性功能进行初步筛选研究。本实验完成了20个橡胶树白粉菌预测Y/F/WxC效应蛋白基因序列扩增及表达载体构建。为橡胶树白粉菌效应蛋白的毒性或无毒性功能的初步筛选及进一步功能验证奠定良好的基础。

关键词：橡胶树白粉菌；效应蛋白；瞬时表达

---

* 基金项目：海南省重点研发计划项目（ZDYF2016208）；海南自然科学基金创新研究团队项目（2016CXTD002）；973计划前期研究专项（20 11C B 1116 12）

** 第一作者：何其光，在读博士，研究方向为分子植物病理学；E-mail：935497300@qq.com

*** 通讯作者：缪卫国，博士，教授，研究方向为分子植物病理等；E-mail：weiguomiao1105@126.com

基于 RNA-Seq 的水稻纹枯病菌菌核发育过程的转录组分析*

舒灿伟**，王陈骄子，江绍锋，周而勋***
（华南农业大学农学院植物病理学系，广州　510642）

摘　要： 水稻纹枯病是水稻生产上的重要真菌病害，菌核在该病的病害循环中起了重要作用，但目前国内外对该病菌菌核发育的分子机制研究甚少。本研究分别以水稻纹枯病菌（*Rhizoctonia solani* AG 1－IA）菌核发育过程中 3 个连续生长阶段（MS＝36h，SI＝72h 和 SM＝14d）的总 RNA 为研究材料，以 Illumina Hi-seq 2000 平台技术进行转录组测序，并进行生物信息学分析。结果表明：过滤低质量 reads 后，MS、SI 和 SM 阶段分别保留了 55 051 284对、52 983 023对和 51 391 446对高质量 reads 用于 DEGs 筛选，其中 77. 22%、72. 58% 和 74. 49% 的 reads 能准确比对到参考序列上，说明 RNA-Seq 结果和参考序列可靠。当从 MS 阶段发育到 SI 阶段时，有 1 675 个基因显著上调，而显著下调的基因则有 1 491个；当 MS 阶段和 SM 阶段相比较时，显著上调的基因达 1 231个，而显著下调的基因有 1 081个；而从 SI 阶段发育到 SM 阶段时，上调的基因有 744 个，而下调的基因则为 742 个。GO 分析表明，DEGs 涉及 36 个细胞组分类群、12 个分子功能类群和 18 个生物过程类群。KEGG 注释表明，有 4 907个 Unigene 富集在 47 条信号通路上，有大量基因涉及过氧化酶体途径和酪氨酸酶代谢途径。qRT-PCR 分析结果表明，10 个 DEGs 的表达模式与 RNA-Seq 分析结果一致。本研究共鉴定出 4 353个参与水稻纹枯病菌菌核发育的基因，明确了 DEGs 富集的分子功能与代谢途径，为水稻纹枯病菌菌核发育机制的深入研究奠定了基础。

关键词： 水稻纹枯病菌；菌核发育；转录组；荧光定量 PCR

---

* 基金项目：国家自然科学基金项目（31271994）

** 第一作者：舒灿伟，博士，讲师，主要从事植物病原真菌及真菌病害研究；E-mail：shucanwei@scau. edu. cn

*** 通讯作者：周而勋，教授，博士生导师，主要从事植物病原真菌及真菌病害研究；E-mail：exzhou@scau. edu. cn

花生网斑病病原菌生物学特性及全基因组测序研究

迟玉成，许曼琳，吴菊香，王　磊，董炜博，鄢洪海，张茹琴
（山东省花生研究所，青岛　266100）

摘　要：花生网斑病是我国北方花生产区危害最重的一种叶部病害，为了有效防治该病害，系统研究花生网斑病病原菌生物学特性，并对分离的莱西菌株进行全基因组测序。该菌在马铃薯培养基（PDA）、燕麦培养基（OA）、Czapek 培养基、花生叶煎汁培养基、水琼脂培养基（WA）上均能生长，以 PDA 培养基上生长最快；生长最适温度为 25℃，超过 35℃ 不再生长；最适 pH 值为 7。在黑暗、光暗交替（12h/12h）、全光照条件下培养 20d，均未检测到分生孢子。花生网斑病莱西分离株的全基因组大小约为 57Mb，G + C 含量约为 47.95%。

Genetic Divergence of Isolates Originated from Singleaecial Cups of *Puccinia striiformis* f. sp. *tritici* Through Selfing on *Berberis shensiana*

LU Xia, Angkana Rotprachak, TIAN Yuan, WANG Jie-rong,
JIANG Shu-chang, ZHAN Gang-ming, HUANG Li-li, KANG Zhens-heng
(*Key State Laboratory of Crop Stress Biology for Arid Areas and College of Plant Protection*, *Northwest A&F University*, *Yangling*, *Shaanxi* 712100, *China*)

Abstract: A single uredospore isolate of *Puccinia striiformis* f. sp. *tritici* was selfed on *Berberis shensiana* to study phenotypic and genotypic characteristics of progeny isolates that separated from single aecial cups. Sixty isolates from 3 aecial cups (named SAC 8, SAC19, and SAC58) were obtained in this study, containing 18, 12, and 30 isolates, respectively. Parental and progeny isolates were tested on 23 *Yr* near isogenic lines (NILs) of wheat for virulence/avirulence. There were 10, 11, and 18 phenotypes in the three aecium cup, respectively. The ratio were 55.6%, 91.2%, and 60%. These results indicated that isolates separated from one aecial cup can be different through sexual reproduction. Additionally, genotype of parental and progeny isolates were studied by 92 simple sequence repeat (SSR) markers. Of the 92 loci analyzed, 13 loci were heterozygous in the parental isolate; 4 loci revealed segregation in progenies of aecial cup SAC8; 3 loci and 5 loci revealed segregation in aecial cup SAC 19 and SAC58. There were 5, 5 and 9 multilocus genotypes (MLGs) in the three aecial cups, respectively. The results showed that highly genetic divergence of isolates existed within each aecial cup by selfing under control condition. Both phenotypic and genotypic anlysis showed that isolates separated from one aecial cup appeared diverse changes, such results were different from previous deduce, and the reason why isolates from one aecial cup have genetic divergence need further study.

苹果炭疽叶枯病原菌的遗传多样性*

刘照涛**，练　森，董向丽，王彩霞，李保华***
（青岛农业大学农学与植物保护学院/山东省植物病虫害综合防控重点实验室，青岛　266109）

摘　要：炭疽叶枯病是由炭疽病菌（*Glomerella cingulate*）在苹果叶部引起的一种新病害，近年来在我国苹果主产区暴发流行。该病主要为害叶片，造成叶片的大量脱落，严重年份落叶率可达95%。炭疽叶病于 2010 年首次在江苏丰县和安徽砀山发现报道，迅速向周边苹果产区扩散，至 2015 年已蔓延至山东文登、甘肃庄浪、辽宁绥中等地，严重威胁我国的苹果生产。本研究拟利用分子标记分析病原菌遗传多样性，明确炭疽病菌（*G. cingulate*）在我国的遗传多样性及病原菌的区域间传播路径。目前，利用 ISSR 分子标记，已对分离自山东栖霞两个果园和江苏丰县一个果园的 90 株菌株的遗传多样性进行了分析。10 个 ISSR 引物在 90 株病原菌中共扩增出 113 条带，其中 112 条具有多态性，占总数的 99.11%。栖霞西口、栖霞寺口和江苏丰县 3 个种群扩增条带多态性比率分别为 54.87%，66.37% 和 60.06%；Shannon 指数分别为 0.2715，0.3429 和 0.2894；Nei's 指数分别为 1.3005，1.3882 和 1.3217。根据 Nei 法对炭疽叶枯菌 3 个种群遗传分析，表明种群的遗传多样性是 *Dst* 为 0.1094，基因分化指数 Gst 为 0.3543，基因流系数 *Nm* 为 0.9119。结果表明，炭疽叶枯菌 3 个种群间变异仅为 35.43%，而种群内部的变异则达 64.57%，群体内多样性大于群体间多样性。基因流系数接近于 1，说明种群间基因交流频繁。为了解炭疽叶枯菌在全国范围内的遗传多样性，我们采集了其他流行区域内菌株，并正对这些菌株的基因多样性分析。

关键词：炭疽叶枯菌；ISSR-PCR 扩增；遗传多样性

---

* 基金项目：国家苹果产业技术体系（CARS－28）
** 第一作者：刘照涛，山东临沂人，硕士研究生，研究方向为植物病害流行学；E-mail：zhaotaoliu@126.com
*** 通讯作者：李保华，教授，主要从事植物病害流行和果树病害研究；E-mail：baohuali@qau.edu.cn

2015 年河北省部分地区苜蓿根腐病病情调查和病原鉴定*

孔前前**，秦　丰，马占鸿，王海光***
（中国农业大学植物病理学系，北京　100193）

摘　要：为了解河北省苜蓿根腐病的发病情况和病原种类，在 2015 年对河北省黄骅市和宣化县部分苜蓿种植地进行了苜蓿根腐病发病情况调查。调查结果表明，在大多数苜蓿种植地块该病害呈零星发生状态，个别地块病株率为 2% ~3%。对调查中采集的苜蓿根腐病植株样品，采用常规组织分离方法进行了微生物分离、纯化和培养，共获得 105 个菌株。采用 CTAB 法提取菌株 DNA，选取引物 ITS4/ITS5 和延伸因子 EF-1H/EF-2T 进行 PCR 扩增，并对扩增产物进行测序，经 BLAST 序列比对，结果表明，所分离微生物共有 17 种，除拟茎点霉（*Phomopsis* spp.）、间座壳菌（*Diaporthe* spp.）、木霉菌（*Trichoderma* spp.）、淡色丛赤壳（*Bionectria ochroleuca*）等外，还有 65 株为镰孢菌（*Fusarium* spp.），包括 22 株木贼镰孢（*F. equiseti*）、21 株茄镰孢（*F. solani*）、16 株尖孢镰孢（*F. oxysporum*）、3 株锐顶镰孢（*F. acuminatum*）、1 株层出镰孢（*F. proliferatum*）、1 株芳香镰孢（*F. redolens*）和 1 株 *F. commune*。采用培养皿法对获得的镰孢菌各个菌株进行了致病性测定，结果表明，多数镰孢菌对萌发的苜蓿种子具有致病性，但致病性强弱有差异。除部分木贼镰孢菌株致病性较弱外，大多数镰孢菌菌株对苜蓿幼苗表现出强致病性，其中，木贼镰孢菌株 QZ3、锐顶镰孢菌株 N12-1、尖孢镰孢菌株 QD13-2 的致病性最强，其病情指数分别为 90.83、82.92、77.13。本研究利用分子生物学技术进行了河北省部分地区苜蓿根腐病病原的鉴定，并对所分离获得的镰孢菌致病性强弱进行了测定，这为了解苜蓿根腐病病原和该病害的及时防治提供了一定基础。

关键词：苜蓿根腐病；病原分离；病原鉴定；ITS；延伸因子；致病性测定

---

* 基金项目：公益性行业（农业）科研专项经费项目（201303057）
** 第一作者：孔前前，硕士研究生；E-mail：1439918864@qq.com
*** 通讯作者：王海光，副教授，主要从事植物病害流行学和宏观植物病理学研究；E-mail：wanghaiguang@cau.edu.cn

灰葡萄孢侵染草莓植株代谢组的分析

代　探*，常旭念，胡志宏，刘盼晴，刘鹏飞**
（中国农业大学植物保护学院植物病理系，北京　100193）

摘　要：草莓灰霉病是草莓大棚种植中的主要病害，它常造成草莓采后和储运过程中大量烂果，引发严重经济损失。其病原菌灰葡萄孢菌寄主范围广，能侵染包括茄科、葫芦科、蔷薇科、豆科、葡萄科等 200 余种植物。有学者发现该病原侵染葡萄后，寄主防御相关代谢物和病原菌增殖相关代谢物含量均上调，但是有关草莓植株受灰葡萄孢侵染后小分子代谢物的变化鲜有报道。本文采用代谢组学法，研究了灰葡萄孢菌侵染不同时间草莓叶片组织中小分子代谢物的变化，目的是寻找灰葡萄孢侵染草莓植株的生物标志物，为探索灰葡萄孢—草莓植株的互作机制提供参考。

在温室条件下，采用孢子悬浮液喷雾对草莓植株接种灰葡萄孢菌，2d、5d 和 7d 后获得病原菌侵染草莓植株叶片。以未接菌的草莓叶片作为对照，进行代谢组提取、衍生化及气相色谱 - 质谱检测。获得了草莓感病组织代谢组的指纹图谱，根据标准物质的保留时间和质谱特征对色谱峰进行定性，并结合 NIST 谱库检索结果对其中匹配度大于 80% 的代谢物进行指认，草莓叶片代谢组中确认了 41 种代谢物，其中有机酸类有 18 种、糖类有 12 种、醇类有 7 种，其他类别有 4 种。ANOVA 分析显示，在 2d、5d 和 7d 采集的样品中，接菌植株与健康植株相比，含量发生显著性变化的代谢物分别为 54 个、44 个和 40 个。根据 OPLS 分析获得与灰葡萄孢侵染密切相关的潜在的生物标志物：病原侵染初期（2d）一些代谢物含量显著上调，包括十六烷酸、十八烷酸、莽草酸、D - 木糖、D - 果糖、半乳糖、蔗糖、来苏糖、β - D - 吡喃葡萄糖、肌醇、蜜二糖、红景天苷，而甘油酸和 2，3 - 丁二醇显著下调，这可能与病原侵染初期对寄主植物的刺激有关；一些代谢物则在侵染一段时间后含量水平发生调整，如 5d 后阿拉伯糖酸、阿拉伯糖醇、半乳糖醛酸、D - 核糖、β - D - 吡喃甘露糖、叶绿醇上调，7d 后苹果酸上调，分析这些代谢物可能处于某些被侵染扰动的代谢途径下游，经过侵染这种外界信号刺激一段时间后发生了累积变化。

关键词：灰葡萄孢菌；代谢组；气相色谱—质谱；生物标志物

---

* 第一作者：代探，硕士研究生，植物病理学；E-mail：daitan@ cau. edu. cn
** 通讯作者：刘鹏飞，副教授，博士，植物病理学；E-mail：pengfeiliu@ cau. edu. cn

重要植物病原真菌：葡萄座腔类真菌的研究进展及问题展望*

李文英**，李　夏，解开治，孙丽丽
（广东省农业科学院农业资源与环境研究所，广州　510640）

摘　要： 葡萄座腔菌目（Botryosphaeriales）隶属子囊菌门（Ascomycota），子囊菌亚门（Pezizomycotina），座囊菌纲（Dothideomycetes），世界性分布，包含一类形态各异的真菌，生于单子叶、双子叶植物或裸子植物等各种各样的植物寄主，主要为木本寄主，也有生于草本植物的叶部、茎部或秆部，在树枝和地衣体和小枝上，营腐生至寄生或内生生活，在长势衰弱、濒临死亡的木本组织上较为常见，大多数种类可以在人工培养基上生长。根据最新的《国际藻类，真菌及植物命名法则》和阿姆斯特丹宣言的要求及提议，采用“一种真菌，一个名称”的最新分类命名方法，目前世界上已知葡萄座腔菌目真菌 6 科 39 属 2 000余种，我国对上述类群的研究始于20 世纪 30 年代，至今已报道该目 2 科分别为：葡萄座腔菌科 15 属 31 种，叶点霉科 3 属 22 种。

葡萄座腔菌目的许多成员属于弱寄生菌，条件合适时会引起重要的植物病害，属重要的潜在植物病原真菌，对农林生产具有重要的经济影响。特别是葡萄座腔菌，新壳梭孢属，毛色二孢菌，小穴壳孢属等重要属的优势类群会引起流胶、茎枯、梢枯、枝枯、溃疡、蒂腐、根腐、炭腐等多种林木、果树及农作物病害，或腐生于重要作物果实，引起果腐、环腐、蕉腐、轮纹类等果实病害或采后货架期病害。

近年来真菌分子系统学的研究进展与菌物命名法规的变革引起真菌分类系统的巨大变化，Wijayawardene（2014）等对葡萄座腔菌类真菌提出较为合理的分类系统，但目前该类重要植物病原真菌研究还存在许多悬而未决的问题，如分类地位和命名的清理与订正，生活史中全型特征关联的证据，病原菌与寄主或基物的辩证关系等。因此，深入开展该类真菌属种多样性研究是认识该类重要病害的基础，可为进一步深入研究其致病机制、病害侵染循环，提出有效防治策略提供基础信息和科学依据。

关键词： 植物病原真菌；葡萄座腔类真菌；葡萄座腔菌目；葡萄座腔菌（*Botryosphaeria*）

* 基金项目：科技部基础性工作专项（2013FY110400）；真菌学国家重点实验室开放课题（SKLMKF201404）
** 第一作者：李文英，博士，副研究员，主要从事微生物多样性资源及环境修复利用研究；E-mail：liwenying2006@126.com

辽宁省稻瘟病菌无毒基因型鉴定分析*

李思博**，魏松红***，刘志恒，李　帅，罗文芳，张　优，王海宁
（沈阳农业大学植物保护学院，沈阳　110866）

摘　要：水稻与稻瘟病菌间存在广泛而特异的相互作用，是研究寄主与病原物互作的重要模式系统，符合 Flor 的“基因对基因”学说。水稻在抵抗稻瘟病菌侵染的过程中，其抗病基因与病菌无毒基因间编码产物可相互识别，引发过敏性坏死反应（HR），抑制病菌在植株中扩展，从而产生抗病效果。在稻瘟病菌中已鉴定出 40 多个稻瘟病菌无毒基因，但目前仅有 9 个无毒基因被克隆和初步分析，其中包括 *PWL1*、*PWL2*、*Avr1 - CO39*、*Avr-pii*、*Avr-pia*、*AvrPiz-t*、*Avr-pik*、*ACE1* 和 *Avr-pita*。本试验采用 7 个稻瘟病菌无毒基因的特异性引物对 2014 年分离自辽宁省沈阳、铁岭、抚顺、盘锦、丹东、大连、营口、锦州共计 120 株稻瘟病菌的 7 种不同无毒基因 *Avr1 - CO39*、*Avr-pii*、*Avr-pia*、*AvrPiz-t*、*Avr-pik*、*ACE1* 和 *Avr-pita* 进行扩增测序分析。结果表明 *Avr1 - CO39*、*Avr-pii* 和 *Avr-pia* 在辽宁省的稻瘟病菌中未被发现。而 *AvrPiz-t*、*ACE1*、*Avr-pik* 和 *Avr-pita* 则以不同频率出现。*ACE1* 在 98% 的菌株中被检测到，*Avr-pita* 在 97% 的菌株中被检测到，*Avr-pik* 在 94% 的菌株中被检测到，*AvrPiz-t* 在 90% 的菌株中被检测到，说明 4 种基因在辽宁省稻瘟病菌中遗传稳定。通过 CDS 测序比对分析发现稻瘟病菌的 *AvrPiz-t* 中存在 2 处点突变导致碱基序列翻译的氨基酸错义突变，对该基因突变所产生的致病力变化有待接种试验的进一步证实。鉴定稻瘟病菌无毒基因，有助于抗病基因的水稻品种合理布局，能够为有效控制稻瘟病的发生提供指导。

关键词：辽宁省；稻瘟病菌；无毒基因型；鉴定分析

---

* 基金项目：国家水稻产业技术体系项目（CARS - 01）；辽宁水稻产业体系专项资金项目（辽农科〔2013〕271 号）

** 第一作者：李思博，硕士研究生，植物病理学专业；E-mail：1967833856@qq. com

*** 通讯作者：魏松红，博士，教授，主要从事植物真菌病害及水稻病害研究；E-mail：songhongw125@163. com

玉米大斑病菌GATA基因家族的生物信息学分析及其表达规律研究*

刘星晨**，冯胜泽，赵　洁，赵立卿，郑亚男，
巩校东，韩建民，谷守芹***，董金皋***
（河北农业大学真菌毒素与植物分子病理学实验室，保定　071001）

摘　要：玉米大斑病（Northern corn leaf blight）是由大斑刚毛座腔菌（*Setosphaeria turcica*）引起的一种威胁玉米生产的重要叶部真菌病害，常给玉米生产造成严重的经济损失。深入研究调控病菌侵染及致病的分子机制对玉米大斑病的有效防治具有重要意义。

研究发现，GATA为一类转录调控因子基因家族，该类基因参与调控植物病原真菌的生长发育和致病过程，在禾谷镰刀菌（*Fusarium graminearum*）、香蕉枯萎病菌（*Fusarium oxysporum* f. sp. *cubense*）中均有对该基因家族的研究报道，但在玉米大斑病菌（*S. turcica*）中尚未见报道。本研究利用生物信息学方法，利用HMMER软件搜索玉米大斑病菌全基因组数据库（JGI，http://genome.jgi-psf.org/settul/settu1.home.html），鉴定了病菌中GATA转录调控因子家族的所有成员，并对其进行了基因组定位、蛋白及保守基序分析、蛋白理化性质分析。结果表明，在玉米大斑病菌基因组中存在5个GATA转录调控因子家族基因，分别命名为*StGATA*1、*StGATA*2、*StGATA*3、*StGATA*4和*StGATA*5；该家族成员均为断裂基因，内含子数量为1～3个；家族蛋白序列及保守基序分析表明，StGATA1含有两个锌指结构，其余4种蛋白均含有一个锌指结构，StGATA3和StGATA4均含有PAS结构域，StGATA4还含有两个PAC结构；理化性质分析发现4个GATA转录因子均呈碱性，仅有*StGATA*3呈酸性（PI＝6.03），且5种蛋白都属于不稳定蛋白（不稳定系数＞40考虑为不稳定）。

本研究进一步利用Real-time PCR技术分析了这5个GATA基因在玉米大斑病菌菌丝发育、分生孢子形成、芽管萌发、附着胞及侵入丝形成等5个时期的基因相对表达水平。结果发现，*StGATA*3、*StGATA*4和*StGATA*5在菌丝形成时期的相对表达水平显著高于其他生长时期；*StGATA*4和*StGATA*5在分生孢子时期相对表达水平显著高于其他生长时期；*StGATA*1和*StGATA*2在各个时期的相对表达量都保持在较为稳定的相对低水平状态。结果表明，*StGATA*3、*StGATA*4和*StGATA*5主要调控病菌的菌丝发育，*StGATA*4和*StGATA*5主要参与调控病菌分生孢子的发育。该研究不仅阐明了玉米大斑病菌GATA基因家族的结构特征及其表达规律，也为进一步研究该家族调控病菌的生长发育及其致病性的分子机制奠定了基础。

关键词：玉米大斑病菌；GATA基因家族；生物信息学分析；表达规律

---

* 基金项目：国家自然科学基项目（31271997；31371897）

** 第一作者：刘星晨，硕士研究生，研究方向为植物分子病理学；E-mail：1552788293@qq.com

*** 通讯作者：谷守芹，教授，博士生导师；E-mail：gushouqin@126.com
董金皋，教授，博士生导师；E-mail：dongjingao@126.com

环境和植物因子诱导禾谷镰刀菌DON毒素合成及运输的分子机制

江　聪*

（西北农林科技大学，杨凌　712100）

1. 禾谷镰刀菌的DON毒素合成受到多种环境因子的刺激。为了揭示相关分子调控机理，我们鉴定了AreA和Skn7这两个转录因子，通过对其功能分析后发现，这两个转录因子分别介导了禾谷镰刀菌对氮源信号以及活性氧信号的响应，并负责了对DON合成的调控。其中，AreA通过与DON合成的核心调控因子之一Tri10的直接互作发挥作用，而Skn7则是通过结合*TRI*基因启动子以刺激*TRI*基因转录，从而达到调控DON合成的目的。

2. cAMP是重要的第二信使，介导了真菌对外界环境信号的感知并在侵染植物的过程中发挥了重要的作用。我们的研究发现，直接用外源cAMP处理禾谷镰刀菌可以大幅度提高DON的合成量，而且同时还可以刺激DON合成相关的细胞分化。我们鉴定了两个*PDE*基因，负责了对胞内cAMP水平的调控，并且这两个*PDE*基因在菌丝营养生长、有性生殖、产孢以及植物侵染过程中均有功能的重叠或冗余。然而在DON合成的调控上，*PDE*2发挥了主要作用。进一步的研究发现，cAMP介导的信号通路对Tri6和Tri10这两个DON合成相关的核心转录因子的调控是截然不同的。

3. 由于Tri6和Tri10在对cAMP的响应上出现了明显的功能分化，我们对Tri6和Tri10的关系和功能的差异进行了进一步研究。研究发现，Tri6和Tri10是可以直接互作的，Tri10对*TRI*基因的调控依赖于Tri6的存在。对Tri10进行过表达会刺激产毒相关的细胞分化同时大量诱导DON的生物合成。然而Tri10的亚细胞定位研究显示，Tri10不同于传统的转录因子，其并非定位于细胞核，而是定位于核周围膜结构上。且在产毒诱导条件下，定位会发生变化。进一步研究发现，Tri10与负责DON合成的产毒小体的定位是互补的，这表明Tri10可能具有双重功能，即调控产毒小体的组装以及作为Tri6的辅因子参与*TRI*表达的调控。

4. Tri6是DON生物合成的核心调控因子之一，研究发现，在禾谷镰刀菌*CAP*1基因的启动子区域有Tri6的结合位点。Cap1可以与Fac1直接互作，介导了对Fac1的激活以调控胞内的cAMP含量和Pka的活性。这说明在禾谷镰刀菌毒素合成过程中存在一个反馈调控的机制。禾谷镰刀菌*cap*1敲除突变体生长受限、孢子小，致病力和毒素合成能力严重降低。不仅如此，Cap1的定位模式与Actin类似。进一步研究显示，Cap1还参与了调控Actin不同形式之间的动态转换。Actin作为结构蛋白可能参与了DON的运输。因此，Cap1可能作为Tri6的下游靶基因，同时调控了DON的生物合成及运输。

---

* 第一作者：江聪，副研究员，从事小麦赤霉病致病真菌禾谷镰刀菌的致病及产毒分子机理

The obtaining and analyzing of *StLAC*4 gene mutant of *Setosphaeria turcica*[*]

MA Shuang-xin[**], CAO Zhi-yan[***], DONG Jin-gao

(*College of Life Science, Agricultural University of Hebei/Key Laboratory of Hebei Province for Molecular Plant-Micrbe Interaction, Baoding* 071000)

Abstract: *Setosphaeria turcica* (anamorph *Exserohilum turcicum*, formerly known as *Helminthosporium turcicum*) is a fungal pathogen that causes northern corn leaf blight (NCLB) in maize around the world, has been aggravating year by year. NCLB is a serious, omnipresent foliar disease. Infections of maize with NCLB before silking can cause grain yield losses of more than 46%, particularly under conditions of moderate temperature and high humidity. *S. turcica* has becoming the urgent problem to be solved in the maize production region.

In order to identify the function of the *StLAC*4 of the *Setosphaeria turcica*, the gene knockout vector has been built followed the Homologus Recombination (HR) strategy . The vector was introduced into the wild type of *S. turcica* by PEG-mediated transformation, two fungal transformant had been selected by the resistance gene Bar and identified by PCR with specific primers successfully, named Δ*StLAC*4 - 1 and Δ*StLAC*4 - 1 as the mutant of *StLAC*4 laid the foundation for the further study.

Key words: Laccase; *Setosphaeria turcica*; Gene mutant; Homologus recombination

---

* 基金项目：国家自然科学基金（31101402）；河北省高等学校科学技术研究项目（ZD2014053）

** 第一作者：马双新，硕士研究生，研究方向为植物真菌病害；E-mail：msx0214@126. com

*** 通讯作者：CAO Zhi-yan；E-mail：caoyan208@126. com

DONG Jin-gao；E-mail：dongjingao@126. com

转 $hpa1_{Xoo}$ 基因棉花对棉花炭疽病的抗性研究*

刘文波**，蔡加挺，林春花，靳鹏飞，孙茜茜，缪卫国***，郑服丛***
（海南省热带生物资源可持续利用重点实验室/海南大学环境与植物保护学院，海口 570228）

摘　要：本研究采用棉花炭疽菌（*Colletotrichum gossypii*）侵染转 $hpa1_{Xoo}$ 基因的棉花材料和普通棉花，比较分析转 $hpa1_{Xoo}$ 基因棉花相对普通棉花对棉花炭疽菌的抗性表现。将棉花炭疽菌接种到转基因棉花（出发品种为陆地棉品种 854）和普通棉花（854）上，分别于接种后 0 h、12h、24h、48h、72h 取样，进行叶片 H_2O_2 含量和 H_2O_2 DAB 定位检测、过氧化物酶（POD）和苯丙氨酸解氨酶（PAL）活性测定及对棉花抗性相关基因的表达情况分析。结果表明，经棉花炭疽菌处理后转 $hpa1_{Xoo}$ 基因棉花和普通棉花在不同时间各测定指标结果不同。在 48h 转 $hpa1_{Xoo}$ 基因棉花的 H_2O_2 含量及其定位检测可显示出峰值，随后又下降到和 0h 相当的水平，POD 与 PAL 活性也显示出类似的变化趋势；然而普通棉花叶片在 24h 显示出峰值；POD 酶活性只有在 72h 才略微上升。抗性相关基因荧光定量检测显示，防卫 *hsr*203*J* 基因在 24h 时转 $hpa1_{Xoo}$ 基因棉花表达高于普通棉，72h 两种几乎都没有表达；*NPR*1 基因在 0 ~ 48h 时转基因植株的表达一直低于在普通棉花的表达，但转基因的表达量比普通棉花增加快，72h 两者的表达量相同；*PR* – 1*b* 基因在 0 ~ 12h 就有微弱的表达，24h 表达量最大，后续逐渐减弱，而普通棉花一直低表达量；与植物生长有关的 *NtEXP*2 基因分别在 12h 和 48h 出现相对表达量的高峰而普通棉花低表达。这些研究结果表明，转 $hpa1_{Xoo}$ 基因棉花为某些防卫基因超水平表达提供了分子基础，使受到棉花炭疽菌的侵染时才诱导产生的高水平活性氧及相关基因表达，说明 $hpa1_{Xoo}$ 编码的 Harpin_{Xoo} 表达赋予转基因棉花对棉花炭疽菌的抗性。

关键词：棉花炭疽病；转 $hpa1_{Xoo}$ 基因；活性氧；防卫基因

---

* 基金项目：南繁区生物安全监测预警及控制关键技术研究与示范（No. 201403075）；海南省自然科学基金（No. 20153131）；国家自然基金（No. 31360029）；农业部 2013 年热作农技推广与体系建设项目（No. 13RZNJ – 20）

** 第一作者：刘文波，硕士，研究方向为热带植物病理学；E-mail：saucher@ 163. com
*** 通讯作者：缪卫国，博士，教授，研究方向为分子植物病理等；E-mail：weiguomiao1105@ 126. com
郑服丛，教授，研究方向为植物病理学；E-mail：zhengfucong@ 126. com

麻山药上一种新病害病原初探*

张梦雅**，张　林，闫红飞***，刘大群***
（河北农业大学植物保护学院/国家北方山区农业工程技术研究中心/河北省农作物病虫害生物防治工程技术研究中心，保定　071000）

摘　要：麻山药属于薯蓣科、薯蓣属植物，有很高的药用价值。在其叶片上发现的一种病害能使叶片出现坏死斑甚至干枯，造成一定的经济损失。本研究利用柯赫氏法则进行了病原鉴定，并利用 PCR 方法进行分子鉴定。初步结果表明：该病原孢子双孢，长形或椭圆形，深褐色，无柄，双孢宽窄相近，着生或分散于有隔菌丝周围。在高湿条件下侵染叶片，有利于该菌的生长和繁殖。该研究初步明确了该病害病原，将为生产实践中该病害的有效防治提供基础。

关键词：麻山药；致病菌；鉴定

* 基金项目：国家重点基础研究发展计划（2013CB127700）；河北省自然基金项目（C2015204105）；河北省中药体系
** 第一作者：张梦雅，硕士研究生，植物病理学；E-mail：1348108060@qq.com
*** 通讯作者：闫红飞，副教授，主要从事植物病害生物防治与分子植物病理学研究；E-mail：hongfeiyan2006@163.com
刘大群，教授，主要从事植物病害生物防治与分子植物病理学研究

Analysis on Resistance of *Pi-gene* to epidemic races of *Mangnaporthe grisea* in Fujian Province of China*

RUAN Hong-chun[1,2**], SHI Niu-niu[1,2], DU Yi-xin[1,2], GAN Lin[1,2], YANG Xiu-juan[1,2], DAN Yu-li[1,2], CHEN Fu-ru[,2***]

(1. *Institute of Plant Protection, Fujian Academy of Agricultural Sciences, Fuzhou* 350013, *China*; 2. *Fujian Key Laboratory for Monitoring and Integrated Management of Crop Peste*, *Fuzhou* 350013, *China*)

Abstract: In order to clarify the blast resistance of 24 monogenic rice lines to epidemic races of Fujian, from 2012 to 2015, 347 isolates of *Magnaporthe grisea* were collected from different parts of Fujian, China. The virulence of these isolates were assayed with the Chinese race differentials and CO39 near-isogenic lines (NILs). The resistance of 24 monogenic lines to epidemic races were determined. The results showed that the 347 isolates could be classified into 6 groups with 36 physiological races by the Chinese race differentials, or 17 pathogenic types by CO39 NILs. The most dominant groups were ZA, ZB and ZC, while the dominant physiological races were ZC15, ZD7 and ZB15, Among the 17 pathogenic types the dominant pathogenic type was I34. 1. The resistance frequency of the tested monogenic lines ranged from 9. 80% to 89. 91%. The resistance frequency of rice varieties with broad-spectrum blast resistance genepathogenic typepathogenic type [*Pi-k*^{m}, *Pi* – 7 (*t*), *Pi* – 9 (*t*), *Pi-k*^{p}, *Pi-k*, *Pi-k*^{h}, *Pi-z*^{5} and *Pi-ta* (1)] were over 70. 00%. In addition, the resistance frequency of 8 major resistance genes to 12 major physiological races and 5 major pathogenic types were also over 70. 00%.

Key words: *Magnaporthe grisea*; Physiological races; Pathotype; *Pi-gene*; Resistance frequency

* 基金项目：国家公益性行业（农业）科研专项（201203014）；福建省种业工程项目（FJZZZY－1526）；福建省属公益类科研院所基本科研专项项目（2015R1024－1，2016R1023－11）

** 第一作者：阮宏椿，助理研究员，主要从事植物真菌病害及其防治研究；E-mail：hcruan@ 126. com

*** 通讯作者：陈福如，研究员，主要从事植物真菌病害及其防治研究；E-mail：chenfuruzb@ 163. com

Differences in Protein Expression Affected by Powdery Mildew in Wheat

LIANG Yin-ping, CHANG Xiao-li, GONG Guo-shu, LUO Li-ya,
HU Yu-ting, LEI Yu, QI Xiao-bo, ZHANG Min*
(*Agricultural college*, *Sichuan Agricultural University*, *Chengdu* 611130, *P. R. China*)

Wheat powdery mildew caused by *Blumeria graminis* f. sp. *tritic* (*Bgt*) is considered as a major wheat leaf disease in the main wheat-producing regions of world. Although many wheat cultivars resistant to this disease have been developed, little is known about their resistance mechanisms. *Pm*40 is an effective powdery mildew resistance gene in wheat line L699, and shows high resistance to *Bgt*. The aim of this study was to investigate protein patterns after *Bgt* infection in wheat line L699, Neimai836 (NM836) and Chuannong26 (CN26), in which NM836 was used as the resistant control while CN26 was susceptible control. Proteins were extracted from the wheat leaf sampled 2, 4, 8, 12 and 24h after inoculation, separated by two-dimensional electrophoresis, and stained with coomassie brilliant blue G－250. Results showed that there were different up-regulated and down-regulated proteins induced in three wheat materials at different indicated time points. As compared to non-inoculated plants the abundance of 62 spots and 71 spots on the wheat L699 gels significantly increased and decreased up on inoculation time respectively, among which 46 up-regulated proteins were identified by mass spectrometry analysis using the NCBInr database of *Triticum*, which not only involves in defense-related proteins such as germin-like protein (PR5), heat shock protein, lipoxygenase, but are also associated with photosynthesis, signal transduction, carbohydrate metabolism, energy pathway, protein turnover as well as cell structure. This indicates that defense proteins cooperate with other cell physiological-related proteins to contribute to *Pm*40-mediated wheat resistance to *Bgt*.

Key words: *Blumeria graminis* f. sp. *tritici*; Protein two-dimensional electrophoresis; Mass spectrometry; *Pm*40

* Corresponding author: ZHANG Min; E-mail: yalanmin@126. com

杧果葡萄座腔菌基因组测序*

李其利**，郭堂勋，黄穗萍，唐利华，莫贱友***
（广西农业科学院植物保护研究所/广西作物病虫害生物学重点实验室，南宁 530007）

摘 要：葡萄座腔菌科（Botryosphaeriaceae）真菌广泛分布于世界各地，其种类繁多，寄主植物多种多样，既可作为病原菌引起树木溃疡病，又能作为内生真菌潜伏在寄主植物组织中。杧果流胶枝枯病是我国杧果种植区的常见病害，各地均有发生。该病可由葡萄座腔菌属的 *Botryosphaeria rhodina*（无性型：可可毛色二孢 *Lasiodiplodia theobromae*，原名可可球二孢 *Botryodiplodia theobromae*），*B. dothidea*（无性型：七叶树壳梭孢 *Fusicoccum aesculi*）和 *B. parva*（无性型：小新壳梭孢 *Neofusicoccum parvum*）引起，其中主要病原是可可毛色二孢。在研究杧果流胶病菌的过程中，我们发现可可毛色二孢对枝条和果实的致病力最强，而七叶树壳梭孢和小新壳梭孢致病力相对较弱，此外，我们还发现这 3 种病原菌均可在人工培养基上产生子囊壳、子囊及子囊孢子。基因组测序有助于深入了解病原菌致病机理及其生活史，而目前未见杧果葡萄座腔菌基因组测序的报道。为此，本研究采用 Illumina 技术对杧果葡萄座腔菌属的可可毛色二孢、七叶树壳梭孢和小新壳梭孢进行全基因组测序，运用 3 种专业软件 ABYSS，SOAP，VELVET 进行组装，结果表明，可可毛色二孢基因组大小为 43.8Mb，N50 为 637kb，七叶树壳梭孢基因组大小为 45.6Mb，N50 为 583kb，小新壳梭孢基因组大小为 43.8Mb，N50 为 503kb。采用 Augustus 基因预测软件对组装的结果进行编码基因预测，分别从可可毛色二孢、七叶树壳梭孢和小新壳梭孢基因组预测出 12 953个、13 144个和 13 258个编码基因。通过同源比对，从杧果流胶病菌的 3 个菌株基因组序列中均可找到 MAT1 －1 和 MAT1 －2 交配型基因，下一步将继续研究杧果葡萄座腔菌交配型基因位点及其功能。

关键词：杧果；葡萄座腔菌；基因组测序

---

* 基金项目：国家农业产业体系广西杧果创新团队建设专项（nycytxgxcxtd－02－08－2）；国家自然科学基金（31560526）；广西作物病虫害生物学重点实验室基金（15－140－45－ST－2）

** 第一作者：李其利，副研究员，主要研究方向植物真菌病害及其防治；E-mail：liqili@ gxaas. net
*** 通讯作者：莫贱友，研究员，主要研究方向为植物真菌病害及其防治；E-mail：mojianyou@ gxaas. net

Etiology and Symptom of Corn Leaf Spot Caused by *Bipolaris* spp. in Sichuan Province

SUN Xiao-fang, QI Xiao-bo, GONG Guo-shu, CHANG Xiao-li, YE Kun-hao
(*Agricultural College*, *Sichuan Agricultural University*, *Chengdu* 611130, *China*)

Abstract: The genus *Bipolaris* is a kind of important plant pathogens causing leaf spot, root rot, and seedling blight in many plants around the world. *Bipolaris* has previously been reported as the potential pathogen caused corn leaf spot, such as southern corn leaf blight and northern corn leaf blight. In recent years, this kind of leaf spot was frequently observed with various complex symptoms and tends to get serious in Sichuan. In all, 747 samples of corn leaf spots were collected from 20 regions of Sichuan Province from 2011 to 2015. Morphological identification along with phylogenetic analysis of the ribosomal internal transcribed spacer (rDNAITS) region, a partial sequence of the β-tubulin gene, transcription elongation factor 1-α gene (EF-1α) and glyceraldehyde 3-phosphate dehydrogenasegene (GPDH) showed that five *Bipolaris* spp. including *B. maydis*, *B. zeicola*, *B. cynodontis*, *B. sorokiniana*, *B. miyabeanus*, were found on corn in Sichuan. Pathogenicity on one major cultivar (cv. Chuandan 28) inoculated with conidial suspension revealed that all these species were pathogenic, among which *B. maydis* and *B. zeicola* were the most pathogenic species, followed by *B. cynodontis*, *B. sorokiniana* and *B. miyabeanus*. As the dominant species, the typical symptoms caused by *B. maydis* are characterized by small fusiform, linear and subround. The characteristic lesions caused by *B. zeicola* are narrow, small point connection into dotted line. Isolates of *B. cynodontis* are also aggressive with small, punctiform lesions. *B. sorokiniana* and *B. miyabeanus* with slightly virulence also causes mall chlorotic spots. These results in this study demonstrated that most of the *Bipolaris* fungi cause corn leaf spot and their virulence vary from high to slight followed by complicated symptoms, respectively. This study provides useful information for disease management for corn leaf spot.

Key words: Corn leaf spot; *Bipolaris*; Identification; Symptom

农杆菌介导的层出镰刀菌的遗传转化及致病性缺陷突变体的筛选*

刘丽媛**，刘力伟，任　洁，王树桐***，曹克强

（河北农业大学植物保护学院，保定　071001）

摘　要：苹果再植病害（Replant disease of apple，ARD）已经广泛发生在世界苹果的主产区。层出镰刀菌是苹果再植病害的重要致病菌之一，但该病菌的致病机制尚缺乏研究。本研究对农杆菌介导的层出镰刀菌的遗传转化的体系进行优化并构建其突变体库，获得了致病力缺陷的单拷贝插入突变菌株，拟为从分子层面解释层出镰刀菌致病机理的研究奠定基础。对影响 H10 农杆菌介导转化效率的主要因子进行单因子条件测验，得到其遗传转化体系为：抑制 H10 菌丝和孢子生长的潮霉素的浓度为 100μg/mL，农杆菌 OD_{600} 的值为 0.3，AS 浓度为 600μg/mL，共培养时间 48h。利用这一体系进行 H10 菌株突变体库的构建，以对潮霉素抗性基因的 PCR 检测呈阳性的转化子作为插入突变体。经过在不含潮霉素的 PDA 培养基上 5 代培养，验证 T-DNA 插入突变体的遗传稳定性。对稳定遗传的转化子进行产孢和致病力试验分析得到产孢能力显著下降的突变体 3 株，致病力显著下降的突变体 8 株。用 Southern blot 技术验证致病力变弱的突变菌株 T-DNA 插入拷贝的拷贝数，证明有 7 个突变体为单拷贝插入并能够稳定遗传。用 TAIL-PCR 技术对产孢量下降且致病力变弱的单拷贝插入的突变体 H10－989 进行了侧翼序列扩增，对扩增到的片段进行回收测序，经过 NCBI-BLAST 分析比，获得的序列与水稻恶苗病菌 58289 基因草图 1 号染色体序列为同源序列。经 tblastx 注释与小麦冠腐病菌 FPCS3096 菌株和禾谷镰刀菌 PH－1 菌株的假定蛋白 mRNA 同源。

关键词：层出镰刀菌；农杆菌介导遗传转化；TAIL-PCR；苹果再植病害

---

* 基金项目：国家苹果产业技术体系（CARS－28）和河北省自然科学基金（C2016204140）

** 第一作者：刘丽媛，硕士研究生，主要从事病原真菌与寄主分子互作研究；E-mail：498799359@qq.com

*** 通讯作者：王树桐，博士，教授，主要从事植物病害流行与综合防治研究；E-mail：bdstwang@163.com

New Hosts of *Corynespora cassiicola* in Sichuan, China

XU Jing[1], CUI Yong-liang[1,2], KONG Xiang-wen[1],
CHANG Xiao-li[1], QI Xiao-bo[1], ZHENG Xiao-juan[1], GONG Guo-shu[1 *]
(1. *College of Agronomy & Key Laboratory for Major Crop Diseases, Sichuan Agricultural University*, *Chengdu* 611130, *P. R. China*; 2. *Sichuan Academy of Natural Resource and Sciences*, *Chengdu* 610041, *Sichuan*, *P. R. China*)

Corynespora cassiicola is the causal pathogen of the most devastating leaf spot diseases in more than 530 plants, mainly including monocotyledon, dicotyledon, ferns, cycads and so on. In China, this fungus has previously been reported in rubber, cucumber, cassava, balsam pear, patchouli, akebia trifoliate as well as cotton. During 2012 to 2016, host range of *C. cassiicola* was investigated in Sichuan, China, and a range of the candidate *C. cassiicola* were isolated from kiwi, cucumber, kidney bean, cowpea bean, hyacinth bean, sweet potato, yams, cotton, ligustrum quihoui, honeysuckle, strawberry and blueberry. The pathogenicity of the representative isolates on these plants was tested based on Koch's postulates. Further morphological and molecular characteristics demonstrated that these candidate pathogens causing leaf spot disease are identified as *C. cassiicola*. This is the first record of Corynespora leaf spot diseases on kiwi, hyacinth bean, sweet potato, ligustrum quihoui, honeysuckle, strawberry and blueberry in China. In addition, kiwi leaves were inoculated separately with each isolate from different hosts of *C. cassiicola*, and in turn, the isolate from kiwi leaves was tested by inoculation on 23 plants including gleguminosae, cucurbitaceae, solanaceae, convolvulaceae, rosaceae, ericaceae, hamamelidaceae, magnoliaceae, malvaceae, euphorbiaceae andoleaceae. Inoculation studies showed that all isolates caused similar leaf spot disease on kiwi, and on the other hand, all tested plants were hosts of the *C. cassiicola* from kiwi, implying that *C. cassiicola* had a wide host range in China. It could be explained by the fact that some of host plants tested above are commonly cultivated in and around kiwi planting gardens. Thus, there is a possibility that *C. cassiicola* on kiwi can spread on to other plant hosts. This will provide some basis for controlling Corynespora leaf spot disease on kiwi.

* Corresponding author: GONG Guo-shu; E-mail: guoshugong@126.com

苹果褐斑病原菌侵染苹果属山定子的转录组学研究*

李海录**，宋艳艳，冯　浩，黄丽丽***，韩青梅
（旱区作物逆境生物学国家重点实验室/西北农林科技大学植物保护学院，杨凌　712100）

摘　要： 苹果褐斑病（Marssonina leaf blotch of apple）是由 *Diplocarpon mali* 引起的东亚地区苹果生产中主要早期落叶病害之一，该病害引起生长季节叶片黄化和早落，进而造成严重的经济损失。前期研究发现山定子对褐斑病具有一定抗性，但病分子机理尚不清楚。本研究采用新一代高通量测序技术 Illumina Hiseq2500 对被侵染的苹果褐斑病菌的苹果属山定子进行转录组双端测序，利用生物信息学方法对基因表达谱研究和功能基因预测。通过测序，获得了 46.86G 包含碱基序列信息。对测序数据进行序列过滤、拼接、组装、去冗余，共获得 48 868 个单基因簇 unigenes，平均长度 1 007bp，序列信息达到了 32Mb。另外，从长度分布、GC 含量、表达水平等方面对 unigenes 进行评估，数据显示测序质量好，可信度高。此外，对组装的 48 868条 unigenes 进行了 NR、Swiss-port、KOG、GO、KEGG 等五大数据库预测，NR 数据库中的序列同源性比较表明，25 797个 unigenes 直接比对到了蔷薇科苹果属苹果（*Malus domestica*）基因上，表明本研究的测序及 unigenes 组装结果可靠。将 unigenes 与 KOG 数据库进行比对，根据其功能大致可分为 25 类。以 KEGG 数据库作为参考，依据代谢途径可将 unigenes 定位到 147 个代谢途径分支，包括核糖体代谢通路、碳水化合物代谢等。利用 RSEM 软件进行差异表达计算，edgeR 软件进行差异表达分析，通过设定阈值 FDR <0.05，logFC >2 或者 logFC <-2 筛选差异基因，共鉴定出上调基因 1 230个，下调基因 1 869个，GO 富集和 Pathway 富集结果表明差异基因参与植物的病原互作、植物信号转导通路等相关代谢通路。通过二代高通量转录组测序技术研究苹果属山定子抗褐斑病中的作用，为明确山定子在抗褐斑病的分子机理奠定理论基础，对于培育抗褐斑病品种具有重要的指导意义。

关键词： 山定子；苹果褐斑病；转录组；高通量测序

---

* 基金项目：国家自然科学基金（31471732 和 31171796）；国家公益性行业（农业）科研专项（201203034）和高等学校学科创新引智计划（B07049）

** 第一作者：李海录，硕士研究生，主要从事苹果褐斑病菌与苹果互作的分子机理研究；E-mail：963694443@ qq. com

*** 通讯作者：黄丽丽，教授，主要从事果树病害病原生物学及病害综合防控方面的研究；E-mail：huanglili@ nwsuaf. edu. cn

稻瘟菌致病因子 MoCon2 的重组表达，纯化和晶体生长

王　超[1]，李国瑞[1,3]，彭友良[1,2]，刘俊峰[1]
（1. 中国农业大学植物病理学系/农业部植物病理学重点开放实验室，
北京　100193；2. 农业生物技术国家重点实验室，北京　100193；
3. 内蒙古民族大学生命科学学院，通辽　028000）

摘　要：水稻作为当今世界主要的粮食作物，稻瘟菌（*Magnaporthe grisea*）是水稻的主要病害，每年造成的粮食损失相当于6 000万人口的粮食供给。稻瘟菌侵染水稻并导致其发病是一个复杂的过程，由大量致病相关因子参与并协同完成。对这些致病因子的研究能够为阐明稻瘟菌的致病机制和稻瘟病的防控提供一定的理论和实验依据。

本课题组前期的研究发现，MoCon2 是稻瘟菌在对水稻的致病过程中必需的致病因子，其缺失将会导致菌体生长缓慢，无分生孢子产生，附着胞能力下降，并完全丧失对宿主的致病性。通过酵母双杂、Co-IP 等技术确定了 MoCon2 与其他的蛋白互作的重要功能结合域。为进一步解析 MoCon2 其致病作用的分子机理，作者利用 x－射线晶体学方法对其进行了结构生物学的研究。通过解析 MoCon2 的蛋白三维结构，探索其执行功能的结构基础，阐述该蛋白的结构与功能关系及推测其调节致病的分子机制。首先选取了 MoCon2 全长和能够与 MoCWF4、MoCWC2 等多个蛋白互作的 100～220AA 区段，利用大肠杆菌表达了 $MoCon2^{100\sim220}$ 重组蛋白，并通过各种层析策略建立了截短体蛋白的纯化流程。采用坐滴法对制备的高纯度重组蛋白进行晶体生长条件的筛选，在 0. 1M Na cacodylate pH 值 5. 5，25% PEG4000 条件下得到蛋白的晶体。正在进一步优化结晶条件，以获得高质量的晶体和收集衍射数据。上述研究结果将为解析 MoCon2 晶体结构和分析其影响剪切的结构机制奠定基础。

关键词：稻瘟菌；MoCon2；纯化；蛋白晶体

苹果树腐烂病菌一个引起植物细胞坏死的效应基因 *VmE02* 的鉴定*

聂嘉俊**，尹志远，李正鹏，康振生，黄丽丽***
（旱区作物逆境生物学国家重点实验室/西北农林科技大学植物保护学院，杨凌 712100）

摘　要：由死体营养型（Necrotrophic）真菌 *Valsa mali* 导致的苹果树腐烂病能够引起苹果树皮严重坏死甚至整株死亡，给苹果产业造成巨大经济损失。引起植物细胞坏死对于死体营养型真菌侵染定殖寄主有重要作用，鉴定该病原菌坏死因子有助于揭示其引起苹果树皮溃烂的分子机制。本研究利用农杆菌介导的瞬时表达体系进行筛选，鉴定到一个能够引起本氏烟草（*Nicotiana benthamiana*）、拟南芥（*Arabidopsis thaliana*）、番茄（*Lycopersicon esculentum*）和辣椒（*Capsicum annuum*）等多种植物细胞死亡的候选效应基因 *VmE*02。序列分析表明，*VmE*02 在其他许多植物病原子囊菌、担子菌和卵菌中均存在同源基因，其中，卵菌普遍含有 3 个拷贝并且是通过基因水平转移的方式从子囊菌获得。烟草瞬时表达试验发现，寄生疫霉（*Phytophthora parasitica*）中 2 个和小麦条锈菌（*Puccinia striiformis* f. sp. *triciti*）中 1 个同源基因也能够引起植物细胞坏死，表明该基因的功能在不同病原菌中是保守的。此外，qRT-PCR 结果表明，该基因在 *V. mali* 侵染早期（6 hpi）显著上调表达。因此，*VmE*02 可能在 *V. mali* 侵染定殖苹果树皮并引起植物细胞坏死的过程中起其重要作用。

关键词：*Valsa mali*；基因水平转移；同源基因；非寄主植物

* 基金项目：国家自然科学基金（31471732 和 31171796）；国家公益性行业（农业）科研专项（201203034）和高等学校学科创新引智计划（B07049）

** 第一作者：聂嘉俊，硕士研究生，主要从事苹果树腐烂病菌效应基因的筛选及功能研究；E-mail：jj. nie@ hotmail. com
*** 通讯作者：黄丽丽，教授，主要从事果树病害病原生物学及病害综合防控方面的研究；E-mail：huanglili@ nwsuaf. edu. cn

芽孢杆菌 EDR2 颉颃苹果树腐烂病菌的相关基因及抗菌蛋白研究*

王娜娜**，黄丽丽***
（旱区作物逆境生物学国家重点实验室/西北农林科技大学植物保护学院，杨凌　712100）

摘　要：为了揭示芽孢杆菌 EDR2 颉颃苹果树腐烂病菌的关键基因及抗菌蛋白，本研究以 SP 转化方法为基础，分别在细胞壁处理方法、质粒和感受态共同孵育时间、转座诱导温度及转座诱导时间 4 个方面进行改进，成功建立了颉颃菌株 EDR2 的转化体系，并利用该体系获得突变体 5 256株；以苹果树腐烂病菌强致病力菌株 WT03 – 8 为筛选靶标，通过平板对峙法从 EDR2 突变体库中筛选到 21 株抑菌活性显著降低的突变体；PCR 扩增和 Southern 杂交结果表明 21 株突变体全部是转座子 TnYLB 随机插入 EDR2 基因组所得，其中，19 株突变体是转座子单拷贝插入突变；通过 SiteFinding-PCR 方法共扩增获得 18 个转座子插入突变的基因序列，其中，7 个基因与颉颃活性相关。同时，利用 TCA – 丙酮法提取菌株 EDR2 及其突变体的胞外蛋白，通过双向电泳技术、PQuest 8. 0 2D Analysis 软件和质谱技术，对部分差异蛋白点进行分析和鉴定；并对其中 1 个差异蛋白成功进行了外源表达、纯化和颉颃活性测定，得到了一种新的颉颃苹果树腐烂病菌的抗菌蛋白 AP-His。

关键词：突变体库；SiteFinding-PCR；双向电泳；重组蛋白

---

* 基金项目：公益性行业（农业）科研专项（nyhyzx201203034）；“111” 高等学校科研创新引智计划（B07049）

** 第一作者：王娜娜，陕西杨凌人，博士研究生，主要从事植物病害生物防治的相关研究；E-mail：wangnaxiaoyu@ nwsuaf. edu. cn

*** 通讯作者：黄丽丽，教授，主要从事果树病害病原生物学及病害综合防控方面的研究；E-mail：huanglili@ nwsuaf. edu. cn

吉林省波叶大黄一新病害——白粉病*

王博儒[1]，刘淑艳[1,2]，唐淑荣[1]，管观秀[1]
（1. 吉林农业大学农学院，长春 130118；2. 吉林农业大学
食药用菌教育部工程研究中心，长春 130118）

摘 要：波叶大黄（*Rheum rhabarbarum* L.）是蓼科，大黄属多年生草本植物。国内主要分布在黑龙江西部、吉林及内蒙古锡东郭勒盟东部。国外主要分布在俄罗斯（东西伯利亚）、蒙古。2015 年 8 月在吉林省长春市吉林农业大学中药园（43. 61°N，125. 41°E）发现波叶大黄上发生白粉菌病害，主要发生在叶片的正面，发病率为 60% ~70%。本文主要结合形态学特征和分子系统学方法对其病原菌进行鉴定。观察发现：发病初期菌丝体覆盖在叶片正面，成片状，后消失；附着器裂片状，单生或对生，直径（3 –）5 ~10μm；分生孢子梗直立，圆柱形，大小为（7 ~12）μm×（77 ~122）（ –132）μm；脚胞直立，圆柱形，大小为（6 ~12）μm×（47 ~80）μm，其上连接 1 ~3 个短细胞；分生孢子椭圆形，大小为（15 ~23）μm×（30 ~46）μm；芽管末端分枝，生于分生孢子顶端。以上形态特征与 *Erysiphe polygoni* DC. 相似。采用 Chelex – 100 法提取 DNA，测序获得 ITS 序列总长 586bp，其中，ITS1 222bp，5. 8S 154bp，ITS2 185bp，与 *Erysiphe polygoni*（KP076437）的相似度为 99. 82%（585/586）。通过相关报道可知，由 *Erysiphe polygoni* 引起的波叶大黄白粉病为吉林省新病害。

关键词：白粉菌；波叶大黄；*Erysiphe polygoni*；新病害

* 基金项目：戈洛文白粉菌族真菌的分类学与分子系统学研究（2015. 01—2018. 12）

*TaMCA*1, a regulator of cell death, is important for the interaction between wheat and *Puccinia striiformis*

HAO Ying-bin[1], WANG Xiao-jie[1], WANG Kang[1],
LI Hua-yi[1], DUAN Xiao-yuan[2], TANG Chun-lei[1], KANG Zhen-sheng[1]
(1. *State Key Laboratory of Crop Stress Biology for Arid Areas and College of Plant Protection, Northwest A&F University, Yangling, China*;
2. *State Key Laboratory of Crop Stress Biology for Arid Areas and College of Life Science, Northwest A&F University, Yangling, China*)

Abstract: Metacaspase orthologs are conserved in fungi, protozoa and plants, however, their roles in plant disease resistance are largely unknown. In this study, we identified a *Triticum aestivum* metacaspase gene, *TaMCA*1, with three copies located on chromosomes 1A, 1B and 1D. The *TaMCA*1 protein contained typical structural features of type I metacaspases domains, including an N-terminal pro-domain. Transient expression analyses indicated that *TaMCA*1 was localized in cytosol and mitochondria. *TaMCA*1 exhibited no caspase-1 activity in vitro, but was able to inhibit cell death in tobacco and wheat leaves induced by the mouse *Bax* gene. In addition, the expression level of *TaMCA*1 was up-regulated following challenge with the *Puccinia striiformis* f. sp. *tritici* (*Pst*). Knockdown of *TaMCA*1 via virus-induced gene silencing (VIGS) enhanced plant disease resistance to *Pst*, and the accumulation of hydrogen peroxide (H_2O_2). Further study showed that *TaMCA*1 decreased yeast cell resistance similar to the function of yeast metacaspase, and there was no interaction between *TaMCA*1 and *TaLSD*1. Based on these combined results, we speculate that *TaMCA*1, a regulator of cell death, is important during the compatible interaction of wheat and *Pst*.

Key words: *Puccinia striiformis*; Metacaspase; Overexpression; VIGS; Wheat

腐烂病菌在苹果枝条木质部内的扩展动态及其影响因子*

王晓焕**，王彩霞，董向丽，练　森，李保华***

（青岛农业大学农学与植物保护学院/山东省植物病虫害综合防控重点实验室，青岛　266109）

摘　要：苹果腐烂病（*Valsa mali*）是我国苹果上一种重要的枝干病害，主要造成死枝、死树，甚至毁园。腐烂病菌主要从剪锯口和各种伤口侵染，且能在木质部内生长扩展。腐烂病菌在木质部内生长扩展是导致腐烂病连年复发和剪锯口发病的主要原因。为了明确腐烂病菌在苹果枝条木质部内的周年生长扩展动态及其影响因子，本研究采用菌饼接种离体和活体富士苹果枝条的方法，测试温度、枝条龄期、木质部的含水量及枝干组分对腐烂病菌在木质部内生长扩展的影响。结果表明：自然条件下，12 月至次年 2 月，腐烂病菌在活体的富士枝条内扩展速度很慢，3～11 月扩展较快。腐烂病菌在离体富士枝条木质部内的生长扩展温度范围为 0～35℃，最适温度为 30℃。枝条龄期对腐烂病菌丝在枝条内生长扩展速度有一定影响，在 25℃下，接种到当年生离体富士枝条木质部内腐烂病菌，在侵染初期平均每天生长 1.03cm，而在一年生离体枝条上平均每天生长 0.83cm，在两年生离体枝条上平均每天生长 0.80cm。腐烂病菌在木质部内的扩展速度与枝条含水量相关。当枝条的含水量很高时，菌丝在木质部内生长较慢；随枝条含水量降低，扩展速度加快；当枝条的含水量低于一定水平后，菌丝扩展速度减缓。腐烂病菌在木质部和树皮培养基上的生长速度基本一致，在木质部浸出液和残渣中的生长速率虽有差异，但不显著。树体冬季休眠后，腐烂病菌在活体枝条上的生长扩展速度显著快于在离体枝条上。本试验为深入了解苹果腐烂病的发生流行规律和病害防治提供参考依据。

关键词：苹果腐烂病；扩展动态；影响因子

* 基金项目：国家苹果产业技术体系（CARS－28）

** 第一作者：王晓焕，山东聊城人，硕士研究生，研究方向为植物病害流行学；E-mail：15865521984@163.com

*** 通讯作者：李保华，教授，主要从事植物病害流行和果树病害研究；E-mail：baohuali@qau.edu.cn

稻瘟菌海藻－6－磷酸合成酶 Tps1 的重组表达、纯化及晶体生长

王珊珊*，易　龙，赵彦翔，彭友良，刘俊峰**
（中国农业大学植物保护学院植物病理学系，北京　100193）

摘　要：海藻糖普遍存在于除哺乳动物外的各种生物体内，在生物体的抗逆等生理过程中起着重要的作用。因而其合成及加工途径的各种酶一直是很有潜力的药物靶标。研究发现稻瘟菌（*Magnaporthe oryzae*）中海藻糖合成过程中的关键酶——海藻糖－6－磷酸合成酶（MoTps1）参与碳源和氮源代谢的调控，影响稻瘟菌侵染、附着胞形成等致病过程，具有重要的生物学作用。但 MoTps1 在致病过程中发生作用的结构机制尚不明确。本研究尝试通过结构生物学的方式，对稻瘟菌中的 MoTps1 的三维结构进行深入研究，以期为其生物学功能的研究提供一定的结构依据，同时为基于靶标的绿色高效杀菌剂的设计提供结构参考。

首先，笔者构建了 MoTps1 全长和截断体的不同表达载体，并将构建好的载体分别热击转化到大肠杆菌 BL21（DE3）、Rosetta-gami（DE3）2 等表达菌株中进行表达，通过试亲和筛选到了可溶且表达好的载体及表达菌株组合及相应的表达条件。在此基础上，笔者通过亲和层析、离子交换层析、凝胶过滤层析等方法对目标蛋白进行进一步的纯化，制备了 MoTps1 全长和截断体符合晶体生长要求的蛋白样品。笔者利用气相扩散法对纯化好的目标蛋白开展了大量的晶体生长条件的筛选，筛选到了截断体较好的晶体生长条件，获得的晶体 X-ray 衍射分辨率达到 2.7 Å，目前正在根据收集到的数据利用同源模型进行结构解析。与此同时，笔者正在继续进行晶体生长条件的优化，以期获得更高分辨率的衍射数据。并进一步开展 MoTps1 与底物类似物等小分子复合物的结构解析和相互作用机制研究，为探究其在影响碳源和氮源代谢的调控机制和开展基于结构的新型农药设计奠定基础。

关键词：稻瘟菌；MoTps1；截断体；蛋白纯化；晶体生长

---

* 第一作者：王珊珊，硕士研究生，植物病理学专业；E-mail：15600912737@163.com
** 通讯作者：刘俊峰，教授；E-mail：jliu@cau.edu.cn

小麦条锈菌在甘肃省不同地区间的传播关系推断*

王翠翠**，谷医林[1]，骆 勇[2]，马占鸿[1]***

（1. 中国农业大学植物病理系，北京 100193；

2. 美国加州大学 Kearney 农业研究中心，Parlier CA 93648）

摘 要：小麦条锈病由小麦条锈菌（*Puccinia striiformis* f. sp. *tritici*，*Pst*）引起，是小麦生产中一种重要的流行性病害。甘肃陇南地区被认为是我国小麦条锈菌的重要越夏区以及新小种的策源地，明确小麦条锈菌在甘肃省内不同地区间的传播方向以及潜在的传播路线，为小麦条锈病的流行预测和集中防治提供重要的参考依据。本研究于 2014 年秋季以及 2015 年春季在甘肃省 11 个县共采集 355 个单孢系样本，利用 12 对 SSR 引物扩增，分析各地区亚群体的遗传多样性、遗传分化程度、有效群体大小，以及 *Pst* 地区亚群体间历史迁移率（*M*）。结果表明：两个季节甘谷县的 Shannon 信息指数最高（0.69），西和县次之（0.63）。PCA 和 Structure 遗传分化显示 2014 年秋季 3 个地区亚群体间遗传差异显著（$P = 0.001$）；2015 年春季东西方向 6 个县和南北方向 5 个县分别属于两个不同的遗传类群。由 Migrate-n 计算的 2 个季节有效群体大小和历史迁移率结果显示，甘谷和文县有效群体最大（$\Theta > 0.5$），康县和徽县有效群体最小（$\Theta < 0.4$）。文县和徽县之间迁移率最大（8.06），西和与康县间迁移率最小（3.11）。经两两群体似然比检验，陇南和陇中地区为对称迁移，其余多为不对称迁移（$P < 0.05$），主要由陇南传到陇东和陇西。综上，陇南地区为重要的基因交流的贡献者，小麦条锈菌在甘肃省整体的传播趋势为由南向北传播，东西方向主要为陇中分别向陇东和陇西传播。

关键词：小麦条锈菌；遗传分化；对称迁移

---

* 基金项目：973 计划项目（2013CB127700）和国家自然科学基金项目（31371881）

** 第一作者：王翠翠，山东潍坊人，在读博士，主要从事小麦条锈病分子流行学研究工作；E-mail：wangcuicui1111@126.com

*** 通讯作者：马占鸿，宁夏海原人，教授、博士生导师，主要从事植物病害流行与宏观植物病理学研究工作；E-mail：mazh@cau.edu.cn

香蕉枯萎病菌 ISSR 遗传多样性分析*

黄穗萍[1,2**]，莫贱友[1]，郭堂勋[1***]，李其利[1]，唐利华[1]，陈 军[3]，吴玉东[3]，魏 晴[4]

（1. 广西农业科学院植物保护研究所/广西作物病虫害生物学重点实验室，南宁 530007；2. 广西大学，南宁 530004；3. 钦州市植保植检站，钦州 535000；4. 南阳师范学院，南阳 473000）

摘 要：为探索香蕉枯萎病菌（*Fusarium oxysporum* f. sp. *cubense*）遗传多样性与生理小种类型、地理来源和致病力的关系，本研究从 26 条 ISSR 引物中筛选出 14 条多态性较好的引物，优化其退火温度，对来自广西、广东、海南、福建和云南等地香蕉产区的 50 个菌株进行 ISSR 分析。同时采用伤根灌淋法测定香蕉枯萎病菌的致病力。结果显示，14 条特异引物扩增出 237 个条带，多态性条带 161 个，多态性比例为 67.93%。聚类分析显示，在阈值为 0.80 时菌株被分为 8 个类群，所占比例分别为 4%、10%、60%、16%、4%、2%、2%、2%。第三类群全部为香蕉枯萎病菌 4 号生理小种，1 号生理小种集中在第一、二、四和五类群。1 号生理小种的遗传变异较 4 号生理小种大。广西香蕉枯萎病菌与其他省份香蕉枯萎病菌的遗传变异较大。ISSR 聚类划分与生理小种类型有明显相关性，与地理来源有一定相关性，与致病力无关。本研究为该病害流行和防治技术提供参考。

关键词：香蕉枯萎病菌；ISSR 指纹分析

---

* 基金项目：广西农业科学院基本科研业务专项（2015YM01，2015YT39）；广西农业科学院科技发展基金（2015JZ40）；钦州市科学技术局科技创新能力与条件建设计划（2015270204）

** 第一作者：黄穗萍，助理研究员，主要从事植物真菌病害研究工作；E-mail：hsp@gxaas.net

*** 通讯作者：郭堂勋，副研究员，主要从事植物病害防治研究工作；E-mail：guotangxun@gxaas.net

橡胶树白粉菌真核表达启动子的预测、克隆及功能鉴定*

王　义**，梁　鹏，刘文波，林春花，缪卫国***
（海南省热带生物资源可持续利用重点实验室/海南大学环境与植物保护学院，海口　570228）

摘　要：启动子是最重要的基因调控序列之一，正确定位基因转录起始位点和核心启动子是理解转录调控机制的重要方面。本实验室从 2010 年开始橡胶树白粉病菌分子致病机理的研究，以橡胶树无性系热研 7－33－97 为寄主材料，单斑菌株 HO－73 为 *O. heveae* 靶标菌，首次基本完成了橡胶树白粉菌 HO－73 基因组的测序工作，基因组组装大小为 53.4M，已得到橡胶树白粉菌主要基因编码区域。Gene Ontology（GO）注释基因 3 890个，其中涉及 311 条 KEG 代谢途径，低于大麦白粉菌 5 854个基因的水平。在 GO level 3 水平上预测致病相关基因 30 个，少于稻瘟病菌的 136 个，仅占注释基因的 0.6%。0hpi、24hpi、3dpi、30dpi、rubber leaves 5 个转录组文库，GO 注释基因 6 910个，共筛选出橡胶树白粉菌差异表达基因 2 169个。利用 Promoter Scan 等在线启动子预测软件预测启动子，通过 TATA 框、CAAT 框、TSS 以及 CTF、HNF1、NF1、Oct、SRF、TFIID、MLTF、ATF、EIIF、Sp1、NFI、TFIID、GCF 等显著信号的定位和分析我们预测得到疑似启动子 62 个，筛选后现已成功构建重组载体 7 个，三亲杂交转入农杆菌 LBA4404 后以叶盘法转入烟草进行了瞬时表达。

关键词：橡胶白粉菌；启动子；预测；克隆；功能鉴定

---

* 基金项目：海南省重点研发计划项目（ZDYF2016208）；海南自然科学基金创新研究团队项目（2016CXTD002）；973 计划前期研究专项（20 11C B 1116 12）

** 第一作者：王义，在读博士，研究方向为分子植物病理学；E-mail：935497300@ qq. com

*** 通讯作者：缪卫国，博士，教授，研究方向为分子植物病理等；E-mail：weiguomiao1105@126. com

层出镰孢菌转化子的获得及侵染分析*

许苗苗**，戴冬青，刘 俊，马双新，杨 阳，王 宽，曹志艳***，董金皋***
（河北农业大学，真菌毒素与植物分子病理学实验室，保定 071001）

摘 要：玉米鞘腐病是近年来我国主要玉米产区频繁出现的一种新型病害，并且呈现逐年加重的趋势。该病害主要是由层出镰孢（*Fusarium proliferatum*）引起的，严重时可导致玉米倒伏，产量下降。本实验通过农杆菌介导遗传转化技术，将带有绿色荧光蛋白 GFP 基因和抗性基因 HPH 的质粒 pBHt1 导入农杆菌感受态 AGL－1 中，再将转化成功的农杆菌与层出镰孢野生型共同培养，经过 HPH 抗性筛选得到疑似转化子。然后进行荧光显微观察和 T－DNA 插入转化子的分子检测，确定得到100多株转化子并对其进行表型分析和产孢量测定。在玉米开花期，分别用层出镰孢野生型和转化子孢悬液接种玉米叶鞘，10d 以后用植物病斑扫描仪测量叶鞘病斑面积，确定转化子致病性强弱。以上实验结果如下。

（1）根据转化子的形态将其分为三类：第一类，生长速率与野生型相近，其气生菌丝和营养菌丝较发达，菌落质地较疏松，呈绒毛状，边缘呈放射状，菌落正面呈白色、浅粉色、浅紫色等，反面呈粉紫色和黄色。第二类，生长速率比野生型变慢，气生菌丝和营养菌丝较短，菌落质地较稀松，成棉絮状，边缘不规则，菌落正反面都呈白色。第三类，生长速率与野生型相比明显缓慢，气生菌丝和营养菌丝既短又细，菌落质地黏稠，成地毯状，边缘较整齐，菌落正反面都呈白色。

（2）部分转化子的产孢量和致病性发生了明显变化，FP62 和 FP80 的产孢量和致病性较野生型显著降低，FP106 的致病性较野生型显著升高。

关键词：玉米鞘腐病；层出镰孢；农杆菌介导的遗传转化

---

* 基金项目：现代农业产业技术体系（CARS－02）
** 第一作者：许苗苗，硕士研究生，研究方向为鞘腐病致病性的研究；E-mail：18232185202@163.com
*** 通讯作者：曹志艳；E-mail：caoyan208@126.com
董金皋；dongjingao@126.com

RNA 编辑调控禾谷镰刀菌 *RID*1 基因功能的研究*

贺　怡[1**]，江　聪[1]，刘慧泉[1]，许金荣[1,2***]

（1. 西农—普度大学联合研究中心/旱区作物逆境生物学国家
重点实验室/西北农林科技大学植物保护学院，杨凌　712100；
2. 美国普渡大学植物及植物病理系，印第安纳州　IN47907）

摘　要：基因组稳定性对丝状真菌的生长发育至关重要。为了限制重复 DNA 片段在基因组中的积累，一种重复序列诱导的点突变机制（RIP）起到了关键的作用，该机制特异发生于丝状真菌的有性阶段。有研究表明，在粗糙脉孢菌中，一个 DNA 甲基转移酶 Rid1 对于 RIP 发生是必须的，这也是目前已知的唯一直接影响 RIP 机制的蛋白。我们在禾谷镰刀菌中鉴定了 *RID*1 的同源基因 *FgRID*1。分析发现 *FgRID*1 本身是一个假基因，在基因的开放阅读框内部存在一个提前终止密码子，因而仅能表达部分截短的蛋白。对有性阶段 RNA-seq 数据进行分析发现该终止密码子在有性生殖阶段会发生 A 到 I 的编辑，从而使基因发生通读并编码出完整的蛋白。为了研究 RNA 编辑对禾谷镰刀菌 *FgRID*1 基因功能乃至 RIP 机制的调控，我们首先对 *FgRID*1 基因进行了敲除，*Fgrid*1 突变体在营养菌丝生长、产孢和致病力等方面与野生型 PH－1 没有明显差异。在有性生殖阶段，*Fgrid*1 突变体可以产生正常形态的子囊壳，但子囊孢子喷发存在缺陷。该突变体在有性生殖后期会产生大量畸形且形态不一的子囊孢子，表明 *FgRID*1 基因在子囊孢子形态建成和喷发过程中发挥了重要作用。此后，我们构建提前终止的互补载体并转入 *Fgrid*1 突变体中，转化子仍表现为突变体的表型。而将终止密码子突变为编辑后状态则可以完全互补突变体的功能缺陷。这表明 RNA 编辑直接影响了 *FgRID*1 基因的功能，并可能参与对 RIP 机制的调控。研究有助于揭示 RNA 编辑和 RIP 这两大表观遗传机制之间的关系。

关键词：禾谷镰刀菌；重复序列诱导的点突变机制；RNA 编辑

---

* 基金项目：西北农林科技大学青年英才之“卓越新星 ”培育计划

** 第一作者：贺怡，硕士研究生

*** 通讯作者：许金荣，国家千人计划特聘教授，主要从事病原真菌功能基因组学研究；E-mail：jinrong@ purdue. edu

真菌中 A－to－I RNA 编辑的适应性进化研究*

王秦虎[1**]，刘慧泉[1***]，许金荣[1,2]

(1. 西农—普度大学联合研究中心/旱区作物逆境生物学国家重点实验室/
西北农林科技大学植物保护学院，杨凌　712100；2. 美国普渡大学
植物及植物病理系，印第安纳州　IN47907)

摘　要： 由 ADAR 酶催化介导的 A-to-I RNA 编辑是动物中普遍存在的一种重要的 RNA 修饰机制，通过水解脱氨基作用将 RNA 分子中的腺苷（A）转变成肌苷（I），而肌苷在翻译过程中被识别为鸟苷（G），因此，发生在 mRNA 上的 A-to-I RNA 编辑有可能导致所编码的蛋白发生氨基酸替代。随着高通量测序和生物信息学技术的发展，目前已在人和动物中鉴定出大量 A-to-I RNA 编辑位点，但是，绝大多数都发生在非编码区，位于编码区内可导致氨基酸改变的非同义编辑位点非常有限。由于 ADAR 酶是动物中特异进化出来的一类蛋白，动物以外的生物不存在 ADAR，因而被认为不具有 A-to-I RNA 编辑。但是，课题组近期研究发现禾谷镰刀菌等子囊真菌中存在不依赖于 ADAR 的 A-to-I RNA 编辑机制，并且该机制特异发生在有性生殖过程中，总共鉴定到 2.6 万多个编辑位点，其中，绝大部分为非同义 RNA 编辑。本研究通过比较非同义 RNA 编辑和同义 RNA 编辑（不导致氨基酸改变）发生的频率和编辑水平，研究真菌中 A-to-I RNA 编辑的适应性进化问题。研究发现，非同义 RNA 编辑发生的频率和编辑水平普遍比同义 RNA 编辑高，非同义 RNA 编辑偏爱编辑功能更加重要、进化上更加保守的基因，但是，偏爱编辑不保守的位点，而且编辑具有密码子特异性，倾向于编辑导致氨基酸理化性质发生改变的位点。上述研究表明真菌中的非同义 RNA 编辑是有重要作用的，是正选择的结果。此外，研究还发现同义 RNA 编辑可增加密码子偏好性，对蛋白翻译效率具有重要作用。

关键词： RNA 编辑；适应性；赤霉菌；正选择

---

* 基金项目：西北农林科技大学青年英才之“卓越新星”培育计划
** 第一作者：王秦虎，博士后，主要从事小麦赤霉菌和生物信息学研究
*** 通讯作者：刘慧泉，副研究员，主要从事小麦赤霉菌和比较基因组学研究；E-mail：liuhuiquan@ nwsuaf. edu. cn

禾谷镰刀菌腺苷脱氨酶基因功能研究*

陈凌峰[1**]，刘慧泉[1]，江　聪[1]，许金荣[1 2***]
（1. 西农—普度大学联合研究中心/旱区作物逆境生物学国家
重点实验室/西北农林科技大学植物保护学院，杨凌　712100；
2. 美国普渡大学植物及植物病理系，印第安纳州　IN47907）

摘　要：RNA 编辑是一种改变 DNA 编码的 RNA 核苷酸序列的机制。近些年在多种生物体中报道了不同类型的 RNA 编辑，其中，最有名的是 A-to-I RNA 编辑，其将 RNA 分子中的腺苷（A）通过水解脱氨基作用转变为肌苷（I）。A-to-I RNA 编辑可发生在 mRNA 和 tRNA 上，发生在 tRNA 上的 A-to-I 编辑由 ADAT 酶催化，广泛存在于各种生物中，包括真核生物、细菌和古细菌，而发生在 mRNA 上的 A-to-I 编辑由 ADAR 酶催化，之前只在动物中报道。课题组近期研究中发现禾谷镰刀菌有性时期产生的子囊壳中也存在 mRNA 上的 A-to-I 编辑现象，由于 ADAR 酶是动物中特异进化出来的，真菌不具有 ADAR 酶，因此，禾谷镰刀菌中的这种 A-to-I 编辑很可能由其他酶负责。为了鉴定真菌中负责 A-to-I 编辑的酶，本研究通过保守结构域分析，共鉴定到 7 个具有腺苷脱氨酶结构域的基因。基因敲除研究发现 *Fgamd*1 敲除突变体子囊壳产生数量、大小，子囊孢子形态、数量、喷发方面均存在显著缺陷，而其他 6 个基因的敲除突变体无明显表型。根据酵母中的报道，AMD1 主要参与催化 AMP 形成 IMP 和氨的脱氨基作用。在禾谷镰刀菌 *FgAMA*1 是否与 A-to-I RNA 编辑有关，还需要进一步研究。

关键词：禾谷镰刀菌；编辑；基因

---

* 基金项目：西北农林科技大学青年英才之“卓越新星 ”培育计划

** 第一作者：陈凌峰，硕士研究生

*** 通讯作者：许金荣，国家千人计划特聘教授，主要从事病原真菌功能基因组学研究；E-mail：jinrong@ purdue. edu

禾谷镰刀菌 *FgAMA*1 基因在有性发育中的调控作用*

郝超峰[1**]，刘慧泉[1]，江　聪[1]，许金荣[1,2***]
（1. 西农—普度大学联合研究中心/旱区作物逆境生物学国家重点实验室/西北农林科技大学植物保护学院，杨凌　712100；2. 美国普渡大学植物及植物病理系，印第安纳州　IN47907）

摘　要：由禾谷镰刀菌引起的小麦赤霉病是小麦上为害最为严重的病害之一。禾谷镰刀菌有性生殖产生的子囊孢子在小麦扬花期侵染小麦穗部的颖壳，是小麦赤霉病的初侵染源。研究禾谷镰刀菌有性生殖调控机理有望为小麦赤霉病防控新策略的制定提供理论指导。课题组前期研究发现禾谷镰刀菌中减数分裂后期促进复合物激活基因 *FgAMA*1 编码框中具有一个提前终止密码子，在营养生长阶段表达不出完整的蛋白，而在有性生殖阶段提前终止密码子经过 RNA 编辑后转变为氨基酸密码子，从而表达出完整的蛋白。本研究通过基因敲除方法获得了 *FgAMA*1 基因的缺失突变体，研究发现基因缺失突变体的菌落形态、生长速率，以及分生孢子形态、数量和萌发均与野生型无显著差异，表明 *FgAMA*1 在禾谷镰刀菌营养生长和无性繁殖中不具有明显作用。而在有性生殖过程中，缺失突变体虽能产生正常数量和形态的子囊壳，但子囊孢子呈球形，大部分单细胞、具有两个细胞核，部分子囊孢子可分裂形成新的细胞，导致一个子囊内具有 8 个以上的子囊孢子。上述研究表明 *FgAMA*1 特异性调控禾谷镰刀菌子囊孢子的形成和发育。目前正在进一步揭示 *FgAMA*1 的调控机理。

关键词：小麦赤霉菌；有性生殖；子囊孢子；减数分裂；RNA 编辑

---

* 基金项目：西北农林科技大学青年英才之“卓越新星”培育计划

** 第一作者：郝超峰，博士研究生

*** 通讯作者：许金荣，国家千人计划特聘教授，主要从事病原真菌功能基因组学研究；E-mail：jinrong@ purdue. edu

禾谷镰刀菌组蛋白乙酰转移酶基因的功能分析*

吴春兰[1**]，江　聪[1]，刘慧泉[1]，许金荣[1,2***]

（1. 西农—普度大学联合研究中心/旱区作物逆境生物学国家重点实验室/西北农林科技大学植物保护学院，杨凌　712100；
2. 美国普渡大学植物及植物病理系，印第安纳州　IN47907）

摘　要：小麦赤霉病是重要的小麦真菌病害。该病害由禾谷镰刀菌引起，会导致小麦产量下降。更为重要的是，禾谷镰刀菌产生的真菌毒素 DON 会在小麦及小麦制品中大量残留，严重危害人畜健康，造成食品安全问题。组蛋白乙酰化修饰作为一种重要的表观遗传调控机制，与基因表达密切相关。为了探究组蛋白乙酰化在禾谷镰刀菌致病产毒阶段的作用。我们鉴定了禾谷镰刀菌中的几个组蛋白乙酰转移酶基因：*HAT*1、*GCN*5、*TAF*1、*ESA*1、*SAS*3、*SAS*2、*ELP*3、*HPA*2/*HPA*3 和 *RTT*109。并对它们进行了敲除。由于多次尝试无法获得 TAF1 和 ESA1 的敲除突变体，判定这两个基因为致死基因。在获得的敲除突变体中，*gcn*5、*sas*3、*elp*3 突变体严重影响菌丝生长速率及气生菌丝生长。*gcn*5 基因敲除突变体不产孢。*sas*3 和 *elp*3 突变体可以正常产孢，但是所产孢子较野生型长，*rtt*109 突变体所产孢子则普遍粗短。*GCN*5 和 *SAS*3 的缺失完全阻断了有性发育过程。*rtt*109 缺失突变体可以形成子囊壳，但是，较野生型子囊壳明显偏小，且子囊无法正常发育。*elp*3 突变体子囊壳和子囊发育均无明显异常，但是，产生子囊孢子多为一个隔膜。*gcn*5、*sas*3、*elp*3 和 *rtt*109 突变体致病力显著下降，其中，*gcn*5、*sas*3 和 *elp*3 突变体的 DON 合成量严重降低。综合以上结果，*GCN*5、*SAS*3、*ELP*3 和 *RTT*109 这几个组蛋白乙酰转移酶对禾谷镰刀菌的生长发育、有性生殖和致病产毒至关重要。

关键词：禾谷镰刀菌；组蛋白；乙酰化修饰

---

* 基金项目：973 计划子课题（2013CB127702）

** 第一作者：吴春兰，硕士研究生

*** 通讯作者：许金荣，国家千人计划特聘教授，主要从事病原真菌功能基因组学研究；E-mail：jinrong@ purdue. edu

禾谷镰刀菌 *FgSNU66* 基因的功能研究*

孙蔓莉[1**]，金巧军[1]，许金荣[1,2***]
（1. 西农—普度大学联合研究中心/旱区作物逆境生物学国家
重点实验室/西北农林科技大学植物保护学院，杨凌　712100；
2. 美国普渡大学植物及植物病理系，印第安纳州　IN47907）

摘　要：由剪接体催化完成的前体 mRNA 剪接是真核生物中重要的生物过程。作为 U4/U6 · U5 tri-snRNP 中的一个组成蛋白，*SNU66* 首先在啤酒酵母（*Saccharomyces cerevisiae*）中被发现。它在剪接体从复合体 B 转为激活态的复合体 B 的过程中发挥着重要作用。实验室前期研究发现蛋白激酶基因 *Fgprp4* 敲除突变体生长速度异常缓慢，但培养 2 周后，接近 10% 的 *Fgprp4* 突变体菌落边缘会产生快速生长的自发抑制突变体。本研究通过测序分析发现，23 个自发抑制突变体中鉴定到了与 *FgSNU66* 有关的突变，且这些突变发生在 8 个不同的位置。其中，603RLKKIEDEK611 和 R644*突变能够回复 *Fgprp4* 突变体的生长缺陷，说明 *FgSNU66* 突变具有抑制 *Fgprp4* 突变体生长缺陷的功能。亚细胞定位研究发现，*FgSNU66* 定位于细胞核，而 R477H 和 R477C 突变并不影响 *FgSNU66* 的亚细胞定位。我们推测鉴定得到的 8 个突变位点在一定程度上影响了 U4/U6 · U5 tri-snRNP 蛋白复合体不同组分蛋白之间的互作。

关键词：剪接体；SNU66；自发抑制突变

---

* 基金项目：973 计划子课题（2013CB127702）

** 第一作者：孙蔓莉，博士研究生

*** 通讯作者：许金荣，国家千人计划特聘教授，主要从事病原真菌功能基因组学研究；E-mail：jinrong@ purdue. edu

禾谷镰刀菌 *FgSAD*1 基因在前体 mRNA 剪接中的功能研究*

李晓平[1**]，宋超妮[1]，金巧军[1***]，许金荣[1,2]

（1. 西农—普度大学联合研究中心/旱区作物逆境生物学国家重点实验室/西北农林科技大学植物保护学院，杨凌 712100；
2. 美国普渡大学植物及植物病理系，印第安纳州 IN47907）

摘 要：在酿酒酵母及人类中，Sad1 蛋白参与了剪接过程中 U4/U6 di-snRNP 的装配，有助于维持 U4/U6 · U5 tri-snRNP 的完整性。然而，我们对 Sad1 蛋白在剪接过程中具体的功能知之甚少。据之前的研究报道，在禾谷镰刀菌 *Fgprp*4 突变体恢复生长速率的角变子中测到基因 *FgPRP*6、*FgPRP*8、*FgBRR*2、*FgPRP*31 上出现角变位点。本实验对 310 个角变子的 *FgSAD*1 基因测序，共测到 14 个角变子，7 个角变位点。这 7 个位点分别位于基因的 3 个区域。尽管 14 个角变子均恢复了菌丝的生长速率，但它们在有性生殖及致病方面仍存在缺陷。我们验证了 FgSad1 中角变位点 L512P 能够恢复 *Fgprp*4 缺失突变体的生长速率。与酿酒酵母中的同源基因相比，*FgSAD*1 基因具有一段特殊的富含丝氨酸与精氨酸的 N 端序列。本实验对 N 端的 70 个氨基酸进行敲除，突变体具有生长速率变缓并容易角变的特征，说明 N 端序列对于 FgSad1 的功能具有重要作用，FgPrp4 的磷酸化位点可能位于该区域。

关键词：Sad1；剪接；角变；恢复生长速率

---

* 基金项目：973 计划子课题（2013CB127702）
** 第一作者：李晓平，硕士研究生
*** 通讯作者：金巧军，副教授，主要从事小麦赤霉病菌研究；E-mail：jqiaojun@ yahoo. com

禾谷镰刀菌中与 FgTub2 互作的蛋白鉴定及其功能分析*

王 欢[1**]，王晨芳[1]，许金荣[1,2***]
(1. 西农—普度大学联合研究中心/旱区作物逆境生物学国家
重点实验室/西北农林科技大学植物保护学院，杨凌 712100；
2. 美国普渡大学植物及植物病理系，印第安纳州 IN47907)

摘 要：禾谷镰刀菌是引起小麦赤霉病的主要病原菌，其产生 DON 等毒素能严重危害人畜健康。微管是目前我国生产上广泛使用的苯并咪唑类杀菌剂的作用靶标，同时也是重要的细胞骨架，在真菌极性生长，物质运输、细胞器定位、有丝分裂和维持细胞形态等众多生物学过程中具有重要功能。禾谷镰刀菌具有 2 个 β－微管蛋白基因（*FgTUB*1，*FgTUB*2），课题组前期研究过程中发现，*FgTUB*1 与有性生殖密切相关，而 *FgTUB*2 与营养生长相关。微管发挥作用可能依赖于其微管相关蛋白（microtubule associated proteins），这种功能分化可能与两个微管蛋白各自的微管相关蛋白调控有关。*FgTUB*2 突变体生长不稳定，容易形成角变子。角变子的产生可能与 *FgTUB*2 相关基因上发生突变有关。通过对 *FgTUB*2 角变子进行重测序分析，鉴定到一个候选基因 *Kar*9 上发生无义突变。后续将对该基因敲除突变体进行表型分析，与 FgTub2 以及其他已知微管相关蛋白的互作关系进行研究。

关键词：禾谷镰刀菌；基因敲除；有性生殖；微管相关蛋白；微管辅因子

* 基金项目：973 计划子课题（2013CB127702）
** 第一作者：王欢，硕士研究生
*** 通讯作者：许金荣，国家千人计划特聘教授，主要从事病原真菌功能基因组学研究；E-mail：jinrong@ purdue. edu

禾谷镰刀菌候选效应蛋白筛选*

徐　妍[1**]，王晨芳[1]，许金荣[1,2***]
（1. 西农—普度大学联合研究中心/旱区作物逆境生物学国家
重点实验室/西北农林科技大学植物保护学院，杨凌　712100；
2. 美国普渡大学植物及植物病理系，印第安纳州　IN47907）

摘　要：禾谷镰刀菌是小麦赤霉病的主要致病菌。在病原菌侵入寄主植物的早期阶段，会分泌一些效应蛋白进入植物体内，从而诱发一系列的抗病反应。因而，对这一时期的分泌蛋白进行研究，可以筛选到参与致病过程的效应蛋白，从而有助于揭示病原菌与寄主植物互作的机制。本研究以实验室所构建的 RNA-seq 数据库为基础，对预测的 296 个未知功能分泌蛋白基因进行了初筛，获得了 44 个只在侵染早期特异上调表达的基因，分别对这些基因进行了敲除，将获得的突变体在小麦穗和玉米须上进行了致病力测试，目前已初步筛到了一个候选效应蛋白 *FgSP1*。后续将对该蛋白在侵染过程中的定位以及寄主植物互作靶标进行研究，以期对该基因的功能有更明确的认识。

关键词：禾谷镰刀菌；分泌蛋白；效应蛋白

* 基金项目：973 计划子课题（2013CB127702）
** 第一作者：徐妍，硕士研究生
*** 通讯作者：许金荣，国家千人计划特聘教授，主要从事病原真菌功能基因组学研究；E-mail：jinrong@ purdue. edu

禾谷镰刀菌 *FgSRP*1 基因的基本功能鉴定*

张艺美[1**]，许金荣[1,2***]
（1. 西农—普度大学联合研究中心/旱区作物逆境生物学国家重点实验室/西北农林科技大学植物保护学院，杨凌　712100；
2. 美国普渡大学植物及植物病理系，印第安纳州　IN47907）

摘　要：禾谷镰刀菌（*Fusarium graminearum*）是引起小麦和大麦赤霉病的主要病原菌，该病是一种世界性的麦类病害，发病的麦粒中含有对人和牲畜有害的毒素。人类和其他模式生物基因组的全测序揭示了可变剪接的存在，而 SR 蛋白家族在前体 mRNA 的组成型剪接和可变剪接中起着重要的作用。SR 蛋白具有保守结构域，在 N 端含有一个或两个 RNA 结合结构域（RBD），C 端含有不同数量的 RS 二肽结构域。裂殖酵母中 *SRP*1 是 SR 蛋白家族的一员，*SRP*1 的 RBD 结构域过量表达会抑制前体 mRNA 的剪接。通过序列比对发现，禾谷镰刀菌中存在保守的 *SRP*1 的同源基因，但至今它的功能还没被报道。

本研究通过 split-PCR 和 PEG－介导的原生质体转化方法将 *FgSRP*1 基因从禾谷镰刀菌的野生型 PH－1 基因组中敲除，得到目的基因缺失突变体，并对突变体进行了基因互补实验，获得基因的互补转化子；通过对基因缺失突变体和互补转化子的菌落形态、产孢量和致病力等表型研究发现，和 PH－1 相比，突变体的生长速率稍微减慢，几乎不能产生分生孢子，对小麦和玉米须的侵染能力显著下降。通过亚细胞定位发现 *FgSRP*1 和只含 *FgSRP*1 C 端部分的突变体定位于细胞核，而只含 *FgSRP*1 N 端的突变体在细胞核和细胞质中均有分布，这和前人的研究 SR 蛋白的 RS 结构域起核定位信号作用相一致。敲除 147～149 位保守的 SRS 氨基酸发现，突变体和 PH－1没有显著的差别。通过双分子荧光互补实验发现 *FgSRP*1 和 *FgPRP*4，*FgDSK*1 存在互作，推测 *FgSRP*1 可能是 *FgPRP*4 和 *FgDSK*1 的磷酸化底物。

关键词：禾谷镰刀菌；*SRP*1；无性生殖；磷酸化

---

* 基金项目：973 计划子课题（2013CB127702）

** 第一作者：张艺美，博士研究生

*** 通讯作者：许金荣，国家千人计划特聘教授，主要从事病原真菌功能基因组学研究；E-mail：jinrong@ purdue. edu

玉米圆斑病病原的快速检测*

马庆周**，李　跃，马乐乐，武海燕，耿月华***，张　猛***
（河南农业大学植物保护学院，郑州　450002）

摘　要：玉米生平脐蠕孢菌（*Bipolaris zeicola*）是引起玉米圆斑病的病原菌。本研究通过对玉米生平脐蠕孢菌及其近似种的延长因子 1α 基因（elongation factor 1α，EF－1α）部分序列进行比对，设计出玉米生平脐蠕孢菌的特异性引物 Y-EF-F 和 Y-EF-R，可以从对应的目的菌株中扩增出 137bp 的特异片段，而其余的 17 个参试菌株扩增结果为阴性。灵敏度实验证明该对引物可以检测到目标 DNA 的浓度为 1pg/μL。用玉米生平脐蠕孢菌接种玉米叶片，以接种发病的病组织 DNA 为模板，利用引物 Y-EF-F 和 Y-EF-R 进行 PCR 扩增，同样可以扩增出 137bp 的特异性条带，而健康玉米组织 DNA 中未能扩增出任何条带，表明该方法可用于快速、准确和灵敏地检测玉米组织中的玉米生平脐蠕孢菌，为玉米圆斑病快速检测，及早采取防治措施提供积极的指导意义。

关键词：玉米生平脐蠕孢；EF－1 α 序列；特异性引物；PCR 检测

* 基金项目：河南省高校科技创新人才支持计划（13HASTIT007）；国家自然科学基金（No. 31171804，30970016）
** 第一作者：马庆周，河南南阳人，研究生，主要从事植物菌物病原研究；E-mail：15237171177@163. com
*** 通讯作者：张猛，教授，主要从事植物菌物病害研究；E-mail：zm2006@126. c0m
耿月华，讲师，主要从事植物菌物病害研究；E-mail：gengyuehua@163. com

小麦近等基因系 *TcLr*19 及其感病突变体与叶锈菌互作的转录组分析

张维宏，范学锋，安　哲，杨文香，刘大群

（河北农业大学植物保护学院，河北省农作物病虫害生物防治工程技术研究中心，保定　071000）

摘　要：小麦抗叶锈病基因 *Lr*19 具有很强的抗叶锈性，目前很少发现对其有毒力的小麦叶锈菌，克隆该基因，研究其抗病机制对于该基因的高效利用具有重要意义。利用基因突变群体进行分析是分离鉴定功能基因的重要途径。为了解 *Lr*19 的结构和功能，本研究以近等基因系 *TcLr*19 以及其 EMS 诱导的感叶锈病突变体 MuTcLr19 M_2 –6 为材料，利用转录组测序技术分析和比较了野生型和突变型 *TcLr*19 在叶锈菌侵染 0、24、144h（hpi）的差异表达基因。借助 Gene Ontology（GO）数据库、KEGG pathway 数据库对差异表达基因的功能和可能参与的分子调控途径进行了分析。主要结果如下：每个小麦材料平均测出 26 498 522条 clean reads，67. 64% 的测序序列能够定位到小麦基因组上；表达基因（RPKM > 1）占总数的 44. 13%。小麦受叶锈菌诱导后，在 MuTcLr19 M_2 –6 与 *TcLr*19 中共筛选出差异表达基因（DEGs）10 671个。在 0hpi 筛选到 3 908个，其中上调 2 503个，下调 1 405个；在 24 hpi 筛选到 2 011个，其中上调 1 147个，下调 864 个；在 144hpi 筛选到 4 752个，其中上调 1 781个，下调 2 971个。在 3 个时间点均表达的共同差异基因有 510 个，其中 190 个（37. 25%）位于抗叶锈基因 *Lr*19 所在的 7DL 染色体上。对小麦与叶锈菌互作的关键点 24hpi 的差异表达基因进行了 GO 功能富集，共注释 1 341个 DEGs，富集到 28 个 term，分布在分子功能、生物过程和细胞组分三大类中。其中，分子功能部分有 18 个，包括：果糖 1，6 – 二磷酸 1 – 磷酸酶活性、碳水化合物磷酸酶活性、糖磷酸酶活性、甘氨酸脱氢酶（脱羧基）活性以及氧化还原酶等；生物进程部分有 8 个：包括了甘氨酸（分解）代谢、丝氨酸家族氨基酸（分解）代谢、α – 氨基酸分解代谢等；细胞组分部分有 2 个，分别为光系统 I 反应中心和光系统 I。金属离子结合、阳离子结合、钙离子结合、特异 DNA 序列结合、致病机理等 term 涉及的 DEGs 最多。对 24 hpi KEGG 通路富集分析发现，差异基因除显著富集在了光合固碳、碳代谢、二羧酸代谢、卟啉与叶绿素代谢、磷酸戊糖途径、光合作用、氮代谢、果糖与甘露糖代谢等基础代谢通路外，126 个基因富集到次生代谢产物的合成途径，23 个差异基因富集在植物与病原物互作通路上，包括 PR 蛋白（PR1、PRS2）、钙依赖型蛋白激酶（CDPK）、钙结合蛋白（CML31）、HSP90、转录因子（WRKY24，WRKY8）等几类与植物抗病反应相关的基因。用 qRT-PCR 方法验证部分差异表达的基因，其结果与转录组测序基本一致。本研究通过对 *TcLr*19 抗病野生型与感病突变型在叶锈菌诱导下的转录组分析，为深入了解小麦与叶锈菌的互作及 *TcLr*19 抗病、感病机理奠定了基础。

关键词：小麦；抗叶锈病基因；突变体；转录组；差异表达基因

玉米大斑病菌 *HLH* 基因家族在其各个生长时期表达规律分析[*]

赵 洁[**]，冯胜泽，赵立卿，刘星晨，郑亚男，
巩校东，董金皋[***]，谷守芹[***]，韩建民[***]
（河北农业大学真菌毒素与植物分子病理学实验室，保定 071001）

摘 要：玉米大斑病（Northern corn leaf blight）是由大斑刚毛座腔菌（*Setosphaeria turcica*）引起的一种威胁玉米生产的重要叶部真菌病害，常给玉米生产造成严重的经济损失。深入研究调控病菌侵染及致病的分子机制对玉米大斑病的有效防治具有重要意义。

研究发现，在植物病原真菌中 *HLH*（helix-loop-helix）为一类转录调控因子基因家族，该类基因参与调控植物病原真菌的生长发育和致病过程。本研究利用生物信息学方法，在玉米大斑病菌基因组中鉴定了 14 个 *HLH* 基因，分别命名为 *StHLH*1、*StHLH*2、*StHLH*3、*StHLH*4、*StHLH*5、*StHLH*6、*StHLH*7、*StHLH*8、*StHLH*9、*StHLH*10、*StHLH*11、*StHLH*12、*StHLH*13 和 *StHLH*14；并利用 Real time PCR 技术，分析了这 14 个 HLH 基因在玉米大斑病菌菌丝发育、分生孢子形成、芽管萌发、附着胞及侵入丝形成等 5 个时期的基因相对表达水平。结果发现，*StHLH*2、*StHLH*4、*StHLH*5、*StHLH*8、*StHLH*10 和 *StHLH*12 在菌丝形成时期的相对表达水平显著高于其他生长时期；*StHLH*3 在芽管时期的相对表达量显著低于其他生长时期；*StHLH*13 在芽管和侵入丝时期都显著低于其他生长时期；*StHLH*1、*StHLH*9 在侵入丝时期的相对表达水平显著高于其他生长时期；*StHLH*11、*StHLH*14 在侵入丝时期的相对表达量要显著低于其他生长时期；而 *StHLH*6、*StHLH*7 在各个时期都比较活跃，相对表达量均较高。结果表明，不同的 *HLH* 基因在玉米大斑病菌的不同生长时期具有不同的调控作用，本研究为进一步明确调控玉米大斑病菌生长发育及其致病性的分子机制鉴定了基础。

关键词：玉米大斑病菌；HLH 基因家族；表达规律

* 基金项目：国家自然科学基项目（31271997）
** 第一作者：赵洁，硕士研究生，研究方向为植物学；E-mail：295277949@qq.com
*** 通讯作者：董金皋，教授，博士生导师；E-mail：gushouqin@126.com
谷守芹，教授，博士生导师；E-mail：gushouqin@126.com
韩建民，教授，硕士生导师；E-mail：hanjmnd@163.com

小麦叶锈菌有性阶段分子鉴定

康　健，张　林，闫红飞，刘大群
（河北农业大学植物保护学院/河北省农作物病虫害生物
防治工程技术研究中心/国家北方山区农业工程技术研究中心，保定　071001）

摘　要：小麦叶锈菌通过有性生殖可产生更多基因型重组的生理小种，在合适的条件下，可造成小麦叶锈病再次大规模流行。20 世纪 90 年代初王焕如先生等通过生物学接种已证实了亚欧唐松草（*Thalictrum minus* L.）和瓣蕊唐松草（*T. petaloideum* L.）是我国小麦叶锈菌的转主寄主。本实验应用小麦叶锈菌特异引物 EST－6 对采集自内蒙古自治区卓资的瓣蕊唐松草（*T. petaloideum*）、河北承德的东亚唐松草（*T. minus*）及河北崇礼的展枝唐松草（*T. squarrosum*）上的锈孢子基因组 DNA 进行 PCR 扩增检测。结果显示在这 3 个地区都检测到小麦叶锈菌的存在，但所占比例不同。其中河北承德锈孢子样品中小麦叶锈菌所占比例最高，为 54.9%；其次为河北崇礼，所占比例为 44.9%；在内蒙古卓资样品中所占比例最低，仅为 4.4%。从分子水平证明了自然条件下小麦叶锈菌可以侵染唐松草，同时唐松草上还有其他锈病存在，但在地区间存在差异。

关键词：小麦叶锈菌；唐松草；有性生殖；分子检测

榛子叶斑病病原菌生物学特性研究*

孙　俊[1**]，刘志恒[2]，刘广平[1]，魏松红[2]
（1. 辽宁省经济林研究所，大连　116031；2. 沈阳农业大学，沈阳　110161）

摘　要：为了对辽宁省近年发生的新病害——榛子叶斑病病原菌的生物学特性进行系统测定。采用已经鉴定并报道的榛子叶斑病致病菌榛叶点霉（*Phyllosticta coryli*）菌株，分别置于设定条件下，采用十字交叉法测量菌落直径，分别测定该病原菌菌丝及分生孢子的生物学特性。对该病害的症状及病原菌形态进行了描述，并对病菌的生物学特性进行了系统测定：病菌菌丝生长以 PSA 培养基为最适；能有效利用多种糖和氮源，分别以麦芽糖和甘氨酸最佳；菌丝适宜生长温度为 20～25℃；最适 pH 值为 7～9；光照可促进菌丝生长；菌丝致死温度为 61℃。病菌分生孢子萌发适宜温度为 15～25℃；最适 pH 值为 5；黑暗条件下对孢子更易萌发；分生孢子致死温度为 58℃。对病原菌菌丝和分生孢子的生物学特性进行了系统测定，不同培养条件下，病原菌菌丝生长和分生孢子萌发均表现出明显差异。

关键词：榛子；叶斑病；榛叶点霉；生物学特性

* 基金项目：辽宁省农业领域青年科技创新人才培养计划项目（2014014）

** 第一作者：孙俊，辽宁省大连市，博士，高级工程师，主要从事干果树种的病害研究工作；E-mail：sunjun3200@163.com

一种通过激活标签筛选稻瘟菌激发子的方法*

房雅丽**，郑晓敏，范 军***
（中国农业大学植物病理学系，北京 100193）

摘 要：在自然界中，非寄主抗性是一种广泛存在的植物抗病现象，它具有广谱和持久的特性。目前尚不清楚非寄主植物的先天免疫系统是如何识别病原菌从而激发下游的防卫反应。因此，本研究以本生烟与稻瘟菌为互作模式，利用经改造的激活标记双元载体构建了稻瘟菌基因组文库，通过农杆菌在烟草上瞬时表达稻瘟菌基因，筛选能够引起植物强烈免疫反应的克隆。本文分析了瞬时表达体系中35S启动子和增强子在稻瘟菌基因激活表达过程中的作用效果，发现35S启动子的作用更为明显。同时，通过筛选大约5 000个克隆，得到1个可以引起烟草细胞死亡的激发子基因。以上结果表明直接从基因功能出发，筛选病原菌潜在激发子的研究策略是可行的，它将有效促进非寄主抗性的遗传解析研究。

关键词：非寄主抗性；先天免疫激发子；基因组文库

---

* 基金项目：教育部高等学校博士学科点专项科研基金博导类资助课题（20130008110005）
** 第一作者：房雅丽，博士研究生，植物病理学方向
*** 通讯作者：范军，教授，植物病理学方向；E-mail：jfan@ cau. edu. cn

玉米大斑病菌 *StKU80* 基因敲除突变体的获得*

郑亚男**，杨晓荣，刘星晨，赵立卿，赵　洁，冯胜泽，巩校东，
韩建民，谷守芹***，董金皋***
（河北农业大学真菌毒素与植物分子病理学实验室，保定　071001）

摘　要：玉米大斑病（Northern Leaf Blight of Corn）是世界各地玉米产区的一种重要真菌病害，引起玉米大斑病的病原菌为大斑凸脐蠕孢菌（*Exserohilum turcicum*），有性态为大斑刚毛座腔菌（*Setosphaeria turcica*）。在生产中该病害的发生常造成玉米严重减产及品质下降，并因此带来巨大的经济损失。虽然目前生产上仍然以使用化学药剂防治玉米大斑病，但由于发病时施药极其困难，因而在生产上防治该病以选育种植抗病品种为主，并辅以栽培措施。近年来，通过分子生物学手段，立足于对病原菌的致病性调控机制的研究，探索更加有效的防治途径，寻找新的药物靶标，研制新型有效的杀菌剂越来越受到研究者的普遍关注。

本研究通过搜索玉米大斑病菌基因组数据库，克隆得到了非同源末端连接途径中的关键基因 *StKU*80，该基因 DNA 全长为 2 439bp，cDNA 为 2 199bp，包含 5 个外显子和 4 个内含子，编码 733 个氨基酸；对其蛋白质序列分析发现，*StKU*80 基因包含 ku70/ku80 特有的 vWa、Ku78 和 Ku-PK-bind 3 个结构域。

本研究进一步在克隆了 *StKU*80 基因的基础上，以 pBS-pUC 载体为基本骨架、草胺磷抗性基因（*BAR*）作为筛选标记，成功构建了 *StKU*80 基因敲除载体；采用 PEG4000 介导的原生质体转化的方法，将 *StKU*80 基因敲除载体转化病菌原生质体，经 0.2‰草胺磷筛选后，得到可正常生长的单菌落。将菌落在含有 0.2‰草胺磷培养基上连续筛选 3 代以上，仍可稳定生长的菌株即为 *StKU*80 基因敲除转化子；然后通过草胺磷抗性筛选、特异引物 PCR、RT-PCR、Southern blotting 验证，最终获得 2 株 *StKU*80 基因敲除突变体，分别命名为 ΔStKU80－1、ΔStKU80－2。该研究不仅可为分析 *StKU*80 基因的功能奠定基础，也为筛选玉米靶基因敲除受体提供研究材料。

关键词：玉米大斑病菌；*StKU*80；突变体

---

* 基金项目：国家自然科学基金项目（31271997，31371897）

** 第一作者：郑亚男，硕士研究生；E-mail：547836287@ qq. com

*** 通讯作者：谷守芹；E-mail：gushouqin@ 126. com
董金皋；E-mail：dogjingao@ 126. com

转录组比较分析稻曲菌（*Villosiclava virens*）子实体发育和产分生孢子阶段的基因表达差异揭示其有性生殖机制*

于俊杰[1**]，俞咪娜[1]，聂亚锋[1]，孙文献[2]，尹小乐[1]，赵　洁[1]，王亚会[1]，丁　慧[1]，齐中强[1]，杜　艳[1]，Li Huang[3]，刘永峰[1***]

（1. 江苏省农业科学院植物保护研究所，南京　210014；2. 中国农业大学植物保护学院，北京　100193；3. 蒙大拿州立大学植物科学与植物病理学系，波兹曼　59717－3150）

摘　要：近年来稻曲病在我国各稻区发生呈总体上升趋势，已成为水稻的重要病害之一。稻曲菌为异宗配合真菌，其有性生殖产生的子囊孢子为稻曲病的重要初侵染源，然而其有性生殖的分子调控机制尚不清楚。本研究以菌株 UV－8b 基因组测序数据为基础，比较分析了稻曲菌在子实体发育和产分生孢子阶段转录组差异。结果表明，有 488 个基因在子实体形成阶段上调表达，通过 KEGG 和 GO 数据库富集分析发现，这些基因涉及交配型基因（*MAT*1－1－2、*MAT*1－1－3）、信息素合成和分泌相关基因（KDB16470.1、KDB16030.1、KDB18386.1、KDB18574.1）、信号传导途径（KDB18934.1、KDB17304.1、KDB15053.1 和 KDB11415.1 等）、转录调控因子（KDB10897.1、KDB11819.1 和 KDB14847.1 等）和有丝分裂相关调控基因（*Ndt*80、*Spo*11 和 *Mei*－3 等）；此外，有 342 个基因在分生孢子形成阶段上调表达，涉及到生物合成和能量代谢相关基因、分生孢子形成调控基因和病原菌侵染相关基因等。本研究深入揭示稻曲菌有性生殖机制奠定了基础，数据在挖据分生孢子形成、致病相关基因方面也有一定潜力。

关键词：稻曲菌；有性生殖；转录组比较

---

* 资助项目：国家自然科学基金（31571961）；江苏省农业自主创新资金项目（CX（12）5005，CX（15）1054）

** 第一作者：于俊杰，副研究员，主要从事水稻病原真菌病理学研究；E-mail：jjyu@ jaas. ac. cn

*** 通讯作者：刘永锋，研究员，主要从事水稻病害及其生防研究；E-mail：liuyf@ jaas. ac. cn

玉米与弯孢叶斑病菌互作过程中的 microRNA 筛选与鉴定*

刘　震**，靳亚忠，崔　佳，曲建楠，左豫虎，刘　铜***
（黑龙江八一农垦大学农学院植物病理与应用微生物所，大庆　163319）

摘　要：由新月弯孢［*Curvularia lunata*（Wakker）Boed］引起的玉米弯孢叶斑病是我国玉米产区的一种重要性病害，给玉米生产带来了极大的损失。培育抗病品种是防止该病大面积发生的经济、有效的途径之一，因此，深入研究该病原菌与玉米互作的分子机制，为抗病育种提供重要理论依据。microRNA（miRNA）是一类非编码小 RNA，在植物抗逆性方面起重要作用。本研究利用高通量测序技术和 miRNA 微阵列芯片技术，筛选与鉴定玉米弯孢叶斑病菌与玉米互作过程中 microRNA，经整合分析从抗病和感病组测序库中获得 454 个已知的 miRNA，其中包括 443 个保守的 miRNA 和 11 个非保守的 miRNA；获得了 77 个新的 miRNA，其中包括 50 个保守的 miRNA，21 个非保守的 miRNA 和 6 个全新的 miRNA。根据 miRNA 测序和芯片杂交数据联合分析了在抗病组和感病组中的差异表达的 miRNA，在感病组中有 14 个上调 miRNA，其中包括 miRNA393、miRNA167、PC－3P－38569 等和 30 个下调的 miRNA，包括 miRNA1859、miR1425、PC－3P－118921_ 8、PC－3P－118921_ 53 等；在抗病组中有 80 个下调的 miRNA 和 59 个上调的 miRNA；在抗病和感病组比较中，有 147 个显著上调的 miRNA 和 148 个显著下调的 miRNA。研究进一步采用降解组技术寻找差异 miRNA 的靶基因，对其靶基因进行 GO 和 GO 富集化分析，比较抗感组中差异 miRNA 主要涉及的功能包括氧化还原酶活性、信号转导、物质转运、水解酶、氨基酸代谢、光合作用、蛋白质翻译加工和降解、激素调节、抗病物质合成与代谢等，发现了 4 个新 miRNA（PC－5p－154581_ 10，PC－3p－284258_ 7，PC－5p－130707_ 14，PC－5p－87546_ 29）可能调控了 PR1 和 PR5 蛋白转录和翻译。在此基础上，目前正开展 miRNA 调控功能研究。该结果将为深入地研究 miRNA 在抗玉米弯孢叶斑病菌的分子机理提供基础。

关键词：玉米弯孢叶斑病菌；miRNA 微阵列；降解组；抗病机理

---

* 基金项目：国家自然基金"基于 microRNA 的抗玉米弯孢叶斑病分子调控机理研究（31272026）"
** 第一作者：刘震，硕士，主要从事分子植物病理学；E-mail：liuzhenhenan@163. com
*** 通讯作者：刘铜，副教授，主要从事植物病理学；E-mail：liutongamy@sina. com

香蕉棒孢霉叶斑病病原鉴定及其生物学特性*

杨佩文**，番华彩，郭志祥，刘树芳，曾　莉***
（云南省农业科学院农业环境资源研究所，昆明　650205）

摘　要：在云南香蕉产区发现一种由多主棒孢霉［*Corynespora cassiicola*（Berk and Curt.）Wei］引起的叶部病害，其病原菌分生孢子和分生孢子梗的形态、大小及色泽均符合棒孢霉属的特征，通过分子生物学手段辅助鉴定了该病原菌。病原菌接种到香蕉和橡胶上，分别于3d和5d后表现致病症状。病原菌部分生物学特性测定结果表明：菌丝生长的最适温度为28～30℃，适宜pH值为6～8。能有效利用各种碳源和氮源，适宜碳源为乳糖和甘露醇，麦芽糖、葡萄糖和蔗糖次之；蛋白胨为最佳氮源，硝酸钠、磷酸氢二铵和酵母菌粉次之，脲最差。菌丝生长适宜培养基为PDA培养基和黑麦培养基，PSA培养基、燕麦培养基、玉米粉培养基次之，胡萝卜培养基最差。在玉米粉培养基上产孢量最高，其次为胡萝卜培养基、燕麦培养基和PDA培养基，在黑麦培养基和PSA培养基上产孢较少。光照处理对菌丝生长速度影响不显著，光照与黑暗交替有利于产孢，全黑暗、全光照的几乎不产孢。菌丝的致死温度是58℃、15min，分生孢子的致死温度是51℃、5min。7种杀菌剂对病原菌敏感性测定结果表明：多抗霉素、丙环唑、苯甲·丙环唑效果最好，其次是氟硅唑、乙蒜素、丙森锌，嘧菌酯效果最差。

关键词：香蕉；多主棒孢；鉴定；生物学特性

* 国家香蕉产业技术体系（CARS－32）；云南省科技富民强县计划（2014EB080）

** 第一作者：杨佩文，研究员，博士，主要从事植物病理学研究；E-mail：pwyang2000@126.com

*** 通讯作者：曾莉，研究员，主要从事香蕉病虫害及综合防控技术研究；E-mail：ynzengli@163.com

bZIP Transcription Factor CgAP1 is Essential for Oxidative Stress Tolerance and Full Virulence of the Poplar Anthracnose Fungus *Colletotrichum gloeosporioides*

SUN Ying-jiao, WANG Yong-lin, TIAN Cheng-ming

(*The Key Laboratory for Silviculture and Conservation of Ministry of Education, College of Forestry, Beijing Forestry University, Beijing, China*)

Abstract: Yeast AP1 transcription factor is a regulator of oxidative stress response. Here, we report the identification and characterization of CgAP1, anortholog of YAP1 in poplar anthracnose fungus *Colletotrichum gloeosporioides*. The expression of *CgAP*1 was highly induced by reactive oxygen species. *CgAP*1 deletion mutants displayed enhanced sensitivity to oxidative stress compared with the wild-type strain, and their poplar leaf virulence was obviously reduced. However, the mutants exhibited no obvious defects in aerial hyphal growth, conidia production, and appressoria formation. CgAP1 :: eGFP fusion protein localized to the nucleus after *tert*-Butyl hydroperoxide treatment, suggesting that CgAP1 also functions as a redox sensor in *C. gloeosporioides*. In addition, CgAP1 prevented the accumulation of ROS during early stages of biotrophic growth. CgAP1 also acted as a positive regulator of several genes (i. e., *Cat*1, *Glr*1, *Hyr*1, and *Cyt*1) involved in antioxidative response. These results highlight the global regulatory role of CgAP1 transcription factor in oxidative stress response and provide insights into the critical role of ROS detoxification in virulence of *C. gloeosporioides*.

稻瘟病菌不同致病力菌株转录组分析*

齐中强**，杜　艳，余振仙，于俊杰，俞咪娜，刘永锋***
（江苏省农业科学院植物保护研究所，南京　210014）

摘　要：稻瘟病菌（*Magnaporthe oryzae*）引起的稻瘟病是生产上一种最具破坏性的真菌病害，严重威胁我国的水稻安全生产；不同致病力的稻瘟病菌共存于田间，是水稻抗性丧失的主要原因之一；稻瘟病菌致病力差异的分子机制仍不清楚。为了对不同致病力稻瘟病菌的致病分子机制进行深入分析，采用转录组测序对致病力强（#248）、中（#235）和弱（#162）的孢子、菌丝和侵染阶段（8h、24h、48h、72h 和 96h）进行分析。差异表达基因分析表明 3 个菌株在侵染阶段 48h 的差异表达基因数量达到最多，强致病力菌株 248 各个阶段的差异表达基因数量均最多。同时，已知的致病相关基因在 248 各个阶段的差异表达基因中也是出现最多的。对上述各个阶段的差异表达基因进行聚类分析可以分为 11 个类别，其中，3 个菌株的孢子和菌丝阶段分别聚为 2 组，而中间类型 235 和弱致病力菌株 162 的侵染早期（8h 和 24h）聚为 1 组，3 个菌株的侵染后期共同聚为 3 组，上述结论表明强致病力菌株的差异基因表达模式和其余两种致病力存在较大差异，但在侵染后期，这种差异逐步缩小。

K-mean 聚类分析可以将侵染 8h 和 24h 两个阶段的差异表达基因各分为 20 个类别，根据 3 个菌株差异基因和致病力的相关性可以分别找到 555 个和 439 个差异表达基因，其中，55 个基因重复出现在两个时间点。通过功能注释可知上述 55 个基因主要参与糖代谢、细胞壁合成、分子转运等过程，对于这些基因在稻瘟病菌中的功能还需进一步深入研究。上述转录组结果加深了对不同致病力菌株的内在分子机制的理解，同时为稻瘟病的防治提供了新思路。

* 基金项目：国家自然科学基金青年基金项目（31401697）；江苏省青年基金项目（BK20140749）；江苏省农业科技自主创新基金项目［CX（15）1054］

** 第一作者：齐中强，助理研究员，主要从事植物病理学研究；E-mail：qizhongqiang2006@126.com

*** 通讯作者：刘永锋，研究员，主要从事植物病理学研究；E-mail：Liuyf@jaas.ac.cn

条锈菌有性菌系的分离与致病力测定分析*

李　巧**，覃剑锋，赵元元，赵　杰***，黄丽丽，康振生***
（西北农林科技大学植物保护学院，旱区作物逆境生物学国家重点实验室，杨凌　712100）

摘　要：小麦条锈病是我国危害最严重的小麦病害之一。培育和种植抗病品种是防治该病害经济有效的途径。但是，小麦生产上面临的严峻问题是，由于条锈菌毒性变异产生新小种导致品种抗病性丧失，造成病害流行。近期，随着小檗被确定为小麦条锈菌的转主寄主，有性生殖在其毒性变异中作用和病害发生中的作用亟待研究。本研究通过对 2015 年采集的自然发病的小檗产生的锈子器，进行接种感病小麦分离，对获得的菌系在鉴别寄主上测定其毒性，并进行分析。结果表明，从 3 种采自不同地点的小檗病害标样中共分离获得小麦条锈菌菌系 8 个，其中 1 个是已知致病类型，其余 7 个为未知小种（致病类型）。93 个单孢子堆菌系共产生了 47 个不同的致病类群，其中，14 个小种（致病类型），占 29.8%；33 个为未知小种（致病类型），占 70.2%。研究结果再次获得了小麦条锈菌在自然条件下进行有性生殖、经有性生殖发生毒性变异产生遗传多样性的证据，证实转主寄主小檗在我国小麦条锈菌的有性循环和病害循环中起重要作用。

关键词：小檗；转主寄主；条锈病；小麦；毒性变异；小种

---

* 基金项目：国家自然科学基金（32171986）；国家“973”项目（2013CB127700）；国家“863”计划（2012AA101503）；“新世纪优秀人才支持计划”

** 第一作者：李巧，硕士，主要从事小麦条锈病研究；E-mail：1013066562@qq.com

*** 通讯作者：赵杰，副教授，主要从事小麦条锈病研究
康振生，教授；E-mail：kangzs@nwsuaf.edu.cn

Histological Observation of Wheat Infected by *Fusarium asiaticum*, the Causal Agent of Wheat Crown Rot Disease*

ZHAO Qin**, ZHANG Xiang-xiang, DENG Yuan-yu, LI Wei, SUN Hai-yan, CHEN Huai-gu***

(*Institute of Plant protection*, *Jiangsu Acedemy of Agricultural science*, *Nanjing* 210014)

Abstract: Wheat crown rot disease caused by multiple soil-born fungus was one of worldwide-spread and devastating disease for wheat production. *Fusarium asiaticum* was the dominant pathogen of wheat crown rot disease in major production of winter wheat in China. It's beneficial to pathogenesis analysis of this fungi and resistance breeding through histological observation of infection process. Gene *gfp* was transferred into strain CF0915 *via* PEG-$CaCl_2$ mediated transformation method. Based on analysis of fluorescence expression, PCR verification, genetic stability, growth parameters and pathogenicity analysis, transformants CF0915 – GFP resembling wild strain was chosen to investigate infection process on wheat cultivars of different resistance. The results showed that transformants CF0915 – GFP showed bright green fluorescence, and *gfp* gene was integrated into the genome of *F. asiaticum* by PCR amplification verification and inherited normally and stably, growth characters and pathogenicity were comparable to wild strain. Numerous conidiaspores attached root hairs and root epidermis of wheat and germinated at 1dpi in susceptible cultivar "Yangmai 158", until 2 dpi conidiaspore germination was observed in moderately-resistant cultivars "CI12633" and "Sumai 3" . The hyphae infected root hairs and grooves along junctions of adjacent epidermal cells and then spread into cortex of roots at 3dpi. The hyphae spread so fast onto leaf shealth from roots at 8 dpi, while at 10 dpi massive hyphae invaded into root cortex cells and epidermal cells of leaf shealth and meanwhile pycidium with abundant macroconidiaspores were observed, brown disease spots appeared on basal stems of susceptible cultivar. Massive hyphae were filled in xylems vessels of roots and leaf shealths at 14dpi, subsequently numerous chlamydospores were observed in the epidermal cells of leaf shealth. Until 25dpi most seedlings appeared wilt and dead. Compared to susceptible cultivar, the infected process was retarded later in moderately-resistant cultivars. In this study the histopathological observation of infection process of *F. asiaticum* in wheat provided an important theoretical basis for disease control and resistance resource utilization.

* Funded by national project (Agriculture) in the public interest

** First author: ZHAO Qin; E-mail: zhaoqin0802@126.com

*** Corresponding author: CHEN Huai-gu; E-mail: huaigu@jaas.cn

甘蔗种苗传播病害病原检测与分子鉴定*

李文凤**，王晓燕，黄应昆***，单红丽，张荣跃，尹　炯，罗志明
（云南省农业科学院甘蔗研究所，云南省甘蔗遗传改良重点实验室，开远　661699）

摘　要：我国蔗区（尤其云南）生态多样化，甘蔗病害病原种类复杂多样，多种病原复合侵染，尤其种苗传播病害病原具有潜育期和隐蔽性，传统方法难以诊断。精准有效地对甘蔗种苗传播病害病原进行诊断检测，明确监测病害致病病原是科学有效防控甘蔗种苗传播病害的基础和关键。本研究针对甘蔗种苗传播病害诊断检测基础薄弱、主要病害病原种类及株系（小种）不明等关键问题，以严重为害我国甘蔗生产的黑穗病、宿根矮化病、病毒病、白叶病等种苗传播病害为对象，系统建立了甘蔗黑穗病、宿根矮化病、白条病、赤条病、花叶病（SCMV、SrMV、SCSMV）、黄叶病、杆状病毒病、斐济病和白叶病9种种苗传播病害11种病原分子快速检测技术，为甘蔗种苗传播病害精准有效诊断、脱毒种苗检测及引种检疫提供了关键技术支撑。通过从不同生态蔗区广泛采集有代表性的甘蔗黑穗病、宿根矮化病、病毒病、白叶病等病害样品，系统分离鉴定明确病害病原种类及主要株系（小种）。结果表明：①云南和广西蔗区甘蔗黑穗病病原为甘蔗鞭黑粉菌（*Ustilago scitaminea* Sydow），病菌存在致病性和生理小种分化，云南蔗区存在生理小种1、小种2及新小种3。②21个蔗区21批1 270个样品宿根矮化病检测、测序分析，致病菌为*Leifsonia xyli* subsp. *xyli*，其核苷酸序列完全一致大小为438bp（GenBank登录号JX424816、JX424817），与GenBank中巴西、澳大利亚（登录号AE016822、AF034641）RSD致病菌相似性为100%；与美国路州（AF056003）的有1个碱基错配和1个碱基插入，相似性为99.54%。③引起甘蔗花叶病病原有甘蔗花叶病毒（Sugarcane mosaic virus，SCMV）、高粱花叶病毒（Sorghum mosaic virus，SrMV）、甘蔗条纹花叶病毒（Sugarcane streak mosaic virus，SCSMV）3种，SCSMV阳性检出率100%、SrMV阳性检出率27.27%、SCMV阳性检出率1.3%，SCSMV为最主要病原（扩展蔓延十分迅速、致病性强），SrMV为次要病原，且存在2种病毒复合侵染。83份SCSMV和34份SrMV克隆及测序分析，SCSMV序列分为3个大类群，中国分离物聚为1个类群，印度分离物聚为2个类群，中国分离物除JX467699外全聚在一起。SrMV形成2个组：I和II组，组间又分为2个亚组。不同来源SrMV在系统树中交叉存在，云南分离物在各个分支中普遍存在，表现出很高的遗传多样性。④甘蔗杆状病毒病病原为甘蔗杆状病毒（Sugarcane Bacilliform virus，SCBV），其核苷酸序列大小为589bp，与SCBV-Australia核苷酸和氨基酸序列相似性为74.0%和84.1%；同SCBV-Morocco核苷酸和氨基酸序列相似性为67.1%和66.7%，可见，SCBV不同分离物间基因组序列存在高度变异特性。⑤89份黄叶病毒阳性样品RT-PCR-RFLP分析结果表明，云南蔗区检测的甘蔗黄叶病毒（Sugarcane yellow leaf virus，SCYLV）全部为BRA基因型。⑥甘蔗白叶病病原为SCWL植原体（Sugarcane white leaf phytoplasma），其核苷酸序列大小为210bp，与

---

* 基金项目：现代农业产业技术体系建设专项资金资助（CARS-20-2-2）；云南省现代农业产业技术体系建设专项资金资助；云南省“人才培养”项目（2008PY087）

** 第一作者：李文凤，云南石屏人，研究员，主要从事甘蔗病害研究；E-mail：ynlwf@163.com

*** 通讯作者：黄应昆，研究员，从事甘蔗病害防控研究；E-mail：huangyk64@163.com

GenBank 中公布的 SCWL 植原体基因组序列同源性在 99.05% ~100%。系统进化分析，SCWL 序列分为 3 个类群，云南保山分离物聚为 1 个类群，与越南、缅甸、泰国聚在一起；云南临沧分离物聚为 1 个类群，与泰国、印度、日本聚在一起。研究结果丰富了种苗传播病害相关理论和技术基础，为监测预警和科学有效防控甘蔗种苗传播病害奠定了重要基础。

关键词：甘蔗；种传病害；病原检测；分子鉴定

小麦叶锈菌 FHPL 在 TcLr19 及其感病突变体上的差异表达分析

范学锋，张维宏，苑　莹，杨文香，孟庆芳，刘大群
（河北农业大学，保定　071000）

摘　要：小麦叶锈菌对不同寄主致病性差异的研究对揭示其致病的分子机制有重要意义。本实验采用对野生型小麦 TcLr19 和其突变体小麦 MuLr19 分别表现侵染型为 0 和 3 的小麦叶锈菌 FHPL 分别接种 TcLr19 和 MuLr19，通过分析其在不同寄主上的表达差异，以期揭示小麦叶锈菌 FHPL 的致病相关基因。我们分别提取未接菌 TcLr19/MuLr19、接种叶锈菌 FHPL 24h 和 6d 后的总 RNA，进行 denove 测序，拼接得到总转录组，扣除寄主小麦的转录组后，得到叶锈菌 FHPL 的 Unigenes 总数 32 168个。对基因表达水平和差异表达进行分析，结果接种 24h 和 6d 后 FHPL 在 TcLr19 和 MuLr19 上差异表达基因分别有 127 个和 179 个，其中，24h 和 6d 共同差异表达的基因 50 个。Gene Ontology 注释成功的 59 个，其中，生物过程 46 个，分子功能 19 个，细胞组分 35 个。KEGG Pathway 显著性富集分析中，接种 24h 后 PtM_ 24h 相对于 Pt19_ 24h 显著富集，苯丙素生物合成及苯丙氨酸代谢基因 c66749_ g3（苯丙氨酸氨裂解酶）下调、RNA 降解及真核生物核糖体生物合成和氧化磷酸化途径中基因 c14572_ g1 XRN2、RAT1 下调、氧化磷酸化途径中基因 c59559_ g1（H^+转运 ATP 激酶）下调；接种 6d 后 PtM_ 相对于 Pt19_ 基因 c36028_ g1 下调（硫胺代谢）、脂肪酸代谢中基因c38476_ g1、c71646_ g1、c32016_ g1（醛脱氢酶）c60322_ g3、c60322_ g1（adhP）下调、戊糖和葡萄糖醛糖相互转换途径中醛脱氢酶相关基因 c38476_ g1、c71646_ g1、c32016_ g1，c13955_ g1 AKR1A1，adh 下调、组氨酸代谢中醛脱氢酶基因 c38476_ g1、c71646_ g1、c32016_ g1 下调、糖酵解途径中醛脱氢酶基因（c32016_ g1、c38476_ g1、c71646_ g1），c13955_ g1（AKR1A1，adh）c60322_ g3、c60322_ g1（adhP）下调、细胞色素 P450 途径中基因 c60322_ g3、c60322_ g1（adhP）下调、玉米素生物合成途径中基因 c47978_ g1（CKX）下调。研究为进一步深入开展差异表达的基因的功能分析，明确其在致病中的作用奠定了基础。

关键词：小麦叶锈菌；转录组测序；差异基因表达

甘蔗锈病病原菌鉴定及系统进化分析*

王晓燕**，李文凤，黄应昆***，单红丽，张荣跃，罗志明，尹　炯
（云南省农业科学院甘蔗研究所，云南省甘蔗遗传改良重点实验室，开远　661699）

摘　要：甘蔗锈病是世界性的甘蔗重要病害之一，常造成巨大的经济损失。根据症状和病原特征，甘蔗锈病分为两种类型，即由黑顶柄锈菌（*Puccinia melanocephala* Syd. et P. Syd.）引起的甘蔗褐锈病和由屈恩柄锈菌（*Puccinia kuehnii* Butler.）引起的甘蔗黄锈病。为明确甘蔗锈病在中国云南蔗区的病原菌种类、发生分布特征、病菌间及其与柄锈菌属其他锈菌的系统进化关系。2014—2015 年，对云南保山、临沧、勐海、孟连、西盟、澜沧和文山等蔗区的甘蔗锈病发生分布状况进行了调查，并采集 57 份甘蔗锈病样品进行形态特征观察及分子鉴定。病原菌鉴定结果表明：来源于云南勐海海引 1 号的 4 份甘蔗锈病样品属甘蔗黄锈病，病原菌为屈恩柄锈菌；其余 53 份甘蔗锈病样品都属甘蔗褐锈病，病原菌为黑顶柄锈菌。从 GenBank 中下载柄锈菌属其他锈菌的 rDNA 序列与本研究获得的序列一起，构建 NJ 树进行系统进化分析。结果表明：本文获得的 4 条屈恩柄锈菌与 *Puccinia polysora*（GenBank 登录号：GU058024）、*Puccinia agrophila*（GenBank 登录号：GU058016）和 *Puccinia physalidis*（GenBank 登录号：DQ354522）聚为 1 组，遗传关系近；53 条黑顶柄锈菌与 *Puccinia miscanthi*（GenBank 登录号：AJ296546）、*Puccinia nakanishikii*（GenBank 登录号：GU058002）、*Puccinia rufipes*（GenBank 登录号：AJ296545）和 *Puccinia coronata*（GenBank 登录号：DQ354526）等柄锈菌属锈菌聚为 1 组，遗传关系近；而两种甘蔗锈病菌的遗传关系相对较远。研究结果为甘蔗锈病的流行预测及综合防治提供了科学依据。

关键词：甘蔗锈病；病原菌鉴定；黑顶柄锈菌；屈恩柄锈菌；系统进化

* 基金项目：现代农业产业技术体系建设专项（CARS－20－2－2）；云南省现代农业产业技术体系建设专项
** 第一作者：王晓燕，云南开远人，助理研究员，主要从事甘蔗病害研究；E-mail：xiaoyanwang402@sina.com
*** 通讯作者：黄应昆，研究员，从事甘蔗病害防控研究；E-mail：huangyk64@163.com

暹罗刺盘孢菌 *Colletotrichum siamense* 群体遗传结构分析

李　杨[1,2]，李　河[1,2]，周国英[1,2]，蒋越西[1,2]，蒋仕强[1,2]，刘君昂[1,2]
（1. 中南林业科技大学，森林有害生物防控湖南省重点实验室，长沙　410004；
2. 中南林业科技大学，经济林培育与保护教育部重点实验室，长沙　410004）

摘　要：炭疽病菌可引起全球范围内多种植物炭疽病害。研究炭疽病菌群体遗传结构可为全面、有效防治植物炭疽病害提供理论依据。本研究对分离自中国海南、江西、湖南、广西、辽宁、云南 6 个地区 11 种寄主植物上共 90 株暹罗刺盘孢菌菌株以及 GenBank 中 76 条暹罗刺盘孢菌的 ITS 序列进行群体遗传结构分析。166 条暹罗刺盘孢菌 ITS 序列可定义为 26 个单倍型，其中，单倍型 H16 及单倍型 H12 为主要单倍型，分布于绝大部分地区。病菌不同地理、寄主种群间的遗传分化较大；AMOVA 分析显示，遗传变异主要发生在种群内；病菌未经历过大规模的种群扩张过程。研究结果表明暹罗刺盘孢菌种群具有丰富的遗传多样性。

关键词：暹罗刺盘孢菌；地理种群；ITS－5. 8S 序列；群体遗传结构

苹果树腐烂病菌果胶裂解酶基因 *Vmpl4* 的致病功能研究*

许春景**，孙迎超，吴玉星，冯　浩，高小宁，黄丽丽***
（旱区作物逆境生物学国家重点实验室/西北农林科技大学植物保护学院，杨凌　712100）

摘　要：苹果树腐烂病是由黑腐皮壳属真菌 *Valsa mali* 引起的严重枝干皮层腐烂病害，*Vmpl4* 是苹果树腐烂病菌侵染过程中上调表达倍数最高的果胶裂解酶（Pectate lyase，PL）基因。本研究通过对 *Vmpl4* 敲除突变体的致病力、果胶利用和基因敲除后家族内其他基因的表达水平变化进行分析，解析 *Vmpl4* 在病菌致病过程中的作用。利用 qRT-PCR 检测 *Vmpl4* 在野生菌株 03－8 与寄主互作过程中的表达水平，发现该基因在病菌侵染过程中上调表达高达 10.20 倍。通过 Double-joint PCR 技术成功构建了基因敲除载体，并通过 PEG 介导原生质体遗传转化获取转化子，进而通过 PCR 及 Southern blot 验证得到 1 个基因敲除突变体。利用离体接种方法对苹果叶片和枝条进行接种检测突变体致病力变化，发现基因突变体在叶片和枝条上的致病力分别降低 16.82% 和 18.59%；将野生型菌株及突变体接种至 PDA 及果胶培养基，发现突变体在果胶培养基上生长速率降低 8.79%。进而利用 qRT-PCR 检测 *Vmpl4* 敲除后 PL 家族内其他基因的表达水平，发现 *Vmpl4* 敲除后家族内 4 个基因在病菌侵染过程中表达水平显著上调。以上结果表明果胶裂解酶基因 *Vmpl4* 通过降解果胶参与致病过程，PL 家族内其他基因与 *Vmpl4* 在 *V. mali* 致病过程中协同发挥作用。

关键词：苹果树腐烂病菌；果胶裂解酶；基因敲除；致病力

---

* 基金项目：国家自然科学基金（31171796 和 31471732）；国家公益性行业（农业）科研专项（201203034）；高等学校博士学科点基金（20120204110002）

** 第一作者：许春景，硕士研究生，主要从事苹果树腐烂病菌果胶酶基因致病机理研究；E-mail：xuchunjing89@163.com

*** 通讯作者：黄丽丽，教授，主要从事果树病害病原生物学及病害综合防控方面的研究；E-mail：huanglili@nwsuaf.edu.cn

菠萝小果芯腐病（*Fusarium ananatum*）在中国的首次发现

谷　会，朱世江，詹儒林，张鲁斌

（中国热带农业科学院南亚热带作物研究所/海南省热带园艺产品采后生理与保鲜重点实验室，湛江　524091）

摘　要：菠萝小果芯腐病是由病原菌 *Fusarium ananatum* 引起的，是一种严重的菠萝［*Ananas comosus*（L.）Merr.］果实病害。该病最初是在 2014 年 3 月份在中国广东省徐闻县“巴厘”菠萝果实上发现，在当地大概有 10% 的菠萝果实采后均受该病侵染。受侵染的果实典型的外部症状表现为小果不能正常褪绿或者延迟成熟，受侵染的小果中部有褐变产生，腐烂有时从小果一直延伸到菠萝果心，褐变的果肉组织比较坚硬，是一种典型的干腐病，病情一般只局限在单个小果内。用组织分离法从 5 个有典型症状菠萝果实上最终分离到 4 个真菌菌株。致病性实验在菠萝植株初花期进行，将分离到的 4 个菌株进行重新培养，并配制浓度为 1×10^6 spores/mL 的真菌孢子悬浮液，在每个菠萝花序盛开的小花上均匀施喷 15mL 悬浮液，对照直接施喷蒸馏水。大概接种 9～10 周后，接种菌株 GH0305 的菠萝果实上均表现出了典型的小果芯腐病症状，而对照和其他 3 个菌株没有病害症状。菌株 GH0305 可以从发病的菠萝植株上重新分离到。为了进一步鉴定该菌株，将菌株 GH0305 置于 PDA 培养基上，在 28℃和自然光周期下培养，7d 后发现有白色的菌丝长出，菌落边缘菌丝比较稀薄，中间菌丝比较浓密，培养 10d 后，在 PDA 培养基中有臧红色的色素产生。将菌株 GH0305 置于合成低营养琼脂（SNA）培养基上进行菌株显微形态观察，在 28℃下培养 5d 后，发现菌株形态特征和 *Fusarium ananatum* 的形态特征完全一致。小型孢子生长在直立的分生孢子梗上，分生孢子梗假头状，单瓶梗，未见链状分生孢子；小型分生孢子卵形，椭圆或长圆形，分生孢子内部有 0～1 个隔膜，大小为（5～15）μm×（2～4）μm。大型分生孢子笔直或镰刀形，内部有 1～3 个隔膜，大小为（15～40）μm×（2～6）μm，有尖锐或圆锥形的顶细胞。为了进一步鉴定致病菌，用编码蛋白基因翻译延长因子 TEF－1α 进行扩增和序列分析，获得长度为 597bp 的片段，与 NCBI 数据库中相关序列进行同源性比较发现，所测的序列（GenBank No. KP751254）与 *F. ananatum*（GenBank No. HE802668）有 100% 的同源性。基于病害症状分析，病原菌致病性测定，病原菌形态分析和 TEF－1α 基因序列分析，该病原菌鉴定为 *F. ananatum*。这是菠萝小果芯腐病在国内的首次报道，本研究为以后进一步防治菠萝小果芯腐病奠定了基础。

关键词：菠萝小果芯腐病；菠萝镰刀菌；*Fusarium ananatum*；病原鉴定

基于转录组测序分析苹果抗炭疽叶枯病的机制*

吴成成**，李保华，李桂舫，董向丽，王彩霞***

（青岛农业大学农学与植物保护学院，山东省植物病虫害综合防控重点实验室，青岛 266109）

摘 要：近几年在我国苹果树上新发现了一种由围小丛壳菌（*Glomerella cingulata* Spauld. & H. Schrenk）引起的炭疽叶枯病（Glomerella Leaf Spot，GLS），该病主要为害叶片和果实，果实受侵染后在果面上形成1～2mm的黑色病斑，高温多雨季节，叶片发病后一周内可造成大量脱落，落叶率一般在90%以上。炭疽叶枯病不仅危害严重，且流行性强，目前已蔓延至山东、陕西、辽宁等苹果主产区，对我国苹果产业构成严重威胁。调查发现，苹果品种间对炭疽叶枯病的抗病性存在显著差异，嘎啦、金冠、乔纳金等品种高度感病，而富士、红星等品种高抗炭疽叶枯病。因此，利用抗病基因育种是克服苹果炭疽叶枯病的理想策略。

为了研究苹果抗炭疽叶枯病的机制及其抗病相关基因，利用高通量测序技术（RNA-Seq）获得了抗病品种富士（*Malus domestica* Borkh. cv. Fuji）和感病品种嘎啦（*Malus domestica* Borkh. cv. Gala）响应炭疽叶枯病菌（*Glomerella cingulata*）侵染的基因表达谱，检测到22 350个富士基因和22 656个嘎啦基因。通过对表达谱进行分析发现，富士和嘎啦有2 731个共有的显著差异表达基因，其中48.92%（1 336）共同上调表达，49.07%（1 340）共同下调表达，1.50%（41）在富士中上调表达在嘎啦中下调表达，0.51%（14）在富士中下调表达嘎啦中上调表达。通过荧光定量PCR（qRT-PCR）对部分基因的表达水平进行了验证。GO富集分析表明这些基因大部分在生物过程和分子功能富集。除此之外，还检测到了几个在抗病品种和感病品种中差异表达的WRKY转录因子。研究结果表明富士特有基因和富士上调嘎啦下调的差异表达基因可能与抗病相关，编码一些抗炭疽叶枯病的物质，通过分子信号途径来激活或抑制感受或抵抗病原菌侵染相关的物质。

关键词：苹果炭疽叶枯病；转录组；基因表达谱；差异表达基因

---

* 基金项目：现代农业产业技术体系建设专项资金（CARS-28）；国家自然科学基金（31272001和31000891）；山东省科技攻关计划（2010GNC10918）

** 第一作者：吴成成，硕士研究生，研究方向果树病理学；E-mail：15610567060@qq.com

*** 通讯作者：王彩霞，教授，主要从事果树病害；E-mail：cxwang@qau.edu.cn

玉米大斑病菌 *StMSN2* 基因的克隆及其敲除载体的构建*

赵立卿**，刘星晨，郑亚男，赵 洁，冯胜泽，
巩校东，韩建民***，谷守芹***，董金皋***
（河北农业大学真菌毒素与植物分子病理学实验室，保定 071001）

摘 要：玉米大斑病是由玉米大斑病菌（*Setosphaeria turcica*）所引起的玉米叶片病害，在流行年份，感病品种可减产高达 50% 以上。由于玉米大斑病菌变异频繁，因此，寻找新的防治方法和预防措施已成为当务之急。近年来，研究植物病原菌的发育与致病性的分子机制、探讨更有效的防治措施已成为植物病理学和植物遗传育种领域最热门的研究课题之一。

大量研究表明，MAPK（mitogen activated protein kinase）信号转导途径对植物病原真菌的抗胁迫反应、生长发育及致病性等均有重要的调控作用。本课题组前期研究表明，在玉米大斑病菌中 StMsn2 很可能为 HOG-MAPK 级联途径下游关键转录因子，参与调控病菌的高渗胁迫反应及致病性。因此，阐明转录因子 StMsn2 的作用机制对于揭示调控病原菌高渗胁迫反应及致病性机理具有重要意义。本研究利用生物信息学方法，从 JGI（http：//genome. jgi-psf. org/）网站上公布的玉米大斑病菌数据库中，搜索并得到了玉米大斑病菌中 *StMSN2* 基因；该基因位于 scaffold_ 9：1301421 - 1303172 （ + ），其 ID 为 100946；该基因包含 1 752bp 的开放阅读框，其中包含 2 个外显子，1 个内含子，编码两个 ZnF_ C2H2 锌指蛋白保守结构域。本研究进一步以该基因为靶序列，设计分别带 *Kpn* I、*Hind* III 和 *Sac* I、*Sma* I 酶切位点的特异引物，克隆了 777bp、788bp 的片段，成功构建含有氯霉素抗性和氨苄抗性的 *StMSN2* 基因的敲除载体 pPZP100 - MSN2。本研究拟通过 ATMT 介导的遗传转化方法，筛选 *StMSN2* 基因敲除突变体，下一步将比较基因敲除突变体与野生型菌株的表型及致病性的差异，以明确该基因的功能及其调控机制，为玉米大斑病致病分子机制的研究提供理论基础。

关键词：玉米大斑病菌；生物信息分析；敲除载体构建

---

* 基金项目：国家自然科学基项目（31171805，31371897）
** 第一作者：赵立卿，硕士研究生，研究方向为植物学；E-mail：370172469@ qq. com
*** 通讯作者：韩建民，教授，硕士生导师；E-mail：hanjmnd@ 163. com
谷守芹，教授，博士生导师；E-mail：gushouqin@ 126. com
董金皋，教授，博士生导师；E-mail：gushouqin@ 126. com

北京、甘肃、山东等地葡萄灰霉病菌 SSR遗传多样性分析*

张艳杰[1**]，李兴红[2]，李亚宁[1***]，刘大群[1,3***]

(1. 河北农业大学，植物保护学院植物病害生物防治与分子植物病理学实验室，河北省农作物病虫害生物防治工程技术研究中心，国家北方山区农业工程技术研究中心，保定　071000；2. 北京市农林科学院植物保护环境保护研究所，北京　100097；3. 中国农业科学院研究生院，北京　100081)

摘　要：以采集自北京昌平、北京延庆、甘肃兰州、甘肃张掖、山东解西、山东龙口、山东蓬莱和辽宁兴城等8个地区的葡萄灰霉病菌为研究对象，经单孢分离和纯化，获得菌株83个，采用灰葡萄孢菌6对SSR引物对8个群体进行扩增，获得多态性位点54个，基因型90个，群体遗传分析发现，群体间遗传距离为0.008 4～0.099 4，遗传相似系数为0.927 8～0.991 6。在物种水平上，得到观察等位基因数为2.000 0，有效等位基因数为1.334 7，Nei's基因多样性指数是0.217 4，Shannon信息指数为0.352 2，多态性位点百分率100%，表明这几个地区的灰葡萄孢菌遗传多样性很丰富。8个群体的总遗传多样性（Ht）为0.215 2，群体内遗传多样性（Hs）为0.175 6，群体间遗传多样性（Dst）为0.039 6，遗传分化系数（Gst）是0.184 0，群体内的遗传多样性远远大于群体间遗传多样性，说明其群体内和群体间存在着一定的遗传分化。8个群体间的基因流强度为2.216 8，表明群体间的菌源交流并不是太强。聚类结果显示，山东解西、山东蓬莱、山东龙口聚在一支，山东的群体与北京群体的距离较近，与辽宁兴城次之，与甘肃兰州群体距离较远，可以看出，菌株的亲缘关系与地理来源具有一定的相关性，地理距离较近的菌株亲缘关系一般也较近。

关键词：葡萄；灰葡萄孢菌；SSR；遗传多样性；多态性

---

* 基金项目：现代农业产业技术体系专项资金（CARS－30）；国家自然科学基金（31572050）

** 第一作者：张艳杰，在读博士，从事植物病害生物防治研究；E-mail：love2006hope@126.com

*** 通讯作者：李亚宁，教授，博导，从事植物病害生物防治研究；E-mail：yaning22@163.com

刘大群，教授，博导，从事植物病理学研究；E-mail：ldq@hebau.edu.cn

稻曲病菌T-DNA插入突变体B1812侧翼基因克隆*

薄惠文[1,2**]，俞咪娜[1]，王亚会[1]，于俊杰[1]，尹小乐[1]，
丁　慧[1,2]，周俞辛[1]，齐中强[1]，杜　艳[1]，宋天巧[1]，张荣胜[1]，刘永锋[1***]
（1. 江苏省农业科学院植物保护研究所，南京　210014；
2. 南京农业大学生命科学学院，南京　210095）

摘　要：水稻稻曲病是发生在水稻穗部的一种真菌病害，由病原菌稻曲病菌（*Ustilagrnoidea virens*）侵染引起。近年来，随着水稻杂交品种的大面积推广、农田氮肥施用量增加等因素，水稻稻曲病在我国各稻区已从次要病害逐渐成为主要病害。稻曲病菌的发生对水稻产量和质量均有严重影响，危害我国粮食安全。本研究室利用农杆菌介导的转化技术构建了稻曲病菌P1的T-DNA插入突变体库，并通过田间接种筛选致病力变化突变体。本研究对筛选得到一株致病力减弱突变体B1812进行研究，观察并分析了突变体B1812的菌落形态、菌落直径、孢子形态、产孢能力等生物学特性，克隆了致病相关基因，并初步探究了该基因的功能。

通过生物学性状研究发现，突变体B1812在固体培养基PSA和TB3上的菌落形态和生长速率与P1相比无显著差异，而在MM培养基上的生长速率下降。突变体液体摇培后产生的分生孢子大小明显小于P1，且其产孢能力也显著下降。Southern杂交表明，T-DNA在B1812的基因组中为单拷贝插入。利用HiTail-PCR和RACE-PCR，克隆了T-DNA插入位点全长为2 196bp的侧翼基因*UvHac*1，T-DNA插入在基因的5′UTR区。荧光定量PCR结果显示，基因的表达量显著下降。侧翼基因*UvHac*1含有一个高度保守的bZIP结构域，与真菌转录激活因子*Hac*1同源，且与UV－8b菌株中的*UV8b*_1075同源。Hac1p是一种在真菌的非折叠蛋白响应（UPR）中起关键作用的转录因子，UPR是真核生物进化中保守的一种机制。稻曲病菌中Hac1p的调控功能目前正在进一步研究，本研究为解析稻曲病菌中转录因子Hac1的功能，及其在致病过程中的作用提供了研究基础。

关键词：稻曲病菌；T-DNA；致病力；基因克隆

---

* 基金项目：江苏省农业科技自主创新基金（CX（15）1054）；国家自然科学基金（31401700）
** 第一作者：薄惠文，山东烟台人，硕士研究生，研究方向为水稻稻曲病致病基因；E-mail：bhwwhb422@163. com
*** 通讯作者：刘永锋，博士，研究员，主要从事水稻病害病理学及其生物防治技术研究；E-mail：liuyf@jaas. ac. cn

稻曲病菌 T-DNA 插入突变体 B2510 的插入位点分析*

丁 慧[1,2]**，俞咪娜[1]，王亚会[1]，
于俊杰[1]，尹小乐[1]，薄惠文[1,2]，黄 星[2]，刘永锋[1]***
（1. 江苏省农业科学院植物保护研究所，南京 210014；
2. 南京农业大学生命科学学院，南京 210095）

摘 要：水稻稻曲病是由稻曲病菌（*Ustilagrnoidea virens*）侵染引起的水稻穗部真菌病害，近年来已逐渐成为水稻的主要病害之一。深入了解稻曲病的致病机理，可为抗病水稻的育种和新靶标药剂的开发提供理论依据。本研究以稻曲病菌 T-DNA 插入突变体库中致病力减弱突变菌株 B2510 为材料，通过分析 T-DNA 插入位点的侧翼序列和突变基因，克隆在稻曲病菌致病过程中起作用的基因。

通过测定突变菌株 B2510 的生长速率、产孢能力及致病力发现，与野生型菌株 P1 相比，B2510 田间接种表现为致病减弱；在 MM 培养基上生长速率下降，而在 PSA 和 TB_3培养基中生长速率与野生型没有显著差异，但丧失产孢能力。Southern 杂交显示 T-DNA 在突变菌株 B2510 中以双拷贝形式插入，利用 TAIL-PCR 技术扩增紧邻 T-DNA 两侧的侧翼序列，经过比对分析发现，T-DNA 分别插在基因 *UVC6TF*（与稻曲病菌 Uv8b 菌株的 *UV8b*_1412 基因同源）的启动子区域和 *UVRASGAP*（与稻曲病菌 Uv8b 菌株的 *UV8b*_1386 基因同源）的下游 3′端。半定量 RT-PCR 分析基因的表达情况，显示两个基因在突变体 B2510 的表达量较 P1 均显著下降，推测 T-DNA 插入位点处的基因与稻曲病菌致病性相关，参与调控稻曲病菌在水稻上的致病过程。为进一步了解基因的功能，我们构建基因 *UVC6TF* 的沉默载体，并通过 ATMT 的方法导入稻曲病菌。获得的 2 个沉默转化子中 *UVC6TF* 的表达分别被抑制了 50% 和 90%。分析它们的产孢量、生长速率等生物学性状和致病力检测，结果发现该基因不影响菌株正常的生长发育，但产孢能力降低，致病能力减弱。此外，T-DNA 插入还破坏了基因 *UVRASGAP* 的表达，后续研究则是对单个基因或是两个基因进行基因敲除、回补等方法，进一步研究基因功能，研究结果将对稻曲病菌在水稻上的致病过程解析具有重要意义。

关键词：稻曲病菌；T-DNA 插入突变；ATMT 转化；致病力；侧翼序列

* 基金项目：江苏省农业科技自主创新基金（CX（15）1054）；国家自然科学基金（31301624）
** 第一作者：丁慧，江苏南通人，硕士研究生，研究方向为水稻稻曲病菌功能基因；E-mail：dinghui90@126.com
*** 通讯作者：刘永锋，博士，研究员，主要从事水稻病害病理学及其生物防治技术研究；E-mail：liuyf@jaas.ac.cn

稻曲病菌特异性的 SSR 标记筛选*

俞咪娜**，王亚会，于俊杰，尹小乐，刘永锋***
（江苏省农业科学院植物保护研究所，南京 210014）

摘 要：水稻稻曲病是由稻曲病菌（*Ustilaginoidea virens*）侵染引起的一种水稻穗部真菌病害。近年来，稻曲病菌呈现发生范围扩大的趋势，逐渐由水稻次生病害发展为主要病害。而且，稻曲病的发生会影响稻米的产量和品质，严重影响粮食安全。稻曲病菌具有潜伏侵染的特性，在病菌侵入水稻早期不易被发现，难以界定初侵染的时期和规模，直至水稻开花后在稻穗上出现稻曲球症状，才说明该水稻在开花之前被稻曲病菌侵染，而对于未显症的水稻植株还无法判断是否存在稻曲病菌侵染。另一方面，稻曲病菌生长十分缓慢，采用植物组织内生真菌分离法对未显症状水稻植株体内稻曲病菌分离检测及稻株内分布的研究比较困难。发展稻曲病菌监测和鉴定技术，对提高稻曲病菌的预测预报和探索稻曲病菌的侵染循环具有重要意义。目前，已经有报道应用于稻曲病菌特异性分子诊断的特异性靶标位点，主要有 ITS、rDNA 和 SCAR，但国内外尚未见基于 SSR 方法获得特异性引物检测稻曲病菌的报道。

SSR 具有数量丰富、多态信息含量高、按照孟德尔方式遗传、呈共显性等特点。而且 PCR 检测 SSR 标记易于操作，对 DNA 质量要求不高，并且用量不大，甚至 DNA 轻度降解后，仍能扩增出目的 SSR 序列。目前 SSR 标记已经应用于小麦秆锈病菌（*Puccinia graminis* f. sp. *tritici*）、小麦条锈病菌（*Puccinia striiformis*）等病原菌的特异性检测。本研究利用稻曲病菌全基因组信息，基于 MISA 和 Primer 3 设计了 127 条 SSR 序列的扩增引物。通过 PCR 检测引物在不同病原菌中［稻曲病菌 134 个（126 个来自江苏省赣榆、南京、扬州、金坛、徐州、淮安等 7 个稻区；8 个来自长沙）；水稻纹枯病菌 1 个；小麦赤霉病菌 1 个；稗草黑穗病菌 1 个；水稻稻瘟病菌 6 个；小麦纹枯病菌 1 个；水稻恶苗病菌 3 个；西瓜枯萎病菌 1 个；油菜菌核病菌 1 个；甘蓝黑斑病菌 1 个］的扩增结果，筛选得到了 3 对能在稻曲病菌中稳定、且特异性扩增的 SSR 序列引物。本研究为稻曲病菌高效、快速、特异性检测提供了新的特异性靶标。

关键词：稻曲病菌；特异性；SSR 标记

* 基金项目：江苏省自然科学基金（BK20151368）；江苏省农业科技自主创新基金（CX（15）1054）
** 第一作者：俞咪娜，浙江萧山人，助理研究员，研究方向为水稻稻曲病菌功能基因研究；E-mail：zjpsyu@163.com
*** 通讯作者：刘永锋，博士，研究员，主要从事水稻病害病理学及其生物防治技术研究；E-mail：liuyf@jaas.ac.cn

腐烂病菌 *Valsa mali* 侵染过程中苹果抗性相关基因的鉴定*

尹志远**，柯希望，康振生，黄丽丽***
（旱区作物逆境生物学国家重点实验室/西北农林科技大学植物保护学院，杨凌　712100）

摘　要：由死体营养真菌 *Valsa mali* 引起的苹果树腐烂病（Apple valsa canker）作为苹果的主要病害，严重威胁着我国苹果产业的可持续发展。目前，由于病菌主要为害韧皮部并且对苹果的抗性机制知之甚少，导致病害很难得到有效防控。本研究利用转录组学方法，分析鉴定病菌侵染过程中苹果（*Malus domestica* Borkh. cv. Fuji）的抗性相关基因。对 2 713个显著上调表达基因进行 GO 和 KEGG 富集分析发现，抗性相关通路如几丁质、激素、细胞坏死等显著富集，表明其可能参与抗菌过程。两个几丁质受体激酶，MDP0000136494 和 MDP0000169047，可能是响应病菌几丁质降解产物的关键受体。茉莉酸和水杨酸信号通路是响应病菌侵染的主要激素通路。此外，11 个可能参与植保素合成或真菌毒素降解的细胞色素 P450 基因也显著上调表达。本研究鉴定的候选抗腐烂菌相关基因，为进一步鉴定抗病基因及创制抗病材料提供了理论依据。

关键词：果树病害；细胞色素 P450；几丁质；植物激素

---

* 基金项目：国家自然科学基金（31471732 和 31171796）；国家公益性行业（农业）科研专项（201203034）和高等学校学科创新引智计划（B07049）

** 第一作者：尹志远，博士研究生，主要从事苹果树腐烂病菌与苹果互作机制的生物信息学研究；E-mail：zhiyuan. yin@hotmail. com

*** 通讯作者：黄丽丽，教授，主要从事果树病害病原生物学及病害综合防控方面的研究；E-mail：huanglili@ nwsuaf. edu. cn

Regulation of Methyltransferase LAE1 on Growth and Metabolic Characteristics in *T. atroviride*

LI Ya-qian, SONG Kai, YU Chuan-jin, GAO Jin-xin, CHEN Jie*

(*School of Agriculture and Biology, Shanghai Jiao Tong University, Shanghai, P. R. China; Key Laboratory of Urban Agriculture (South), Ministry of Agriculture, P. R. China; State Key Laboratory of Microbial Metabolism, Shanghai Jiao Tong University, Shanghai, P. R. China*)

The biocontrol potential of *Trichoderma* has been the most fascinating and interesting topic of research for agricultural scientists. *Trichoderma* spp. secretes a chemically diverse range of secondary metabolites, of which broad-spectrum antimicrobial properties have been demonstrated well in many in vitro assays. The methyltransferase LAE1 is a global transcriptional regulator that affects the expression of multiple secondary metabolite gene clusters by modifying heterochromatin structure in *Trichoderma*.

This study chose *Trichoderma atroviride* T23 as the initial strain, *agrobacterium-mediated transformation* (ATMT) technique was used to construct knockout strain T23Δ*lae*1 and over expression strain T23O*lae*11. The growth phenotype (mycelia and spores) was assayed comparing T23 an T23Δ*lae*1 & T23O*lae*11. The results showed that the mutant T23Δ*lae*1 had a slower mycelia growth, spore production delayed, by contrast, over-expression strain of T23O*lae*1 increased the growth and sporulation. 18 non-ribosomal peptide synthase (NPRSs) and 16 polyketide synthase (PKSs) are involved in the biosynthesis of secondary metabolites in *T. atroviride*. QRT- PCR analyzed the regulation of *lae*1 on the expression of NPRSs and PKSs, and showed that *lae*1 promote the expression of most of the NRPS and PKS enzymes and inhibit of a few of NRPS and PKS expressions. Therefore, the transcription factor Lae1 is closely related to the production of secondary metabolites. The GC-MS and LC-MS analysis of metabolites spectrum of three strains (T23, T23Δ*lae*1 and T23O*lae*1) including the volatile and non-volatile metabolites synthesis. The results showed that the quality and quantities of metabolites in T23Δ*lae*1 are significantly reduced, while the kinds of secreted metabolites were obvious increased in T23O*lae*1 and had much higher yields of some metabolism. Overall, Lae1 promote synthesis of amino acids metabolites and induce certain antimicrobial peptaibols compounds, which were not detected in wild strain T23. Therefore, the results verified that Lae1 can activate partial silencing of the NPRs or PKSs gene cluster, and lead to mine some novel metabolites unreported by over-expression *lae*1 gene.

In summary, the putative methyltransferase Lae1 is a global regulator that affects the cell growth and expression of multiple secondary metabolite gene clusters in *T. atroviride* T23. This study will shed light on the development effective biocontrol agents that take secondary metabolites-based formulation in agriculture application.

* Corresponding author: CHEN Jie, professor; E-mail: Jiechen59@ sjtu. edu. cn

我国大麦新病害——冠锈病*

赵　杰[1**]，赵元元[1]，覃剑锋[1]，左淑霞[1]，
郑　丹[1]，姚　强[2]，康振生[1***]
（1. 西北农林科技大学植物保护学院，旱区作物逆境生物学国家重点实验室，
杨凌　712100；2. 青海省农林科学院植物保护研究所，西宁　810016）

摘　要：锈病（条锈、秆锈、叶锈和冠锈）是世界范围内大麦及其他禾谷类作物生产上的毁灭性真菌病害，该病害流行常常可致严重的产量损失。其中，冠锈是禾谷类作物及其他杂草上最重要的病害之一，在世界上局部地区（北美和欧洲部分国家）发生，但是，在我国尚未有发生的相关报道。本研究利用形态学方法、分子生物学方法结合致病力测定，对发生于我国青海省大麦上的冠锈病进行了研究。研究结果表明，该病害病原菌的冬孢子顶端生长有较长的数目不等的指状突起（分枝），颜色栗褐色；基因组 DNA 的 ITS（Internal transcribed spacer）区序列与已知大麦冠锈菌（*Puccinia coronata* var. *hordei*）高度同源；对燕麦、小黑麦和大多数小麦的幼苗不致病，对 83 份测试的大麦的 35 个品种、黑麦和极少数小麦品种的幼苗致病。因此，认为此病害是大麦冠锈病，是我国大麦上的新病害，对我国大麦生产具有严重的潜在威胁，应当引起足够的重视。本研究为我国大麦病害的防治和安全生产提供了科学依据。

关键词：大麦；冠锈菌；冬孢子；致病力；ITS 区

---

* 基金项目：高等学校学科创新引智计划资助项目（B07049）；国家科技支撑计划（十二五）（2012BAD19B04）；“新世纪优秀人才支持计划资助”

** 第一作者：赵杰：副教授，主要从事植物病理学研究；E-mail：jiezhao@ nwsuaf. edu. cn

*** 通讯作者：康振生，教授；E-mail：kangzs@ nwsuaf. edu. cn

An Extracellular Zn-only Superoxide Dismutase from *Puccinia striiformis* Confers Enhanced Resistance to Host-derived Oxidative Stress

LIU Jie [1*], GUAN Tao[1*], ZHENG Pei-jing[1*], CHEN Li-yang[1], YANG Yang[1], HUAI Bao-yu[1], LI Dan[1], CHANG Qing[2], HUANG Li-li[2], KANG Zhen-sheng[2**]

(1. *State Key Laboratory of Crop Stress Biology for Arid Areas and College of Life Sciences, Northwest A&F University, Yangling, PR China*; 2. *State Key Laboratory of Crop Stress Biology for Arid Areas and College of Plant Protection, Northwest A&F University, Yangling, PR China*)

Abstract: The accumulation of reactive oxygen species (ROS) following plant-pathogen interactions can trigger plant defence responses and directly damage pathogens. Thus, it is essential for pathogens to scavenge host-derived ROS to establish a parasitic relationship. However, the mechanisms protecting pathogens from host-derived oxidative stress remain unclear. In this study, a superoxide dismutase gene, *PsSOD*1, was cloned from a wheat-*Puccinia striiformis* f. sp. *tritici* (*Pst*) interaction cDNA library. Transcripts of *PsSOD*1 were up-regulated in the early *Pst* infection stage. Heterologous mutant complementation and biochemical characterization revealed that *PsSOD*1 encoded a Zn-only SOD. The predicted signal peptide was functional in an invertase-mutated yeast strain. Furthermore, immunoblot analysis of apoplastic proteins in *Pst*-infected wheat leaves and bimolecular fluorescence complementation suggested that PsSOD1 is a secreted protein that forms a dimer during *Pst* infection. Overexpression of *PsSOD*1 enhanced *Schizosaccharomyces pombe* resistance to exogenous superoxide. Transient expression of *PsSOD*1 in *Nicotiana benthamiana* suppressed Bax-induced cell death. Knocking down *PsSOD*1 using a host-induced gene silencing system reduced the virulence of *Pst*, which was associated with ROS accumulation in HIGS plants. These results suggest that *PsSOD*1 is an important pathogenicity factor that is secreted into the host-pathogen interface to contribute to *Pst* infection by scavenging host-derived ROS.

---

* These authors contributed equally to this work

** Corresponding author: KANG Zhen-sheng; E-mail: kangzs@nwsuaf.edu.cn

16种杀菌剂对新疆红枣黑斑病的室内毒力测定*

刘础荣[1**]，阮小珀[1]，董　玥[1]，罗来鑫[1]，朱天生[2]，李志军[2]，李健强[1***]
（1. 中国农业大学植物病理学系/种子病害检验与防控北京市重点实验室，北京　100193；
2. 塔里木大学，阿拉尔　843300）

摘　要：新疆是我国红枣的主产区，红枣黑斑病是该地区红枣生产中常见的病害之一；该病害主要为害红枣果实，导致果实黑斑，严重影响红枣的商品品质，也可侵染叶片和花导致叶片黑斑和落花现象。自2010年新疆阿克苏地区红枣黑斑病大暴发以来，新疆各地区该病害均有报道，细极链格孢（*Alternaria tenuissima*）是引起新疆地区红枣黑斑病的主要病原菌之一。

目前生产中防控主要以化学防治为主，筛选防控红枣黑斑病的安全高效药剂以及研究施药技术具有重要意义。本研究以抑制呼吸作用、信号转导、细胞膜甾醇合成和氨基酸及蛋白质合成的4种不同作用机制的16种杀菌剂原药为供试药剂，采用平皿菌丝生长抑制法进行了室内毒力测定。结果表明，吡唑醚菌酯、啶酰菌胺、嘧菌环胺、嘧霉胺、咯菌腈和戊唑醇等15种杀菌剂对红枣黑斑病菌表现出较高抑菌活性，其中，吡唑醚菌酯和嘧菌环胺对供试的细极链格孢（*A. tenuissima*）室内抑菌效果最好，其EC_{50}值分别为0. 033 7μg/mL和0. 100 6μg/mL。

关键词：红枣黑斑病；细极链格孢（*Alternaria tenuissima*）；杀菌剂；室内毒力

* 基金项目：国家科技支撑计划专题（2014BAC14B04－3）
** 第一作者：刘础荣，硕士研究生；E-mail：x8281288@163. com
*** 通讯作者：李健强，教授，主要研究方向为种子病理与杀菌剂药理学；E-mail：lijq231@cau. edu. cn

越南黄檀锈病及其天敌昆虫研究

王　妏，周国英，苏圣淞，何苑皞，董文统，刘君昂
（中南林业科技大学经济林培育与保护教育部重点实验室/
中南林业科技大学森林有害生物防控湖南省重点实验室，长沙　410004）

摘　要：越南黄檀是世界珍稀名贵木材，海南省是我国主要栽培种植区，目前国内越南黄檀病虫害的报道较少。2015 年，对海南省澄迈县越南黄檀人工林锈病的发生为害进行了调查，同时对该锈病的病原及天敌昆虫进行研究。结果表明，锈病是越南黄檀人工林发生最为严重的病害，该锈病病原夏孢子几乎全年均可发生，每年有两个暴发期，冬孢子在天气转凉时偶见，通过对夏孢子、冬孢子形态学分析确定了越南黄檀锈病的病原菌为紫檀无眠单胞锈菌（*Maravalia pterocarpi*）；通过收集越南黄檀人工林中的昆虫，并对所有收集到的昆虫进行饲养，根据各昆虫的取食特性成功筛选出了一种取食无眠单胞锈菌夏孢子的天敌昆虫，并对该昆虫的形态学与生物学特性进行了观察与研究，经鉴定该昆虫是一种锈菌瘿蚊（*Mycodiplosis* sp.）。本研究在详细描述该锈病发生过程与其天敌昆虫的形态学、生物学特性的同时，探索出天敌昆虫的最适饲养条件为：光照周期 L：D＝14h：12h、温度（27±0.5）℃、湿度（70±5）%，为越南黄檀的进一步研究与防治奠定了基础，也丰富了菌食性天敌昆虫的理论知识。

关键词：越南黄檀；锈病；锈菌瘿蚊

重庆地区玉米穗腐病致病镰孢菌种类及产毒研究*

周丹妮[1,2]**，王晓鸣[1]，陈国康[2]，杨　洋[1,2]，孙素丽[1]，朱振东[1]，段灿星[1]***
（1. 中国农业科学院作物科学研究所/国家农作物基因资源与基因改良重大科学工程，北京　100081；2. 西南大学植物保护学院，重庆　400715）

摘　要：玉米穗腐病，属世界性病害，也是我国玉米生产上的重要病害，可由多种镰孢菌和其他真菌引起。镰孢菌引起的玉米穗腐病不仅造成产量损失，也会因病菌产生的多种毒素，对人畜健康造成极大威胁。因此，开展玉米穗腐致病镰孢菌种群组成及毒素的研究，具有重要意义。本研究于2014—2015年在重庆市及周边地区32个区县的98个乡镇采集玉米穗腐病样品，经组织分离和单孢纯化共鉴定出111个镰孢菌分离物，采用分子检测方法对不同地区的分离物进行种级和产毒基因型鉴定，获得以下结果：

（1）基于形态学和分子检测，在获得的111个镰孢菌分离物中，共鉴定出10种致病镰孢，分别为拟轮枝镰孢（*Fusarium verticillioides*），禾谷镰孢复合种（*F. graminearum* species complex），层出镰孢（*F. proliferatum*），尖镰孢复合种（*F. oxysporum* species complex），藤仓镰孢（*F. fujikuroi*），木贼镰孢（*F. equiseti*），黄色镰孢（*F. culmorum*），变红镰孢（*F. incarnatum*）、九州镰孢（*F. kyushuense*）和茄镰孢（*F. solani*），其中，拟轮枝镰孢、禾谷镰孢复合种、层出镰孢、尖镰孢复合种分离频率较高，为重庆地区玉米穗腐病优势致病菌。

（2）采用Fum5F/Fum5R特异性引物对43个拟轮枝镰孢分离物和19个层出镰孢分离物的产毒关键基因*fum*1进行检测，发现上述所有分离物均能扩增出该基因的特异性条带，表明拟轮枝镰孢和层出镰孢均具有产生伏马菌素的能力；采用Tril3P1/Tril3P2特异性引物对禾谷镰孢复合种进行了产毒化学型检测，发现19个禾谷镰孢复合种均为NIV毒素类型。

关键词：玉米；穗腐病；镰孢菌；产毒化学型

* 基金项目：国家现代农业（玉米）产业技术体系（CARS－02）；中国农业科学院科技创新工程（谷物质量安全与风险评估团队）

** 第一作者：周丹妮，硕士研究生，从事植物病理学研究

*** 通讯作者：段灿星，副研究员，从事作物种质资源抗性与玉米病害研究；E-mail：duancanxing@caas.cn

First Report of Powdery Mildew Caused by *Erysiphe alphitoides* on *Euonymus japonicas* and Its Natural Teleomorph Stage in China

CHI Meng-yu[1], QIAN Heng-wei[1], DUAN Xue-lan[1], ZHAO Ying[1], HUANG Jin-guang[1,2]*

(1. College of Agronomy and Plant Protection, QingDao Agriculture University, Qingdao, Shandong 266109, China; 2. Key Lab of Integrated Crop Disease and Pest Management of Shandong Province, Qingdao 266109, China)

Abstract: Powdery mildew is one of the major diseases on *Euonymus japonicus*, which occurred in Japan, America and Europe. In China, the powdery mildew on *E. japonicas* was spreading rapidly and the pathogen types causing powdery mildew on *E. japonicas* are as follows: it was identified as *Oidium euonymi-japonicae* (Arcang.) Sacc in Sichuan, Shandong, Jiangxi, and Henan province, but latterly *Oidium* was classified as *Erysiphe*. And the pathogen caused powdery mildew on *E. japonicas* was identified as *E. euonymi-japonicae* in Zhoukou in 2010. During the summer of 2015, severe *E. japonicas* plants exhibiting symptoms typical of powdery mildew was observed in several gardens of Qingdao city (Northern China). In all locations, 80% ~ 90% of foliage was visibly affected. The infected leaves covered powdery mildew colonies on both sides of leaves, ultimately caused leaves discoloration and crinkles.

According to both the morphological characteristics and rDNA-ITS sequence analysis, the powdery mildew pathogen on *E. japonicas* was identified as *Erysiphe alphitoides*. And the pathogenic teleomorph stage was first found in field in November. To our knowledge, this is the first case of *E. alphitoides* natural teleomorph stage on *E. japonicas* and is the first report of *E. alphitoides* infection on *E. japonicas* in China. The occurrence of powdery mildew is a potential threat to the widespread ornamental plantings of *E. japonicas*. The information about the morphological characteristics and genetic characteristics of this disease could help gardeners to monitor and control the disease best.

Key words: *Euonymus japonicas*; Powdery mildew; *Erysiphe alphitoides*; Morphological characteristics; Molecular identification

* Corresponding author: HUANG Jin-guang; E-mail: jghuang@ qau. edu. cn

我国玉米小斑病调查、O 小种不同致病力菌株的侵染规律和侵染过程中的转录本发掘*

王　猛**，陈　捷***
（上海交通大学农业与生物学院，上海　200240）

摘　要：玉米小斑病（southern corn leaf blight）是我国和世界其他玉米产区的主要叶部病害，一般可造成减产 10% ~20%。病原为异旋孢腔菌（*Cochliobolus heterostrophus*）［无性态为玉蜀黍平脐蠕孢（*Bipolaris maydis*）］，分为 O、T、C 三个小种。根据过去 5 年对黄淮海、东北、江浙等玉米产区的调查、采样和鉴定（每年分离菌株不少于 200 株），发现我国玉米小斑病优势小种为 O 小种，仅在 2015 年度夏季石家庄市分离到 1 株 T 小种菌株。采用 4 组玉米自交系对分离到的 O 小种菌株进行致病力鉴定，并对病斑面积量化，发现我国 O 小种致病力有分化现象。为了探索小斑病 O 小种下致病力分化机理，我们挑选了 1 株强致病力菌株和 1 株弱致病力菌株为研究材料，对两株菌进行的单细胞纯化和绿色荧光蛋白标记。对标记好的两株小斑病菌 O 小种进行侵染规律观察，发现强致病力菌株在孢子萌发率和芽管伸长速度明显高于弱致病力菌株，同时强致病力菌株比弱致病力菌株更早完成在玉米叶肉细胞内的定殖。同时，采用激光捕获显微切割将侵入玉米叶肉细胞的病原分离，抽取 total RNA，并进行后续的建库和测序。测序数据初步分析发现，除 micro RNA 和无意义的 rRNA 之外的全部转录本信息（mRNA、lnc RNA、circ RNA）被成功挖掘到，对病原菌在侵染寄主细胞过程中的转录本差异分析有助于我们发现我国玉米小斑病 O 小种的致病力分化分子机理，对我国玉米小斑病的致病力变异预测提供科学理论依据。另外，对两株致病力差异菌株转录本的共性分析还有助于回答一个悬而未决的问题：玉米小斑病菌 O 小种的致病因子有哪些？

关键词：玉米；玉米小斑病菌；O 小种；侵染；转录本

* 基金项目：十二五国家玉米产业技术体系（CARS－02）
** 第一作者：王猛，博士研究生，主要研究方向：分子植物病理；E-mail：wangmeng407@ hotmaii. com
*** 通讯作者：陈捷，教授，主要研究方向：分子植物病理学；E-mail：jiechen59@ sjtu. edu. cn

DNA 提取法对 PCR 扩增检测进境油菜籽携带茎基溃疡病菌的影响

宋培玲[1]，皇甫海燕[1]，郝丽芬[1]，燕孟娇[1]，
石　嵘[1,2]，贾晓清[1]，皇甫九茹[1]，吴　晶[1,2]，李子钦[1]
（1. 内蒙古农牧业科学院植物保护研究所，呼和浩特　010031；
2. 内蒙古大学生命科学学院，呼和浩特　010020）

摘　要：2009 年，我国将油菜茎基溃疡病列入进境植物检疫性有害生物名录，为建立适于进境油菜籽携带茎基溃疡病菌的 DNA 提取及检测方法，比较分析了 5 种 DNA 提取法对 PCR 扩增检测的影响。采用 PVP、CTAB、Kit 以及 PVP + Kit、CTAB + Kit 5 种方法对带菌种子进行 DNA 提取。特异性扩增结果显示，以 PVP、CTAB、Kit 3 种方法所得 DNA 为模板时，检出率为零；以 PVP + Kit 法所得 DNA 为模板时，有特异性条带出现但较弱，且检出率不稳定；以 CTAB + Kit 法提取的 DNA 为模板时，检出率高，且重复率好，该提取法不仅保证了油菜籽带菌量较低时，所得总 DNA 中茎基溃疡病菌 DNA 的微量存在，且其含量可满足进一步 PCR 检测的要求，特异性扩增可有效检出 1 000粒或10 000粒种子中 1 ~2 粒种子所携带的病菌。即 CTAB + Kit 这种大量与微量相结合的 DNA 提取法，适于进境油菜籽携带茎基溃疡病菌的检测。

关键词：油菜籽；茎基溃疡病；DNA 提取；PCR 检测

全基因组预测禾谷炭疽菌碳水化合物酶类蛋白及其理化性质分析*

韩长志**，许　僖

（西南林业大学林学院/云南省森林灾害预警与控制重点实验室，昆明　650224）

摘　要：碳水化合物酶类蛋白（Carbohydrate-Active Enzymes，CAZymes）作为分泌蛋白中的一大类，是病原菌侵染过程中突破寄主细胞第一道屏障植物细胞壁的关键因素。对其开展深入研究，有助于解析植物病原菌侵入、操控植物的作用机制。禾谷炭疽菌［*Colletotrichum graminicola*（Cesati）Wilson］作为炭疽菌属重要的植物病原菌，可以侵染玉米、小麦、高粱等禾本科作物而引起炭疽病，给各国农业生产造成巨大的经济损失。自20世纪70年代以来，由该菌引起的玉米炭疽病在美国、印度等国就非常普遍。随着该菌全基因组序列的释放，目前，关于该菌MAPK途径蛋白预测、分泌蛋白预测以及14－3－3蛋白和RGS生物信息学分析等已见报道。前期本研究小组利用SignalP、ProtComp、TMHMM、big-PI Fungal Predictor和TargetP等预测程序对*C. graminicola*蛋白数据库中的12 006条序列进行分析预测，明确该菌中含有630个分泌蛋白。后续通过CAT分析，明确上述蛋白中含有267个CAZymes，然而，对于上述蛋白的性质和功能一直缺少较为深入的后续研究报道。目前学术界关于禾谷炭疽菌中CAZymes的数量有相关报道，但尚未见有关该菌中CAZymes理化性质分析报道。本研究基于前期所获得的267个CAZymes蛋白序列，通过生物信息学分析方法解析该菌上述蛋白的基本理化性质，结果表明，近2/3的蛋白理论等电点小于6，属于酸性蛋白，同时，通过随机数软件对上述蛋白进行10%抽样分析，明确了不同理论等电点类别的分泌蛋白所具有的分子质量、分子式、原子数量、半衰期、不稳定性系数、脂肪族氨基酸指数、总平均亲水性等性质。该研究为深入解析禾谷炭疽菌CAZymes蛋白的功能奠定了坚实的理论基础，也为深入解析该病原菌CAZymes蛋白在侵染操控寄主机制方面提供重要的理论基础。

关键词：禾谷炭疽菌；碳水化合物酶；分泌蛋白；理化性质；预测程序

---

* 基金项目：国家自然科学基金项目（31560211）；云南省森林灾害预警与控制重点实验室开放基金项目（ZK150004）；云南省优势特色重点学科生物学一级学科建设项目（项目编号：50097505）；云南省高校林下生物资源保护及利用科技创新团队（2014015）

** 第一作者：韩长志，河北石家庄市人，博士，讲师，研究方向：经济林木病害生物防治与真菌分子生物学；E-mail：hanchangzhi2010@163.com

环渤海湾和黄河故道苹果产区轮纹病成灾机制及防控策略*

李保华
（青岛农业大学农学与植物保护学院，青岛　266109）

在环渤海湾和黄河故道两大苹果产区，轮纹病已成为苹果树的第一大病害，其危害已超过腐烂病。苹果轮纹病菌侵染枝干形成轮纹病瘤、马鞍状病斑、干腐病斑、粗皮等症状，严重削弱树势，造成死枝、死树等。侵染果实形成轮纹烂果症状，造成严重经济损失。轮纹病菌来源广，侵染期长，侵染量大，侵染后能长期存活，属于典型的积年流行病害。

感病品种富士的大面积栽培是轮纹病菌积累和繁殖的基础。环渤海湾和黄河故道苹果产区的多雨气候为纹病菌的产孢、孢子传播和孢子侵染提供了有利的条件。在病瘤内生长扩展，以及在皮孔和伤口内长期潜伏的病菌可导致果园内轮纹病菌逐年积累；干腐病斑和马鞍状病斑上所产孢子的大量侵染，扩大了病菌在果园内的种群数量；树势衰弱、树体受旱受涝、枝条枯死等，导致潜伏在病瘤、皮孔和伤口内的病菌迅速扩展，形成干腐病斑和马鞍状病斑，并产生大量孢子，进行再侵染，加重了轮纹病的为害。轮纹病菌除产生大量分生孢子随雨水传播外，遇连续阴雨还能产生大量子囊孢子；子囊孢子随气流进行远距离传播，扩大病菌的分布范围，导致苗圃、新建果园发病；随苗木传带是轮纹病菌的重要传播方式。轮纹病菌的侵染条件容易，2mm 的降雨可导致病菌孢子释放与传播，持续 3h 的降雨可使病菌在寄主表面定殖，随后侵入寄主的活体组织，或潜伏于枝干表层的死组织内。轮纹病菌侵入寄主组织后难以铲除。防治轮纹病的唯一有效措施是防止轮纹病菌侵入寄主组织；增强树势防止侵染病菌扩展致病，可减少轮纹病菌的产孢量，降低轮纹病的为害。

对苹果轮纹病需自苗期和幼树期开始，保护枝干和剪锯口不受病菌侵染；病菌侵染后及时治疗或铲除，防止病菌逐年积累，导致积年流行；同时加强果园管理、增强树势，以防止已侵染的病菌扩展致病和产孢。在实际的病害防治中，应综合运用“清、防、健、治”四项措施，即“清”除侵染菌原，减少侵染菌原量；采用物理、化学、生物等方法保护枝干、剪锯口和果实，“防”止病菌侵染；加强栽培、肥水、花果管理和病虫害防治，保持“健”旺树势，防止已侵染病菌在树体内生长、扩展、致病和产孢；适时适度“治”疗病斑，铲除潜伏病菌，刮除病瘤，防止已侵染的病菌进一步扩展为害。四项措施综合运用，坚持数年，可有效控制轮纹病。

* 基金项目：国家苹果产业技术体系（CARS－28）

利用 RNA-seq 研究玉米弯孢叶斑病菌黑色素和毒素协同表达机理*

高金欣[1,2]** 高士刚[1,2] 李雅乾[1,2] 陈　捷[1,2]***

（1. 上海交通大学农业与生物学院，都市农业（南方）重点开放实验室，上海　200240；2. 上海交通大学微生物代谢国家重点实验室，上海　200240）

摘　要：玉米弯孢叶斑病菌（*Curvularia lunata*）作为一种主要的玉米叶部病害在我国玉米种植区广泛分布。黑色素和毒素是玉米弯孢叶斑病菌次生代谢产生的两种主要的致病因子。尽管玉米弯孢叶斑病菌 CX－3 全基因组序列已经测通，但是对于黑色素和毒素的合成机制及二者的协同表达机理尚未完全研究透彻。前期研究获得了一株 PKS18 基因敲除突变株 ΔPKS18 和一株产毒缺陷型菌株 T806。ΔPKS18 完全丧失了产黑色素的能力，同时降低了产毒素能力；而 T806 完全不能产毒素，同时产黑色素能力降低。因此，我们对 ΔPKS18 和 T806 进行了转录组测序，以便发现黑色素和毒素合成基因，以及二者的协同表达模式。通过 RNA-seq 分析，我们发现了大量差异表达基因，发现了黑色素合成过程的相关基因，并且对毒素合成基因簇进行预测。另外，我们进行了可变剪接分析，发现 76.15% 的可变剪接属于 Intron retention（IR）。

关键词：玉米弯孢霉叶斑病菌；黑色素；毒素；RNA-seq；基因表达

---

* 基金项目：国家自然科学基金（31471734）；中国农业研究体系（CARS－02）

** 第一作者：高金欣，博士生，主要从事分子植物病理学研究；E-mail：jinxingao@ yeah. net

*** 通讯作者：陈捷，教授，博士生导师，主要从事植物病理学和环境微生物工程研究；E-mail：jiechen59@ sjtu. edu. cn

Screening and Analysis of Effector Proteins from *Puccinia triticina* by Transcriptome Sequencing

LI Jian-yuan, WANG Zhi, ZHANG Yu-mei, YANG Wen-xiang, LIU Da-qun

(*Department of Plant Pathology, Agricultural University of Hebei/ Biological Control Center of Plant Diseases and Plant Pests of Hebei Province/National Engineering Research Center for Agriculture in Northern Mountainous Areas, Baoding* 071001)

Abstract: Leaf rust caused by the fungus *Puccinia triticina* (*Pt*) is a major constraint to wheat production worldwide. The molecular events that underlie *Pt* pathogenicity are largely unknown. Like all rusts, *Pt* forms a specialized infection structure called haustoria to take nutrients and water from host plant tissue, and to secrete pathogenicity factors called effector proteins. We extracted the total RNA of urediniospores, germinated urediniospores and infected wheat leaves harvest at 6d from three different virulence strains of *Pt* and used Illumina Hiseq sequencing platforms to assemble the transcriptomes and then we got 80392 Unigenes. Candidate effectors were screened through Signal P 4.1, Target P 1.1, TMHMM 2.0 and Effector P, and 325 genes which constitute candidate effectors were identified. About two thirds of these were up-regulated in the harvest leaves compared to urediniospores and germinated spores. 42 of the genes were predicted to code for proteins of known function like a variety of metabolism processes, nutrient transport, and biosynthesis by using a BLAST2GO platform and manual annotation. Pfam was used to predict the frameworks and 61 proteins were predicted with Pfam notes. The remaining 264 secreted proteins were screened and predicted the de novo structure by using MEME software. The results showed that 3 conserved domains were found in 11 effector proteins. RT-PCR analysis confirmed the expression patterns of 52 effector candidates and four genes showed a special increase in infected leaves which may be associated with the formation of urediniospores. This analysis of candidate effectors is an initial step towards functional research and established the foundation for clarify the pathogenic mechanism of leaf rust pathogens.

Key words: *Puccinia triticina*; Effector proteins; Transcriptome sequencing

Heterologous Expression and Homology Modelling of Sterol 14α-demethylase of *Fusarium graminearum* and Its Interaction with azoles*

QIAN Heng-wei[1], CHI Meng-yu[1], ZHAO Ying[1], HUANG Jin-guang[1,2]**

(1. *College of Agronomy and Plant Protection*, *QingDao Agriculture University*, *Qingdao* 266109, *China*; 2. *Key Lab of Integrated Crop Disease and Pest Management of Shandong Province*, *Qingdao* 266109, *China*)

Abstract: *Fusarium graminearum* is the main pathogen of *Fusarium* Head Blight (FHB), a worldwide plant disease and one of the major wheat diseases in China. The infected wheat grain could product the mycotoxins and posed a serious threat to human and animal health. The control of the FHB is main dependent on the application of fungicides, Demethylase inhibitors (DMI fungicides, DMIs), such as tebuconazole and diniconazole, which inhibit the biosynthesis of ergosterol in fungi because of sterol 14α-demethylase (CYP51) binding. Ergosterol is essential for cell membrane fluidity and permeability, so ergosterol depletion will result in the inhibition of fungal growth. Due to the intensive use of DMI fungicides in agriculture, resistance has been detected in infected crops and become more serious in recent years.

F. graminearum has three CYP51 genes, CYP51A, CYP51B and CYP51C, which share 61.65% amino acid sequence identity. CYP51B is expressed constitutively, while gene CYP51A is reported to be inducible. Fg CYP51B, as the most conserved CYP51 gene in all fungi, encodes the enzyme primarily responsible for sterol 14α-demethylation. In order to explore the molecular mechanism of drug resistance of CYP51, we amplified the FgCYP51B (FGSG_01000) fragments by PCR and cloned into *Escherichia coli* expression vector pETM - 30, the constructed plasmids were transformed into *E. coli* strain BL21 (DE3). After IPTG induced, it was not expressed in *E. coli* verified by SDS-PAGE. According to TM-HMM Server V. 2. 0 prediction, the *N*-terminal 49aa (transmembrane domains) of FgCYP51B was truncated (T-FgCYP51B). T-FgCYP51B was successfully expressed in *E. coli* after IPTG induced, but the protein mainly existed in the form of inclusion body after affinity chromatography.

Fungal CYP51 proteins are membrane bound, making it more difficult to study their structural compared with the solube bacterial CYP51. So homology modeling and molecular docking with ligand has been used extensively to explain possible protein-drug interactions. The FgCYP51B structure was modeled based on the crystal structure of AfCYP51B (PDB ID: 4UYM) as the template using the Swiss-Model server, then energy minimization was performed to eliminate improper contacts for the modeled structure in Amber. The structures of five small molecular inhibitors, including Diniconazole (DIC), Tebuconazole (TEC), Triadimenol (TRL), Triadimefon (TRN) and Propiconazole (PRC), were conducted geometry optimization by Gaussian programe. AutoDock vina was then used to dock the inhibitors into the

* This work was supported by National Natural Science Foundation of China (NSFC) No. 31471735

** Corresponding author: HUANG Jin-guang; E-mail: jghuang@ qau. edu. cn active cavity of the enzyme respectively. The structure of complexes were visualized using PyMOL. Modeled structure exhibiting a high degree of sequence identity with template is in consistant with the view and outlines a highly hydrophobic cavity consisting of a number of hydrophobic residues, mainly including Tyr123, Phe131, Tyr137, Phe230, Leu510 and Phe511, which are conserved. Tyr123, Phe131 and Tyr137 of FgCYP51 are highly conserved across all biological kingdoms, while Phe230 is conserved across fungal species, Leu510 and Phe511 varies in different phyla. Thus, the non-conservative and Phe230, Leu510 and Phe511 of FgCYP51 are supposed to be key residues interacting with azole fungicides. Elucidation of molecular mechanisms of protein-drug interactions is underway. In summary, our results offer new insights into resistance molecular mechanisms of *F. graminearum*.

Key words: *Fusarium graminearum*; CYP51; Expression; Homology modeling; Molecular docking

First Report of *Colletotrichum scovillei* Causing Anthracnose Fruit Rot on Pepper in Anhui Province, China

ZHAO Wei, WANG Tao, CHEN Qing-qing, CHI Yuan-kai, QI Rrede

(*Institute of Plant Protection and Agro-products Safety*, *Anhui Academy of Agricultural Sciences*, *Hefei* 230031, *China*)

Anthracnose fruit rot is a major disease of pepper (*Capsicum annuum*), which causes severe damage and enormous economic loss in pepper production. *Colletotrichum acutatum*, *C. coccodes*, *C. boninense*, and *C. capsici* have been reported as causal agents for anthracnose on pepper in China, Since 2014, dark, circular, sunken zones with concentric rings of orange conidial masses, the typical anthracnose symptoms, have been appearing on pepper fruits in Shitai, Anhui Province, China. Small tissue pieces from the edges of lesions on fruits were disinfected in 70% (v/v) ethanol for 30s and in 2% (v/v) hypochlorous acid solution for 2 min, rinsed twice in sterilized water, air-dried, then incubated on potato dextrose agar (PDA) at 25℃. Eight isolates of the pathogen were obtained from different diseased plants. Colony morphology on Spezieller Nährstoffarmer Agar (SNA) was pale gray to pale orange fluffy aerial mycelium. The conidia were hyaline, aseptate, smooth, straight, subcylindrical to clavate, with one end round and one end round or acute, 11.8 to 16.5μm long and 3.2 to 4.5μm wide, L/W ratio = 3.6. These morphological characters are consistent with the description of *Colletotrichum scovillei*. To confirm this identification, the internal transcribed spacer (ITS) rDNA regions, Glyceraldehyde-3-phosphate dehydrogenase (GADPH) gene and β-Tubulin-2 (TUB2) gene were amplified with the primers ITS1/ITS4, GDF/GDR and TUBT1/TUBT2, respectively. Compared with the sequence of strain CBS 126529 of *C. scovillei* in GenBank, the amplification products showed 100% homology with the ITS sequence (GenBank Accession No. JQ948267), the GADPH sequence (GenBank Accession No. JQ948597) and the TUB2 sequence (GenBank Accession No. JQ949918), respectively. A pathogenicity test was performed by depositing 10μL droplets of a suspension (10^5 conidia/mL) on the surfaces of 5 artificially wounded fruits (prick a small hole on pepper fruit using needle). Five wounded fruits were inoculated with sterilized, distilled water as the control. After 5 days in a moist chamber at 25℃, symptoms similar to those observed on the original pepper plants had developed on the five wounded fruits that had been inoculated, and *C. scovillei* was re-isolated from the lesions. The control fruits remained healthy. Recently, *C. acutatum* has been considered a species complex. *C. scovillei* as one of the species belongs Clade D3 within *C. acutatum* species complex. *C. scovillei* causing anthracnose disease on pepper was previously reported in Japan, Brazil. However, to our knowledge, this is the first report of occurrence of anthracnose fruit rot caused by C. *scovillei* on pepper in China. This disease causes severe damage both in the field and postharvest and needs to develop effective control strategies in the pepper production in these regions.

Development of New EST-SSR Markers of Wheat Leaf Rust Based on *Puccinia triticina* Transcriptome

YANG Wen-xiang, WANG Zhi, MENG Qing-fang, LIU Da-qun

(*Department of Plant Pathology, Agricultural University of Hebei/Biological Control Center of Plant Diseases and Plant Pests of Hebei Province/National Engineering Research Center for Agriculture in Northern Mountainous Areas, Baoding* 071001)

Abstract: Wheat leaf rust caused by *Puccinia triticina* has different degree of threat to wheat production in the important wheat diseases in almost all wheat producing areas in the world. Due to the variation of wheat leaf rust and the same disease resistant varieties of single plant, pathogen is constantly changing. Pathogen monitoring of virulence trend and population polymorphism are important task of the wheat leaf rust population dynamic transformation research and virulence polymorphism research. Development of new molecular markers of *P. triticina* can make us deeply understand the molecular pathogenetic mechanism, survay virulence and control wheat leaf rust. At present, only 31 pairs of wheat leaf rust EST-SSR markers have been reported, so it is an urgent need to develop new EST-SSR perimers. The objective of this research was to develop new polymorphic EST-SSR markers of wheat leaf rust based on transcriptome sequencing for reveal the genetic relationship and genetic background. We used Illumina Hiseq 2000 to assemble the transcriptomes and then we got 46 008 Unigenes. MISA software was then used to screen the EST-SSR molecular markers from these and 3 085 unigenes containing 3 729 EST-SSR loci were obtained. A total of 3 065 EST-SSR primers on 613 loci were designed by using Primer3 software and 180 of these were synthesized randomly. Among them 141 pairs of the primers could be amplified in *P. triticina*, and 37 of these produced polymorphic bands which can be applied to detect the genetic diversity in a natural population of *P. triticina*. The total allele number was 85, with average allele number of 2. 3 per locus. Then we used the 37 EST-SSRs to judge the polymorphism of different 48 stains of *P. triticina* collected from Hebei province and Xinjiang. The results showed that most of the *P. triticina* stains in the two regions were clustered into different groups which indicated that the EST-SSR polymorphism and geographical distribution of *P. triticina* stains have certain correlation.

Key words: Wheat leaf rust; Transcriptome; EST-SSR molecular marker; Polymorphism

Comparative Genome Analyses Reveal Distinct Modes of Host-specialization Between Dicot and Monocot Powdery Mildew

WU Ying [1], MA Xian-feng[1], PAN Zhi-yong[2], KALE Shiv[3], SONG Yi[1], XIAO Shun-yuan[1]

(1. *Institute of Biosciences and Biotechnology Research & Department of Plant Science and Landscape Architecture*, *University of Maryland College Park*, *Rockville MD* 20850, *USA*; 2. *Key Laboratory of Horticultural Plant Biology of Ministry of Education*, *Huazhong Agricultural University*, *Wuhan* 430070, *PR China*; 3. *Virginia Bioinformatics Institute*, *Virginia Tech*, *Blacksburg*, *Virginia*, *USA*)

Abstract: Monocot powdery mildew (PM) fungi such as *Blumeria graminis* f. sp. *hordei* (*Bgh*) infectious on barley and *B. graminis* f. sp. *tritici* (*Bgt*) infectious on wheat exhibit extremely high-level of host-specialization. By contrast, many dicot PM fungi display rather broad host ranges. To understand why different PM fungi adopt distinct modes of host-adaption, we sequenced the genomes of four dicot PM strains (three *Golovinomyces* and one *Oidium*) and conducted comparative sequence analyses along with the genomes of *Bgh*, *Bgt* and *Erysiphe necator* (infectious on grapevine) available from the Genbank. While 3 129 genes are found in all seven genomes (thus they are conserved PM genes), ~50% are lineage-specific, of which 116 ~ 534 are genes encoding candidate secreted effector proteins (CSEPs). Interestingly, while the two monocot PM fungi possess up to 534 CSEPs with several families containing up to 21 members, all the five dicot PM fungi have only 116 ~ 178 CSEPs with limited gene amplification. These two evolutionary patterns concerning PM CSEPs coincide with the contrasting modes of their host-adaption: the monocot PM fungi show a high-level of host specialization, likely reflecting an advanced host-pathogen arms-race, whereas the dicot PM fungi tend to practice polyphagy, which has presumably lessened the selective pressure for escalating an arms-race with a particular host. Strikingly, the tomato PM *Oidium neolycopersici* has a genome only half the size of other PM genomes, despite having a similar number of genes, suggesting that *O. neolycopersici* has followed a unique genome evolutionary path.

Cloning, Expression and Function Analysis of *TaSGT*1 in *TcLr*19

YUAN Ying[1], HU Ya-ya[2], LI Jan-yuan[1], YANG Wen-xiang[1], LIU Da-qun[1]

(1. *Department of Plant Pathology, Agricultural University of Hebei/Biological Control Center of Plant Diseases and Plant Pests of Hebei Province/National Engineering Research Center for Agriculture in Northern Mountainous Areas, Baoding* 071001; 2. *Institute of Cereal and Oil crops, Hebei academy of Agriculture and Forestry Sciences, Shijiazhuang* 050035)

Abstract: Wheat leaf rust, caused by *Puccinia triticina* is an airborne fungal disease in wheat worldwide, which can cause significant yield losses to the wheat production in the world. Using resistant cultivars is the most economical and environmentally sound method to control the disease. A leaf rust resistance gene *Lr*19 could provides resistance to all races of *P. triticina* currently found in China, and has shown high potential in application at home and abroad. The objectives of this work were to clone a *TaSGT*1 gene from the TcLr19 wheat line and understanding its function in against the wheat leaf rust. A spliced sequence with the length of 1 690bp was obtained by rapid amplification cDNA ends (RACE), and the full length cDNA of the *TaSGT*1 gene isolated from TcLr19 was 1 372bp. The TaSGT1 protein, containing TPR, VR1, CS, VR2 and SGS domains, was predicted by using the SMART software. *TaSGT*1 gene was located on plasma membrane and nucleus by subcellular localization. Silencing of *TaSGT*1 in the leaf rust-resistant plants by virus induced gene silencing (VIGS) resulted in increased susceptibility to *P. triticina*. Further histological observation found that mycelium of *P. triticina* grew with small area hypersensitive cell death at 120 hpi in plants with silenced gene *TaSGT*1. Temporal and spatial expression profile of *TaSGT*1 was detected by real-time quantification PCR. The results showed that *TaSGT*1 transcript was up-regulated inoculated with leaf rust and *TaSGT*1 could be induced by salicylic acid (SA) but not induced by abscisic acid (ABA). All those results indicated that *TaSGT*1 is an important role involved in *Lr*19 resistance expression and SA induced resistance signal channel.

Key words: Wheat leaf rust; *TaSGT*1; Rapid amplification cDNA ends (RACE); Virus induced gene silencing (VIGS); Subcellular localization; Real-time quantitative PCR

Transcriptome Analysis of *Ganoderma pseudoferreum* of *Hevea brasiliensis* *

TU Min, CAI Hai-bin, CHENG Han, HUANG Hua-sun
(*Rubber Research Institute CATAS*, *Danzhou* 571737, *China*)

Abstract: *Ganoderma pseudoferreum* was the pathogen of *hevea* red root disease, which generally occurred in the rubber tree planting areas in China and brought huge losses to the rubber industry. In order to explore the pathogenic mechanism of *G. pseudoferreum*, cDNA library was sequenced and analyzed with the mycelium pure cultured by PDA medium and infected rubber tree root for 30d respectively, by the Illumina hiseq 2 500 high-throughput sequencing platforms. Sequencing results showed that the valid data were 46 024 635 after pretreatment. There were 6 892 differently expressed genes ($q < 0.05$ & | log2 (Fold-change) | >1), 3 030 up-regulated genes, 3 862 down-regulation of gene. To up-regulated genes, there were 32 genes which log2 (Fold-change) greater than 6 and 13 genes which greater than 7. Significantly up-regulated genes mainly included cellulose, hemicellulase, oxidase, dehydrogenase and transcription factor. The results lay a foundation for further research on the pathogenic molecular mechanism of *G. pseudoferreum*.

Key words: *Ganoderma pseudoferreum*; Transcriptome; Infection; Cellulose

* Funding: Supported by the Fundamental Research Funds for Rubber Research Insistute, CATAS (1630022015002)

辣椒棒孢叶斑病在广东首次报道*

蓝国兵[1,2]**，汤亚飞[1,2]，佘小漫[1,2]，何自福[1,2]***
（1. 广东省农业科学院植物保护研究所，广州　510640；
2. 广东省植物保护新技术重点实验室，广州　510640）

摘　要：辣椒是广东省主要冬种蔬菜作物之一，主要种植在湛江和茂名等粤西地区，年种植面达 100 万亩以上。2015 年 12 月，广东省湛江市冬种辣椒产区调查时首次发现了辣椒棒孢叶斑病。病株田间症状主要表现为叶片、茎秆和果柄处密布近圆形病斑，病斑中间白色、边缘浅褐色，叶部病斑不穿孔。田间病株率为 15% ~50%，发病严重地块病株率达 100%，并造成落叶和落果现象。通过对发病植株的组织进行病原分离和病菌致病性测定，获得了辣椒棒孢叶斑病的病原菌。在 PDA 培养基上，该病菌的菌落平展，灰色至浅墨绿色，表面菌丝毡毛状；分生孢子梗由菌丝衍化而来，无子座。分生孢子圆柱状或棍棒状，直立或稍弯，单生或链状，基部平截，顶部钝圆，浅灰色，有 0 ~8 个假隔膜。分生孢子大小差异较大。根据形态学观察和相关文献报道，将引起辣椒棒孢叶斑病的病原鉴定为多主棒孢霉（*Corynespora cassiicola*）。本文是多主棒孢霉在广东引起辣椒棒孢叶斑病的首次报道。目前，在广东已观察到多主棒孢霉可为害黄瓜、茄子、辣椒、番茄等蔬菜作物，引起病叶斑病害，并造成较严重损失。

关键词：辣椒；多主棒孢霉；棒孢叶斑病

* 基金项目：广东省科技计划项目（2014B070706017，2014B020203002）；广东省农作物病虫害绿色防控技术研究开发中心建设

** 第一作者：蓝国兵，助理研究员，硕士，植物病理学，主要从事蔬菜病害研究；E-mail：languo020@163.com
*** 通讯作者：何自福，博士，研究员；E-mail：hezf@gdppri.com

苹果树腐烂病菌 β－葡糖苷酶基因的克隆与原核表达*

李　婷**，史祥鹏，李保华，王彩霞***

（青岛农业大学农学与植物保护学院/山东省植物病虫害综合防控重点实验室，青岛　266109）

摘　要： 苹果树腐烂病是对苹果产业威胁最大的一种毁灭性病害，在我国各苹果产区发生普遍，主要为害主枝主干，严重时可造成整树死亡甚至毁园，成为制约我国苹果生产的主要限制因子。苹果树腐烂病由黑腐皮壳菌（*Valsa mali*）引起，研究表明该病原菌在对寄主植物的侵染致病过程中可产生一系列的细胞壁降解酶：木聚糖酶、纤维素酶、β－葡糖苷酶、多聚半乳糖醛酸酶、果胶甲基半乳糖醛酸酶。

为了解析苹果树腐烂病菌 β－葡萄糖苷酶的分子特征和表达，明确其在病原菌侵染致病过程中的作用，通过 RT-PCR 结合 RACE 技术从苹果树腐烂病菌中克隆了一条 β-glucosidase 基因（GenBank 登录号为 KX013493），命名为 *Glucosidase* Ⅰ，构建重组表达载体 pET32a-*glucosidase* Ⅰ，将该重组载体转入大肠杆菌 Rosetta 菌株，用 IPTG 诱导表达。结果表明：成功克隆了 *Glucosidase* Ⅰ基因，全长 1 677bp，开放阅读框长度为 1 491bp，编码 496 个氨基酸，分子量约为 54. 56kDa，5′端和 3′端非编码区长度分别为 54bp 和 132bp。融合蛋白的可溶性检测表明，低温诱导下该蛋白主要以可溶性蛋白的形式存在，Western-blot 印迹也证明大肠杆菌中表达了一个分子量约为 72. 56kDa 的蛋白质，与预测分子量大小一致。本研究成功从苹果树腐烂病菌中克隆得到一个 β-葡萄糖苷酶基因，并成功实现了其编码蛋白的原核表达，为后期探索其在苹果树腐烂病菌侵染致病中的功能提供理论基础。

关键词： 苹果腐烂病；β-glucosidase；基因克隆；原核表达

---

* 基金项目：国家自然科学基金（31272001 和 31371883）；现代农业产业技术体系建设专项资金（CARS－28）

** 第一作者：李婷，硕士研究生，研究方向果树病理学；E-mail：m18960878069@163. com

*** 通讯作者：王彩霞，教授，主要从事果树病害研究；E-mail：cxwang@qau. edu. cn

北沙参锈病病原夏孢子萌发现象观察*

高　颖**，张　林，张梦雅，闫红飞***，刘大群***
（河北农业大学植物保护学院，国家北方山区农业工程技术研究中心，
河北省农作物病虫害生物防治工程技术研究中心，保定　071000）

摘　要：北沙参锈病是一种严重为害北沙参的真菌病害。近几年在各北沙参种植区均有发生且呈逐年加重的趋势，2013 年、2014 年和 2015 年在河北省安国、承德、张家口等沙参种植区发生较为严重。因此，不断加强对北沙参病原的生物学研究可以对该病害的防控提供一定的理论指导。本试验通过人工接种北沙参锈菌，发病后收集新鲜夏孢子，对其萌发温度进行实验，并对芽管形态特征进行观察。结果显示，夏孢子在 25℃ 条件下萌发率最高，且在 0 ~ 6h 的时间段内夏孢子的萌发率会随时间的延长而不断增加，6 ~ 24h 内逐渐趋于稳定。夏孢子萌发时一般只产生 1 个芽管，但有时会出现 2 个或 3 个，且个别芽管顶端会出现分叉现象，但目前只观察到 2 个分叉，且这两个分支生长速度不同，对于产生该现象的原因及其是否都具有侵染功能目前尚不能确定，有待进一步探索。

关键词：北沙参锈病；夏孢子；萌发

* 基金项目：国家重点基础研究发展计划（2013CB127700）；河北省自然基金项目（C2015204105）
** 第一作者：高颖，硕士研究生，研究方向为分子植物病理学；E-mail：873223823@qq.com
*** 通讯作者：闫红飞，副教授，主要从事植物病害生物防治与分子植物病理学研究；E-mail：hongfeiyan2006@163.com
刘大群，教授，主要从事植物病害生物防治与分子植物病理学研究

灰葡萄孢 *Bcspp* 基因功能的初步研究*

申德民[1,2]**，谢甲涛[1]，陈　桃[1]，付艳萍[1]，姜道宏[1,2]，程家森[1]***
（1. 湖北省作物病害监测和安全控制重点实验室（华中农业大学），武汉　430070；
2. 农业微生物学国家重点实验室（华中农业大学），武汉　430070）

摘　要：灰葡萄孢（*Botrytis cinerea*）是一种死体营养型病原真菌，在世界范围内可侵染200多种植物，造成严重的经济损失。灰葡萄孢主要以菌核或菌丝作为下一个病害循环的初侵染源，菌核在灰葡萄孢的生活史中发挥重要作用。筛选灰葡萄孢菌核形成相关基因并探讨其基因功能，不仅有助于认识灰葡萄孢的菌核形成机理，也有助于提供新的防控靶标或为灰霉病的防治提供新的方法和思路。本研究首先通过灰葡萄孢在不同生长发育阶段的RNA-Seq数据筛选到了一个在菌丝生长和菌核形成时期表达量差异较大的基因*Bcspp*。该基因在菌核形成时期的表达明显下调，在培养16d后几乎检测不到该基因的表达。*Bcspp*基因的开放阅读框架由2 135bp组成，包含3个外显子和2个内含子，编码一个含有677aa的蛋白质。BLAST比对发现该蛋白质含有一个酸性鞘磷脂酶和金属磷酸酶结构域。酸性鞘磷脂酶能在酸性条件下水解鞘磷脂得到神经酰胺，而神经酰胺是一种有生物活性的次级信号分子，在鞘磷脂信号途径中发挥作用，能影响细胞的凋亡、增殖、免疫和代谢等。为了进一步研究该基因的功能，我们通过融合PCR的方法构建了基因敲除片段，并通过同源重组获得了该基因的敲除转化子。敲除*Bcspp*基因后灰葡萄孢菌丝的生长速度和尖端形态无明显变化，在菌核形成早期发育延缓，但最终的菌核形成能力并无明显差异，同时敲除转化子的产孢能力显著下降（平均每皿产孢量是出发菌株的70%左右）。研究结果初步揭示了*Bcspp*基因在灰葡萄孢的产孢和早期菌核发育上发挥了某种程度的作用，具体机制正在深入研究之中。

关键词：灰葡萄孢；菌核形成；*Bcspp*

---

* 资助基金：国家公益性行业（农业）科研专项（201303025）
** 第一作者：申德民，硕士研究生，主要从事分子植物病理学相关研究；E-mail：sdsdm01@163.com
*** 通讯作者：程家森，副教授，主要从事分子植物病理学及生物防治的相关研究；E-mail：jiasencheng@mail.hzau.edu.cn

重寄生真菌通过寄生获得真菌病毒的研究*

刘　四[1,2**]，谢甲涛[1]，程家森[1]，陈　桃[1]，付艳萍[1]，姜道宏[1,2***]
（1. 湖北省作物病害监测和安全控制重点实验室（华中农业大学），武汉　430070；
2. 农业微生物学国家重点实验室（华中农业大学），武汉　430070）

摘　要：真菌病毒可侵染真菌并在寄主细胞中复制，且广泛存在于真菌种群中。真菌病毒主要通过菌丝融合在营养亲和型菌丝体之间进行水平传播或者通过孢子进行垂直传播，目前尚未报道其他有效的传播方式。

我们从发病的油菜上分离得到灰葡萄孢菌株10HN454，该菌株感染真菌病毒BfTV/10HN454、BfHV/10HN454和BfRV/10HN454。其中BfTV/10HN454属于Totiviridae的病毒，BfHV/10HN454属于Fusariviridae的病毒。BfRV/10HN454在菌株10HN454中不能被检测到，但在脱去病毒BfHV/10HN454后，则可以检测到BfRV/10HN454的dsRNA，结果提示BfHV/10HN454和BfRV/10HN454可能存在相互抑制的现象。

以往研究表明盾壳霉是核盘菌的专性重寄生真菌，不能寄生与核盘菌亲缘关系较近的灰葡萄孢。但是将菌株10HN454与盾壳霉菌株ZS-1对峙培养时，该菌株可以被盾壳霉寄生。从对峙培养菌落分离出盾壳霉菌株，检测发现这些菌株同时获得了病毒BfTV/10HN454和BfHV/10HN454或者病毒BfTV/10HN454和BfRV/10HN454，部分菌株只获得了病毒BfTV/10HN454。继代培养5代后发现这些盾壳霉菌株中BfHV/10HN454和BfRV/10HN454无法被检测到，而病毒BfTV/10HN454依然存在，说明该病毒在盾壳霉体内可以稳定地复制遗传给后代菌株。

本研究说明重寄生真菌可以通过寄生直接获得真菌病毒，研究结果揭示了一种新的真菌病毒传播途径，对利用真菌病毒防治植物真菌病害提供了一种新的思路。

关键词：真菌病毒；重寄生真菌；病毒传播

---

* 资助基金：国家自然科学基金（31371895、31570136）和现代农业产业技术体系建设专项资金（CARS-13）

** 第一作者：刘四，博士研究生，主要从事真菌病毒学相关研究；E-mail：hzauls1987cn@126.com

*** 通讯作者：姜道宏，教授，主要从事分子植物病理学及生物防治的相关研究；E-mail：daohongjiang@mail.hzau.edu.cn

核盘菌病毒 SsPV1 对盾壳霉生物学特性影响的研究*

王　涌[1,2]**，付艳萍[1]，谢甲涛[1]，程家森[1]，姜道宏[1,2]***
（1. 湖北省作物病害监测和安全控制重点实验室（华中农业大学），武汉　430070；
2. 农业微生物学国家重点实验室（华中农业大学），武汉　430070）

摘　要：随着人们对自身健康及环境安全的日益关注，植物病害的生物防治相关研究愈发受到研究者的重视。其中在由核盘菌引起的菌核病的防治方面，真菌病毒和盾壳霉作为有效的生防资源正不断被开发利用。本实验室前期从核盘菌弱毒力菌株 WF-1 中发现了双分病毒 SsPV1，研究发现 WF-1 的弱毒现象是由 SsPV1 引起的。本研究拟向盾壳霉中转入有弱毒特性的双分病毒 SsPV1，探讨利用该病毒提高盾壳霉对菌核病防治功效的可行性。提取 SsPV1 病毒粒子后以盾壳霉 ZS-1 的原生质体为受体，通过 PEG 介导转染法，最终得到了 8 株候选转染子。提取这 8 株转染子的总 RNA，用 SsPV1 的 *CP* 基因和 *RdRp* 基因上的特异性检测引物进行 RT-PCR，结果发现只有 4 号转染子有预期大小的扩增片段，对扩增产物进行测序分析，发现分别为 SsPV1 的 *CP* 基因和 *RdRp* 基因的信息；同时对 8 株转染子进行了 dsRNA 提取，并用 S1 核酸酶处理，发现大约 2300bp 处有明亮的核酸条带。结果证明通过原生质体转染的方法可以获得携带 SsPV1 的盾壳霉菌株，将该 4 号转染子命名为 ZS-1V。菌株 ZS-1V 的菌落形态及其产孢能力与出发菌株 ZS-1 无显著差异，但 ZS-1V 的孢子萌发率与 ZS-1 存在显著差异。后期我们将进一步研究 ZS-1V 分生孢子携带病毒的效率以及 ZS-1V 向核盘菌传毒的效率，同时将探讨感染病毒后盾壳霉 ZS-1V 的核盘菌寄生能力等生物学特性。

关键词：盾壳霉；核盘菌；弱毒病毒；SsPV1

---

* 基金项目：国家自然科学基金（31371982）和产业技术体系建设专项资金（CARS-13）
** 第一作者：王涌，湖北咸宁人，博士研究生，主要从事分子植物病理学研究；E-mail：13429883936@163. com
*** 通讯作者：姜道宏，教授，主要从事分子植物病理学及生物防治的相关研究；E-mail：daohongjiang@mail. hzau. edu. cn

核盘菌 SsNEP2 在盾壳霉重寄生中功能的初步研究*

赵会长[1,2]**，胥月丽[1]，程家森[1]，谢甲涛[1]，陈　桃[1]，姜道宏[1,2]，付艳萍[1]***
（1. 湖北省作物病害监测和安全控制重点实验室（华中农业大学），武汉　430070；
2. 农业微生物学国家重点实验室（华中农业大学），武汉　430070）

摘　要： 盾壳霉是植物病原真菌核盘菌的重寄生真菌，其重寄生作用是利用盾壳霉防治核盘菌的主要机制之一，然而其机制尚不清晰。本研究从核盘菌与盾壳霉互作不同时期的 RNA_ Seq 数据库中筛选到一个表达显著上升的核盘菌基因，编码一个坏死和乙烯诱导蛋白 SsNEP2（*Sclerotinia sclerotiorum* necrosis and ethylene-inducing protein 2），并初步研究了该基因在核盘菌应答盾壳霉寄生过程中的作用。

核盘菌基因 *SsNEP*2 全长 857bp，含有两个内含子，编码一种含有 244 个氨基酸的坏死与诱导蛋白，预测含有 21 个氨基酸的一段信号肽，不含细胞膜脂锚定结合位点和跨膜结构域，推测为分泌蛋白。在盾壳霉寄生核盘菌时，*SsNEP*2 基因的表达量随寄生时间的延长而逐渐上升。核盘菌基因 *SsNEP*2 沉默转化子与出发菌株相比，它们的菌丝生长速度、菌落形态、菌核形状和致病力无显著差异。选取其中不同沉默效率的 6 个沉默转化子分别与盾壳霉 ZS-1 对峙培养，盾壳霉寄生 *SsNEP*2 沉默转化子的菌丝能力受到显著的抑制，并且分生孢子的产量显著降低，而且这种影响与 *SsNEP*2 基因沉默的效率显著正相关，初步推测 *SsNEP*2 基因沉默可以一定程度上抑制盾壳霉寄生。进一步研究发现盾壳霉在添加了 *SsNEP*2 基因沉默转化子发酵液的 PDA 培养基中的产孢受到显著抑制，但生长速度没有明显区别；而基本培养基中添加沉默转化子发酵液总蛋白对盾壳霉生长和产孢影响不显著。由此推定，盾壳霉可能通过诱导表达寄主核盘菌的 *SsNEP*2 基因而创造适合其寄生核盘菌和生长的环境，详细机理研究正在进一步研究。

关键词： 核盘菌；盾壳霉；坏死和乙烯诱导蛋白；重寄生

---

* 基金资助：国家自然科学基金（31572048）

** 第一作者：赵会长，博士研究生，主要从事盾壳霉基因组及其与核盘菌互作研究；E-mail: huizhanghzau@ webmail. hzau. edu. cn

*** 通讯作者：付艳萍，教授，主要从事植物病理学相关研究；E-mail：yanpingfu@ mail. hzau. edu. cn

稻瘟菌双分病毒 MoPV1 的研究*

陈伟博[1,2**]，姜道宏[1,2]，程家森[1]，谢甲涛[1]，陈　桃[1]，付艳萍[1***]
（1. 湖北省作物病害监测和安全控制重点实验室（华中农业大学），武汉　430070；
2. 农业微生物学国家重点实验室（华中农业大学），武汉　430070）

摘　要：稻瘟病是水稻上最重要的病害之一，每年都会造成水稻大量的损失，目前已经发现真菌病毒可以作为植物病害生物防治的重要资源之一。本实验从发病的水稻叶片上采用单胞分离的方法获得纯化培养的稻瘟菌株。通过对所得菌株进行 dsRNA 的检测，选取 YC-13 作为研究对象。通过随机克隆获和末端克隆获得了病毒序列的全长；在 NCBI 上用 blastx 进行比对发现，病毒与 *Penicillium stoloniferum* virus F（PsV-F）具有一定的相似性，命名为 *Magnaporthe oryzae* partitivirus virus 1（MoPV1）。MoPV1 含有两条 dsRNA：dsRNA 1 编码依赖于 RNA 的 RNA 聚合酶（RNA-dependent RNA polymerase，RdRp），dsRNA 2 编码外壳蛋白（Coat Protein，CP）。根据 MoPV1 的 CP 及 RdRp 序列进行的系统进化分析，该病毒与主要侵染丝状真菌的 *Gammapartitivirus* 属的病毒亲缘关系更近，因此，MoPV1 归属于 *Partitiviridae* 科 *Gammapartitivirus* 属，是一种新的真菌病毒。为了研究 MoPV1 对稻瘟菌的影响，通过病毒粒子转染试验，将 MoPV1 转到稻瘟菌株 RB11 中；试验证明，MoPV1 对 RB11 的菌落形态、菌丝尖端形态、生长速度、分生孢子形态和致病力均无明显的影响，但可以使 RB11 的分生孢子产生量明显降低。本研究发现了新的真菌病毒，进一步丰富病毒学的内容，而且可以为稻瘟病的防治提供新的思路。

关键词：稻瘟病菌；真菌病毒；dsRNA；双分病毒

---

* 基金资助：农业公益性行业科研专项（201203014）

** 第一作者：陈伟博，硕士研究生，主要进行稻瘟菌中真菌病毒的相关研究；E-mail：weibochen@ mail. hzau. edu. cn

*** 通讯作者：付艳萍，教授；E-mail：yanpingfu@ mail. hzau. edu. cn

水稻根际真菌多样性及其生防潜能的初步研究*

李松鹏[1,2**]，程家森[1]，付艳萍[1]，姜道宏[1,2]，谢甲涛[1***]
（1. 湖北省作物病害监测和安全控制重点实验室（华中农业大学），武汉 430070；
2. 农业微生物学国家重点实验室（华中农业大学），武汉 430070）

摘　要：纹枯病是水稻种植耕作过程中普遍发生的一类重要真菌病害，给农业生产造成不可挽回的经济损失。近年来内生真菌群体多样性研究热度日益提高，对其生防潜能评估同样具有理论意义和应用价值。实验室前期通过对水稻根际土和根系组织采样分离，共获得 64 株真菌菌株，经过 ITS 序列分析初步鉴定归类于 14 个属，其中，以曲霉属、木霉属和青霉属真菌居多，共占总数的 50% 以上。通过平板对峙培养的方法筛选对纹枯病菌有颉颃效果的菌株，有 5 株菌株对纹枯病菌抑菌率在 60% 以上，其中，3S1 - 13 和 4S2 - 46 两株真菌菌株对纹枯菌抑菌率在 75% 以上，后经鉴定均为哈茨木霉。将 5 株有明显颉颃效果的真菌菌株进行液体摇培（PDB，20℃，3d），将发酵液离心取上清并用细菌过滤器处理后，以 100μL（发酵液）/10mL（PDA）的体积比混合均匀，接纹枯菌 WH-1 菌株进行发酵液平板抑菌评估。发现 5 株真菌菌株发酵液均具有良好的抗生作用，抑菌率可达 60%，其中，菌株 3S1 - 13 和 4S2 - 46 最为明显（抑菌率在 85% 以上）。水稻离体叶片生防评估，初步表明筛选获得的 5 株真菌发酵液均可抑制病斑扩展，抑制率超过 30%。筛选到具有生防潜能的生防机制及其对水稻等植物的影响正在作进一步的研究。

关键词：水稻纹枯病；水稻根际真菌；生物防治

* 资助基金：长江学者创新团队基金（IRT1247）

** 第一作者：李松鹏，硕士研究生，主要从事水稻纹枯病生物防治的相关研究；E-mail：905157826@ qq. com
*** 通讯作者：谢甲涛，副教授，主要从事分子植物病理学及生物防治的相关研究；E-mail：jiataoxie@ mail. hzau. edu. cn

水稻纹枯菌上发现的全新 Ophiovirus 病毒的初步研究*

李阳艺[1,2**]，谢甲涛[1]，程家森[1]，陈　桃[1]，付艳萍[1]，姜道宏[1,2***]
（1. 湖北省作物病害监测和安全控制重点实验室（华中农业大学），武汉　430070；
2. 农业微生物学国家重点实验室（华中农业大学），武汉　430070）

摘　要：水稻纹枯菌（Rice Sheath blight）作为世界性水稻三大病害之一，在 20 世纪 90 年代中期成为我国南方部分稻产区的第一大病害，具有广泛的地理分布。水稻纹枯菌中的病毒种类繁多，能引起水稻纹枯病菌弱毒特性的真菌病毒在生物防治上有具有重要的实用价值。本研究经过对实验室前期从湖北省各地区采集分离的 500 个水稻纹枯菌样本的深度测序，得到了样本中大量的病毒信息，选取未被在真菌中报道过的 Ophiovirus 病毒的核酸部分序列，设计特异性引物定位得到水稻纹枯菌菌株 JZ56，将该病毒命名为 RsOPV。

这个带病毒菌株在生长速度、菌丝尖端，菌落形态均与野生不含病毒菌株 190 有明显差异，特别在致病力方面能引起水稻叶片过敏性坏死。现已在植物上发现并报道的 Ophiovirus 病毒基因组总长 11 ~ 12kb，分为 7.5 ~ 9.0kb、1.6 ~ 1.8kb 和 1.5kb 三个 RNA 片段，病毒粒子呈现假线性。现阶段通过末端克隆获得 RsOPV 病毒 RNA1 的完整序列，分析发现 RNA1 包含两个完整的 ORF，分别编码一个 RdRp 和一个未知功能的蛋白。进化树分析，RsOPV 病毒与以往报道在植物上发现的 Ophiovirus 病毒属于不同分支，怀疑是一个新的病毒种类。通过原生质体脱毒的方法，获得了脱去 RsOPV 病毒且与原始菌株 JZ56 具有相同遗传背景的菌株 JZ56-VF3，该菌株与原始菌株 JZ56 在生长速度、菌丝尖端，菌落形态上有明显差异，而与野生不含病毒菌株 190 在生长速度、菌丝尖端、菌落形态等方面差异不显著，并且在生长后期产生的菌核在形态上也恢复为更加接近球形的菌核。关于病毒基因组其他节段的克隆以及引起纹枯菌致病性改变的相关基因的研究仍在继续。

* 资助基金：公益性行业（农业）科研专项（201103016）、教育部科技研究重大项目（313024）
** 第一作者：李阳艺，硕士研究生，主要从事分子植物病理学相关研究；E-mail：13667229734@163.com
*** 通讯作者：姜道宏，教授，主要从事分子植物病理学及生物防治的相关研究；E-mail：daohongjiang@mail.hzau.edu.cn

灰葡萄孢 *BcMito*1 基因影响盾壳霉对其寄生作用的探究*

梁克力[1,2**]，刘　四[1,2]，程家森[1]，谢甲涛[1]，陈　桃[1]，姜道宏[1]，付艳萍[1***]
（1. 湖北省作物病害检测和安全控制重点实验室（华中农业大学），武汉　430070；
2. 农业微生物学国家重点实验室（华中农业大学），武汉　430070）

摘　要：灰葡萄孢（*Botrytis cinerea*）是引起作物灰霉病的病原菌，与油菜上重要病原菌核盘菌（*Sclerotinia sclerotiorum*）同属于核盘菌科。两者亲缘关系虽然较近，但生防真菌盾壳霉（*Coniothyrium minitans*）能够寄生核盘菌，却不能寄生灰葡萄孢，表明灰葡萄孢与盾壳霉之间存在着非寄主识别的相互作用。

实验室前期从油菜罹病植株的菌核中分离获得一株葡萄孢新种-拟蚕豆葡萄孢菌株（10HN-454），发现该菌株能被盾壳霉高效寄生。本研究以此出发，以盾壳霉孢子分别侵染核盘菌菌丝、灰葡萄孢菌丝、拟蚕豆葡萄孢菌丝 0h、12h 为 RNA-seq 测序材料，筛选盾壳霉与寄主和非寄主互作过程中表达差异的基因。发现灰葡萄孢基因 *BcMito*1 在盾壳霉与灰葡萄孢的互作中表达明显上调，而该基因的同源基因在盾壳霉与核盘菌及拟蚕豆葡萄孢互作中的表达无显著差异，提示 *BcMito*1 基因在灰葡萄孢与盾壳霉的非寄主识别过程中可能发挥重要作用。

*BcMito*1 基因全长 897bp，含 1 个外显子，不含内含子，推定编码含有 298 个氨基酸的蛋白质，含有的保守结构域 Mito-carr 属于 Mito-carr 超级家族，蛋白为线粒体转运蛋白。利用分割标记法对 *BcMito*1 基因进行了敲除，敲除转化子 ΔBcMito1 在 PDA 培养基上菌丝茂盛，生长速度无明显变化，但产孢量显著减少，不形成菌核且对番茄叶片的致病力显著降低。将转化子 ΔBcMito1 与盾壳霉菌株 ZS-1（hyg 标记）对峙培养，结果显示 ΔBcMito1 可以明显被盾壳霉寄生，进一步证明该基因在灰葡萄孢抵抗盾壳霉寄生过程中发挥重要作用，*BcMito*1 基因参与灰葡萄孢与盾壳霉互作的分子机理正在进一步深入研究。

关键词：灰葡萄孢；盾壳霉；*BcMito*1；重寄生

---

* 资助基金：国家自然科学基金（31371895）
** 第一作者：梁克力，硕士研究生，主要从事分子植物病理学相关研究；E-mail：keliliang@webmail.hzau.edu.cn
*** 通讯作者：付艳萍，教授，主要从事分子植物病理学及生物防治的相关研究；E-mail：yanpingfu@mail.hzau.edu.cn

盾壳霉产孢缺陷突变体的分析*

罗晨薇[1,2]**，程家森[1]，谢甲涛[1]，陈　桃[1]，姜道宏[1,2]，付艳萍[1]***
（1. 湖北省作物病害监测和安全控制重点实验室（华中农业大学），武汉　430070；
2. 农业微生物学国家重点实验室（华中农业大学），武汉　430070）

摘　要：盾壳霉（*Coniorhyrium minitans*）是核盘菌（*Sclerorinia sclerotsorum*）的重寄生菌，其产孢和寄生能力对于菌核病的防治至关重要。为了从分子水平上研究盾壳霉产孢和寄生菌核的机制，实验室在前期构建了盾壳霉 ZS-1 菌株的 T -DNA 标记插入突变体库。我们对其中三个产孢异常的突变体 ZS-1TN15322、ZS-1TN16231 和 ZS-1TN24363 进行了深入研究。

通过 Hi-tail PCR 扩增出了 ZS-1TN15322 的 T-DNA 插入侧翼 DNA 片段，根据该序列在盾壳霉基因组中进行比对，发现该突变体中 T-DNA 标记插入一个基因的启动子区域，该基因 *CMZS*1_ 06889 全长为 896bp，无内含子，推定编码一个 297aa 的蛋白。对推定的氨基酸序列进行功能预测，发现该序列含有保守区域 GAT1_ cyanophycinase。该氨基酸序列与其他真菌的藻青素酶具有 66% 的同源性。通过 Hi-tail PCR 扩增出了 ZS-1TN16231 的 T-DNA 插入破坏的编码 F-Box 蛋白基因，该基因 *CMZS*1_ 04523 与其他真菌中的 F-Box 蛋白基因只有 31% 的同源性，基因全长 2 435bp，包含 3 个内含子和 4 个外显子，推定编码一个 504 aa 的蛋白，T-DNA 标记插入该基因的启动子区域，距离起始密码子 353bp 处。通过 Hi-tail PCR 确定了 ZS-1TN24363 突变体的 T-DNA 插入破坏了一个编码转录因子 CMR1 的基因，该基因 *CMZS*1_ 076383 与其他真菌的转录因子编码基因 *CMR*1 具有 72% 的同源性，基因全长 3 167bp，包含两个内含子和 3 个外显子，推定编码 1 011个氨基酸。同时我们利用 Split-Marker 重组技术构建了这些基因的重组片段，并通过 PEG-原生质体方法对盾壳霉 ZS-1 菌株进行转化，获得了相关基因敲除的转化子，并利用 PCR 对转化子进行了初步鉴定，转化子的生物学特性正在研究中。

关键词：盾壳霉；产孢；T-DNA 插入突变体

---

* 资助基金：国家自然科学基金（31572048）
** 第一作者：罗晨薇，硕士研究生，主要从事分子植物病理学相关研究；E-mail：410406816@ qq. com
*** 通讯作者：付艳萍，教授，主要从事分子植物病理学及生物防治的相关研究；E-mail：yanpingfu@ mail. hzau. edu. cn

核盘菌弱毒菌株 DT-8 菌剂研发及其田间应用研究*

曲　正[1,2**]，谢甲涛[1]，付艳萍[1]，陈　桃[1]，程家森[1]，姜道宏[1,2***]
（1. 湖北省作物病害监测和安全控制重点实验室，华中农业大学植物科学技术学院，武汉　430070；2. 华中农业大学农业微生物学国家重点实验室，武汉　430070）

菌核病是由核盘菌引起的重要病害，严重影响了油菜、大豆和向日葵等作物及蔬菜的生产安全。本课题组前期从感染油菜的菌株 DT-8 中分离和鉴定了一种核盘菌低毒衰退相关 DNA 病毒 1（SsHADV-1），该 DNA 病毒不仅可导致核盘菌毒力衰退，而且可以在核盘菌不同营养亲和型菌株间进行有效传播，具有很大的生防价值，有潜力发展成一种新型生物“杀菌剂”。

本研究在前期研究的基础上，通过优化发酵条件，发现在 100r/min，不调节 pH 值，初始接种浓度为 9% 的条件下，获得的发酵产物干重最高，可达（5.329 ±0.750）g/L。产物经研磨，添加防腐剂 0.05% 硫酸链霉素、防霉剂液体石蜡油、保护剂 4% 海藻糖、防冻剂甘油、紫外保护剂核黄素后，得到了活菌数达 10^5CFU/mL 的水悬浮剂。制剂的悬浮率达 91.83%，消泡性优良，在 4℃下贮藏 28d 后带毒率达 65%，且贮藏期间没有腐败发霉现象。这些发酵条件、助剂和剂型的优化，为 DT-8 菌株的规模化生产和商品制剂开发奠定了基础。

田间防效实验在湖北省鄂州市、江陵县试验田进行。设置蕾薹期喷施、始花期喷施、盛花期喷施 3 个单独生长期喷施处理，蕾薹期、始花期连续喷施和蕾薹期、始花期、盛花期连续喷施 2 个不同生长期连续喷施处理。每次以每亩施用 10% 的发酵液 15L 为标准进行喷施。在收获期调查各处理发病情况，证明我们研制的 DT-8 制剂在油菜蕾薹期、始花期、盛花期连续喷施后，油菜菌核病的发病程度均显著降低，可以起到明显的防病效果，具有一定的应用前景。

关键词：弱毒菌株；DT-8；剂型

* 资助基金：现代农业产业技术体系建设专项资金（CARS-13）

** 第一作者：曲正，山东潍坊人，在读硕士研究生，主要从事植物病害生物防治
*** 通讯作者：姜道宏，教授；E-mail：daohongjiang@ mail. hazu. edu. cn

核盘菌4种分子寄生物传播特性的初步研究*

吴松松[1,2]**，付艳萍[1]，谢甲涛[1]，程家森[1]，姜道宏[1,2]***
（1. 湖北省作物病害监测和安全控制重点实验室（华中农业大学），武汉 430070；
农业微生物学国家重点实验室（华中农业大学），武汉 430070）

病毒作为一种严格的细胞内寄生分子，其传播方式有两种：水平传播和垂直传播。垂直传播是以有性孢子为介质直接由亲代传播给子代的传播方式，由于复杂的生殖屏障机制隔离了大部分病毒的垂直传播。对于占有绝大部分植物病原真菌种类的子囊菌，在其携带的真菌病毒中，目前报道仅有部分线粒体病毒能以高频率进入有性孢子子囊孢子，而应用真菌病毒对植物病原真菌的生物防治最主要的局限是其传播能力。因此，寻找在子囊菌中能通过有性传播的真菌病毒将为扩展弱毒相关真菌病毒的传播能力及研究真菌有性生殖隔离机制提供可能。本研究于2009年在分离自青海省的核盘菌菌株DT47中发现，其携带有4种类型的核酸分子寄生物，分别命名为SsNSRV2-QH47，SsRV-QH47，SsMV6-QH47，Sat-RNA。SsNSRV2-QH47为-ssRNA病毒，基因组全长约10 000nt；SsRV-QH47为+ssRNA病毒，全长4 474nt；SsMV6-QH47为线粒体病毒；Sat-RNA为一类未知的核酸分子，大小约750bp。通过DT47菌株子囊孢子单孢分离获得的有性后代中，100%携带SsMV6-QH47及Sat-RNA，62.5%携带SsRV-QH47，SsNSRV2-QH47仅有部分菌株携带。因此，这4种核酸分子寄生物均能通过子囊孢子进行传播。通过PEG介导及菌丝对峙培养发现SsRV-QH47，SsMV6-QH47，Sat-RNA单独对核盘菌的生物学特征无明显影响，而SsNSRV2-QH47，SsRV-QH47混合侵染的菌株会促使核盘菌菌核形成时间提前1～2d。4种分子寄生物分子特性及其传播机制有待进一步研究。

关键词：核盘菌；真菌病毒；传播

* 基金项目：教育部科技重大项目（313024）和现代农业产业技术体系建设专项资金（CARS-13）

** 第一作者：吴松松，湖北武汉人，博士研究生，主要从事分子植物病理学（真菌病毒学）研究；E-mail：wusong425@gmail.com

*** 通讯作者：姜道宏，教授，主要从事分子植物病理学及生物防治的相关研究；E-mail：daohongjiang@mail.hzau.edu.cn

芝麻茎点枯病菌（*Macrophomina phaseolina*）致病相关基因分析*

赵　辉**，刘红彦***，刘新涛，倪云霞，文　艺，
刘玉霞，杨修身，王　飞，高素霞
（河南省农业科学院植物保护研究所，农业部华北南部作物有害生物综合治理重点实验室，
河南省农作物病虫害防治重点实验室，郑州　450002）

摘　要：芝麻茎点枯病是芝麻生产中最具威胁性的病害之一，其病原菌菜豆壳球孢（*Macrophmina phaseolina*，简称 Mp）是一种非专性寄生的土传病原真菌，寄主范围很广，可侵染除芝麻以外的 500 多种植物。为明确 Mp 与芝麻互作过程中参与致病的基因种类，系统研究 Mp 与芝麻互作中致病基因的总体表达模式，初步推测该病原菌致病的分子机理，本研究在 Mp 全基因组测序的基础上，以芝麻茎点枯病菌——芝麻互作 0h、8h、16h、24h、32h 材料为研究对象，采用高通量测序技术进行转录组测序，通过与 Mp 全基因组测序结果和数据库比对，共获得表达基因 15 233个。与 Mp 全基因组数据比较，发现了新转录本结构域 3 608个，注释新基因 340 个，并对原 Mp 全基因组预测的 7 710个基因的起始和终止位置进行校正，补充校正了 Mp 全基因组的测序结果。与病原——寄主互作数据库（PHI-base）比对结果表明，有 844 种 1 531个致病相关基因参与整个致病过程。经 RT-PCR 验证，在互作 8 ~ 32h 差异表达的致病相关基因 611 个，其中，8h 差异表达的致病相关基因 12 个，16h 差异表达的致病相关基因 67 个，24h 差异表达的致病相关基因 52 个，32h 差异表达的致病相关基因 60 个，4 个时间点共同表达的致病相关基因 73 个。通过数据分析和 RT-PCR 试验验证，在芝麻茎点枯病菌——芝麻互作的不同阶段，参与致病过程的致病相关基因种类、数量和表达量是不同的，随着互作时间的增加参与互作的致病相关基因数量和代谢途径也增多，在互作 16h ~ 24h 时参与致病过程的致病相关基因最多，此时参与的代谢途径也最多。

关键词：芝麻茎点枯病；*Macrophomina phaseolina*；致病基因；转录组；互作

---

* 基金项目：国家自然科学基金资助项目（31301631）；农业部现代农业产业技术体系专项（CARS-15-03B）

** 第一作者：赵辉，辽宁省昌图县人，副研究员，博士，主要从事植物病理学和植物病害生物防治研究；E-mail：zhaohui_ 0078@126. com

*** 通讯作者：刘红彦，研究员，主要从事植物病理学和植物病害生物防治研究；E-mail：liuhy1219@163. com

稻曲病菌侵染机制研究*

宋杰辉[1]，罗朝喜[1,2**]
（1. 华中农业大学植物科学技术学院，武汉　430070；
2. 湖北省作物病害监测和安全控制重点实验室，武汉　430070）

摘　要：稻曲病已发展成为我国水稻的三大病害之一，其不仅引起水稻产量损失，稻曲病菌还产生毒素污染谷物，对我国粮食安全及食品安全造成严重威胁。随着稻曲病菌人工接种技术的改善，世界范围内对稻曲病菌的侵染机制进行了大量研究，极大地提高了对稻曲病菌侵染的认识。然而，由于稻曲病菌人工接种效率受限，研究者不易对相同批次的侵染谷粒在不同侵染阶段进行系统观察，导致对病菌侵染位点及侵入后的发育过程存在较大争议。本研究利用高效的稻曲病菌人工接种技术，使用绿色荧光标记的菌株接种敏感水稻品种晚籼98，在接种后逐日观察病菌的侵染进程及水稻各花器的发育情况，初步揭示了稻曲病菌对水稻的侵染是一个逐步侵染的模式。稻曲病菌在水稻花器发育成熟前通过颖片表面缝隙进入谷粒后，首先侵染水稻花丝，阻止花粉成熟并阻断子房受精。随后病菌侵入柱头和花柱，偶尔能通过花柱到达子房。此时病菌通过启动水稻籽粒灌浆相关基因模拟水稻受精，欺骗水稻持续地向子房输送营养，病菌通过花柱与子房的连接劫持大量的营养物质供应病菌形成稻曲球。本研究揭示了稻曲病菌对水稻花器的逐步侵染模式以及阻止子房受精后又模拟子房的受精来获取持续的营养供应从而成功地形成稻曲球，代表了一种较新颖的寄主—病原菌互作生物学过程。

关键词：稻曲病菌；水稻；侵染机制

---

* 第一作者：宋杰辉，博士研究生；E-mail：370178295@ qq. com

** 通讯作者：罗朝喜，教授，主要从事稻曲病与水稻互作及病原真菌抗药性分子机理研究；E-mail: cxluo@ mail. hazu. edu. cn

Studies on the Function of SsCut from *Sclerotinia sclerotiorum* in Elicitor Activity and Signaling Transduction of Plant Resistance

ZHANG Hua-jian*

(Anhui Agricultural University, Hefei 230036)

In this study, we report the cloning of the *SsCut* gene (GenBank Accession No. XM_001590986.1) encoding cutinase from *Sclerotinia sclerotiorum*. We isolated a 609-bp cDNA encoding a polypeptide of 202 amino acids with a molecular weight of 20.4 kDa. Heterologous expression of SsCut in *Escherichia coli* (His-SsCut) caused the formation of lesions in tobacco that closely resembled hypersensitive response lesions. His-SsCut was caused cell death in Arabidopsis, soybean (*Glycine max*), oilseed rape (*Brassica napus*), rice (*Oryza sativa*), maize (*Zea mays*), and wheat (*Triticum aestivum*), indicating that both dicot and monocot species are responsive to the elicitor. Furthermore, compounds related to disease resistance including hydrogen peroxide, phenylalanine ammonia-lyase, peroxides, and polyphenol oxidase increased markedly after His-SsCut treatment. His-SsCut-treated plants exhibited enhanced resistance as indicated by a significant reduction in the number and size of *S. sclerotiorum*, *Phytophthora sojae*, and *P. nicotianae* lesions on leaves relative to controls. Real-time PCR results indicated that the expression of defense-related genes and genes involved in signal transduction were induced by His-SsCut. Our results demonstrate that SsCut is an elicitor that triggers defense responses in plants and will help to clarify its relationship to downstream signaling pathways that induce defense responses.

Additionally, we utilized *Nicotiana benthamiana* and virus-induced gene silencing to individually decrease the expression of over 2 500 genes. Using this forward genetics approach, several genes were identified that, when silenced, compromised SsCut-triggered cell death based on a cell death assay. A C_2H_2-type zinc finger gene was isolated from *N. benthamiana*. Sequence analysis indicated that the gene encodes a 27-kDa protein with 253 amino acids containing two typical C_2H_2-type zinc finger domains; this gene was named *NbPIF*1. We found that SsCut-induced cell death could be inhibited by virus-induced gene silencing of *NbPIF*1 in *N. benthamiana*. In addition, SsCut induces stomatal closure, accompanied by reactive oxygen species (ROS) production by NADPH oxidases and nitric oxide (NO) production. *NbPIF*1-silenced plants showed impaired SsCut-induced stomatal closure, decreased SsCut-induced production of ROS and NO in guard cells, and reduced SsCut-induced resistance against *Phytophthora nicotianae*. Taken together, these results demonstrate that the NbPIF1-ROS-NO pathway mediates multiple SsCut-triggered responses, including stomatal closure, hypersensitive responses, and defense-related gene expression. This is the first report describing the function of a C_2H_2-type zinc finger protein in tobacco.

* Corresponding author: ZHANG Hua-jian; E-mail: hjzhang@ahau.edu.cn

cAMP 信号在甘蔗鞭黑粉菌细胞形态转变中的作用研究*

李玲玉**，蔡恩平，梅　丹，颜梅新，常长青***，姜子德***
（华南农业大学群体微生物创新团队/微生物信号与病害防控重点实验室，广州　501642）

摘　要：甘蔗鞭黑粉菌（*Sporisorium scitamineum*）引致的黑穗病是甘蔗生产上的重要病害。该菌为二态型真菌，其单倍体酵母状菌体以腐生方式生活，只有具亲和性的单倍体经配合形成双核菌丝体才能侵染甘蔗，因此，菌体配合及细胞形态转换是侵染致病和完成生活史过程的关键步骤。本研究从以甘蔗鞭黑粉菌单倍体菌株 WT18 为材料构建的 T-DNA 插入突变体库中筛选到一个在固体培养基上持续生长白色菌丝的突变体 G17。实验发现外源加入 cAMP 可抑制该突变体在固体培养基上形成白色菌丝，使其恢复到酵母状细胞菌落性状，而 5d 以上的培养时间则该酵母状菌落又逐渐长出白色菌丝；液体培养条件下突变体形态为酵母状细胞，与野生型并无差异；以上研究结果表明 cAMP 信号在甘蔗鞭黑粉菌不同细胞形态中维持一个动态平衡，其浓度对甘蔗鞭黑粉菌细胞形态转变起着调控作用。通过 Southern blot 分析已确定 G17 是 T-DNA 单拷贝插入，经 HITAIL-PCR 鉴定所突变的基因是一个功能未知的锌指结构基因。为明确该基因在 cAMP 信号合成和传导中的作用机制，今后拟对 G17 突变体进行细胞内源性 cAMP 信号检测、致病性测定及转录组分析等研究。

关键词：甘蔗鞭黑粉菌；二态型；cAMP；锌指结构基因

---

* 基金项目：国家 973 计划项目（2015CB150600）和广东省群体微生物基础理论与前沿技术创新团队项目（2013S034）

** 第一作者：李玲玉，河南人，博士研究生
*** 通讯作者：常长青，副教授；E-mail：changcq@ scau. edu. cn
姜子德，教授；E-mail：zdjiang@ scau. edu. cn

The Complete Genome Sequence of a Novel *Fusarium graminearum* RNA Virus in a New Proposed Family With the Order Tymovirales*

CHEN Xiao-guang**, HE Hao***, YANG Xiu-fen,
ZENG Hong-mei, QIU De-wen, GUO Li-hua***
(*State Key Laboratory for Biology of Plant Diseases and Insect Pests, Institute of Plant Protection, Chinese Academy of Agricultural Sciences, Beijing* 100193, *China*)

Abstract: The complete nucleotide sequence of *Fusarium graminearum* deltaflexivirus 1 (FgDFV1), a novel positive single-stranded (+ss) RNA mycovirus, was sequenced and analyzed. The complete genome of FgDFV1/BJ59 was shown to be 8 246 nucleotides (nts) long excluding the poly (A) tail. FgDFV1/BJ59 was predicted to contain a large open reading frame (ORF 1) and four smaller ORFs (2 ~ 5). ORF1 encodes a putative replication-associated polyprotein (RP) of 2 042 amino acides (aa) and contains three conserved domains, viral RNA meylthtransferase (Mtr), viral RNA helicase (Hel) and RNA-dependent RNA polymerase (RdRp). ORFs (2 ~ 5) encode four putative small hypothetical protein (12 ~ 18kDa) with unknown biological functions. Phylogenetic analysis based on RP sequences indicated that FgDFV1 is phylogenetically related to soybean leaf-associated mycoflexivirus 1 (SlaMyfV1) and *Sclerotinia sclerotiorum* deltaflexivirus 1 (SsDFV1), a cluster of an independent group belonging to a newly proposed family *Deltaflexiviridae* in the order *Tymovirales*. However, FgDFV1 is markedly different from SsDFV1 and SlaMyfV1 in genome organization and nucleotide sequence. FgDFV1 may represent additional species in the new genus *Deltaflexivirus*or possibly new different genera in the family *Deltaflexiviridae*.

Key words: *Fusarium graminearum*; FgDFV1; Mycovirus

---

* Acknowledgments: This work was supported by the National Natural Science Foundation of China (31171818) and Science and Technology Plan Project of Beijing (No. D151100003915003).

** First author: CHEN Xiao-guang; E-mail: chenxg0306@126.com

*** Corresponding author: GUO Li-hua; E-mail: guolihua72@yahoo.com

侵染禾谷镰刀菌的芜菁黄花叶病毒科新病毒的鉴定*

李鹏飞**，林艳红，章海龙，王双超，邱德文，郭立华***
（中国农业科学院植物保护研究所，植物病虫害生物学国家重点实验室，北京　100193）

小麦赤霉病是小麦穗期“三病三虫”中较为严重的病害之一，其优势病菌是禾谷镰刀菌（*Fusarium graminearum*）。该病在全球范围内普遍发生，也一直是我国淮河以南及长江中下游麦区发生最为严重的病害之一，它能够造成小麦严重减产，甚至绝收，更重要的是它危害小麦后，可产生多种真菌毒素，尤其是脱氧雪腐镰刀菌烯醇（deoxynivalenol，DON），影响人畜健康，恶化籽粒品质，降低种用价值。

目前，在禾谷镰刀菌中分离出来7个真菌病毒，分别是中国报道的FgHV1和FgHV2，韩国报道的FgV1、FgV2、FgV3和FgV4，和德国报道的FgV-ch9。在这些病毒中，只有FgHV2和FgV1与禾谷镰刀菌的低毒力有关。因此，我们有必要寻找新的能降低禾谷镰刀菌致病力的真菌病毒，作为生防资源以防治小麦赤霉病。

芜菁黄花叶病毒科主要是一个侵染植物的病毒科，有3个病毒属组成，分别是芜菁黄花叶病毒属（*Tymovirus*）、玉米雷亚朵非纳病毒属（*Marafivirus*）及葡萄斑点病毒属（*Maculavirus*）。目前，在芜菁黄花叶病毒属、玉米雷亚朵非纳病毒属和葡萄斑点病毒属分别鉴定26，4和1个病毒。近来，许多未归类的病毒种也陆续报道出来，但至今为止还没有从植物病原真菌分离出与芜菁黄花叶病毒科相关的真菌病毒。

我们从陕西省禾谷镰刀菌菌株SX64中分离到一个全长7863nt的新病毒，通过对该病毒的基因组结构比较和序列及系统发育树全面分析，我们确定该病毒是芜菁黄花叶病毒科（*Tymoviridae*）的一个新种，但是，与该病毒科已确定的3个病毒属（*Tymovirus*，*Marafivirus*，*Maculavirus*）有明显的不同，因此，我们建议在*Tymoviridae*病毒科中成立一个新属*Mycotymovirus*，该病毒命名为*Fusarium graminearum* mycotymovirus 1（FgMTV1/SX64），FgMTV1/SX64是第一个从植物病原真菌中鉴定到的与芜菁黄花叶病毒科病毒相关的真菌病毒。对芜菁黄花叶病毒科病毒FgMTV1/SX64的生物学功能进行了系统鉴定，尽管FgMTV1/SX64侵染对孢子产量、生物产量和致病力有微弱影响，但是，该病毒的侵染对真菌寄主SX64的生长速度、菌落直径和脱氧雪腐镰刀菌烯醇（DON）产量有显著的影响。

关键词：真菌病毒；禾谷镰刀菌；芜菁黄花叶病毒科；*Mycotymovirus*；FgMTV1/SX64

* 基金项目：国家自然科学基金面上项目（31171818）；北京市科技计划（No. D151100003915003）
** 第一作者：李鹏飞，硕士，主要从事禾谷镰刀菌真菌病毒学的研究；E-mail：li_ pengfei2014@163. com
*** 通讯作者：郭立华，副研究员，主要从事镰刀菌真菌病毒学的研究；E-mail：guolihua1972@126. com

Transcriptome-based Discovery of *Fusarium graminearum* Stress Responses to FgHV1 Infection*

WANG Shuang-chao[1,2]**, ZHANG Jing-ze[1]**,
LI Peng-fei[1], QIU De-wen[1], GUO Li-hua[1]***
(1. *State Key Laboratory for Biology of Plant Disease and Insect Pests, Institute of Plant Protection, Chinese Academy of Agricultural Science, Beijing, China*; 2. *Walloon Centre of Industrial Biology, Gembloux Agro-Bio Tech, University of Liège, Passage des Déportés, 2, Gembloux, Belgium*)

Abstract: *Fusarium graminearum* hypovirus 1 (FgHV1), which is phylogenetically related to Cryphonectria hypovirus 1 (CHV1), is a virus in the family *Hypoviridae* that infects the plant pathogenic fungus *F. graminearum*. Although hypovirus FgHV1 infection results in defects in mycelial growth and spore production, it does not attenuate the virulence of the host (hypovirulence). We now report that the vertical transmission rate of FgHV1 through asexual spores reached 100% and that hypovirus FgHV1 infection did not alter phenotypes such as pigment production and sexual development. Using RNA deep sequencing, we performed genome-wide expression analysis to reveal phenotype-related genes with expression changes in response to FgHV1 infection. A total of 248 geneswere differentially expressed, suggesting that hypovirus infection causes a significant alteration of fungal gene expression. Nearly two times as many genes were up-regulated as were down-regulated. A differentially expressed gene enrichment analysis identified a number of important pathways. Metabolic processes, cellular redox regulation, the ubiquitination system and transcription factors were the most affected categories in *F. graminearum* challenged with FgHV1. Papain-like protease, p20, encoded by FgHV1 could cause H_2O_2 accumulation in *Nicotiana benthamiana* leaves and programmed cell death in *N. tabacum cv.* Samsun NN leaves. Moreover, hypovirus FgHV1 may trigger the RNA silencing pathway in *F. graminearum*.

Key words: *Fusarium graminearum*; FgHV1; Transcriptome; Stress responses; Cellular redox regulation

---

* Acknowledgments: This work was supported by the National Natural Science Foundation of China (31171818) and the Science and Technology Plan Project of Beijing (No. D151100003915003). We thank the University of Liège-Gembloux Agro-Bio Tech and more specifically the research platform AgricultureIsLife for the funding of the scientific stay in Belgium that made this paper possible

** First authors: These authors contributed equally to this work: WANG Shuang-chao; E-mail: yuxinren2006@163.com; ZHANG Jing-ze; E-mail: jingzezhang0820@gmail.com

*** Corresponding author: GUO Li-hua; E-mail: guolihua72@yahoo.com

Transcriptome Analysis of Resistant and Susceptible Kiwifruit Varieties Upon *Botryosphaeria dothidea* Infection*

WANG Yuan-xiu**, LIU Bing, HUANG Chun-hui, SONG Shui-lin, JIANG Jun-xi***
(*College of Agronomy*, *Jiangxi Agricultural University*, *Nanchang* 330045, *China*)

Abstract: Ripe rot of kiwifruit, caused by *Botryosphaeria dothidea*, is one of the most important diseases in Fengxin county of Jiangxi province. In our previous study, it was showed that two kiwifruit varieties, "Hongyang" and "Jinyan", are susceptible (S) and resistant (R) to *B. dothidea* respectively. To understand resistant mechanism of kiwifruit to ripe rot, we use RNA-seq technology for comparatively analyzing the transcriptome profiling of both Jinyang and Hongyang in response to *B. dothidea* infection. A total of 2 034 850 196 clean reads were generated from the thirty-six libraries, of which 863 163 054 clean reads came from S and 1 171 687 142 came from R. From all the libraries, 44 656 genes including 39 041 reference genes and 5 615 novel transcripts were identified, and 13 898 differentially expressed genes (DEGs) were screened. We employed GO terms and KEGG pathway to analyze functions of these DEGs. 2 252 potential defense-related genes mainly involved in calcium regulating, mitogen-activated protein kinase, cell wall modification, resistance proteins, COBRA protein, transcription factors, pattern-recognition receptors and pathogenesis related proteins were identified. Especially, DEGs involved in calcium regulating and cell wall modification were induced to higher levels in Rvariety than in S variety. Plant-pathogen interaction pathways were enriched at all three time points in Rvariety, but only at one (2nd) time point in S variety. The results have partly disclosed the internal reasons of differences in kiwifruit ripe rot resistance between Rvariety and S variety.

Key words: *Ripe rot of* kiwifruit; *Botryosphaeria dothidea*; Transcriptome

---

* Funding: National Natural Science Foundation of China (31460452), The Science and Technology Department of Jiangxi Province (20141BBF60019) and the Education Department of Jiangxi Province (GJJ13269)

** First author: WANG Yuan-xiu, Ph, Dstudent, major in molecular plant pathology; E-mail: wangyuanxiu2002@163. com

*** Corresponding author: JIANG Jun-xi, professor, engaged in control of plant diseases; E-mail: jxau2011@126. com

Biological Characteristics of *Bipolaria maydis* Race O in Fujian Province*

DAI Yu-li[1**], GAN Lin[1], RUAN Hong-chun[1], LIAO Lei[2],
SHI Niu-niu[1], DU Yi-xin[1], CHEN Fu-ru[1], YANG Xiu-juan[1***]
(1. *Fujian Key Laboratory for Monitoring and Integrated Management of Crop Pests, Institute of Plant Protection, Fujian Academy of Agricultural Sciences, Fuzhou* 350013; 2. *College of Plant Protection, Fujian Agriculture and Forestry University, Fuzhou* 350002)

Abstract: To determine the biological characteristics of *B. maydis* race O in Fujian Province, three race O isolates, which were collected from Jian' ou, Shaxian and Fuzhou areas, were selected as representatives for research objective. The effects of temperature, moisture, pH, light and nutrition on the mycelial growth, sporulation and conidia germination were tested in this work. The results showed that the optimum temperature and pH for mycelium growth were 30℃ and 6; the optimum temperature and pH for sporulation were 25℃ and 6 ~ 7; and the optimum temperature and pH for conidia germination were 25 ~ 28℃ and 6. Fluorescent light could significantly inhibit the sporulation. When relative humidity for 85%, 95% or on water agar, the rate of conidia germination reached 50%, 88% and 90%, respectively. Glucose, lactose, mannitol and soluble starch were suited for mycelial growth and sporulation. However, urea, ammonium sulphate, ammonium nitrate, ammonium chloride could significantly inhibit the mycelial growth, sporulation and conidia germination. The lethal temperature for the mycelium was 65℃ remaining 10 min or 60℃ remaining 30 min, while the lethal temperature for the spores was 60℃ remaining 10 min or 55℃ remaining 30 min. Three isolates were obviously different in spoulation characters, lethal temperature of the mycelium and the pathogenicity.

Key words: *Bipolaria maydis*; Mycelium; Conidia; Biological characteristics

---

* 基金项目：福建省自然科学基金项目（2016J05073）；福建省农业科学院博士启动基金（2015BS－4）；福建省农业科学院青年科技英才百人计划项目（YC2016－4）

** 第一作者：代玉立，博士，助理研究员，研究方向：真菌学及植物真菌病害；E-mail：dai841225@126.com

*** 通讯作者：杨秀娟，硕士，研究员，研究方向：植物病害防治；E-mail：yxjzb@126.com

Comparison of Cell Wall Degrading Enzymes from Three Different *Fusarium oxysporum* Formae Speciales of Solanaceae

DONG Zhang-yong[1], LUO Mei[1], WANG Zhen-zhong[2], XIANG Mei-mei[1 *]

(1. *Department of Plant Protection*, *Zhongkai University of Agriculture and Engineering*, *Guangzhou* 510225, *China*; 2. *Department of Plant Protection*, *South China Agricultural University*, *Guangzhou* 510642, *China*)

Abstract: *Fusarium* species are among the most important phytopathogenic fungi, causing economically important wilts. Within these, *F. oxysporum*, has a broad host range, infecting both monocotyledonous and dicotyledonous plants. To understand the molecular underpinnings of pathogenicity in the different formae speciales of F. oxysporum, we compared the cell wall degrading enzymes, including polygalacturonase (PG), pectin methyl-galacturonase (PMG), polygalacturonic acid transeliminase (PGTE), pectin methyl transeliminase (PMTE) and cellulose (Cx), from three different *Fusarium oxysporum* formae speciales of Solanaceae as *F. oxysporum* f. sp. *lycopersici* (FOL), *F. oxysproum* f. sp. *melongenae* (FOM), and *F. oxysporum* f. sp. *capsicum* (FOC). We found that the higher to lower enzyme activity order were PG, PMG, PGTE, Cx and PMTE in inoculated eggplant or tomato *in planta*, while an order of PG, PGTE, PMG, PMTE and Cx in inoculated *Capsicum*. We also found that PG activity of FOC was higher than that of FOM or FOL *in vitro*. There were no significant difference in the activity of PMG, PMTE and Cx in these three fungal *in vitro*. Analysis of PG isoenzymes by IEF-PAGE showed that FOL, FOM and FOC could secrete a total of 13 PGs, with 5, 5, and 3 PG isoenzymes, respectively. Each fungus had a specific PG band. These results will provide useful information for managing *Fusarium* wilt and insights into Pathogenic factors for phytopathology.

Key words: Solanaceae; *Fusarium* Wilt disease; *Fusarium oxysporum*; Formae speciale; Cell wall degrading enzymes

* Corresponding author: XIANG Mei-mei; mm_ xiang@ 163. com

Carbamoyl Phosphate Synthetase subunit MoCpa2 Controls Development and Pathogenicity by Modulating Arginine Biosynthesis in *Magnaporthe oryzae*

LIU Xin-yu, CAI Yong-chao, ZHANG Xi,
ZHANG Hai-feng*, ZHENG Xiao-bo, ZHANG Zheng-guang
(*College of Plant Protection/Key Laboratory of Integrated Management of Crop Diseases and Pests, Ministry of Education, Nanjing Agricultural University, Nanjing* 210095, *China*)

Abstract: Arginine is a semi-essential amino acid that affects physiological and biochemical functions. The *CPA*2 gene in yeast encodes a large subunit of arginine-specific carbamoyl phosphate synthetase and is involved in arginine biosynthesis. Here, an ortholog of yeast *CPA*2 was identified in the rice blast fungus *Magnaporthe oryzae*, and was named *MoCPA*2. MoCpa2 is a 1180-amino acid protein which contains an ATP grasp domain and two CPSase domains. Targeted deletion of *MoCPA*2 showed that it was involved in *de novo* arginine biosynthesis in *M. oryzae*. The Δ*Mocpa*2 mutant exhibited defects in asexual development and pathogenicity but not appressorium formation. Further examination revealed that the invasive hyphae of the Δ*Mocpa*2 mutant were restricted mainly to the primary infected cells. In addition, the Δ*Mocpa*2 mutant was unable to induce a plant defense response and had the ability to scavenge ROS during pathogen-plant interactions. Exogenous rice leaf extracts could partially suppress the growth defect, but not conidiogenesis or pathogenicity defects. Structure analysis revealed that the ATP grasp domain and each CPS domain were indispensable for the full function of MoCpa2. In summary, our results demonstrate that MoCpa2 plays an important role in arginine biosynthesis, and affects growth, conidiogenesis, and pathogenicity. These results suggest that research into metabolism and processes that mediate amino acid synthesis are valuable for understanding *M. oryzae* pathogenesis.

Key words: *Magnaporthe oryzae*; Carbamoyl phosphate synthetase; Arginine biosynthesis; Conidiation; Pathogenicity

* Corresponding author: ZHANG Hai-feng; E-mail: hfzhang@ njau. edu. cn

MoDnm1 Dynamin Mediating Peroxisomal and Mitochondrial Fission in Complex with MoFis1 and MoMdv1 is Important for Development of Functional Appressorium in *Magnaporthe oryzae*

ZHONG Kai-li[1], LI Xiao[1], LE Xin-yi[1], KONG Xiang-yi[1], ZHANG Hai-feng[1], ZHENG Xiao-bo[1], WANG Ping[2], ZHANG Zheng-guang[1 *]

(1. *College of Plant Protection/Key Laboratory of Integrated Management of Crop Diseases and Pests, Ministry of Education, Nanjing Agricultural University, Nanjing* 210095 *China*; 2. *Department of Pediatrics, Louisiana State University Health Sciences Center, New Orleans, Louisiana, United States of America*)

Abstract: Dynamins are large superfamily GTPase proteins that are involved in various cellularprocesses including budding of transport vesicles, division of organelles, cytokinesis, and pathogen resistance. Here, we characterized several dynamin-related proteins from the rice blast fungus *Magnaporthe oryzae*. We found that MoDnm1 is required for normal functions, including vegetative growth, conidiogenesis, and full pathogenicity. In addition, we found that MoDnm1 co-localizes with peroxisomes and mitochondria, which is consistent with the conserved role of dynamin proteins. Importantly, MoDnm1-dependent peroxisomal and mitochondrial fission involves the functions of the mitochondrial fission protein MoFis1 and the WD-40 repeat protein MoMdv1. These two proteins display similar functions as well as cellular localizations to MoDnm1, and are required for full pathogenicity. Further studies showed that MoDnm1, MoFis1 and MoMdv1 are in complex to regulate peroxisomal and mitochondrial fissions. This is required for not only pexophagy and mitophagy progression but also appresorium function and host penetration. In summary, our studies provide insights into how dynamin-related proteins interact with their partners to mediate peroxisomal and mitochondrial function and how such regulatory events may link to differentiation and pathogenicity in the rice blast fungus *M. oryzae*.

Key words: Peroxisome; Mitochondria; Fission; Pexophagy; Mitophagy; Autophagy; Pathogenicity

* Corresponding author: ZHANG Zheng-guang; E-mail: zhgzhang@ njau. edu. cn

Molecular Detection and Sequence Analysis of Polygalacturonase *Szpg*6 to *Szpg*10 Gene of Zebra Disease of Sisal*

WU Wei-huai[1], ZHENG Jin-long, XI Jin-gen, ZHENG Xiao-lan, LIANG Yan-qiong, LI Rui, HE Chun-ping**, YI Ke-xian**

(*Environment and Plant Protection Institute*, *CATAS* /*Hainan Key Laboratory for Detection and Control of Tropical Agricultural Pests*, *Haikou* 571101, *China*)

Abstract: Sisal zebra disease caused by *Phytophora nicotianae* var. *parasitica* is a kind of main diseases which can cause serious damage to the sisal. In the present study, five specific primer pairs were designed in coding region based on the reference sequence of *pppg*6 to *pppg*10 gene from *Phytophora parasitic*. Five specific primer pairs were used for the molecular detection and gene cloning from sisal zebra pathgoen DNA. The results showed that *Szpg*6 (sisal zebra polygalacturonase 6) to *Szpg*10 genes were conserved in all the tested sisal zebra isolates. Sequence alignment with the reference sequences of *pppg*6 to *pppg*10 gene from *Phytophora parasitic* revealed that 1, 3, 1, 6 and 4 SNPs were identified in the *Szpg*6 to *Szpg*10 genes coding regions, respectively. There, five and three of these mutations were nonsynonymous in *Szpg*7, *Szpg*9 and *Szpg*10, and the rest were synonymous. Thus we speculated that there were some differences in function when *Szpg*7, *Szpg*9 and *Szpg*10 genes were compared to *pppg*7, *pppg*9 and *pppg*10. The results have laid a solid foundation for the further study of the role of sisal zebra germs polygalaturonase in the pathogenic process.

Key words: Zebra disease of sisal; Polygalaturonase; Molecular detection

---

* This study was sponsored by the Natural Science Foundation of Hainan Province (314105)

** Corresponding authors: HE Chun-ping; E-mail: hechunppp@163.com

YI Ke-xian; E-mail: yikexian@126.com

Comparative Genome Analyses Reveal Distinct Modes of Host-specialization Between Dicot and Monocot Powdery Mildew

WU Ying[1], MA Xian-feng[1], PAN Zhi-yong[2], Shiv Kale[3], SONG Yi[1], XIAO Shunyuan[1]

(1. *Institute of Biosciences and Biotechnology Research & Department of Plant Science and Landscape Architecture*, *University of Maryland College Park*, *Rockville MD* 20850, *USA*; 2. *Key Laboratory of Horticultural Plant Biology of Ministry of Education*, *Huazhong Agricultural University*, *Wuhan* 430070, *PR China*; 3. *Virginia Bioinformatics Institute*, *Virginia Tech*, *Blacksburg*, *Virginia*, *USA*)

Abstract: Monocot powdery mildew (PM) fungi such as *Blumeriagraminis* f. sp. *hordei* (*Bgh*) infectious on barley and *B. graminis* f. sp. *tritici* (*Bgt*) infectious on wheat exhibit extremely high-level of host-specialization. By contrast, many dicot PM fungi display rather broad host ranges. To understand why different PM fungi adopt distinct modes of host-adaption, we sequenced the genomes of four dicot PM strains (three *Golovinomyces* and one *Oidium*) and conducted comparative sequence analyses along with the genomesof *Bgh*, *Bgt* and *Erysiphenecator* (infectious on grapevine) available from the Genbank. While 3 129 genes arefound in all seven genomes (thus they are conserved PM genes), ~50% are lineage-specific, of which 116 ~ 534 are genes encoding candidate secreted effector proteins (CSEPs). Interestingly, while the two monocot PM fungi possessup to 534 CSEPs with several families containing up to 21 members, all the five dicot PM fungi have only 116 ~ 178 CSEPs with limited gene amplification. These two evolutionary patterns concerning PM CSEPs coincide with the contrasting modes of their host-adaption: the monocot PM fungi show a high-level of host specialization, likely reflecting an advanced host-pathogen arms-race, whereas the dicot PM fungi tend to practice polyphagy, which has presumably lessened the selective pressure for escalating an arms-race with a particular host. Strikingly, the tomato PM *Oidium neolycopersici* has a genome only half the size of other PM genomes, despite having a similar number of genes, suggesting that *O. neolycopersici* has followed a unique genome evolutionary path.

URP Analysis of *Fusarium pseudograminearum* that Is the Dominant Pathogen Causing Crown Rot of Wheat*

HE Xiao-lun, ZHOU Hai-feng, DING Sheng-li, CHEN Lin-lin, YUAN Hong-xia, LI Hong-lian**

(*College of Plant Protection, Henan Agricultural University, Zhengzhou* 450002, *China*)

Crown rot of wheat is a newly discovered soil-borne disease in China, which is caused by a diverse group of Fusarium species. It occurs generally in Huang-Huai winter wheat area and shows a tendency of expansion and aggravation, which is a serious threat to wheat production. Li et al. reported the *F. pseudograminearum* causing crown rot for the first time in China in 2012. The aim of the present study was to examine the population structure of eight populations of *F. pseudograminearum* collected from different regions in Huang-Huai by using URP-PCR, and to characterize the genetic differentiation within and between these populations. An understanding of the genetic structure of *F. pseudograminearum* may provide theoretic foundation and technique measures for the management of crown rot.

Nine URP primers which can amplify more polymorphic loci were screened from 12 URP primers. The genetic diversity of 261 *F. pseudograminearum* isolates was analyzed with 9 URP primers. The amplification results showed that 132 fragments were amplified and 128 fragments displayed polymorphism which accounted for 96.97% in the total amplified fragments. The average number of bands amplified per primer was 14.67. By analysis of the genetic similarity coefficient, the relationship of *F. pseudograminearum* populations of Northern Henan and Middle of Henan were closest and that of Shandong and Southern Henan were the farthest. The coefficient of the population genetic differentiation (Gst) among geographical groups was 0.1499, while it reached 0.8501 within the populations indicating more divergent genetic diversity within the populations. The number of migrants per generation among 8 geographical populations was 2.8362, which indicated that genetic information exchange was frequent. At the similarity coefficient of 0.962, all isolates were clustered into two groups. The group Ⅰ included five regions, i. e. Northern Henan, Middle of Henan, Eastern Henan, Western Henan and Southern Henan. The group Ⅱ included three regions, i. e. Hebei, Shandong and Shanxi populations.

* Funding: This study was funded by the Special Fund for Agro - scientific Research in the Public Interest (201503112)

** Corresponding author: Prof. LI Hong-lian; E-mail: honglianli@sina.com

玉米大斑病菌（*Setosphaeria turcica*）非核糖体肽合成酶（NRPS）基因结构域预测*

陈　楠**，肖淑芹，闫丽斌，王　芬，孙玉鑫，薛春生***
（沈阳农业大学植物保护学院，沈阳　110866）

非核糖体肽合成酶参与的次生代谢途径与植物病原真菌的无毒因子、毒素和铁离子代谢密切相关，其基础模块由一个腺苷酸化结构域（A）、巯基化结构域（T）和缩合结构域（C）组成，A结构域特异性识别底物，T结构域固定中间产物，C结构域负责中间产物肽键的形成。

根据 *S. turcica* 全基因组序列信息分析预测 NRPS 的结构域。以 *Cochliobolus heterostrophus*、*Fusarium* spp.、*Aspergillus* spp. 的 NRPS 为参考，利用 DNAMAN 6.0 和 MEGA 5.2 等软件分析发现，*S. turcica* 预测 ID65284、ID179280 和 ID18754 等 NRPS 分别与 *C. heterostrophus* 的 NRPS3、NRPS4 和 NRPS12 相似性高，利用 NRPS-KPS 在线软件分析它们的结构域分别为 NM-T-C、T-E-C-A-T-C-A-T-E-C-A-T-C-A-T-E-C-T-T 和 A；而 ID54477 与 *C. heterostrophus*、*Fusarium* spp.、*Aspergillus* spp. 中所有的 NRPS 同源性低，结构域为 A-T-C。

* 基金项目：国家自然科学基金（31271992）
** 第一作者：陈楠，在读博士研究生，从事玉米病害研究
*** 通讯作者：薛春生；E-mail：cshxue@sina.com

香蕉枯萎病菌 MAP 激酶信号途径假定转录因子的功能分析*

丁兆建**，杨腊英，郭立佳，王国芬，汪　军，梁昌聪，刘　磊，黄俊生***
（中国热带农业科学院环境与植物保护研究所，海口　571101）

摘　要：尖孢镰刀菌古巴专化型（*Fusarium oxysporum* f. sp. *cubense*，Foc）是一种重要的土传病原真菌，其引起的香蕉枯萎病是世界性的毁灭性病害。可是，该病原菌的分子致病机理仍不清晰。丝裂原活化蛋白（MAP）激酶信号途径在调控真菌胞外的多种信号转导、生长和分化过程起着重要的作用。为了更好地理解香蕉枯萎病菌的致病机理，我们深入研究了香蕉枯萎病菌 MAP 激酶信号途径假定转录因子 FoSwi6 和 FoRlm1 的生物学功能。通过 PEG 介导的基因定点敲除，获得了突变体 Δ*FoSwi*6 和 Δ*FoRlm*1。致病性测试表明该2个突变体对香蕉苗的致病性均减弱。细胞壁敏感性测试表明突变体 Δ*FoSwi*6 和 Δ*FoRlm*1 均对细胞壁抑制因子荧光增白剂（CFW）和细胞壁降解酶不敏感，但对细胞壁抑制因子刚果红（CR）的抗性增加。平板测试表明突变体 Δ*FoSwi*6 在 PDA 平板上生长缓慢，气生菌丝明显减少；然而突变体 Δ*FoRlm*1 仅气生菌丝稍微减少。以上结果表明该2个 MAP 激酶信号途径转录因子具有重叠的生物学功能。总之，研究结果揭示香蕉枯萎病菌 MAP 激酶信号途径假定转录因子 FoSwi6 和 FoRlm1 均调控 Foc 的生理学特性和致病性，相关研究工作正在进行中。

关键词：香蕉枯萎病菌；MAP 激酶信号途径；转录因子；致病性

* 基金项目：中央级公益性科研院所基本科研业务费专项（2015hzs1J013）；海南省自然科学基金面上项目（20163103）和海南省重点研发计划（ZDYF2016039）

** 第一作者：丁兆建，助理研究员，主要从事植物病原真菌的致病机理研究；E-mail：dingzhaojian@163.com

*** 通讯作者：黄俊生，研究员，从事热带作物病害防控技术研究；E-mail：H888111@126.com

香蕉枯萎病菌 Foc1 和 Foc4 对香蕉果胶甲酯酶影响的差异分析*

范会云**，李华平***
（华南农业大学农学院，广州 510642）

摘 要：香蕉枯萎病是由尖孢镰刀菌古巴专化型（*Fusarium oxysporum* f. sp. *cubense*，Foc）引起的香蕉上的一种毁灭性病害。该病原共有 3 个生理小种，其中，1 号生理小种（Foc1）只侵染粉蕉类香蕉，而不能侵染矮香蕉，但 4 号生理小种（Foc4）却能侵染包括矮香蕉在内的所有的香蕉栽培品种。为了明确 2 个小种在侵染不同香蕉寄主时寄主所表现的差异，本研究利用化学法和免疫荧光标记法比较了 Foc1 和 Foc4 分别接种巴西蕉后，巴西蕉根部的果胶甲酯酶活性与果胶甲酯化程度的时空变化。结果表明，2 个小种在接种巴西蕉后，其果胶甲酯酶活性和果胶甲酯化程度具有显著性差异；进一步利用不同果胶酶类单克隆抗体标记 2 个小种侵染巴西蕉根部后的果胶含量，结果表明，在接种 Foc4 后，巴西蕉根部被标记的高甲酯化的果胶含量较 Foc1 低；而被标记的低甲酯化的果胶含量较 Foc1 高。这进一步表明果胶甲酯酶可能是 Foc1 和 Foc4 在巴西蕉上产生致病差异的因子之一。

关键词：香蕉枯萎病菌；香蕉；果胶甲酯酶；果胶甲酯化程度

---

* 基金项目：现代农业产业技术体系建设专项（CAR－32－05）
** 第一作者：范会云，博士研究生，植物病理学；E-mail：fhy1262006@126.com
*** 通讯作者：李华平，教授；E-mail：huaping@scau.edu.cn

柑橘褐斑病菌 *FTR*1 基因生物学功能初步研究*

张　倩**，王洪秀，唐科志***
（西南大学柑桔研究所，中国农业科学院柑桔研究所，重庆　400712）

摘　要：由链格孢菌（*Alternaria alternata*）引起的柑橘褐斑病（Alternaria brown spot，ABS）是严重为害柑橘的的一种真菌病害。近年来在我国柑橘各主产区迅速蔓延，造成巨大的经济损失。目前，国内外已经相继开展病原菌分离鉴定、生物学特性、侵染机制、品种抗性及病害流行发生规律等试验且取得一定成果。其中，病原菌致病基因的挖掘当属研究热点。*A. alternata* 作为典型的死体营养型真菌，在侵入之前即依赖其产生的毒素杀死寄主细胞。ACT 毒素作为主要致病因子，其致病机理已有文献报道。Chen 等发现，该病菌对 ROS 的解毒机制也是致病所必需的。柑橘褐斑病菌存在一套对 ROS 的高效解毒机制。外界刺激下 Nox 合成少量 ROS，其中 H_2O_2 可以作为信号分子，激活 *YAP*1、*HOG*1、*SKN*7 的表达，启动对 ROS 的解毒反应。研究发现铁离子在解毒 ROS 过程中也发挥重要作用。*NPS*6 诱导合成铁载体，介导低铁环境下铁离子的摄取，继而通过下游的系列氧化还原反应解毒 ROS。其中 *YAP*1 与 *HOG*1 可能调控依赖于铁的 ROS 解毒系统，表明铁离子的摄取对 ROS 的解毒能力有重大影响。

为建立适合于柑橘褐斑病菌的遗传转化体系，本研究首先优化原生质体制备条件。结果表明，以 0.7mol/L NaCl 为稳渗剂，在液体 PDB 中摇培 36 h 的菌丝，以单一酶 1% Kitalase 30℃ 条件下消化 2.5h 后所释放原生质体数量最多，达 7.5×10^8 个/mL。PEG 介导下，不同基因转化该方法所获得的原生质体均可获得足量的转化子。因此，基于该方法，进一步对柑橘褐斑病菌 *FTR*1（Fe transporter permease）基因展开研究。*FTR*1 编码一种对铁离子具有高亲和性的渗透酶，它具有三价铁结合位点 Glu-Xaa-Xaa-Glu 的结构单元。承担的作用主要是将胞外的铁转入细胞，供生物进一步改造利用。借助融合 PCR，通过 Split Marker 重组技术构建 *FTR*1 基因敲除盒，PEG 介导下转化原生质体，获得柑橘褐斑病菌 *Δftr*1 突变体。表型分析结果显示与野生型相比，突变体生长速度、菌丝形态、产孢量及孢子形态均无明显差异，表明该基因并不参与调控其生长发育。下一步将开展突变菌株致病力、胁迫应答及铁代谢相关基因分析。

关键词：柑橘褐斑病菌；原生质体；*FTR*1；生长发育

* 基金项目：教育部创新团队（IRT0976）；重庆市科技攻关项目（CSTC2012GG－YYJS0475）
** 第一作者：张倩，在读硕士，研究方向：分子植物病理学；E-mail：1335032933@qq.com
*** 通讯作者：唐科志，副研究员，主要从事柑橘病害研究；E-mail：tangkez@163.com

稻瘟病菌 Mo*Duo*1 基因酵母双杂交诱饵载体的构建和自激活检测

吕　哲[1*]，许雨晨[1]，陈保善[2]，彭好文[1,2**]

（1. 广西大学农学院，南宁　530004；2. 广西大学亚热带生物资源保护与利用国家重点实验室，南宁　530004）

摘　要：由稻瘟病菌引起的水稻稻瘟病是水稻生产中最重要的病害，常造成水稻大幅度减产，甚至绝收。围绕稻瘟病菌致病机理的研究主要集中在与信号传导相关的基因中，Dam1 复合体在芽殖酵母中首次发现，参与酵母细胞有丝分裂过程动粒与微管的正确连接。我们在前期研究发现，与芽殖酵母 Sc*Duo*1 同源的稻瘟病菌 Mo*Duo*1 基因是一个致病相关基因，在稻瘟病菌中缺失该基因导致病菌营养生长减缓，分生孢子形态异常，萌发率大幅下降，为了进一步研究该基因的功能及其产物参与的细胞进程，我们利用酵母双杂交技术筛选其互作蛋白。本研究首先利用传统基因克隆技术构建了酵母双杂交诱饵质粒 pGBKT7 – Duo1，之后将其转化到酵母 AH109 感受态中，涂 SD/-Trp 和 SD/-Trp/X-α-gal 培养基，30℃倒置培养 3 ~ 5d，观察到两培养基均有菌落生长，且 SD/-Trp/X-α-gal 培养基上的菌落没有变蓝，而 SD/-Leu/-Trp/X-α-gal 平板上的阳性对照为蓝色，因此诱饵质粒在酵母 AH109 中并没有自激活。本研究证明 pGBKT7-Duo1 载体可以作为酵母双杂交方法中的诱饵载体筛选与 Duo1 蛋白相互作用的蛋白。

关键词：稻瘟病菌；Mo*Duo*1；酵母双杂交

---

* 第一作者：吕哲，硕士研究生，研究方向为真菌及植物真菌病害防治；E-mail：lvzhelz@ 163. com

** 通讯作者：彭好文，副教授，主要从事真菌及植物真菌病害防治方面的研究；E-mail：phwxx@ gxu. edu. cn

广西杧果炭疽菌致病力、抗药性及遗传多样性分析*

郭堂勋[1,2**]，赵　广[3]，李其利[1,2]，黄穗萍[1,2]，唐利华[1,2]，莫贱友[1,2***]
（1. 广西农业科学院植物保护研究所，南宁　530007；2. 广西作物病虫害生物学重点实验室，南宁　530007；3. 广西大学农学院，南宁　530004）

摘　要：杧果是世界五大水果之一，其广泛种植于热带、亚热带地区，目前在我国主要种植于海南、广西、广东、云南、福建、四川、贵州等地。炭疽病是杧果种植和采后的主要病害之一，严重影响杧果的产量和品质。该病的发生与病原种类及其致病力关系密切，而广西杧果炭疽菌的种群分布、致病力分化及其对吡唑嘧菌酯的抗药性尚不明确。本研究在马铃薯葡萄糖琼脂培养基（PDA）观察了29个供试菌株的形态特征，结果表明，菌落大部分边缘规则，菌丝体呈绒毛状，培养后期多变成灰黑色，并有粉红色分生孢子堆形成。平均生长速率在6.74～12.43mm/d，产孢量在2.96×10^5～7.93×10^7个/mL，孢子大小在（12.05～20.2）μm×（3.72～6.54）μm。将29个菌株在离体叶片上进行致病性测定，结果发现其中弱致病菌株4株，占13.8%；中等致病菌株11株，占37.9%；强致病菌株14株，占48.3%。将29个菌株在离体果实上进行致病性测定，其中，弱致病菌株14株，占48.3%；中等致病菌株7株，占24.1%；强致病菌株8株，占27.6%。对比杧果叶片和果实上菌株的致病力发现：TD3、TD6-5、TD4均表现弱致病性；WM01-2、WM52、NN133均表现中等致病性；WM03-1、NKY-1、QZ-3、TD6-7均表现强致病性。

采用菌丝生长速率法测定了广西杧果炭疽菌对吡唑醚菌酯的抗药性，结果表明，从南宁市区采集的杧果炭疽病菌对吡唑醚菌酯的敏感性较高，其EC_{50}在0.14～0.4μg/mL；而从百色田东县、田阳县，武鸣华侨投资区采集的杧果炭疽菌对吡唑醚菌酯的敏感性显著降低，EC_{50}最高可达4.64μg/mL。采用ISSR-PCR分析了广西杧果炭疽病菌遗传多样性，结果表明，选用15条ISSR随机引物共计扩增出277个条带，多态性条带有174个，多态率为62.81%，每个引物均能扩增出2～13个条带，条带大小在200～2 500bp。UPGMA聚类分析在遗传相似系数0.78水平，供试菌株共划分为6个大的类群。29个菌株的遗传多样性与地理位置、形态多样性、对吡唑醚菌酯的敏感性均无显著相关性。

关键词：杧果炭疽菌；致病力；抗药性；遗传多样性

---

* 基金项目：国家农业产业体系广西杧果创新团队建设专项（nycytxgxcxtd－02－08－2）；国家自然科学基金（31560526）；广西作物病虫害生物学重点实验室基金（15－140－45－ST－2）

** 第一作者：郭堂勋，副研究员，主要研究方向为植物真菌病害及其防治；E-mail：guotangxun@ gxaas. net
*** 通讯作者：莫贱友，研究员，主要研究方向为植物真菌病害及其防治；E-mail：mojianyou@ gxaas. net

Comparative Transcriptome Analysis of Two *Valsa pyri* Strains

HE Feng, ZHANG Xiong, QIAN Guo-liang, LIU Feng-quan, DOU Dao-long
(*Department of Plant Pathology, Nanjing Agricultural University, Nanjing, China*)

Abstract: *Valsa pyri* belongs to ascomycetes of Valsaceae family (Sordariomycetes, Diaporthales) and causes pear canker disease, resulting in great economic losses of pear yield worldwide especially in eastern Asia. The fungus invades the host from injury sites of trunks and fruits and then forms cankers. We have collected hundreds of *V. pyri* isolates in China and found that they exhibited different phenotypes in growth, development and pathogenesis, implying that the fungi have evolved diversity pathotypes. In order to dissect the mechanism of evolution between different pathotypes, we analyzed the differential expressed genes between two strains which were different in growth, fruiting body production and pathogenicity through RNA-seq. Using sequencing, de novo assembly, and annotation of differentially expressed genes in two different pathogenic *V. pyri* strains (Vp14 and Vp297), we found that, similar to other necrotrophic pathogenic fungi, two isolates contained a large number of genes encoding transporters, protein kinases, transcription factors and especially the heat shock proteins. Interestingly, the Vp14 isolate has higher number of cell wall degrading enzymes and sugar transporters than Vp297, indicating the Vp14 isolate is especially suitable for sugar utilization. The number of metallopeptidases and cysteine peptidases which are important for acquiring nutrition and enhancing pathogenicity was also expanded in Vp14. However, signal transduction-related genes are slightly more representative in Vp297. Taken together, these results accord with the phenotypes of the two strains indicating that Vp14 can effectively infect the host and take up nutrition from host.

核盘菌（*Sclerotinia sclerotiorum*）抗多菌灵β-微管蛋白基因在禾谷镰孢菌（*Fusarium graminearum*）中的表达研究[*]

杨　莹[**]，赵东磊，段亚冰，周明国[***]
（南京农业大学植物保护学院，南京　210095）

摘　要：植物病原真菌β-微管蛋白基因是多菌灵等苯并咪唑类杀菌剂的作用靶标，其特异位点的突变可导致植物病原真菌对苯并咪唑类杀菌剂产生不同水平的抗药性。禾谷镰孢菌（*Fusarium graminearum*）具有 β_1 和 β_2 两个微管蛋白基因，其中 β_2-微管蛋白基因特异位点突变可导致该菌对多菌灵产生不同水平的抗性。而其他植物病原真菌仅有1个β-微管蛋白基因，且该基因特异位点突变调控着对多菌灵的抗性。同源比对分析发现植物病原真菌β-微管蛋白基因与禾谷镰孢菌 β_1-微管蛋白基因的同源性要远高于 β_2-微管蛋白基因。为了进一步揭示植物病原真菌对多菌灵的抗性机制，本实验室刘圣明博士将灰葡萄孢菌（*Botrytis cinerea*）对多菌灵抗性的突变基因型（E198A）分别同源置换禾谷镰孢菌的 β_1-和 β_2-微管蛋白基因后，所得突变体对多菌灵均不表现抗性，但能发挥其他生物学功能。在本研究中，我们将核盘菌（*Sclerotinia sclerotiorum*）对多菌灵抗性的突变基因型（SstubE198A）分别同源置换禾谷镰孢菌的 β_1-和 β_2-微管蛋白基因。结果表明，核盘菌抗多菌灵β-微管蛋白（SstubE198A）无论单独替换禾谷镰孢菌的 β_1 还是 β_2 微管蛋白基因，所得突变体均对多菌灵不表现抗性。这可能是由于禾谷镰孢菌中另一微管蛋白的存在与核盘菌抗多菌灵β-微管蛋白（SstubE198A）相互作用造成的。为了验证这个假设，我们以得到的突变体为出发菌株，分别用核盘菌抗多菌灵β-微管蛋白（SstubE198A）将另一个β-微管蛋白基因进行同源置换，得到 β_1 和 β_2 微管蛋白基因均被核盘菌抗多菌灵β-微管蛋白基因（SstubE198A）替换的禾谷镰孢菌突变体。结果发现突变体对多菌灵表现高水平抗性。

关键词：禾谷镰孢菌；核盘菌；β-微管蛋白；多菌灵

* 基金项目：公益性行业（农业）科研专项“农作物重要病原菌抗药性监测及治理技术研究与示范”（项目编号：201303023）

** 第一作者：杨莹，在读硕士，主要从事杀菌剂毒理及抗药性研究；E-mail：2014102132@ njau. edu. cn

*** 通讯作者：周明国，教授，博士生导师，主要从事植物病害化学防治研究；E-mail：mgzhou@ njau. edu. cn

核盘菌蛋白激发子同源物 SsPemG1 的基因功能分析*

徐玉平**，姚传春，魏君君，高智谋，潘月敏***
（安徽农业大学植物保护学院，合肥 230036）

摘　要：核盘菌（*Sclerotinia sclerotiorum*）是一种广泛分布的产菌核的死体营养型真菌，造成巨大经济损失。由于传统防治手段弊端凸显，从基因水平研究核盘菌致病机理，寻找新一代环保、高效的微生物农药刻不容缓。植物在长期进化过程中形成了复杂的防卫反应以应对病原微生物的入侵，而蛋白激发子可以刺激植物快速形成程序性细胞死亡（Programmed Cell Death，PCD）阻止病原菌的进一步扩散。因此，研究蛋白激发子对挖掘新型安全、高效的微生物农药意义重大。

PemG1（GenBank accession number：EF062504）是一个分离自稻瘟病菌（*Magnaporthe grisea*）的蛋白激发子，对诱导水稻抵御稻瘟病菌的侵染有较好作用。本研究使用稻瘟病菌 PemG1 的氨基酸序列对核盘菌全基因组数据库进行 Blast_ p，找到一个具有相同功能域的蛋白，编码基因为 SS1G_ 07345，将其命名为 *SsPemG1*。故可推测 SsPemG1 可能也是蛋白激发子。

本实验扩增了核盘菌中 SsPemG1 编码基因序列，构建出 *SsPemG1*-pSilent-1 的沉默表达载体，利用 PEG 介导的核盘菌原生质体转化实验获得了该基因的沉默突变菌株。通过潮霉素 B 抗性和荧光定量 PCR 筛选出具有良好的沉默效果的突变菌株 M6 和 M9。通过分析沉默突变菌株的表型差异和生物信息学信息，明确 SsPemG1 在核盘菌中的功能。

表型分析结果表明：沉默突变体的致病性明显较野生型强。通过进一步研究发现，沉默突变体比野生型和 Mock 菌株具有更快的生长速度；对 NaCl 和 SDS 的耐受性变的更强；形成更多侵染垫；细胞壁降解酶类的活性变的更强。SsPemG1 通过调控这 4 个因素共同导致其致病力变强。生物信息学预测结果表明，主要功能是结合 RNA，调节转录和翻译，这与表型实验的结果一致。

本实验的开展有助于进一步了解核盘菌 SsPemG1 的功能，为研究激发子提供一个新的方法，为核盘菌的防治提供理论基础。

关键词：核盘菌；SsPemG1；沉默；致病性；激发子

* 基金项目：国家公益性行业（农业）科研专项（201103016）
** 第一作者：徐玉平，安徽安庆人，在读硕士生，主要从事真菌学及植物真菌病害研究；E-mail：1204903469@ qq. com
*** 通讯作者：潘月敏，副教授，主要从事真菌学及植物真菌病害研究；E-mail：panyuemin2008@ 163. com

核盘菌磷酸泛酰巯基乙胺基转移酶编码基因的预测与生物信息学分析*

李秀丽[1,2]**，张茜茹[1]，刘腾飞[1]，吴　亚[1]，高智谋[1]***
（1. 安徽农业大学植物保护学院，合肥　230036；2. 安徽宿州学院生物与食品工程学院，宿州　234000）

核盘菌［*Sclerotinia sclerotiorum*（Lib.）de Bary］是一种重要的植物病原真菌，可以侵染 400 种以上的植物，使多种重要的农作物产生重大经济损失。磷酸泛酰巯基乙胺基转移酶（Phosphopantetheinyl transferases，PPTases）蛋白在各生物的代谢过程中发挥着重要作用，它参与调节初生代谢和次生代谢过程，进而影响生物的生长发育和致病性等过程。在脂肪酸、聚酮化合物以及非核糖体多肽合成酶复合体中，载体蛋白添加一个来自于辅酶 A 的磷酸泛酰巯基乙胺基后才可形成具活性的全蛋白，而这一重要的修饰过程需要 PPTases 的催化作用。有关文献报道证实核盘菌存在 PPTases 基因，且 PPTase 蛋白对生命过程的影响是多方面的，除了已揭示部分的功能，可能还有一些未知的作用机理等待我们去探索，这对我们深入研究核盘菌的致病机理具有重要意义。为了解磷酸泛酰巯基乙胺基转移酶（PPTases）在核盘菌生理代谢与致病过程中的作用，在核盘菌全基因组中寻 PPTases 同源物。利用同源比对的方法，预测核盘菌中 PPTases 基因，进而分析其编码蛋白的基本理化性质、疏水性、跨膜区、信号肽、亚细胞定位、结构域和二级、三级结构，并对 PPTases 编码氨基酸进行遗传进化分析。结果表明，在核盘菌全基因组中预测存在 3 个 PPTases 基因同源物，分别命名为 Ss-Ppt1、Ss-Ppt2 和 Ss-Ppt3，3 个蛋白均为亲水性的非分泌型蛋白，在其二级和三级结构中 α 螺旋结构和无规则卷曲占主要部分，且三者都含有 PPTases 的保守结构域 SCOP 和（或）ACPS。

* 基金项目：国家公益性行业（农业）科研专项（201103016）和宿州学院优秀青年人才基金项目（2013XQRL06）资助
** 第一作者：李秀丽，博士，研究方向：植物病原真菌学；E-mail：lixiuli2187@163.com
*** 通讯作者：高智谋，教授；研究方向：真菌学与植物真菌病害；E-mail：gaozhimou@126.com

黑龙江省水稻褐变穗病原鉴定及分子检测技术的研究*

韩雨桐，常　浩，牟　明，李云鹏，范　琳，钟庆燕，张俊华**
（东北农业大学农学院，哈尔滨　150030）

摘　要：水稻是世界重要的粮食作物之一，中国是世界最大的水稻生产国和消费国。黑龙江省是全国主要的水稻产区。近年来，随着黑龙江省水稻种植面积增加、品种单一以及环境条件等因素的影响，水稻褐变穗病已经成为黑龙江省主要水稻病害之一，影响着水稻的产量和品质。为了控制该病害的发生，减少水稻褐变穗病对水稻生产的危害，开展黑龙江省水稻褐变穗病病原鉴定、生物学特性、分子检测的研究迫在眉睫。本研究对采自黑龙江省 14 个市县的典型症状的水稻褐变穗病穗进行病原菌的分离，利用形态学及分子生物学技术相结合进行病原菌的鉴定，测定了病原菌生物学特性，建立病原菌分子检测体系，主要研究结果如下。

（1）明确了引起黑龙江省水稻褐变穗病的病原为链格孢［*Alternaria alternata*（Fr.）Keissl］。

（2）该病菌的最适培养基为大米马铃薯煎汁培养基，最适培养温度为 27℃，最适 pH 值为 6.5，最适碳源为果糖和蔗糖，最适氮源为胰蛋白胨和牛肉膏，最适光照条件为 24h 黑暗。该病菌分生孢子萌发的最适温度为 30℃，最适 pH 值为 7，最适光照条件为 24h 黑暗培养。

（3）通过 ITS 序列分析，设计了一对检测水稻褐变穗病菌的特异性引物 HBS-S/HBS-A。利用该特异性引物对所有供试菌株进行 PCR 扩增，只有水稻褐变穗病菌扩增出了一条 170bp 的条带，其他供试菌株均未扩增出目的条带，证明了本研究所设计的引物对水稻褐变穗病菌（*A. alternata*）具有特异性。该对引物在水稻褐变穗病菌 DNA 水平上，PCR 反应的最小检测浓度为 1.0ng/μL。在病原菌接种水稻穗粒 48h 时，便可检测到病原菌。

关键词：水稻褐变穗病；链格孢；分子检测

* 基金项目：哈尔滨市应用技术研究与开发项目（2014AB6BN036）；黑龙江省粮食作物新品种选育及持续增产栽培技术创新平台项目（2011 计划项目）

** 通讯作者：张俊华

黄淮麦区小麦根腐病菌的遗传多样性及致病力分析*

胡艳峰，王利民，陈琳琳，袁虹霞，丁胜利，李洪连**
（河南农业大学植物保护学院，郑州 450002）

黄海麦区是我国的主要小麦生产区，根腐病是小麦生产上一种重要的土传病害。近年来，随着耕作制度的改变，小麦根腐病在该地区的发生日益严重，对小麦产量造成巨大损失。小麦根腐病的主要病原菌为麦根腐离蠕孢菌（*Bipolaris sorokiniana*），国内对该菌的遗传多样性及致病力分化还缺乏系统研究。本实验于2014—2015年在小麦灌浆期，在黄淮麦区河南、河北、山东、山西、安徽及江苏等地进行广泛采样调查，并对病菌的遗传多样性及致病力分化进行了初步研究。

通过多态性引物（URP）对两年分离得到的麦根腐离蠕孢菌菌株进行PCR扩增，在优化反应条件的基础上筛选出5条引物用于分子标记，统计扩增结果，构建0、1矩阵。利用popgene 1.32软件分析矩阵，结果显示2014年供试菌株各地理群内基因多样度Hs平均值为0.1002，种群基因分化系数Gst平均值为0.0867，基因流Nm为5.2676；2015年供试菌株各地理群内基因多样度Hs平均值为0.1131，种群基因分化系数Gst平均值为0.1601，基因流Nm为2.6235。结果表明，黄淮麦区不同地理群体的供试菌株存在一定的基因多样性，地理群间不仅存在着一定的遗传分化，还存在着较大的基因流动。通过比较供试菌株的地理来源和遗传相似性之间的关系，两年的结果都表明：发现遗传相似性和地理来源之间存在着负相关性，即地理位置越远，遗传相似性越低。

利用小麦品种矮抗58，采用室内接种鉴定方法分别对2014年采集的74个菌株和2015年来自不同群体的46个代表性菌株进行致病力测定。试验结果表明：2014年和2015年供试菌株的致病力都存在显著性的差异（$P<0.001$），致病力分化现象十分明显。对不同地理来源的病原菌群体进行分析表明，2014年的河北群体、豫南群体、豫西群体、豫北群体和豫中东群体5个地理群体致病力十分接近，平均病情指数分别为62.16、59.85、58.37、57.61和57.26，无显著性差异（$P>0.05$）；2015年河南群体、河北群体、山西群体、山东群体和安徽群体5个地理群体的致病力也基本一致，平均病情指数分别为42.10、44.47、41.83、40.98和44.51，不存在显著性差异（$P>0.05$）。但江苏群体平均病情指数达到61.43，与其他各省致病力差异显著。通过比较菌株的致病力和地理来源之间的关系，两年的结果均表明致病力的分化和地理位置之间不存在着显著的相关性。

* 基金项目：本研究受到国家公益性行业科研专项（201503112）资助

** 通讯作者：李洪连，教授，主要从事植物土传病害研究；E-mail：honglianli@sina.com

柑橘褐斑病菌信号转导基因研究进展*

张　倩**，王洪秀，唐科志***

（西南大学柑桔研究所，中国农业科学院柑桔研究所，重庆　400712）

摘　要：由链格孢菌（*Alternaria alternata*）引起的柑橘褐斑病（*Alternaria* brown spot，ABS）是一种严重的真菌病害，近年来在我国柑橘主产区迅速蔓延，造成巨大的经济损失。目前，国内外已经相继开展病原菌分离鉴定、生物学特性、侵染机制、品种抗性及病害流行发生规律等试验且取得一定成果。研究最热的当属病原菌致病基因的挖掘，其中毒素基因和信号转导基因研究最为集中。相对于前者，信号转导基因的研究仅围绕单个基因生物学功能的鉴定，对其上游及下游靶标基因的确定仍有大量工作待完成。本文综述了柑橘褐斑病菌主要信号转导基因的研究进展，包括 MAPK 基因、TCST 基因、G 蛋白基因、cAMP-PKA 基因和 Ga^{2+} 信号途径基因。分析了部分通路间的交叉调控，探讨了未来的研究热点，旨在为进一步研究柑橘褐斑病菌分子致病机制提供参考。

关键词：柑橘褐斑病菌；信号转导；基因；调控

* 基金项目：教育部创新团队（IRT0976）；重庆市科技攻关项目（CSTC2012GG－YYJS0475）

** 第一作者：张倩，在读硕士，研究方向：分子植物病理学；E-mail：1335032933@qq.com

*** 通讯作者：唐科志，副研究员，主要从事柑橘病害研究；E-mail：tangkez@163.com

基于 RNA-Seq 的水稻纹枯病菌菌核发育过程的转录组分析*

舒灿伟**，王陈骄子，江绍锋，周而勋***
（华南农业大学农学院植物病理学系，广州 510642）

摘　要：水稻纹枯病是水稻生产上的重要真菌病害，菌核在该病的病害循环中起了重要作用，但目前国内外对该病菌菌核发育的分子机制研究甚少。本研究分别以水稻纹枯病菌（*Rhizoctonia solani* AG 1-IA）菌核发育过程中 3 个连续生长阶段（MS = 36h，SI = 72h 和 SM = 14d）的总 RNA 为研究材料，以 Illumina Hi-seq 2000 平台技术进行转录组测序，并进行生物信息学分析。结果表明：过滤低质量 reads 后，MS、SI 和 SM 阶段分别保留了 55 051 284、52 983 023和 51 391 446对高质量 reads 用于 DEGs 筛选，其中 77. 22%、72. 58% 和 74. 49% 的 reads 能准确比对到参考序列上，说明 RNA-Seq 结果和参考序列可靠。当从 MS 阶段发育到 SI 阶段时，有 1 675个基因显著上调，而显著下调的基因则有 1 491个；当 MS 阶段和 SM 阶段相比较时，显著上调的基因达 1 231个，而显著下调的基因有 1 081个；而从 SI 阶段发育到 SM 阶段时，上调的基因有 744 个，而下调的基因则为 742 个。GO 分析表明，DEGs 涉及 36 个细胞组份类群、12 个分子功能类群和 18 个生物过程类群。KEGG 注释表明，有 4 907个 Unigene 富集在 47 条信号通路上，有大量基因涉及过氧化酶体途径和酪氨酸酶代谢途径。RT-qPCR 分析结果表明，10 个 DEGs 的表达模式与 RNA-Seq 分析结果一致。本研究共鉴定出4 353个参与水稻纹枯病菌菌核发育的基因，明确了 DEGs 富集的分子功能与代谢途径，为水稻纹枯病菌菌核发育机制的进一步深入研究奠定了基础。

关键词：水稻纹枯病菌；菌核发育；转录组；荧光定量 PCR

---

* 基金项目：国家自然科学基金项目（31271994）

** 第一作者：舒灿伟，博士，讲师，主要从事植物病原真菌及真菌病害研究；E-mail：shucanwei@ scau. edu. cn
*** 通讯作者：周而勋，教授/博导，主要从事植物病原真菌及真菌病害研究；E-mail：exzhou@ scau. edu. cn

假禾谷镰刀菌 Endo G 同源核酸酶的鉴定与功能分析*

陈琳琳，侯　莹，李洪连**
（河南农业大学植物保护学院，郑州　450002）

假禾谷镰刀菌（*Fusarium pseudograminearum*）引起的小麦茎基腐病是一种严重威胁小麦生产的世界性病害，先后已有 10 多个国家报道该病害的发生和危害。2012 年，Li 等在我国河南省小麦产区首次报道假禾谷镰刀菌的为害，罹病植物茎基部腐烂、呈蜜褐色病斑，染病地块的田间发病率达 70% 以上。由于常年实施秸秆还田，造成菌源积累，品种抗性差等原因，小麦茎基腐病的发生呈现不断加重和蔓延趋势。前期我们对假禾谷镰刀菌的鉴定、遗传多样性以及致病性等方面已进行大量的研究报道，但对其致病机制了解的还非常少。

核酸酶 Endo G 是物种中非常保守的核酸酶，在人、布氏锥虫和酿酒酵母等物种中参与调控细胞凋亡的发生。本研究通过 GenBank 获得编码 Endo G 蛋白的已知的基因序列，选择的物种包括人（*Homo sapiens*）、秀丽隐杆线虫（*Caenorhabditis elegans*）、酿酒酵母（*Saccharomyces cerevisiae*）和稻瘟菌（*Magnaporthe grisea*）。通过 BLASTn 方法，在假禾谷镰刀菌基因组中鉴定到 1 个保守的编码 Endo G 的同源基因，命名为 *FpNUC*1。通过 PCR 扩增获得 *FpNUC*1 的基因序列和编码序列，利用 SMART 数据库预测其编码的蛋白含有保守的 Endo G 蛋白结构域。对不同物种的 Endo G 蛋白构建进化树，结果显示假禾谷镰刀菌的 Endo G 蛋白与稻瘟菌亲缘关系最近，其次是酿酒酵母，与人和秀丽隐杆线虫亲缘关系较远，其进化关系符合生物体总体进化过程，说明 Endo G 在物种中出现的比较早。利用 qRT-PCR 和转录组测序方法检测 *FpNUC*1 在假禾谷镰刀菌侵染阶段的表达，发现 *FpNUC*1 在侵染初期阶段上调表达，且在与抗性小麦周麦 24 的互作中，诱导表达程度高于病原菌与感病小麦国麦 301 的亲和互作，表明 *FpNUC*1 可能参与假禾谷镰刀菌的致病过程。

利用 PEG 介导的原生质体转化方法在假禾谷镰刀菌中敲除 *FpNUC*1 基因，基因敲除转化子菌丝生长减慢，而致病性却有明显增强。已有研究报道植物与病原菌互作过程中，双方都通过复杂的网络试图去操纵对方的细胞死亡以最大限度的提高自身的生存概率。植物可以通过局部细胞死亡来抵抗病原菌的侵入，而最近的研究发现多种病原真菌也在经历着细胞凋亡的发生，且在病原菌生长、产孢和侵染过程中起至关重要的作用。所以，在假禾谷镰刀菌与植物互作过程中，病原菌可能也经历着植物诱导的其细胞凋亡的发生，病原菌需要抑制自身细胞凋亡的发生，才能保证更多的病原菌得以存活，以实现有效的侵染和定殖。

* 基金项目：国家自然科学基因项目（30601221）和国家公益性行业科研专项（201503112）
** 通讯作者：李洪连，教授，主要从事植物土传病害研究工作；E-mail：honglianli @ sina. com

假禾谷镰孢侵染小麦幼苗过程中的基因差异表达分析和部分基因功能验证*

丁胜利，胡艳峰，王利民，李永辉，袁虹霞，邢小萍，陈琳琳，李洪连**
（河南农业大学植物保护学院，郑州 450002）

2012 年笔者所在课题组首次报道发现假禾谷镰孢（*Fusarium pseudograminearum*）在我国的为害。近年来的小麦茎基腐病调查和病原鉴定结果显示，在河南、河北、山西和山东都发现有假禾谷镰孢，且为优势菌群，该菌造成的为害呈现不断加重趋势。假禾谷镰孢起初为禾谷镰孢（*Fusarium graminearum* Group 2）内的一个菌群（*Fusarium graminearum* Group 1），因为其有性生殖为异宗配合，从禾谷镰孢（*F. graminearum*，有性态 *Gibberella zeae*）中分出来并定名为假禾谷镰孢（*F. pseudograminearum*，有性态 *Gibberella coronicola*）。关于假禾谷镰孢的侵染机制，除了类似于禾谷镰孢的 DON 毒素作为致病因子参与早期侵染过程以外，目前了解甚少。2012 年假禾谷镰孢全基因组序列公布后，为研究单倍体假禾谷镰孢侵染发病过程的分子机理提供了有利条件。

本课题组用假禾谷镰孢强致病菌株 WZ2-8AW，通过接种从小麦高感品种国麦 301 和中抗品种周麦 24 的根部开始侵染，收集发病的小麦和茎基部组织，采用 RNAseq 技术，在小麦发病组织样中，获得百万数量级假禾谷镰刀菌的 reads，然后进行假禾谷镰孢接菌后 5d 和 15d 差异表达基因的 GO 和 KEGG 分析，尽管大部分差异表达基因编码功能未知蛋白，也发现在其他菌中已知功能的基因，部分初步结果如下：

（1）Trichothecene（单端孢酶烯族毒素）合成途径基因的总体上高度诱导表达，在抗感品种间没有明显差别，其中包括 Tri1，Tri3，Tri4，Tri5，Tri16，Tri8，Tri10，Tri11，Tri12，Tri14 和 Tri15。可能同禾谷镰孢相似，Tri 毒素的合成参与早期的侵染定殖过程。

（2）与发育和致病相关的次生代谢相关基因的差异表达：如多聚酮合成酶基因（*PKS*）和非核糖体合成酶途径基因（*NRPS*），*PKS* 包括 *PKS*2，*PKS*3，*PKS*4，*PKS*5，*PKS*6，*PKS*7，*PKS*8，*PKS*9，*PKS*10，*PKS*11，*PKS*12，*PKS*13，*PKS*14 和 *PKS*15，其中，除了 *PKS*4 和 *PKS*9 在两个品种中整体下调表达，其他主要是上调诱导表达。*NRPS* 包括 *NRPS*1，*NRPS*2，*NRPS*3，*NRPS*4，*NRPS*5，*NRPS*6，*NRPS*9，*NRPS*10，*NRPS*11，*NRPS*12，*NRPS*13，*NRPS*14，*NRPS*15，*NRPS*16 和 *NRPS*19，除了 *NRPS*4，*NRPS*10 和 *NRPS*13 在两个品种中整体下调，其余都是诱导上调表达。选择性挑选了几个基因进行验证非核糖体合成酶在假禾谷镰孢侵染定殖方面的功能验证正在进行中。

（3）MAPK 途径基因的诱导上调表达：其中包括 *STE*11，*STE*7，*MPS*1，*RLM*1，*FST*12 和 *HOG*1 同源的基因，尽管没有测到 *PMK*1 同源基因表达，但 PMK1 敲除突变体丧失了致病性。说明 MAPK 途径在丝状真菌中的功能很保守，也参与调控假禾谷镰孢的致病过程。其他几个基因的功能验证也在进行当中。

（4）分泌蛋白：除了一些细胞壁降解酶诱导高表达，还有许多具有编码 N 端信号肽的经典

* 基金项目：国家公益性行业（农业）科研专项（201503112）；河南省科技计划项目（152300410073）；河南农业大学人才引进项目启动基金（30600861）

** 通讯作者：李洪连，教授；E-mail：honglianli@ sina. com

分泌蛋白高诱导表达，在早期侵染诱导表达量最高的也是具有信号肽的分泌蛋白。选择性对个别细胞壁降解酶和小分泌蛋白进行功能验证。

假禾谷镰孢侵染小麦的转录组测序数据中绝大多数来自寄主小麦，关于小麦诱导抗性表达的内容也在分析当中，其中，防卫反应相关的 JA，SA 和 ET 几个主要途径的基因和一些特异差异表达的定量 PCR 正在验证当中。

苜蓿（*Medicago sativa* L.）叶片应答假盘菌（*P. medicaginis*）早期侵染的蛋白质组学分析*

李　杨**，史　娟***，张会会
（宁夏大学农学院，银川　750021）

苜蓿（*Medicago sativa* L.）作为优质牧草，在世界各国的畜牧业发展中占据着非常重要的地位。近年来，随着种植面积的增加，苜蓿病害日趋严重。由苜蓿假盘菌（*Pseudopeziza medicaginis*（Lib.）Sacc.）引起的褐斑病是苜蓿最常见和破坏性很大的病害之一，感病后会引起苜蓿减产、品质下降，诱导生成香豆雌酚等类黄酮物质，致使家畜采食后流产、不育。为了解析苜蓿假盘菌的侵染机制，寻找出致病关键蛋白，本试验采用2-D双向电泳、质谱鉴定技术，通过人工接种对*P. medicaginis*侵染苜蓿叶片12h、24h、36h三个不同侵染时期的差异蛋白进行蛋白质组学分析。结果表明，*P. medicaginis*早期侵染叶片后与健康叶片相比共有57个差异蛋白质点，33个上调点，24个下调点。其中，侵染12h时，有29个差异蛋白质点，13个上调点，16个下调点；侵染24h时，有29个差异蛋白质点，16个上调点，9个下调点；侵染36h时，有29个差异蛋白质点，有17个上调点，12个下调点。通过质谱分析及NCBI数据库检索有21个蛋白质点属于皮盘菌科，主要为杨树黑斑病菌的Snf7家族蛋白、核菌病菌蛋白、棘白菌蛋白；有28个蛋白质点在苜蓿中得以鉴定，主要涉及1，5-二磷酸核酮糖羧化酶、叶绿体未知蛋白产物、ATP合成酶β亚基、苹果酸脱氢酶前体。根据上述蛋白质组学分析，在病菌侵染过程中，苜蓿的光合作用及氧化还原系统受到不同程度的影响。这些不同的差异蛋白是否与*P. medicaginis*的成功侵染以及功能性有待于进一步分析和验证。

* 基金项目：苜蓿褐斑病菌［*Pseudopeziza medicaginis*（Lib.）Sacc］侵染过程的蛋白质组学研究及其功能分析（31460033）
** 第一作者：李杨，山西大同人，在读硕士研究生，专业方向：草原保护；E-mail：568867940@qq.com
*** 通讯作者：史娟，教授，主要从事草原保护研究；E-mail：shijuan0@163.com

2015年河北省部分地区苜蓿根腐病病情调查和病原鉴定*

孔前前**，秦　丰，马占鸿，王海光***
（中国农业大学植物病理学系，北京　100193）

摘　要：为了解河北省苜蓿根腐病的发病情况和病原种类，在2015年对河北省黄骅市和宣化县部分苜蓿种植地进行了苜蓿根腐病发病情况调查。调查结果表明，在大多数苜蓿种植地块该病害呈零星发生状态，个别地块病株率为2%～3%。对调查中采集的苜蓿根腐病植株样品，采用常规组织分离方法进行了微生物分离、纯化和培养，共获得105个菌株。采用CTAB法提取菌株DNA，选取引物ITS4/ITS5和延伸因子EF-1H/EF-2T进行PCR扩增，并对扩增产物进行测序，经BLAST序列比对，结果表明，所分离微生物共有17种，除拟茎点霉（*Phomopsis* spp.）、间座壳菌（*Diaporthe* spp.）、木霉菌（*Trichoderma* spp.）、淡色丛赤壳（*Bionectria ochroleuca*）等外，还有65株为镰孢菌（*Fusarium* spp.），包括22株木贼镰孢（*F. equiseti*）、21株茄镰孢（*F. solani*）、16株尖孢镰孢（*F. oxysporum*）、3株锐顶镰孢（*F. acuminatum*）、1株层出镰孢（*F. proliferatum*）、1株芳香镰孢（*F. redolens*）和1株*F. commune*。采用培养皿法对获得的镰孢菌各个菌株进行了致病性测定，结果表明，多数镰孢菌对萌发的苜蓿种子具有致病性，但致病性强弱有差异。除部分木贼镰孢菌株致病性较弱外，大多数镰孢菌菌株对苜蓿幼苗表现出强致病性，其中，木贼镰孢菌株QZ3、锐顶镰孢菌株N12-1、尖孢镰孢菌株QD13-2的致病性最强，其病情指数分别为90.83、82.92、77.13。本研究利用分子生物学技术进行了河北省部分地区苜蓿根腐病病原的鉴定，并对所分离获得的镰孢菌致病性强弱进行了测定，这为了解苜蓿根腐病病原和该病害的及时防治提供了一定基础。

关键词：苜蓿根腐病；病原分离；病原鉴定；ITS；延伸因子；致病性测定

---

* 基金项目：公益性行业（农业）科研专项经费项目（201303057）
** 第一作者：孔前前，硕士研究生；E-mail：1439918864@qq.com
*** 通讯作者：王海光，副教授，主要从事植物病害流行学和宏观植物病理学研究；E-mail：wanghaiguang@cau.edu.cn

玉米大斑病菌 *StCAK3* 基因敲除载体的构建*

李　盼，赵玉兰，于　波，郝志敏**，董金皋**
（河北农业大学真菌毒素与植物分子病理学实验室，保定　071001）

摘　要：钙/钙调素依赖性蛋白激酶（calcium/calmodulin dependent protein kinases，CaMKs）是 Ca^{2+} 途径中主要的组成成分之一。本实验室前期已经通过克隆方法，获得 CaMKs 的基因，并初步利用简并引物从玉米大斑病菌中克隆到 4 个 CaMKs 基因片段，分别命名为 *CAK*1、*CAK*2、*CAK*3 和 *CAK*4 并对其进行了基因家族的鉴定和表达模式的分析。研究发现 *CAK*3 在功能表达方面主要是起上调作用的。为进一步明确 CaMKs 的分子机制，本试验拟通过创制 *CAK*3 突变体，研究 *CAK*3 基因的功能。

本试验利用 JGI（www. jgi. doe. gov）公布的玉米大斑病菌的基因组序列，查到了 *CAK*3 的基因序列，该基因 DNA 全长 3 125bp。本试验设计了带有 *Xho* Ⅰ、*EcoR* Ⅰ和 *BamH* Ⅰ、*Spe* Ⅰ酶切位点的引物，以玉米大斑病菌基因组 DNA 为模板，分别扩增得到 *CAK*3 基因的同源臂片段Ⅰ（929bp）和同源臂片段Ⅱ（734bp），以本实验室带有草氨磷片段的质粒扩增得到草氨磷片段，将其纯化后分别连接于 pMD19-T Simple Vector 上，送样测序。将测序正确的菌液提取质粒。用相应的酶将 3 个目的片段双酶切并纯化回收。双酶切 pBS 载体后，将 3 个纯化的目的片段按照相应的酶切位点依次连接到 pBS 载体上，并转化到大肠杆菌 DH5α，获得阳性转化子后再进行双酶切验证，产生的目的条带与预期结果一致，说明成功构建了 *CAK*3 基因敲除载体。下一步拟将该重组载体利用 PEG 介导的转化方法，转化玉米大斑病菌原生质体，利用草氨磷抗性筛选基因敲除突变体，通过对突变体与野生型菌株致病性比较，明确 *CAK*3 基因的功能。

---

* 基金项目：国家自然科学基金项目（No. 31301616）；河北省高等学校青年拔尖人才计划项目（No. BJ2014349Y）和河北省自然科学基金项目（No. C2016204160）

** 通讯作者：郝志敏；E-mail：hzm_ 0322@ 163. com
董金皋；E-mail：dongjingao@ 126. com

重要植物病原真菌：葡萄座腔类真菌的研究进展及问题展望[*]

李文英[**]，李 夏，解开治，孙丽丽
（广东省农业科学院农业资源与环境研究所，广州 510640）

摘 要：葡萄座腔菌目（Botryosphaeriales）隶属子囊菌门（Ascomycota），子囊菌亚门（Pezizomycotina），座囊菌纲（Dothideomycetes），世界性分布，包含一类形态各异的真菌，生于单子叶、双子叶植物或裸子植物等各种各样的植物寄主，主要为木本寄主，也有生于草本植物的叶部、茎部或秆部，在树枝和地衣体和小枝上，营腐生至寄生或内生生活，在长势衰弱、濒临死亡的木本组织上较为常见，大多数种类可以在人工培养基上生长。根据最新的《国际藻类，真菌及植物命名法则》和阿姆斯特丹宣言的要求及提议，采用"一种真菌，一个名称"的最新分类命名方法，目前世界上已知葡萄座腔菌目真菌 6 科 39 属 2 000余种，我国对上述类群的研究始于 20 世纪 30 年代，至今已报道该目 2 科分别为：葡萄座腔菌科 15 属 31 种（Li *et al.*，2016），叶点霉科 3 属 22 种。

葡萄座腔菌目的许多成员属于弱寄生菌，条件合适时会引起重要植物病害，属重要的潜在植物病原真菌，对农林生产具有重要的经济影响。特别是葡萄座腔菌（*Botryosphaeria* Ces. & De Not.），新壳梭孢属（*Neofusicoccum* Crous），毛色二孢菌（*Lasiodiplodia* Ellis & Everh.），小穴壳孢属（*Dothiorella* Sacc.）等重要属的优势类群会引起流胶、茎枯、梢枯、枝枯、溃疡、蒂腐、根腐、炭腐等多种林木、果树及农作物病害，或腐生于重要作物果实，引起果腐、环腐、蕉腐、轮纹类等果实病害或采后货架期病害。

近年来真菌分子系统学的研究进展与菌物命名法规的变革引起真菌分类系统的巨大变化，Wijayawardene（2014）等对葡萄座腔菌类真菌提出较为合理的分类系统，但目前该类重要植物病原真菌研究还存在许多悬而未决的问题，如分类地位和命名的清理与订正，生活史中全型特征关联的证据，病原菌与寄主或基物的辩证关系等。因此，深入开展该类真菌属种多样性研究是认识该类重要病害的基础，可为进一步深入研究其致病机制、病害侵染循环，提出有效防治策略提供基础信息和科学依据。

关键词：植物病原真菌；葡萄座腔类真菌；葡萄座腔菌目（Botryosphaeriales）；葡萄座腔菌（*Botryosphaeria*）

---

* 基金项目：科技部基础性工作专项（2013FY110400）；真菌学国家重点实验室开放课题（SKLMKF201404）

** 第一作者：李文英，博士，副研究员，主要从事微生物多样性资源及环境修复利用研究；E-mail：liwenying2006@126. com

铁离子对玉米弯孢叶斑病菌生长发育及致病性的影响*

路媛媛**，高艺搏，肖淑芹，赵丽琨，薛春生***
（沈阳农业大学植物保护学院，沈阳 110866）

玉米弯孢叶斑病是重要的叶部病害，曾多次对我国的玉米安全造成严重威胁，其致病菌——新月弯孢菌［*Curvularia lunata*（Wakker）Boed］致病性变化是影响其流行的主要因素之一。目前，对其致病性影响因素的研究主要集中在细胞壁降解酶、毒素、黑色素、致病性分化及信号调控等方面，但是对于营养元素中铁与其致病力的关系未见研究和报道。

C. lunata CX-3 菌株在含有不同铁离子浓度（0～80μm）的 MM 培养基上生长速度差异不明显，日平均速度在 0.85～0.90cm，但菌丝干重存在显著差异，无铁离子时 CX-3 的菌丝干重为 0.192g，随着浓度的增加，干重增加，在铁离子为 40μm 时达到最大值为 0.759g，随后浓度再增加，菌丝干重下降；观察菌落形态发现无铁离子时菌落颜色浅，且菌丝稀疏，而有铁离子存在时菌落颜色深，菌落致密并呈轮纹状；液体培养时上清液颜色也存在差异，在铁离子浓度为 40～60μm 时液体培养基的颜色较深；CX-3 在不同浓度铁离子培养基培养后产孢量差异显著，随着铁离子浓度的增加，产孢量增加，浓度为 40μm 时，产孢量最大，浓度过高，孢子量下降。孢子萌发、附着胞形成数量与孢子产生的结果相同。致病力检测结果表明，接种 5d 后，铁离子浓度为 30～60μm 时培养的 CX-3 先发病，但 7d 后在各处理获得的病原菌均在自交系叶片上产生明显的病斑。

* 基金项目：国家自然科学基金（31271992）
** 第一作者：路媛媛，在读博士研究生，从事玉米病害研究
*** 通讯作者：薛春生；E-mail：cshxue@ sina. com

麦根腐平脐蠕孢（*Bipolaris sorokiniana*）分泌蛋白基因 *CSSP*1-*CSSP*4 的功能研究*

王利民，张一凡，胡艳峰，丁胜利**，李洪连
（河南农业大学植物保护学院，郑州　450002）

麦根腐平脐蠕孢（*Bipolaris sorokiniana*）是禾旋孢腔菌（*Cochliobolus sativus*）的无性态，是一种世界性的植物病原真菌，能侵染小麦地下和地上部分，引起小麦根腐病、叶斑病和黑胚病等病害，严重为害小麦生产。近年来，由于秸秆还田等因素的影响，该菌引起的病害在黄淮麦区有逐渐加重的趋势。研究病菌和寄主的互作，对于弄清病菌的致病机理和寄主的抗病机理，以及病害的监测和防控工作具有重要意义。在麦根腐平脐蠕孢与寄主小麦的互作过程中，数量众多的分泌蛋白（289 个具有信号肽分泌蛋白，比其他同属真菌多 100 多个）在病原菌—寄主互作系统中可能扮演着十分重要的角色，通过对植物病原分泌蛋白功能及其作用机制的研究，可以极大地促进人们对植物抗病分子机制的了解。分泌蛋白在玉米黑粉病菌（*Ustilago maydis*）、稻瘟菌（*Magnaporthe oryzae*）、大丽轮枝菌（*Verticillium dahliae*）和核盘菌（*Sclerotinia sclerotiorum*）等植物病原真菌上已经有比较深入的研究，麦根腐平脐蠕孢分泌蛋白却鲜有研究报道。

本课题组在麦根腐平脐蠕孢侵染小麦过程的转录组数据中，筛选了 4 个表达模式相似的高表达基因，其编码的蛋白在 N 端都具有经典信号肽序列，且定位在胞外，初步推定为分泌蛋白。前期选择 Gm1. 8922_G 编号 *CSSP*1 通过生物信息学分析含有信号肽 17 氨基酸，基因大小 270bp，预测是胞外分泌蛋白。采用 split-marker 基因敲除方法，通过 PEG 介导原生质体转化，经过 4 对引物 PCR 检测，成功获得 Δ*cssp*1 突变体。转化体性状和野生型相比有较大变化，致病性明显降低，基本不产孢，生长速率变慢。表明 *CSSP*1 参与麦根腐平脐蠕孢的生长发育和致病过程。依据相似表达模式的基因可能具有类似功能的原则，4 个分泌蛋白基因可能作为致病因子在麦根腐平脐蠕孢菌侵染致病过程中单独或协同发挥作用，或者其中有的作为效应因子与寄主互作，调控侵染发病过程，选择 *CSSP*2、*CSSP*3、*CSSP*4 共同作为研究对象。采用同样方法，我们获得 Δ*cssp*3 突变体，*CSSP*2、*CSSP*4 尚未获得阳性转化子，需要再次转化。相关的敲除突变体的生物学性状测、功能互补实验、侵染过程显微观察和编码蛋白的亚细胞定位等实验正在进行中。

---

* 基金项目：国家公益性行业（农业）科研专项（201503112）；河南农业大学人才引进项目（30600861）和河南省科技计划项目（152300410073）

** 通讯作者：丁胜利，河南农业大学特聘教授；E-mail：shengliding@ henau. edu. cn

油菜菌核病菌菌核的菌丝型萌发研究*

母红岩**，郑　露，刘　浩，黄俊斌***
（华中农业大学植物科学技术学院，武汉　430070）

摘　要：核盘菌菌核的菌丝型萌发可引起向日葵根和茎基部发病。为了更好的了解核盘菌复杂的侵染机制，我们对不同地理来源菌株的菌核菌丝型萌发特性进行研究。通过对8个菌株的产菌核能力进行测定，发现这些菌株产菌核的数量以及菌核的平均质量差异显著。菌核在水中吸水较快，2h左右吸水速率放缓，6h达到饱和状态。在基质势为－8kPa、－15kPa和－32kPa的土壤中，菌核到饱和所需时间相同。菌核在土壤中吸水较慢，60h时才达到饱和状态。我们发现无外源营养存在的条件下，正常菌核充分吸水后在第6d时可以观察到菌丝型萌发的开始，仅需4～5d几乎所有菌株的萌发率达到100%。菌核在10～30℃都可以进行菌丝型萌发，25℃时萌发速率最快。酸性的环境更有利于菌核的菌丝型萌发和菌丝的生长。

关键词：核盘菌；菌核；菌丝型萌发

* 基金项目：油菜创新技术体系岗位科学家专项（nycytx-00514）
** 第一作者：母红岩，在读硕士研究生，主要从事植物真菌病害研究
*** 通讯作者：黄俊斌，教授；E-mail：junbinhuang@ mail. hzau. edu. cn

苹果炭疽叶枯病菌致病相关基因 *eg_* 3859.15 的克隆及功能分析*

吴建圆**，周宗山，冀志蕊，张俊祥***
（中国农业科学院果树研究所，兴城　125100）

摘　要：苹果是我国第一大水果，2012 年栽培面积达到 220 万 hm^2，苹果产业已成为我国苹果主产区农民增收的支柱产业。近几年，苹果一种重要的新病害——苹果炭疽叶枯病，在我国山东、河南、江苏、安徽、河北、陕西、辽宁和山西等苹果主产区连年大发生，给苹果产业带来巨额损失。

苹果炭疽叶枯病主要由胶胞炭疽菌（*Colletotrichum gloeosporioides*）［有性世代：围小丛壳菌（*Glomerella cingulata*）］引起。为了解析苹果炭疽叶枯病菌致病的分子机制，笔者所在课题组利用农杆菌介导（ATMT）方法，构建了苹果炭疽叶枯病菌强致病力菌株 W16 的 T-DNA 插入突变体库。以感病品种嘎拉为寄主，采用无伤接种方法，从 T-DNA 插入体库中筛选到一些致病性变异的突变体，其中突变体 M1068 丧失了致病能力。Southern 杂交结果显示，该突变体的 T-DNA 插入为单拷贝插入，*hi*TAIL-PCR 和 Bridge-PCR 结果显示，T-DNA 的插入位点位于一个预测的基因（*eg_* 3859.15）的第一个外显子上。通过对目的基因的上下游核苷酸序列设计引物，构建了潮霉素抗性基因靶标的敲除载体，然后通过 ATMT 技术进行同源重组，敲除 *eg_* 3859.15 基因的敲除突变株。主要结果如下。

（1）*eg_* 3859.15 基因 DNA 全长 2 568bp，包含 5 个内含子，编码 652 个氨基酸。

（2）表型分析结果表明：*eg_* 3859.15 敲除突变体在 PDA 平板上呈现菌丝明显白化，生长速度约为野生型菌株的 2/3。*eg_* 3859.15 敲除突变体不影响分生孢子的形成和萌发。致病性测验结果表明，*eg_* 3859.15 敲除突变体对苹果叶片及果实均无致病性。侵染过程观察结果表明，*eg_* 3859.15 敲除突变体不能形成附着胞，暗示其致病性的缺失与附着胞的形成相关。

（3）*eg_* 3859.15-*egfp* 融合表达结构（construct）导入 *eg_* 3859.15 敲除突变体基因组中，获得的转化子恢复了 *eg_* 3859.15 敲除突变体的表型。GFP 信号观测结果表明，*eg_* 3859.15 主要在附着胞形成时表达（该结果仅由 GFP 信号强度判断）。初步证实基因 *eg_* 3859.15 决定该病菌的附着胞的形成及致病性。

关键词：基因；炭疽；苹果；致病；突变体

---

* 基金项目：中国农业科学院科技创新工程

** 第一作者：吴建圆，硕士研究生，主要从事分子植物病理学研究；E-mail：jianyuanwu920115@163.com

*** 通讯作者：张俊祥，E-mail：zhangjunxiang@caas.cn

上海地区番茄灰叶斑病病原鉴定*

曾 蓉，徐丽慧，高士刚，戴富明**
（上海市农业科学院生态环境保护研究所，上海市园艺设施重点实验室，上海 201403）

摘 要：番茄灰叶斑病是一种世界性的病害。近年来，我国的北京、辽宁、河北、山东等地均有发生报道。2015 年 7 月初，在上海市浦东新区曹路镇某育种基地的番茄发现了一些叶部病害，有的品种（品系）的叶片上出现近圆形至不规则病斑，病斑四周深褐色，中间为灰白色，病斑后期可愈合成大病斑，有的品种（品系）上表现为深褐色不规则或椭圆形坏死斑，边缘有黄色晕圈，有的表现为深褐色病斑，病斑相对较大，同时伴有同心轮纹圈，最后一类的症状与番茄早疫病田间症状非常相似，病斑后期易破裂，田间叶片发病率严重的可达 20% ~30%，严重影响番茄产量和品质。3 种不同症状的叶片病样，通过组织分离和纯化，各分离保存了 1 个分离物，3 个分离物在 PDA 的菌落相似，初期表现为白色或灰色，逐渐变深，后期褐色，同时培养基上能产生橘色色素，这与番茄早疫病菌的菌落特征差别较大。致病性试验表明这 3 个分离物均能导致番茄叶片发病，且能获得与原来接种相同的分离物，确定这些分离物为致病菌。3 个病原菌的分离物在 PDA 培养基上的最适生长温度为 26℃左右，生长速度在 0. 77cm/d 左右，超过 36℃生长缓慢或不生长。菌丝分隔、深褐色。分生孢子顶生、深褐色、粗短，有砖格状分隔，2 ~4 个横隔，多个纵隔，大小为（20 ~40）μm×（5 ~10）μm。分子生物学方法鉴定表明：菌丝用 SDS-CTAB 法提取 DNA，利用通用引物 ITS1/ITS4 扩增 rDNA ITS 序列，将测定序列在 GenBank 上进行核苷酸序列相似性分析（BLASTn），这 3 个分离物序列之间相似性为 100%，与番茄匐柄霉（*Stemphylium lycopersici*）（Accession No. KR911814）的相似性也超过 99%。因此，通过致病性、形态特征、分子生物学等方法鉴定，引起上海市保护地番茄叶片上出现的灰白色、深褐色以及具轮纹圈的深褐色叶斑症状的病原初步诊断为番茄匐柄霉（*Stemphylium lycopersici*），这是番茄灰叶斑病在上海的首次发现。

关键词：番茄；匐柄霉；病原鉴定

---

* 资助项目：上海市科技兴农重点攻关项目［沪农科攻字（2015）第 4 -4 号，沪农科攻字（2012）第 2 -10 号］

** 通讯作者：戴富明，研究员，研究方向：植物病害绿色防控与致病机理研究

Regulation of Methyltransferase LAE1 on Growth and Metabolic Characteristics in *T. atroviride*

LI Ya-qian, SONG Kai, YU Chuan-jin, GAOJin-xin, CHEN Jie*

(*School of Agriculture and Biology, Shanghai Jiao Tong University, Shanghai, P. R. China; Key Laboratory of Urban Agriculture (South), Ministry of Agriculture, P. R. China; State Key Laboratory of Microbial Metabolism, Shanghai Jiao Tong University, Shanghai, P. R. China*)

The biocontrol potential of *Trichoderma* has been the most fascinating and interesting topic of research for agricultural scientists. *Trichoderma* spp. secretes a chemically diverse range of secondary metabolites, of which broad-spectrum antimicrobial properties have been demonstrated well in many in vitro assays. The methyltransferase LAE1 is a global transcriptional regulator that affects the expression of multiple secondary metabolite gene clusters by modifying heterochromatin structure in *Trichoderma*.

This study chose *Trichoderma atroviride* T23 as the initial strain, agrobacterium-mediated transformation (ATMT) technique was used to construct knockout strain T23Δ*lae*1 and over expression strain T23O*lae*11. The growth phenotype (mycelia and spores) was assayed comparing T23 an T23Δ*lae*1 & T23O*lae*11. The results showed that the mutant T23Δ*lae*1 had a slower mycelia growth, spore production delayed, by contrast, over-expression strain of T23O*lae*1 increased the growth and sporulation. 18 non-ribosomal peptide synthase (NPRSs) and 16 polyketide synthase (PKSs) are involved in the biosynthesis of secondary metabolites in *T. atroviride*. QRT- PCR analyzed the regulation of *lae*1 on the expression of NPRSs and PKSs, and showed that *lae*1 promote the expression of most of the NRPS and PKS enzymes and inhibit of a few of NRPS and PKS expressions. Therefore, the transcription factor Lae1 is closely related to the production of secondary metabolites. The GC-MS and LC-MS analysis of metabolites spectrum of three strains (T23, T23Δ*lae*1 and T23O*lae*1) including the volatile and non-volatile metabolites synthesis. The results showed that the quality and quantities of metabolites in T23Δ*lae*1 are significantly reduced, while the kinds of secreted metabolites were obvious increased in T23O*lae*1 and had much higher yields of some metabolism. Overall, Lae1 promote synthesis of amino acids metabolites and induce certain antimicrobial peptaibols compounds, which were not detected in wild strain T23. Therefore, the results verified that Lae1 can activate partial silencing of the NPRs or PKSs gene cluster, and lead to mine some novel metabolites unreported by over-expression *lae*1 gene.

In summary, the putative methyltransferase Lae1 is a global regulator that affects the cell growth and expression of multiple secondary metabolite gene clusters in *T. atroviride* T23. This study will shed light on the development effective biocontrol agents that take secondary metabolites-based formulation in agriculture application.

---

* Corresponding author: CHEN Jie, professor, E-mail: Jiechen59@ sjtu. edu. cn

甘蔗鞭黑粉菌有性配合相关基因 *SsGpa* 的功能分析*

孙　飞**，张小萌，常长青，李敏慧，沈万宽，习平根***，姜子德***
（华南农业大学微生物信号与病害防控重点实验室/群体微生物创新团队，广州　501642）

摘　要：甘蔗鞭黑粉菌（*Sporisorium scitamineum*）引起的黑穗病是甘蔗上一种世界性的病害，严重影响甘蔗的质量和产量。目前，对于甘蔗鞭黑粉菌有性配合的分子机制知之甚少，从基因水平研究该菌的有性配合机理，将可为发展新的防治策略提供理论依据。本实验室前期研究自甘蔗鞭黑粉菌单倍体菌株 WT18 筛选到一株有性配合能力明显减弱的 T-DNA 插入突变体菌株 A9，获得了其插入位点的侧翼序列。为了进一步明确因插入而引起突变的该基因及其在有性配合过程中的功能，本研究预测了该基因结构，并通过基因敲除、功能互补及其表型及致病性等对该基因功能进行了分析。主要结果如下。

（1）分析了突变体 A9 被 T-DNA 插入破坏的基因结构。生物信息学分析表明被 T-DNA 破坏的基因全长 1 365 bp，含有一个 29bp 的内含子，编码 454 个氨基酸；所编码的蛋白含有一个 G-Patch保守结构域，拟命名该基因为 SsGpa；蛋白比对显示该基因与玉米丝轴黑粉菌（*Sporisorium reilianum*）和玉米黑粉菌（*Ustilago maydis*）的 uncharacterized protein 基因分别有 66% 的一致性和 57% 一致性。

（2）分别构建了基于基因 *SsGpa* 的敲除载体和功能互补载体。

（3）基因 *SsGpa* 影响甘蔗鞭黑粉菌有性配合，而不影响其生长。

（4）转录组分析和实时荧光定量检测表明 PCR 显示基因 *SsGpa* 影响甘蔗鞭黑粉菌有性配合过程的信息素调控因子基因的表达，导致有性配合能力的减弱，继而降低了对甘蔗的致病能力。

关键词：甘蔗鞭黑粉菌；有性配合；*SsGpa* 基因；致病性

---

* 基金项目：国家 973 计划项目（2015CB150600）

** 第一作者：孙飞，江西人，硕士研究生
*** 通讯作者：习平根，副教授；E-mail：xpg@ scau. edu. cn
姜子德，教授；E-mail：zdjiang@ scau. edu. cn

农杆菌介导玉米北方炭疽病菌的遗传转化及转化子初步分析*

孙佳莹[1**]，肖淑芹[1]，许瑞迪[1]，薛春生[1***]，陈　捷[2***]
（1. 沈阳农业大学植物保护学院，沈阳　110866；
2. 上海交通大学农业与生物学院，上海　200240）

摘　要：目前关于玉米北方炭疽病的报道多集中于病害发生为害、典型症状特征描述等方面，对于病原菌致病机制的研究还未见报道。利用根癌农杆菌介导转化方法对玉米北方炭疽病菌进行遗传转化，构建大容量的 ATMT 突变体库，筛选致病性相关突变体，为从分子水平上揭示病菌致病机理奠定基础。

以玉米北方炭疽病菌强致病力菌株 SYND-12 作为出发菌株，利用以潮霉素和绿色荧光蛋白基因为标记基因的敲除载体 pPZP100HG，采用 ATMT 技术进行遗传转化。优化了农杆菌浓度、农杆菌与分生孢子的混合比例、共培养时间等遗传转化条件，建立稳定根癌农杆菌介导的玉米北方炭疽病菌遗传转化体系。随机选择 30 个转化子，采用潮霉素特异性引物进行验证，扩增出 1 300bp 条带，证明 T-DNA 已稳定插入基因组。荧光显微镜观察发现到转化子均发出绿色荧光。转化子菌落形态分为以下 5 种类型：Ⅰ型转化子菌落边缘为乳白色，中央为浅绿色或灰黄色，生长速率与野生型相似，占总数的 33%；Ⅱ型转化子菌落为浅粉色，生长速率缓慢，占总数的 4%；Ⅲ型转化子菌落为白色，上面覆盖绒毛状菌丝，占总数的 16%；Ⅳ型转化子菌落与野生型无明显差别，占总数的 38%；Ⅴ型转化子菌落边缘为墨绿色中央浅粉色，生长速率与野生型一致，占总数的 9%。利用本试验建立的遗传转化技术，使在基因组范围内大规模进行玉米北方炭疽病菌功能基因筛选和鉴定成为可能，为玉米北方炭疽病菌的致病分子机制研究奠定基础。

关键词：玉米北方炭疽病菌；ATMT；遗传转化

---

* 基金项目：国家自然科学基金（31271992）；国家现代农业（玉米）产业技术体系（CARS－02）

** 第一作者：孙佳莹，在读硕士研究生，从事玉米病害研究

*** 通讯作者：薛春生；E-mail：cshxue@ sina. com
陈捷；E-mail：jiechen59@ sjtu. edu. cn

牡丹土壤木霉菌鉴定及多态性分析

唐 琳，蔡 冰，吴瑞雪，黄龙梅
（洛阳师范学院生命科学学院，洛阳 471934）

摘 要：木霉菌是作为生物防治、诱导植物抗性和促进植物生长的重要微生物之一，明确不同地区牡丹土壤木霉菌菌株间的系统发育关系和种群分布特点，是进一步开发防控牡丹病害和促进牡丹生长的多功能木霉菌制剂的理论基础。采用土壤稀释分离法和稀释平板法，我们对采集自洛阳、菏泽的 15 个地区的 166 份土壤样品进行分离纯化，共得到 482 株木霉菌菌株。采用形态学鉴定和 EF-1α 序列分析对供试菌株进行鉴定，采用 UP-PCR 方法进行多态性分析。结果表明：供试木霉菌分属于 8 个种，分别为长枝木霉（*Trichoderma longibrachiatum*）、黄绿木霉（*Trichoderma aureoviride*）、拟康氏木霉（*Trichoderma pseudokoningii*）、黏绿木霉（*Trichoderma virens*）、哈茨木霉（*Trichoderma harzianum*）、棘孢木霉（*Trichoderma asperellum*）、非钩木霉（*Trichoderma inhamatum*）和深绿木霉（*Trichoderma atroviride*）。其中，哈茨木霉（*T. harzianum*）在洛阳和菏泽的所有土壤中均有分布，其分离频率均最高，分别为 56.83% 和 42.33%，而非钩木霉（*T. inhamatum*）的分离频率最低，仅在洛阳嵩县、汝阳和菏泽曹县的牡丹土壤中存在。UP-PCR 结果分析表明：5 个引物共扩增出 71 条条谱带，其中，多态性谱带为 65 条，多态性比率为 91.5%，40 个菌株在相似系数为 0.82 时，可被划分为 9 个组。本研究表明不同地区牡丹土壤的木霉菌种群存在差异，UP-PCR 分析比 EF-1α 序列分析更能体现木霉菌种间和种内的亲缘关系及遗传差异性，各个自然种群之间的遗传亲缘关系与地理分布之间无相关性。

关键词：牡丹；木霉；EF-1α；UP-PCR

黄连叶斑病病原鉴定及其生物学特性*

席中刚，刘　浩，郑　露，黄俊斌**
（华中农业大学植物科学技术学院，武汉　430070）

摘　要：黄连（*Coptis chinensis* Franch）又名味连、鸡爪连，为毛茛科多年生草本植物，以根茎入药，是我国重点发展的63种中药材品种之一。黄连叶斑病是近几年来在湖北省恩施地区黄连叶片上发生的一种十分严重的病害，2014年在当地调查中发现黄连植株发病率达30%以上，田间黄连叶片上的病斑近圆形至不规则形，常汇合形成较大病斑，中央淡黄褐色，具黄褐色晕圈，直径3～10mm上生小黑点（分生孢子器），发病植株矮小，严重时会导致植株干枯死亡。

从病区采集典型黄连病叶，对样品进行了组织分离纯化，经柯赫氏法则验证获得了黄连叶斑病病原菌株，通过形态学和分子生物学方法对病原物进行了鉴定，并研究了病原菌的生物学特性。研究结果表明，引起湖北恩施地区黄连叶斑病的病原菌是 *Phoma aquilegiicola*。其在 PDA 培养基上的菌落圆形，黄褐色，绒状，边缘规则；分生孢子无色，单胞，长椭圆状或棍棒状，大小为（2～2.5）μm×（4.1～5.4）μm；菌丝的最适生长温度为20℃，最适 pH 值为6，致死温度为50℃、10min；在所测试的碳源中，*P. aquilegiicola* 对麦芽糖的利用效果最好，对蔗糖和甘露醇的利用效果最差；在所测试的氮源中，病原菌对蛋白胨利用最好，对硝酸铵的利用最差。本实验为湖北省黄连叶斑病病害的深入研究提供了重要的科学依据，具有重要的现实意义。

关键词：黄连；叶斑病；病原鉴定；生物学特性

* 第一作者：席中刚，在读硕士研究生，主要从事植物真菌病害研究
** 通讯作者：黄俊斌，教授；E-mail：junbinhuang@mail.hzau.edu.cn

黄曲霉中 COP9 信号体 CsnE 亚基调控了生长发育和毒素的产生

王秀娜*，汪世华**
（福建农林大学生命科学学院，福州 350002）

COP9 信号体（COP9 signalosome，CSN）是以 Cullins 为基础的泛素化连接酶 E3 的重要调节因子，已知该调节机制之一是通过 CsnE 亚基中 JAMM 结构域的金属蛋白酶活性将 NEDD8/RUB1 从 CUL1 解离而完成。已知在丝状真菌中该复合体的 CsnE 调控了生长发育、DNA 损伤、氧化胁迫应答和次级代谢等多种过程。本研究中我们揭示了黄曲霉 COP9 信号体的 CsnE 亚基在产孢、DNA 损伤和毒素产生等过程中的功能。利用同源比对的方法预测黄曲霉基因组编码一个和构巢曲霉中已知功能的 CsnE 高度同源的基因 *AfcsnE*，进一步分析该基因编码的蛋白包含保守的功能结构域 JAMM。*AfcsnE* 基因的缺失突变株在黑暗条件下不再产生分生孢子，然而光照条件下可以产生分生孢子但是产孢量显著少于野生型。*AfcsnE* 基因的缺失突变株和野生型相比对 DNA 损伤试剂 MMS 和 HU 更敏感。利用黄曲霉野生型和 Δ*AfcsnE* 菌株侵染玉米和花生时，和野生型相比突变株接种的发病率明显降低。同时，侵染花生和玉米时 Δ*AfcsnE* 菌株黄曲霉毒素的产量明显低于野生型。由此可见，黄曲霉 COP9 信号体的 CsnE 亚基正调控了分生孢子的产生、DNA 损伤应答、致病力和毒素产生等过程。

* 第一作者：王秀娜，博士，主要从事黄曲霉次级代谢的调控；E-mail：xiuna0304@163.com
** 通讯作者：汪世华，教授，博士生导师，主要从事黄曲霉功能基因研究；E-mail：wshyyl@sina.com

我国冬小麦主产省三种化学型小麦赤霉病菌适合度研究*

刘杨杨[1**]，夏云磊[1,2]，孙海燕[1]，陈怀谷[1***]
（1. 江苏省农业科学院植物保护研究所，南京　210014；
2. 南京农业大学植物保护学院，南京　210095）

摘　要：由镰孢菌（*Fusarium* spp.）引起的小麦赤霉病已成为世界范围的重要病害。我国小麦赤霉病主要由禾谷镰孢菌综合物种（*Fusarium graminearum* species complex，FGSC）引起。根据产生单族毒素种类的不同，可以分为3种化学型，分别为3-乙酰脱氧雪腐镰孢菌烯醇（3-AcDON）、15-乙酰脱氧雪腐镰孢菌烯醇（15-AcDON）和雪腐镰孢菌烯醇（NIV）。为预测这3种化学型病菌的发展趋势，从每种化学型菌株中随机挑选了20株菌并且对其生物学特性进行了研究。研究结果表明，在15℃、20℃和25℃时，3-AcDON、15-AcDON和NIV化学型菌株的生长速度没有显著的差异，在30℃的生长条件下，NIV化学型菌株生长速度要显著高于其他两种化学型。在4个不同温度下（15℃、18℃、21℃和25℃），15-AcDON产生子囊壳的量均是最多并且产生子囊壳的时间较其他两种化学型菌株提前2~3d。15-AcDON在分生孢子的产量和大小方面也显著高于其他两种化学型。利用菌碟接种法测定3种化学型病菌对小麦苗期的致病力，结果显示，3-AcDON、15-AcDON和NIV化学型菌群的平均病情指数分别为72.71、70.72、48.27，方差分析显示NIV化学型菌群病情指数显著低于3-AcDON和15-AcDON化学型菌株，而3-AcDON和15-AcDON化学型菌群之间的差异不显著。在小麦扬花期采用单花滴注接种法对3种化学型病菌进行了致病力测定，3-AcDON化学菌群的致病力仍为最强。在小麦的扬花期利用单花滴注接种法接种，收集接种病麦穗、脱粒并进行毒素提取，利用高效液相色谱法（HPLC）检测病麦粒中的毒素含量，3-AcDON化学型菌株产生的DON毒素含量高于15-AcDON和NIV化学型菌株。采用大米培养基对3种化学型菌株进行产毒培养并且利用液相色谱质谱法进行毒素含量测定，15-AcDON化学型菌株产生的DON毒素含量最高。

关键词：小麦赤霉病；禾谷镰孢菌；化学型；适合度

* 基金项目：国家小麦产业体系（CARS-3-1-17），公益性行业（农业）科研专项（201303016）

** 第一作者：刘杨杨，研究实习员；E-mail：llyy050@126.com
*** 通讯作者：陈怀谷，研究员；E-mail：huaigu@jaas.ac.cn

希金斯刺盘孢ODC基因的研究

严亚琴[*]，袁勤峰，郑　露，刘　浩，黄俊斌[**]
（华中农业大学植物科学技术学院，武汉　430070）

摘　要：由希金斯刺盘孢（*Colletotrichum higginsianum*）侵染引起的十字花科蔬菜炭疽病是一类重要的世界性植物真菌病害，主要分布于中国、日本、印度、美国以及东南亚等地区，可引起多种十字花科蔬菜上的炭疽病。国内外研究和生产实践证明，抗病品种种植和药剂防治是控制十字花科蔬菜炭疽病的主要措施；但近年来，已陆续出现抗病品种退化及化学药剂失效的实例，致使该病防治面临严峻挑战。希金斯刺盘孢是一种半活养寄生真菌，其侵染过程复杂，如果破坏或终止病原菌侵染的任何一个阶段，可以减轻病害的发生。

为了研究该病原菌的致病机理，本研究通过生物信息学的方法，选择鸟氨酸脱羧酶（ornithine decarboxylase，ODC）基因作为研究对象。鸟氨酸脱羧酶是鸟氨酸循环中的第一个限速酶，鸟氨酸循环与蛋白代谢密切相关。现已报道在人体中鸟氨酸脱羧酶是多胺合成代谢途径中的第一个限速酶，ODC活性的异常会引起包括肿瘤在内的一系列疾病的发生，近年来被作为抗肿瘤分子靶点进行抗肿瘤药物研究。而在真菌中鲜有报道ODC基因，并且其鸟氨酸循环代谢途径并未完全清楚。希金斯刺盘孢中ODC基因全长1 421bp，具有2个外显子，1个内含子，编码455aa的蛋白，含有2个Pyridoxal-dependent decarboxylase保守结构域。目前正通过基因敲除和互补方法研究ODC在希金斯刺盘孢生长发育和致病过程中的作用。

关键词：希金斯刺盘孢；ODC，基因敲除

---

* 第一作者：严亚琴，在读硕士研究生，主要从事植物真菌病害研究
** 通讯作者：黄俊斌，教授；E-mail：junbinhuang@mail.hzau.edu.cn

解淀粉芽孢杆菌与香蕉枯萎病菌互作中基因表达分析*

叶景文**，谢晚彬，李华平***
（华南农业大学农学院，广州　510642）

摘　要：香蕉枯萎病是由尖孢镰刀菌古巴专化型（*Fusarium oxysporum* f. sp. *cubense*）引起的真菌毁灭性病害，共有3个生理小种，其中，以4号生理小种（FOC4）为害最大。本实验室在之前的研究中筛选到对FOC4生长具有明显抑制作用的菌株（ZJ6-6），经鉴定该菌株为解淀粉芽孢杆菌。通过对ZJ6-6与FOC4对峙培养以及非胁迫条件下进行基因表达分析，通过比较共获得差异表达基因4 659个。进行差异表达基因的GO功能显著性富集分析，结果表明，差异表达基因的主要生物学功能为结合肽聚糖、结合多糖、结合碳水化合物。进行差异表达基因的Pathway显著性富集分析，结果表明，差异表达基因参与的最主要生化代谢途径为次级代谢产物生物合成、苯丙氨酸代谢、酪氨酸代谢、色氨酸代谢。其中，Unigene0007246等7个差异表达基因的产物能够识别结合FOC4细胞壁中的肽聚糖和多糖，参与FOC4细胞壁的合成过程，这些差异表达基因下调，FOC4细胞壁的合成过程受阻，细胞生长停滞。Unigene0000253等3个差异表达基因参与细胞组分生物合成过程和细胞组装过程，这些差异表达基因下调，细胞组分生物合成过程和细胞组装过程受阻，细胞生长停滞。进一步选取部分表达量下调的基因，通过同源臂重组的原理，对该部分基因进行了敲除和互补，结果表明，这些基因的敲除，将使得突变菌株生长缓慢，分泌色素更多，气生菌丝减少。进一步的研究正在进行中。

关键词：香蕉；香蕉枯萎病菌；互作；差异基因；基因敲除

---

* 基金项目：现代农业产业技术体系建设专项（CAR－32－05）
** 第一作者：叶景文，硕士研究生，植物病理学；E-mail：369443288@ qq. com
*** 通讯作者：李华平，教授；E-mail：huaping@ scau. edu. cn

乙烯利对胶胞炭疽菌 RC169 生物学特性影响初步研究*

郑肖兰[1**]，郑金龙[1]，李　锐[1]，梁艳琼[1]，吴伟怀[1]，
习金根[1]，胡　飞[2]，贺春萍[1***]，易克贤[1,3***]
（1. 中国热带农业科学院环境与植物保护研究所，海口　571101；
2. 海南大学环境与植物保护学院，海口　570228；
3. 中国热带农业科学院热带生物技术研究所，海口　571101）

摘　要：本研究目的是探讨植物生长调节剂乙烯利对橡胶树胶胞炭疽菌生长速率、孢子萌发率和附着胞形成率的影响，为橡胶树炭疽病防治提供新途径。以橡胶树胶胞炭疽菌 RC169 为参试菌株，利用十字交叉法、孢子萌发测试法等常规的植物病理学方法测定浓度为 50g/L、25g/L、12. 5g/L、6. 24g/L、3. 12g/L、0g/L 乙烯利处理对 RC169 的菌丝生长的影响；以及 0g/L、0. 2g/L、0. 4g/L、0. 8g/L、1. 6g/L、3. 2g/L 乙烯利处理对 RC169 的孢子萌发、附着胞形成等的影响。结果显示不同浓度的乙烯利对炭疽菌的菌丝生长、孢子萌发和附着胞形成均有一定的抑制作用，且结果符合线性回归：其中参试梯度浓度乙烯利对 RC169 生长速率影响的抑菌率分别为 98. 6%、88. 1%、34. 7%、25. 0%、8. 1%、0，其相应回归方程为 $y=2.917x-6.796$，r 为 0. 9752；参试梯度浓度乙烯利对 RC169 孢子萌发率分别为 92. 2%、36. 7%、20. 6%、12. 3%、2. 63%、0，其相应线性回归方程为 $y=-1.3945x+7.0663$，$r=0.8248$；参试梯度浓度乙烯利对 RC169 附着胞萌发率则分别为 90. 5%、5. 37%、3. 13%、1. 10%、0、0，对应线性回归方程为 $y=-1.7383x+6.7965$，$r=0.9216$。综上所述，不同浓度的乙烯利对胶胞炭疽菌 RC169 的菌丝生长、孢子萌发、附着胞形成均能形成抑制作用，而且随着浓度的升高，其抑制作用随之增强，其相应的线性回归方程差异均达显著水平。

关键词：乙烯利；橡胶树胶胞炭疽菌；抑菌率；孢子萌发率；附着胞形成率

---

* 基金项目：国家自然科学基金（31101408）；国家外专局引智项目（20154600004）和中央级公益性科研院所基本科研业务费专项（2015hzs1J002）

** 第一作者：郑肖兰，副研究员，主要从事植物病理学研究工作；E-mail：orchidzh@ 163. com
*** 通讯作者：贺春萍，研究员，主要从事植物病理学研究工作；E-mail：hechunppp@ 163. com
易克贤，研究员，博士，主要从事植物抗病育种和病理学研究工作；E-mail：yikexian@ 126. com

玉米大斑病菌 *PP2A-B'* 基因敲除载体的构建*

于　波，李贞杨，申　珅，郝志敏**，董金皋**
（河北农业大学真菌毒素与植物分子病理学实验室，保定　071000）

PP2A 是生物体内主要的 Ser/Thr 蛋白磷酸酶，位于信号转导通路的下游，并与多种蛋白激酶相互配合，通过磷酸化和去磷酸化的方式参与调节多条信号转导通路。该酶是由结构亚基 A、催化亚基 C 和调节亚基 B 构成的异源三聚体。经研究发现，PP2A 调控多条信号转导通路，其中，调节亚基 B 发挥了重要的调节作用。本课题组前期利用 PP2A 特异性抑制剂斑蝥素证明 PP2A 负调控玉米大斑病菌（*Setosphaeria turcica*）的产孢、黑色素合成及致病性，并克隆了玉米大斑病菌 *Stpp2A-C*、*Stpp2A-B*、*Stpp2A-B'* 基因的全长。

本研究以玉米大斑病菌野生型 DNA 为模板，根据 *Stpp2A-B'* 基因的 DNA 序列和 pBS 质粒的酶切位点设计两对引物以扩增 *Stpp2A-B'* 基因片段，即 PP2A-B'-I 和 PP2A-B'-II。经 PCR 扩增产物的凝胶回收及克隆测序正确后可用于敲除载体的构建。首先，对 pBS 质粒和与重组质粒 pMD19-T-PP2A-B'-I 都进行双酶切及 PCR 验证，分别获得了 2.9 kb 和 0.83 kb 的目的片段，与预期大小一致，回收目的片段，然后进行连接转化获得第一步重组载体 pBS-I。然后，将 pBS-I 和 pMD19-T-Bar 质粒分别进行双酶切及 PCR 验证，获得了 3.73 kb 和 1.0 kb 的目的条带，回收目的片段。将带有草铵膦抗性的基因表达元件的部分与 pBS-I 进行连接，获得第二步重组载体 pBS-I-Bar。最后，将 pBS-I-Bar 和 pMD19-T-PP2A-B'-II 进行双酶切，PCR 验证，得到 4.8kb 和 0.85 kb 的目的条带，回收目的片段，连接转化获得同源重组载体 pBS-I-Bar-II。对载体 pBS-I-Bar-II 再次进行双酶切及 PCR 验证，所得扩增及酶切产物经琼脂糖凝胶电泳检测均与预期结果一致，表明该同源重组载体构建成功，可以用于转化玉米大斑病菌的原生质体，以获得 *Stpp2A-B'* 基因的敲除突变体。

* 基金项目：国家自然科学基金项目（No. 31301616）；河北省高等学校青年拔尖人才计划项目（No. BJ2014349Y）和河北省自然科学基金项目（No. C2016204160）

** 通讯作者：郝志敏；E-mail：hzm_ 0322@163.com
董金皋；E-mail：dongjingao@126.com

玉米大斑病菌中 PDE 表达规律的研究*

李贞杨，于 波，申 珅，郝志敏**，董金皋**
（河北农业大学真菌毒素与植物分子病理学实验室，保定 071000）

摘 要：玉米大斑病是由玉米大斑病菌（*Setosphaeria turcica*）引起的玉米叶部重要病害，在流行年份常造成严重的经济损失。许多植物病原真菌的生长发育都受到细胞信号转导途径的调控，cAMP 作为第二信使在调节多种细胞过程中发挥重要作用，而磷酸二酯酶（PDE）是将 cAMP 水解为 5′-AMP 的酶，因此对 PDE 的研究具有重要意义。前期研究证实在玉米大斑病菌中存在两个磷酸二酯酶基因，分别为高亲和力磷酸二酯酶（HPDE）和低亲和力磷酸二酯酶（LPDE）。

本实验通过收集玉米大斑病菌野生型从孢子萌发、附着胞形成及成熟到侵染的整个过程中各个时期的材料，提取其 RNA，反转录得到 cDNA。进一步利用 Real-time PCR 技术，对两个磷酸二酯酶基因在分生孢子发育形成侵染结构的过程中不同阶段进行相对表达量分析，结果表明，*StHPDE* 基因在附着胞成熟时期 12 hpi（hours post inoculation）显著上调（$P<0.05$）；而 *StLPDE* 在整个侵染过程中均下调（$P<0.05$）。由此表明附着胞时期 *StHPDE* 比 *StLPDE* 具有更重要的调控作用，本研究为深入解析植物病原真菌 PDE 的功能提供了理论依据。

关键词：玉米大斑病菌；磷酸二酯酶；Real-time PCR

---

* 基金项目：国家自然科学基金项目（No. 31301616）；河北省自然科学基金项目（No. C2014204111）、河北省高等学校科学技术研究项目（QN2016071）

** 通讯作者：郝志敏；E-mail：hzm_ 0322@163. com
董金皋；E-mail：dongjingao@126. com

云南省马铃薯疮痂病菌的组成研究

杜魏甫*，巩　晨，张红骥，杨梦平，赵晓松，于德才**
（云南农业大学植物保护学院，农业生物多样性与病虫害控制
教育部重点实验室，昆明　650201）

摘　要：马铃薯疮痂病病原菌为放线菌目，链霉菌属（*Streptomyces* spp.）。该菌侵染后，病部细胞组织木栓化，开裂后病斑边缘隆起，中央凹陷，呈疮痂状。目前研究表明，引起疮痂病链霉菌的组成多样。世界上已报道的病原菌有近 20 种，其中，*S. scabies* 在世界范围内分布较广、报道较早，而其他种类在近些年才有报道。云南省对疮痂病的研究起步较晚，目前报道云南省主要病原菌仅为 *S. scabies*。为了明确云南省疮痂病菌组成，研究自 2013 年对云南省昆明、曲靖、楚雄、普洱、临沧、红河等地带有典型疮痂病病斑的马铃薯进行多点采集，采用组织分离法共分离得到链霉菌 143 株。结合形态学和 16S rDNA 分子鉴定、共鉴定出 7 种链霉菌，分别为 *S. anulatus*、*S. scabies*、*S. acidiscabies*、*S. griseus*、*S. europaeiscabiei*、*S. caviscabies*、*S. enissocaesilis*。其中，*S. anulatus*、*S. caviscabies* 首次在国内发现。通过致病性鉴定发现，该 7 种病原菌均能引起不同程度的疮痂病症状，不同种病原菌致病性存在差异，需进一步深入研究。

关键词：马铃薯；疮痂病菌；16S rDNA

* 第一作者：杜魏甫，河北石家庄人，硕士研究生，主要研究方向为马铃薯有害生物生态治理；E-mail：286683918@qq. com

** 通讯作者：于德才，副研究员，主要研究方向为马铃薯有害生物生态治理；E-mail：459025316@ qq. com

生防菌盾壳霉胞外丝氨酸蛋白酶基因的克隆与功能研究*

余　涵，王永春，吴明德，张　静，李国庆，杨　龙**
（华中农业大学植物科学技术学院，武汉　430070）

摘　要：盾壳霉（*Coniothyrium minitans*）是核盘菌（*Sclerotinia sclerotiorum*）的一种重寄生菌，该菌对核盘菌专性寄生，作用时间久，而且与核盘菌的生物学特性相似，对植物无致病性，因而具有广阔的应用前景。实验室前期研究证实盾壳霉重寄生核盘菌过程中会分泌大量胞外丝氨酸蛋白酶。本研究利用同源序列比对，通过简并引物扩增，获得丝氨酸蛋白酶基因（*CmSp*）片段，比对盾壳霉基因组数据库获得11个盾壳霉丝氨酸蛋白酶基因，经过信号肽筛选和表达模式分析，最终筛选到2个受核盘菌菌核提取物诱导上调表达基因 *CmSp*1 和 *CmSp*2。其中 *CmSp*1 基因全长1 335bp，有4个外显子3个内含子，而 *CmSp*2 全长1 407bp，没有内含子。这两个丝氨酸蛋白酶基因均属于类枯草杆菌蛋白酶S8超家族，活性中心均含有Asp/Ser/His催化三联体。进一步通过Split-Marker技术成功获得 *CmSp*1 和 *CmSp*2 敲除转化子，发现只有 *CmSp*1 敲除转化子的胞外蛋白酶活性显著降低，而 *CmSp*2 敲除转化子胞外蛋白酶活没有显著变化。对各转化子的生物学特性进行研究发现，*CmSp*-1 和 *CmSp*2 对盾壳霉的菌丝生长和产孢均没有显著影响。平板对峙及沙皿寄生核盘菌菌核试验结果表明：*CmSp*1 的缺失会导致盾壳霉对核盘菌菌丝和菌核寄生能力的显著下降，而 *CmSp*2 的缺失对盾壳霉的重寄生能力没有显著影响。可见，*CmSp*1 在盾壳霉的重寄生过程中起重要作用。

关键词：盾壳霉；核盘菌；丝氨酸蛋白酶基因；重寄生

---

* 基金项目：国家自然科学基金（31471813）

** 通讯作者：杨龙，副教授；E-mail：yanglong@ mail. hzau. edu. cn

莲雾炭疽病的病原鉴定

李杨秀，张璐，吴 凡，蒙姣荣，李界秋
（广西大学农学院，南宁 530005）

摘 要：炭疽病是莲雾主要采后病害之一，是引起莲雾果实腐烂的主要因素。为明确莲雾炭疽病病原菌种类，在观察病原菌培养性状、分生孢子及附着胞形态特征的基础上，以核糖体内转录间隔区（internal transcribed spacer，ITS）、微管蛋白（β-tubulin 2，TUB2）、几丁质合成酶（chitin synthase，CHS-1）、肌动蛋白（actin，ACT）和3－磷酸甘油醛脱氢酶（glyceraldehyde-3-phosphate dehydrogenase，GPDH）的多基因系统进化分析进行病原种类鉴定。结果表明，病原菌在马铃薯蔗糖培养基平板培养基中初生菌丝为白色，培养3～4d后变为墨绿色，最后为灰黑色；8～10d后开始产孢，孢子堆为橘红色；分生孢子为单细胞，无色透明，长椭圆形，1.5～4.5μm，有的含有1～2个油球，萌发产生附着胞；病原菌的ITS序列与果生刺盘孢菌（*Colletotrichum fructicola*）和*C. aeschynomenes*相似性最高，达99.7%，与胶孢炭疽菌（*C. gloeosporioides*）的相似性为99.2%；基于*ACT*、*ITS*、*CHS*-1、*GPDH*和*TUB*2基因序列构建的多基因联合系统进化树分析的结果显示，莲雾炭疽病菌与果生刺盘孢菌的不同分离物聚集在同一个分支上，亲缘关系最近，将病原菌鉴定为果生刺盘孢菌。本研究结果为采后莲雾炭疽病的防治研究提供了理论依据。

关键词：莲雾炭疽病；果生刺盘孢；多基因联合系统进化树分析

玉米大斑病菌 CAK2 基因同源重组载体的构建*

赵玉兰，李　盼，申　珅，郝志敏**，董金皋**
（河北农业大学真菌毒素与植物分子病理学实验室，保定　071000）

摘　要：玉米大斑病是由大斑刚毛座腔菌（*Setosphaeria turcica*，俗称玉米大斑病菌）所引起的一种严重为害玉米产量的真菌性病害。研究发现，造成植物病害的病原真菌的生长、发育及致病性受多条不同信号途径调节，主要为 MAPK 信号转导途径、cAMP 信号转导途径和 Ca^{2+} 信号转导途径。Ca^{2+} 途径的主要组成成分为钙调素（calmodulin，CaM）、磷脂酶 C（phospholipase C，PLC）和钙调磷酸酶（calcineurin，CaN）等。钙/钙调素依赖性蛋白激酶（calcium/calmodulin-dependent protein kinases，CaMKs）是钙调素下游的一类重要靶蛋白，能够引起众多代谢的关键酶或转录因子磷酸化，从而完成对细胞代谢活动或某些基因表达的调节。本课题组前期已经从玉米大斑病菌中克隆到 4 个 CaMKs 基因片段并获得其全长 cDNA 序列，分别命名为 CAK1、CAK2、CAK3 和 CAK4。后续研究表明，在附着胞诱导萌发及穿透过程中，CAK2 和 CAK3 基因表达水平明显上调。为进一步明确 CaMKs 基因对玉米大斑病菌发育调控的作用机制，本试验拟通过 CAK2 基因同源重组载体的构建获得 CAK2 基因敲除突变体，研究基因功能。

本研究以野生型玉米大斑病菌 DNA 为模板，根据 CAK2 基因的 DNA 序列、bar 基因序列及 pBS 质粒的酶切位点设计三对引物分别扩增 CAK2 上游（CAK2-I）和下游（CAK2-II）两段同源臂序列以及 bar 基因序列，经 PCR 扩增产物的凝胶回收及克隆测序正确后可用于敲除载体的构建。首先对 pBS 质粒和 CaMK2-1 与 pMD19-T 重组质粒进行双酶切及 PCR 验证，获得与预期大小一致的目的片段并回收，回收产物经连接转化获得第一步重组载体 pBS-I。然后，将 pBS-I 和 pMD19-T-Bar 质粒分别进行双酶切及 PCR 验证，回收目的片段。将带有草铵膦抗性的基因表达元件的部分与 pBS-I 进行连接，获得第二步重组载体 pBS-I-Bar。最后最后，将 pBS-I-Bar 和 pMD19-T-CAK2-II 进行双酶切，PCR 验证，回收目的片段，连接转化获得同源重组载体 pBS-I-Bar-II。对载体 pBS-I-Bar-II 再次进行双酶切及 PCR 验证，所得扩增及酶切产物经琼脂糖凝胶电泳检测均与预期结果一致，表明该同源重组载体构建成功，可以用于转化玉米大斑病菌的原生质体，以获得 CAK2 基因敲除突变体。

* 基金项目：国家自然科学基金项目（No. 31301616）；河北省自然科学基金项目（No. C2016204160）；河北省高等学校科学技术研究项目（QN2016071）

** 通讯作者：郝志敏；E-mail：hzm_ 0322@ 163. com
董金皋；E-mail：dongjingao@ 126. com

Virulence Variation and Genetic Diversity in the Populations of Barley Spot Blotch Pathogen *Bipolaris sorokiniana* in China*

GUO Huan-qiang, YAO Quan-jie, WANG Feng-tao, FENG Jing, LIN Rui-ming**, XU Shi-chang**

(*State Key Laboratory for Plant Diseases and Insect pests, Institute of Plant Protection, Chinese Academy of Agricultural Sciences, Beijing* 100193)

Spot blotch is widely found where barley is grown. It is caused by *Bipolaris sorokiniana* (Sacc.) Shoemaker, whose teleomorph is *Cochliobolus sativus*. Spot blotch usually initiates significant yield losses in regions with a warm, humid climate condition. It is the most destructive disease on barley in the northeast China, including Heilongjiang province and eastern areas of Inner Mongolia. Yield reduction is as high as 20%, even up to 40% in some fields. The most effective control method is to utilize durable resistance cultivars. However, there is no document of virulence variation on *C. sativus* populations in these areas. Through selecting 78 representative major cultivars and backbone parent lines used widely in China with 21 strains collected across northeast China, twenty one candidate cultivars were obtained. Then total 71 strains were used to furtherly evaluate the resistance of candidate cultivars to spot blotch. Finally, a set of differential cultivars was set up, including 12 cultivars such as highly resistant ND B112, Kenpimai 11, 10PJ-24; moderately resistant Kenpimai 9, Mengpimai 3, Tradition, Bowman; moderately susceptible ones Kenpimai 7, Vorunda, Morex; and highly susceptible Zaoshu 3 and ND 5883. Every cultivar is with different genetic background, and ND B112, Bowman and ND 5883 are also used as differentials in the United States. Based on the virulence test results with this set of differentials, 71 stains isolated from barley host were classed into 18 pathotypes, of which 83.1% of the tested strains belong to 11 pathotypes. The most predominant race is C0466, whose appearance frequency is 12.7%, and the second potential dominant races are C0027, C0067 and C0267, all of whose frequency is 9.9%. In this study the genetic diversity of 45 strains of *B. sorokiniana* isolated from host barley and 27 strains from host wheat was characterized with AFLP molecular markers. 261 polymorphic AFLP DNA bands were developed using 23 AFLP primer combinations. Cluster analysis results indicated that significant difference in DNA polymorphism exists between the spot blotch pathogen populations isolated from hosts barley and wheat respectively, and high level genetic diversity was found even among the strains collected from one sampling field.

---

* Funding: Barley Industry Technology System of Modern Agriculture (CARS－05)

** Corresponding authors: Lin Ruiming; E-mail: linruiming@ caas. cn; XU Shichang; E-mail: shichangxu317@ 163. com

胶孢炭疽菌侵染草莓的转录组学研究

张丽勍*，段　可，邹小花，高清华**

（上海市农业科学院，上海　201403）

摘　要：草莓（*Fragaria ananasa* Duch.）具有非常高的经济价值，为我国传统的优秀果品之一。草莓炭疽病由半活体营养型真菌胶孢炭疽菌（*Colletotrichum gloeosporioides*）引起，是我国乃至世界草莓生产上最重要的真菌病害之一，近年来的发生危害有上升趋势。揭示植物与植物病原微生物之间相互作用过程，了解病原微生物致病的分子机制，是提高植物的抗病性和控制植物病害的长期有效的途径。

本研究利用 High-Seq2000 测序平台，对 *C. gloeosporioides* 菌丝及其侵染草莓叶片后三个不同时期样品的转录组进行了分析。采用 FPKM 法计算各基因在不同样本中的差异表达情况，共分析获得差异表达基因 4 766个。对 *C. gloeosporioides* 侵染后 3 个不同时期之间差异表达基因进行交集和并集分析，结果发现三个时期连续性表达的基因有 2 791个，而特异性表达的基因分别为 771 个，986 个和 785 个。对差异表达基因进行 GO 和 KEGG 功能富集，分析结果表明在 *C. gloeosporioides* 侵染过程中，碳水化合物代谢、抗氧化通路及水解酶基因等明显富集；PHI（Pathogen-Host Interaction database）注释共得到 962 个致病相关基因，其中增强致病力的基因 8 个，致死相关基因 71 个，丧失致病力基因 91 个，降低致病力基因 342 个。运用生物信息学工具对效应子进行了预测，共获得 337 个候选效应子基因。本研究结果可为深入了解草莓和胶孢炭疽菌共进化的机制，培育广谱、持久抗病的新品种，提供理论基础和技术支持。

关键词：草莓；胶孢炭疽菌；基因转录

---

* 第一作者：张丽勍，助理研究员，主要从事真菌—寄主互作分子机制研究；E-mail：zlq1985 - 345@ 163. com

** 通讯作者：高清华，研究员，主要从事果树设施栽培生理研究；E-mail：qhgao20338@ sina. com

香蕉枯萎病菌在土壤中的生存状态分析*

邹　杰**，李华平***

（华南农业大学农学院，广州　510642）

摘　要： 由尖孢镰刀菌古巴专化型（*Fusarium oxysporum* f. sp. *cubense*，FOC）侵染引起的香蕉枯萎病是香蕉病害中危害最为严重的土传病害之一，此病的传染性强，对全球香蕉种植业的健康发展造成严重障碍。近年来的研究表明，香蕉枯萎病的发生与土壤中尖孢镰刀菌的数量有关，降低土壤中病原菌数量，抑制病原菌的增长是防治香蕉枯萎病的关键因素。本试验通过常规病理学、土壤微生物学研究技术和荧光定量 PCR 方法等对香蕉枯萎病菌在土壤中的孢子形态、类型及数量等进行了分析，结果表明：病菌在土壤中的生存状态随土壤类型、含水量和时间等出现明显变化。当病菌分生孢子进入土壤初期，一段时间内主要以菌丝体形式存在，随后出现大量的分生孢子和少量的厚垣孢子，病菌最终都发育成厚垣孢子；随着土壤含水量的增加和在土壤中时间的延长，分生孢子及厚垣孢子在土壤中的数量将逐渐减少；同时发现土壤中的病菌与土壤微生物群落种类和数量之间存在密切的关系，不同类型土壤和土壤中的相关微生物群落是影响病菌生存状态的主要因子之一。进一步的研究正在进行中。

关键词： 香蕉枯萎病菌；土壤类型；病菌形态；微生物群落

---

* 基金项目：现代农业产业技术体系建设专项（CAR－32－05）

** 第一作者：邹杰，硕士研究生，植物病理学；E-mail：351201726@ qq. com

*** 通讯作者：李华平，教授；E-mail：huaping@ scau. edu. cn

光周期对灰葡萄孢（*Botrytis cinerea*）生长发育及致病力影响的初步研究*

赵思霁**，王晓莹，张国珍***
（中国农业大学植物保护学院植物病理学系，北京 100193）

摘　要：灰葡萄孢（*Botrytis cinerea*）引起的草莓灰霉病是草莓生产中的一种重要病害。灰葡萄孢的寄主范围广泛，至少可以侵染包括双子叶植物及一些重要经济作物在内的 200 多种植物，造成灰霉病。光是影响真菌生长发育、物质代谢、繁殖传播及周期节律的调节等各种生命活动的重要环境因子。探讨环境中的光信号对灰葡萄孢生长发育和致病力的调控机制，对于控制灰霉病具有重要的意义。

本研究选取了 24 株来自不同省市草莓上的灰葡萄孢菌株，设置持续光照、12h：12h 光暗交替、持续黑暗三种光周期条件，比较了不同光周期条件下培养的灰葡萄孢主要生物学性状的差异；将灰葡萄孢接种的草莓叶片置于不同光周期条件下培养，分析光周期对灰葡萄孢致病力的影响。结果表明，光周期对灰葡萄孢产孢、产菌核及致病力均有明显的影响。大多数菌株表现为在持续黑暗条件下，产孢量最高、产菌核数量及干湿重最高；持续光照条件下，产孢量最低、不能形成菌核。表明持续黑暗能促进灰葡萄孢产生分生孢子，有利于菌核的形成及生长发育，持续光照对产孢及菌核的形成有明显的抑制作用。接种灰葡萄孢的草莓叶片在持续黑暗条件下产生的病斑较光照条件下明显增大，表明持续黑暗有利于灰霉病病斑的扩展。而菌落生长速率在不同光周期条件下，差异不显著，表明光周期对灰葡萄孢菌落生长速率影响不大。试验结果为生产上利用光的调控防治灰霉病提供了科学依据，具有生产实际指导意义。

关键词：光周期；灰葡萄孢；生长发育；致病力

---

* 基金项目：公益性行业（农业）科研专项（201303025）
** 第一作者：赵思霁，在读硕士研究生，植物病理学专业；E-mail：zsj149@126.com
*** 通讯作者：张国珍，教授，主要从事植物病原真菌学的研究；E-mail：zhanggzh@cau.edu.cn

The Pmt2p-mediated Protein *O*-Mannosylation is Required for Morphogenesis, Adhesive Properties, Cell Wall Integrity and Full Virulence of *Magnaporthe oryzae*[*]

GUO Min[**], TAN Le-yong[**], NIE Xiang[**], ZHU Xiao-lei,
PAN Yue-min, GAO Zhi-mou[***]
(*Department of Plant Pathology*, *College of Plant Protection*,
Anhui Agricultural University, *Hefei* 230036, *China*)

Protein *O*-mannosylation is a type of *O*-glycosylation that is characterized by the addition of mannose residues to target proteins, and is initially catalyzed by evolutionarily conserved protein *O*-mannosyltransferases (PMTs). In this study, three members of PMT were identified in *Magnaporthe oryzae*, and the pathogenic roles of MoPmt2, a member of PMT2 subfamily, were analyzed. We found that MoPmt2 is a homolog of *Saccharomyces cerevisiae* Pmt2 and could complement yeast Pmt2 function in resistance to CFW. Quantitative RT – PCR revealed that *MoPmt*2 is highly expressed during conidiation, and targeted disruption of *MoPmt*2 resulted in defects in conidiation and conidia morphology. The *MoPmt*2 mutants also showed a distinct reduction in fungal growth, which was associated with severe alterations in hyphal polarity. In addition, we found that the *MoPmt*2 mutants severely reduced virulence on both rice plants and barley leaves. The subsequent examination revealed that the fungal adhesion, conidial germination, CWI and invasive hyphae growth in host cells are responsible for defects on appressorium mediated penetration, and thus attenuated the pathogenicity of *MoPmt*2 mutants. Taken together, our results suggest that protein *O*-mannosyltransferase MoPmt2 plays essential roles in fungal growth and development, and is required for the full pathogenicity of *M. oryzae*.

Key words: *Magnaporthe oryzae*; *O*-mannosylation; Conidia germination; Appressoria formation; Cell wall integrity; Pathogenicity

---

* 基金项目：国家自然科学基金青年基金项目（31101401）

** These authors have contributed equally to this work.

*** 通讯作者：郭敏，博士，副教授；E-mail：kandylemon@163. com

安徽省烟草根腐病病原鉴定及生物学特性研究

江　寒，叶　磊，王文凤，檀根甲*
（安徽农业大学植物保护学院，合肥　230036）

摘　要：烟草是重要的经济作物，其质量和产量受到烟草病虫害等一些重要因素的影响，烟草根腐病是农业生产上的一种重要根部病害，而耕作制度和栽培方式的改变，更是加重了该病害的发生。本文从烟草根腐病病样根部，采用常规组织分离法分离并纯化获得两种病原，通过形态学和分子生物学方法鉴定并确定两种病原菌分别为尖孢镰刀菌（*Fusarium oxysporum*）和黄色镰刀菌（*Fusarium culmorum*），并探究了两种病原菌的生物学特性，结果表明：两种病原菌菌丝生长和孢子萌发的最适温度分别是 25℃及 30℃，最适 pH 值都是 7；两种病原菌菌丝生长最适碳氮源都分别为蔗糖和硝酸钾；尖孢镰刀菌的致死温度为 55℃，黄色镰刀菌则为 60℃。其结果为烟草根腐病的综合绿色防治提供了科学理论依据.

关键词：烟草；根腐病；病原菌；生物学特性

Identification and Characterization of Tobacco Root Rot Disease Pathogens in Anhui Province

JIANG Han, YE Lei, WANG Wen-feng, TAN Gen-jia
(*School of Plant Protection*, *Anhui Agricultural University*, *Hefei* 230036)

Abstract: Tobacco is an important economic crop, one of the factors that infect the quality and yield of tobacco were diseases and insect pest. Tobacco root rot was one kind of root and stem diseases. The adjustment of planting structure aggravates the degree of this disease. From the selected sample root of Tobacco root rot, two pathogens isolated by tissue separation method was obtained. The pathogens were identified morphologically and molecularlly as *Fusarium oxysporum and Fusarium culmorum*, the biological characteristics of the pathogens was studied. The results indicated that the most suitable temperature for the mycelial growth and spore germination is 25℃ and 30℃ respectively. The most suitable for mycelia growth, spore germination of pH is 7. The optimum carbon and nitrogen source were sucrose and potassium nitrate. The lethal temperature for *Fusarium oxysporum is* 55℃, for *Fusarium culmorum is*60℃. The finding shows theoretical basis for the control of tobacco root rot.

Key words: Tobacco; Root Rot; Pathogens; Biological characteristics

烟草是一种属于管状花目茄科的草本植物，获得高质量、高产量的烟草不仅对经济发展有重要影响，而且能促进医药卫生的发展，提高人们的生活水平，保障人们的身体健康。然而烟草病害却严重影响着烟草的产量；制约着烟草产业的发展；造成了经济的重大损失，防治烟草病害亟待解决。随着社会的不断发展，环境不断恶化，全球温度不断上升，再加上烟草病虫害的发生受

* 通讯作者：檀根甲，博士，教授，博士生导师，主要从事植病流行与绿色防控技术研究；E-mail：tgj63@163.com

到各地地势，耕作方式和气候的影响，所以确定引起烟草病害的病原物种类，摸清烟草病虫害的发生规律，明确引起烟草病害的病原菌的生物学特性十分紧迫。烟草根腐病作为烟草根部腐烂病的一种，在我国部分地区也时有发生。该病害一般病株率在3% ~5%。本文对该病病原菌进行了分离鉴定，并研究了该病菌的生物学特性，为实现该病害的综合绿色防治奠定了科学理论基础[1-2]。

1　材料与方法

1.1　病原菌的分离与鉴定

1.1.1　病样采集：

病样由安徽省农科院提供，病株的采集地均来自于安徽。

1.1.2　供试培养基：

PDA 培养基：马铃薯 200g，葡萄糖 20g，琼脂粉 18g，蒸馏水 1 000mL。

PDB 培养基：马铃薯 200g，葡萄糖 20g，蒸馏水 1 000mL。

1.1.3　病原菌的分离和纯化：

采用常规组织分离法。将病根冲洗干净后进行表面消毒，按组织分离法分离病原菌，待菌落形成后，进行初次镜检鉴定。并采用菌丝块法切取一小块菌丝进一步纯化，获得纯化菌株，保存备用。

1.1.4　病原菌的形态学鉴定

将菌株移置于 PDA 平板恒温培养 5d 后，观察菌株在培养基上的培养性状、质地、色泽等。用0.05%吐温 20 外理下孢子，取分生孢子悬浮液滴在玻片上，在显微镜下观察孢子的形态。

1.1.5　病原菌的分子生物学鉴定

1.1.5.1　病原菌基因组 DNA 的提取

对已分离的菌株进行活化，用灭菌的打孔器（直径 =6mm）打孔，置于 PDB 培养基中，并移置 25℃的摇床上振荡 7d，待其长出菌丝，过滤，弃滤液，并挑出菌碟，滤渣备用。

CTAB 法提取分离菌株的 DNA：

（1）将滤渣置于研钵中，加入液氮，充分研磨，破坏菌丝的组织结构，取适量于 2mL 离心管中。

（2）加入 1mLCTAB 抽提液，混匀。

（3）65℃水浴 1h，每隔 10min 摇晃一次。

（4）加入等体积的酚：氯仿：异戊醇（25：24：1），充分混匀，12 000r/min 离心 10min。

（5）小心吸取上层液于新管子，加入等体积的预冷异丙醇，-20℃静置 30min。

（6）12 000r/min 离心 10min。

（7）弃上清液，加入 1mL75%乙醇洗涤 2 次。

（8）12 000 离心 3min，弃上清。

（9）12 000r/min 离心 15s，用枪头小心吸去残液。

（10）通风橱内干燥 30min，加入 50μLddH_2O[3]。

1.1.5.2　rDNA ITS 区序列扩增

采用通用引物 ITS1 和 ITS4 进行 PCR 扩增，基因组 DNA 做扩增模板，引物序列为：

ITS1：5′-TCCG TAGG TGAA CCTG CGG-3′

ITS4：5′-TCCT CCGC TTAT TGAT ATGC-3′

PCR 反应体系（25μL）：

PCR Buffer	2.5μL	总体系25μL[4]
ddH2O	19.8μL	
DNTP	0.5μL	
ITS1	0.5μL	
ITS4	0.5μL	
Taq 酶	0.2μL	
DNA 模板	1μL	

反应程序：

94℃ 预变性	5min	30个循环
94℃ 变性	45s	
55℃ 退火	45s	
72℃ 延伸	1min	
72℃ 终延伸	1min	

1.1.5.3 PCR产物测序

琼脂糖凝胶电泳检测PCR产物，使用试剂盒回收目的条带，送去生物测序公司测序，将测序结果在BLAST比对测序结果并进行分析，选择相似性较高的菌株序列与供试的菌株序列进行同源性比较。

1.1.6 柯赫氏法则验证与致病性测定

采用人工接种法验证：将分离后的病原菌通过针刺法或灌浇法接种至烟草根部，观察其发病情况。采用常规组织分离法，分离病原物，移入PDA平板培养基上获得纯化的培养菌株，培养一段时间后与原接种菌株进行形态学和生物学特性比较。

致病性测定：将所得病原菌接到PDA培养基上培养7d后，用灭菌水配成分生孢子数为1×(10^5~10^6)个/mL的悬浮液，浸根或灌根，对照用清水浸根或灌根后置于25℃温室中培养。待处理后的烟苗发病后，用组织分离法进行病原菌的再分离，并镜检观察分离物是否与接种物一致。

1.2 病原菌的生物学特性研究

1.2.1 温度对烟草根腐病病原菌菌丝生长和孢子萌发的影响

从培养5d的菌落边缘取直径为6mm的菌碟，接种到到PDA培养基上，并置于10℃、15℃、20℃、25℃、28℃、30℃、35℃的下培养，每处理3重复，5d后用十字交叉法测量菌落生长直径。

孢子萌发：在培养7d的菌落上，用打孔器打成菌碟，将菌碟置于PDB培养基中在摇床3d后，过滤，得孢子悬浮液，用无菌水稀释成浓度为1×(10^5~10^6)个/mL，用悬滴法制成玻片，分别放置于10℃、20℃、30℃、40℃、50℃的不同恒温下培养10h，每隔2h记录一次孢子萌发率[5-9]。

1.2.2 pH对烟草根腐病病原菌菌丝生长和孢子萌发的影响

将PDA培养基用0.1%的NaOH和0.1%HCl调成pH值为5、6、7、8和9。从培养5d的菌落边缘取直径为6mm的菌碟，接种到到不同ph值的PDA培养基上，置于25℃下培养。每处理3重复，5d后用十字交叉法测量菌落生长直径。大致确定适宜菌株生长的pH值范围，再进行后续实验。

孢子萌发：按2.2.1所述方法制成孢子悬浮液，将无菌水配制成pH值为5、6、7、8和9稀释孢子悬浮液，并按上述方法接种病原菌，25℃恒温培养，测定时间与方法同上。

1.2.3　碳氮源对烟草根腐病病原菌菌丝生长的影响

菌丝生长：碳、氮源测定均以 Czapek（$NaNO_3$2.00g、$K_2HPO_4$1.00g、KCl 0.50g、$MgSO_4 \cdot 7H_2O$ 0.50g、$FeSO_4$0.01g、蔗糖 30.00g、蒸馏水 1 000mL）为基础培养基，每 1 000mL 加入 20.00g 琼脂制成固体培养基。其中碳源分别用等量碳元素的葡萄糖、乳糖、麦芽糖、淀粉、代替蔗糖；氮源分别用等量氮元素的尿素、蛋白胨、硫酸铵、硝酸钾代替硝酸钠，配成不同碳源的培养基，无碳氮源培养基和无氮碳源培养基作对照。置于 25℃恒温箱中培养，每个处理 3 个重复，5d 后用垂直十字交叉法测菌落直径[5-9]。

2.2.4　病原菌致死温度的测定

用液体培养基配制分离病原菌孢子悬浮液。取灭菌 1.5mL 离心管，加入 1mL 孢子悬浮液（若浓度过高，可加入无菌水稀释），分别置于 45℃、50℃、55℃、60℃、65℃恒温水浴锅中，水浴 10min，另设室温对照，每处理重复 3 次。后吸取 5μL 孢子悬浮于凹玻片中央，置于 25℃培养，每隔 2h 镜检孢子萌发率；同时吸取 100μL 孢子悬浮液涂于 PDA 平板培养基上，置于 25℃培养，隔 12h 检查菌落情况。

2　结果与分析

2.1　病害症状

该病害主要危害烟草幼苗的根部，成株期也能发病。在发病初期，病害只危害须根和支根，并向主根扩散蔓延，早期植株不表现明显症状，当烟草生长的环境湿度大时，其发病部位可以发现粉红色的霉状物[10-11]。随着发病面积的扩展以及发病程度的加重，烟草根部腐烂加剧，烟草吸收营养物质和水分的能力逐渐减弱，地上部分供氧不足，致使新叶首先变黄。病情严重时，烟草的整株叶片发黄、枯萎，根皮变褐，根系逐渐减少，而且明显变黑，并逐渐与髓部分离，环境湿度大时，根部的发病部位还可以产生白色、粉红色霉层[12]，如图 1 所示。烟草根腐病在大田期间发生时，病株高度较低，叶片逐渐变黄，植株生长缓慢甚至停滞且茎秆瘦小。

图 1　烟草根腐病症状图

2.2　烟草根腐病病原的分离鉴定

2.2.1　病原菌的形态学鉴定

通过组织分离法，分离得到 2 种菌，将这 2 种菌分别接斜面低温保存，并同时转移置 PDA 平板上培养 5d 后，观察该菌在平板上生长的菌落的质地、性状、颜色等。

2.2.1.1　尖孢镰刀菌（*Fusarium oxysporum*）

在 PDA 平板上培养，菌丝繁茂致密，白色。菌落白色，浅粉色或肉色[11]。尖孢镰刀菌可产生大型分生孢子和小型分生孢子，大型分生孢子多胞，无色，镰刀形，略弯曲[12]；小型分生孢子单胞，无色，卵圆形[13]（图 2 ~ 图 4）。

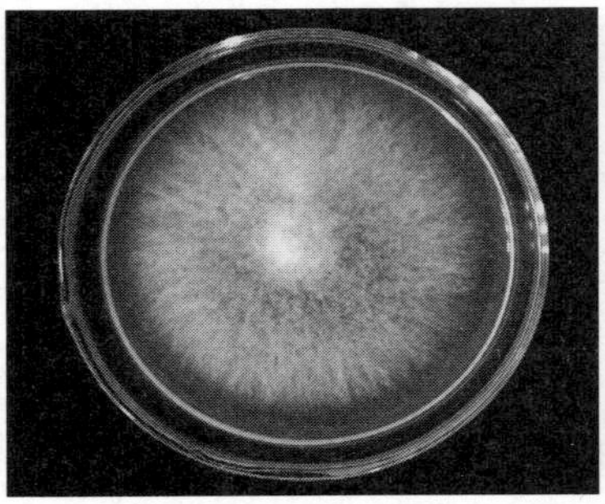

图 2　尖孢镰刀菌（正面）
(*Fusarium oxysporum*)

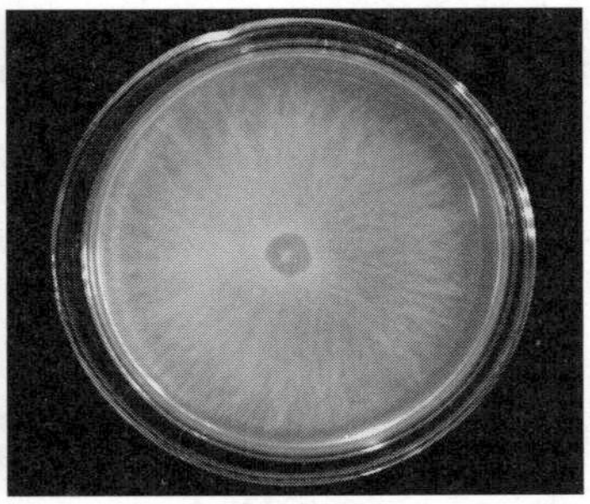

图 3　尖孢镰刀菌（反面）
(*Fusarium oxysporum*)

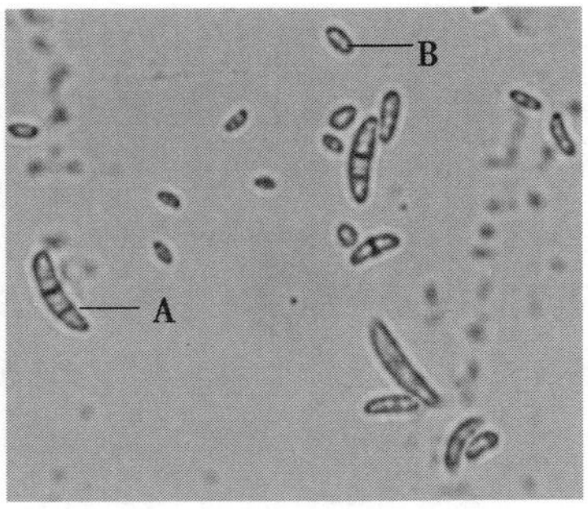

图 4　尖孢镰刀菌孢子图
（A：大型分生孢子；B 小型分生孢子）

2.2.1.2　黄色镰刀菌（*Fusarium culmorum*）

在 PDA 培养基上生长较快，菌丝质密，表面浅黄色，菌落背面产生玫瑰红色素。黄色镰刀菌在也可产生大型分生孢子和小型分生孢子，大型分生孢子多为刀型；小型分生孢子多为柱状或棒状[13]（图 5～图 7）。

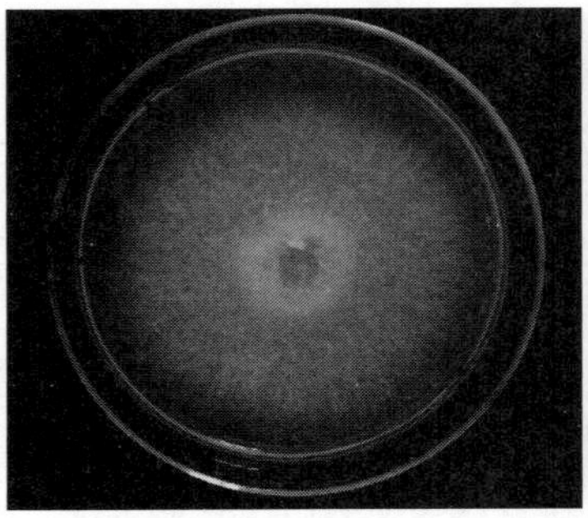

图 5　黄色镰刀菌（正面）
(*Fusarium culmorum*)

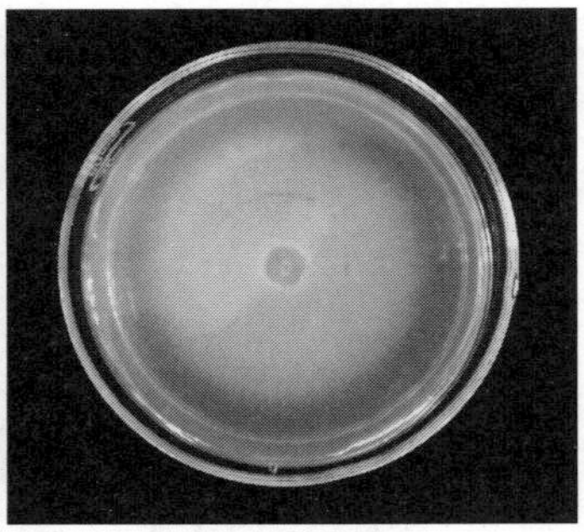

图 6　黄色镰刀菌（反面）
(*Fusarium culmorum*)

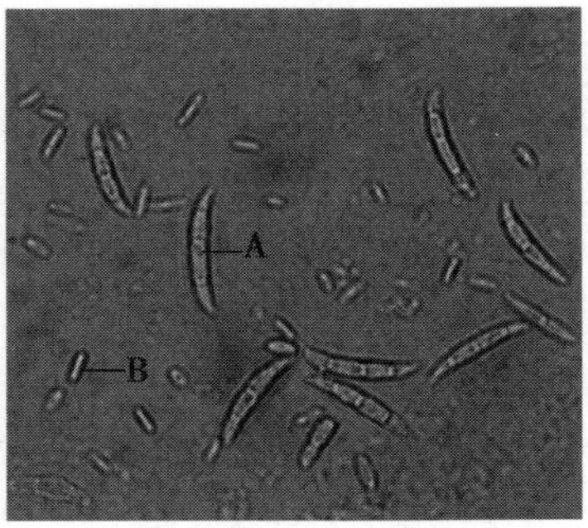

图 7　黄色镰刀菌孢子图
（A：大型分生孢子；B：小型分生孢子）

2.2.2　病原菌分子生物学鉴定

在经过形态学鉴定的基础上，用分子生物学技术辅助鉴定，进一步确定病原。利用 CTAB 法从真菌中提取的 DNA，并以真菌的通用引物 ITS1 和 ITS4 为引物进行 PCR 扩增，其扩增结果经凝胶电泳显示病原菌的 ITS 大小约为 500bp（图 8）。

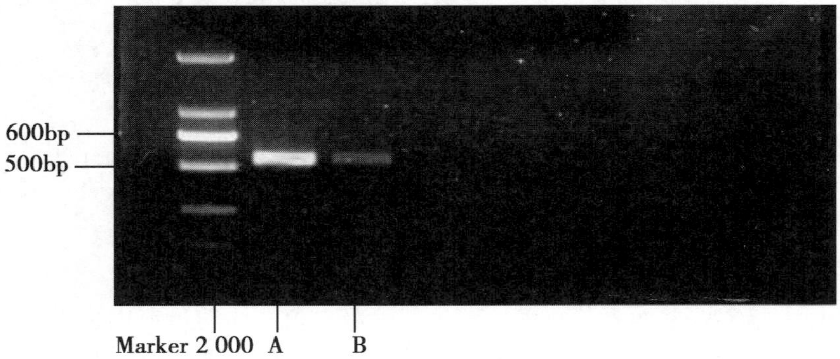

图 8　凝胶电泳显示的病原菌的 ITS 条带
（A：尖孢镰刀菌；B：黄色镰刀菌）

将扩增的得到的片段通过快速DNA产物纯化试剂盒纯化，并将处理后的DNA片段送至生物公司测序，将测定的rDNA-ITS序列与GenBank数据库中已知的rDNA进行同源性比较。结果表明所分离纯化得到两种病原菌与GenBank中*Fusarium oxysporum*及*Fusarium culmorum*菌株序列的同源性最高（图9），且都为100%[14]。分子鉴定的结果与形态学鉴定的结果一致，进而鉴定了引起烟草镰刀菌根腐病的两种病原菌分别为尖孢镰刀菌*Fusarium oxysporum*及黄色镰刀菌*Fusarium culmorum*[15]。

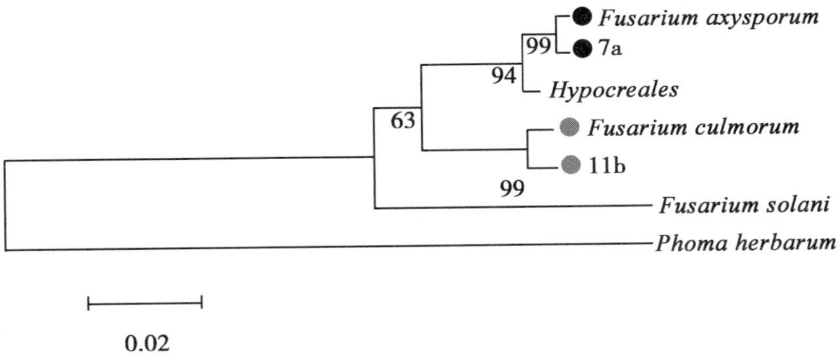

图9　系统发育树

2.2.3　柯赫氏法则验证与病原物致病性

采用常规真菌分离法，分离到的菌株用PDA培养基培养后，通过多次重复挑取菌落边缘未污染的菌丝，获得了纯菌株。根据柯赫氏法则对感病烟草根部进行了活体接种，接种3～5d后的烟草出现了和大田感病烟草相同的症状，根部出现粉红色霉状物。说明分离到的两种病原菌菌株均为烟草根腐病病原菌，且有致病性。

2.3　烟草根腐病病原菌的生物学特性

2.3.1　温度对烟草根腐病病原菌菌丝生长及孢子萌发的影响

两种病原菌均能在10～35℃的温度范围内正常生长，且两种病原菌的最适生长温度都为25℃。低温5℃环境中，两种病原菌菌丝生长极其缓慢，甚至几乎不繁殖[8-9]，如图10。

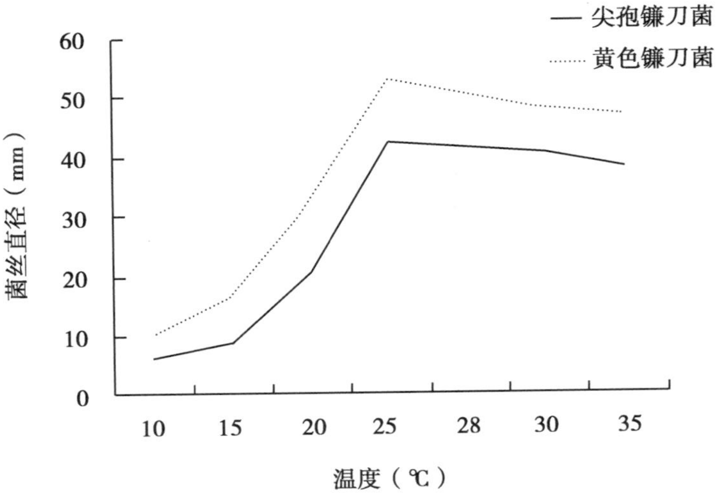

图10　温度对病原菌菌丝生长的影响

在20～30℃范围内，分离得到的两种供试菌株的分生孢子都能萌发，且最适萌发温度都是

25℃，如图11。

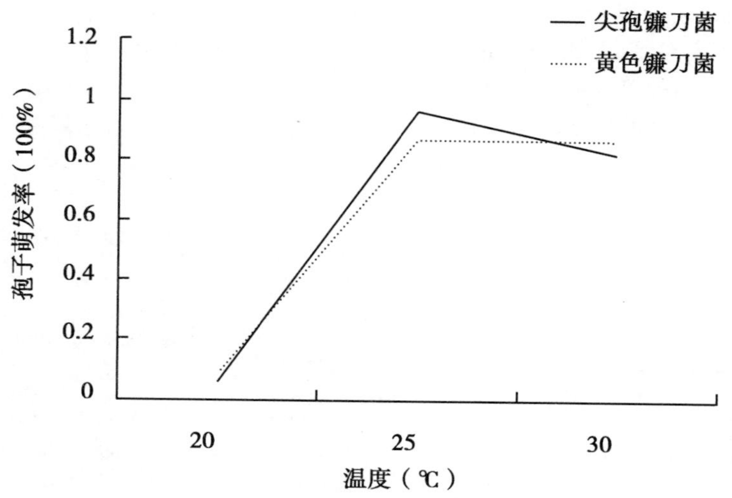

图11 温度对病原菌孢子萌发的影响

2.3.2 pH对烟草根腐病病原菌菌丝生长及孢子萌发的影响

用生长速率法测定分离得到的两种菌株在不同pH值条件下的生长速度，可以发现不同pH值对菌株菌落形态、生长速度有一种病原菌的最定影响：两适生长pH值为7，此时菌株生长速度也最快，如图12。

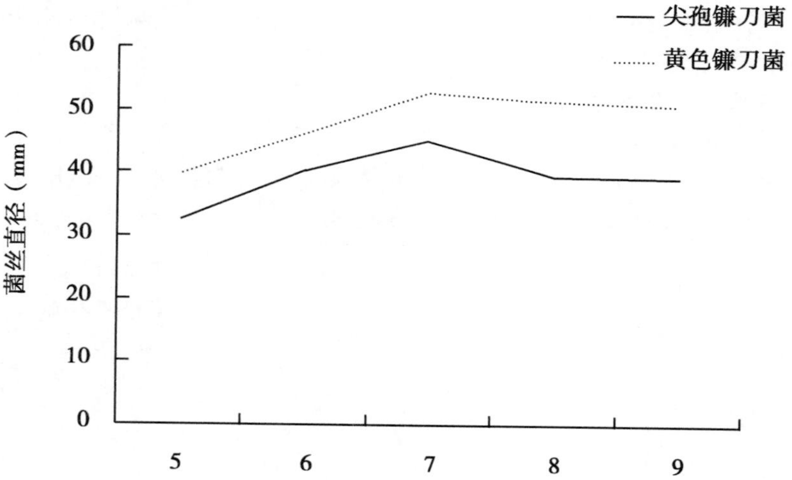

图12 pH对病原菌菌丝生长的影响

孢子：两种病原菌的分生孢子在pH值5~9的范围内均能萌发，且两种菌的分生孢子最适萌发pH值为7[10]，如图13。

2.3.3 碳氮源对烟草根腐病病原菌菌丝生长的影响

2.3.3.1 碳源对烟草根腐病病原菌菌丝生长的影响

由图14可知，病原菌培养在不同的碳源培养基中，当碳源为麦芽糖时，病原菌菌丝生长直径最小，两种供试菌株菌丝生长直径在碳源为蔗糖时最大，表明蔗糖最适于病原菌菌丝生长。虽然菌丝在无碳环境中仍能生长，且生长较快，但能观察到菌落菌丝极其稀疏。

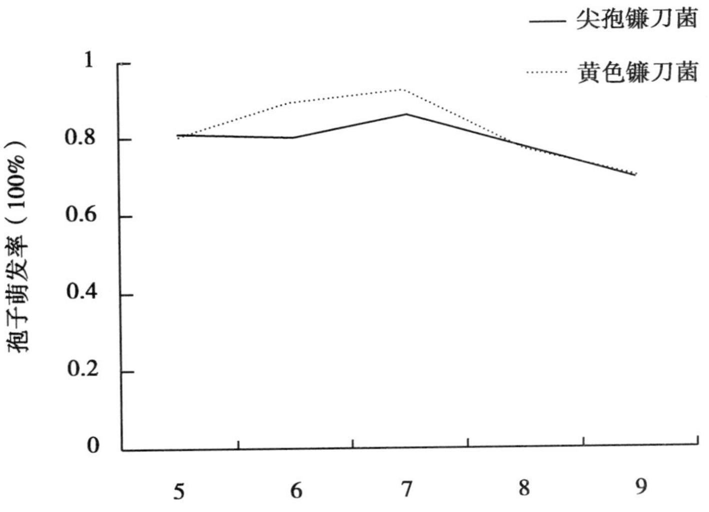

图 13　pH 对病原菌孢子萌发的影响

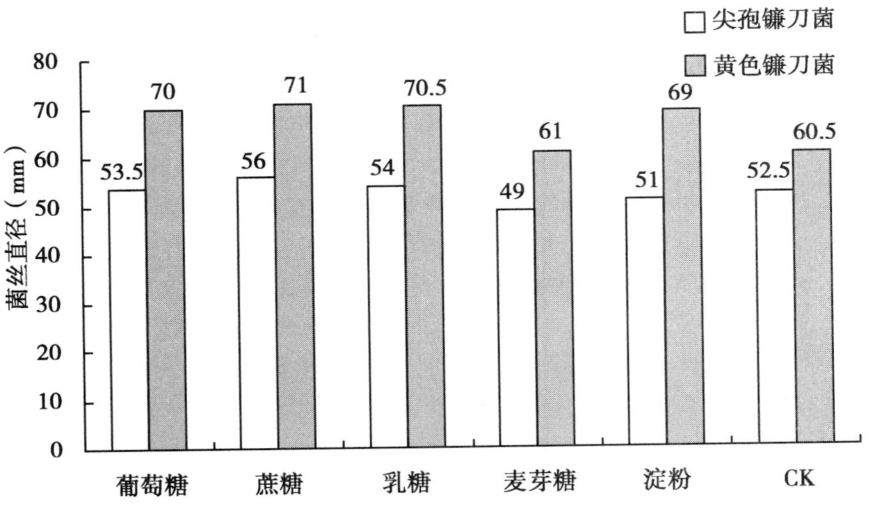

图 14　碳源对病原菌丝生长的影响

2.3.3.2　氮源对烟草根腐病病原菌菌丝生长的影响

如图 15，在病原菌在氮源为硝酸钾的培养基上生长直径最大，而在氮源为硫酸铵的培养基上，菌落生长直径明显较小，且菌落分泌橘黄色色素。虽然于无氮的生长环境中，菌丝生长较快，却能观察到菌落菌丝极其稀疏。

2.3.4　病原菌致死温度的测定

镰刀菌孢子悬浮液在 60℃恒温水浴中加热 10min 后，吸取 100μL 涂布到 PDA 平板上，培养 48h 后，出现菌丝不能生长情况，在 60℃、65℃处理时，两种菌均出现不能生长的情况，实验结果表明黄色镰孢菌（*Fusarium culmorum*）的致死温度为 60℃；而尖孢镰孢菌（*Fusarium oxysporum*）的孢子悬浮液在 55℃水浴中处理 10min 后，涂于 PDA 平板，培养 48h 后，出现菌丝不能正常生长情况，则尖孢镰刀菌（*Fusarium oxysporum*）致死温度是 55℃。这两种镰刀菌的致死温度较高，说明该两种病原菌对环境的生长适应能力强。

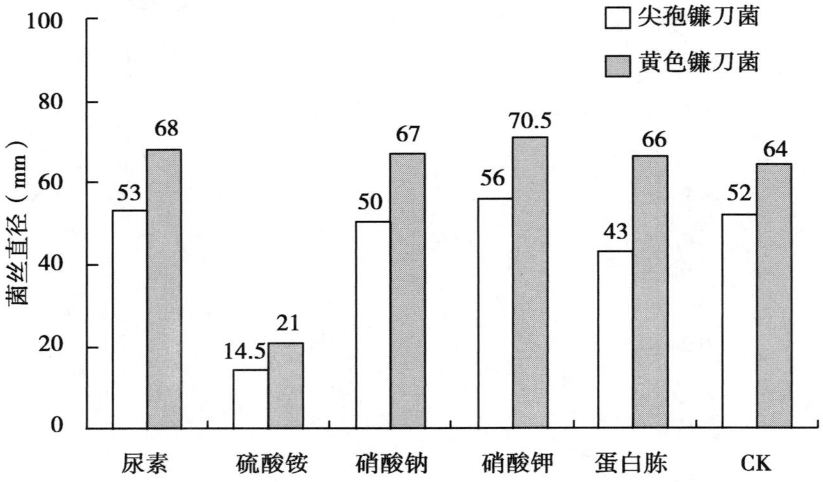

图 15　氮源对病原菌丝生长的影响

3　结论与讨论

本文对安徽省农科院提供的典型烟草根腐病病样的病原进行了分离纯化、鉴定及致病性测定研究，研究结果表明，烟草根腐病的两种病原菌分别是尖孢镰刀菌（*Fusarium oxysporum*）和黄色镰刀菌（*Fusarium culmorum*）。

通过对分离得到的烟草根腐病两种病原菌生物学特性的探究，进一步了解了烟草根腐病两种病原菌菌株生长特点，确定了该病害的发生和发展规律，为该病的防治给予了重要的理论依据。本文明确了温度、pH 和碳氮源对烟草根腐病病原菌菌丝生长和孢子萌发的影响。其中，温度对烟草根腐病菌菌丝生长影响明显，在 10 ~ 40℃ 范围内菌丝都能生长，最适生长温度为 25℃。低于 5℃ 或高于 45℃ 均不能生长。而病原菌在 30℃ 时产生的孢子量最大。在致死温度的测定实验中，结果表明，尖孢镰刀菌和黄色镰刀菌在 45 ~ 50℃ 范围内都能繁殖，但温度达到 55℃ 时，尖孢镰刀菌不能生长；温度达到 60℃ 时，黄色镰刀菌不能繁殖，则尖孢镰刀菌的致死温度为 55℃，黄色镰刀菌的致死温度为 60℃。

参考文献

[1] 陈芷，张绍升．福建四种烟草根茎病害诊断及其病原学研究［D］．福州：福建农林大学，2012.
[2] 胡语婕，冯斗．生物柴油树种石栗繁殖技术研究［D］．南宁：广西大学，2010.
[3] 李锡宏，许汝冰，黎妍妍，等．清江流域烟草病虫害发生与防治技术研究［J］．中国烟草科学，2011，32（05）：112 – 116.
[4] 叶瑞强．戚益军．拟南芥 AGO4/siRNA 复合体细胞质内组装和不依赖 DCL 的小 RNA 介导 DNA 甲基化的机理研究［D］．杭州：浙江大学，2014.
[5] 费丹，檀根甲，罗道宏．安徽省水稻穗腐病病原鉴定及生物学特性研究［J］．安徽农业大学学报，2014，41（5）：777 – 782.
[6] 窦彦霞，肖崇刚，蒋茜，等．重庆烟草根黑腐病菌的生物学特性［J］．烟草科技，2009（11）：56 – 60.
[7] 王佳，杜春梅．大豆根腐病生防菌的鉴定及发酵条件的优化［D］．哈尔滨：黑龙江大学，2010.
[8] 杨继余，王娜，王立事．不同碳、氮源对榆白涩病病菌的影响［J］．北方园艺．2012（19）：153 – 154.
[9] 蒋晓东，杨晓萍．茶叶提取物对桃软腐菌的抑制机理及其活体保鲜研究［D］．武汉：华中农业大学，2014.
[10] 陈高航，侯明生．烟草根腐病病原鉴定及其生物学特性观察［D］．武汉：华中农业大学，2013.

[11] 叶旭红，林先贵，王一明．尖孢镰刀菌致病相关因子及其分子生物学研究进展［J］．应用与环境生物学报．2011，17（5）：759－762．
[12] 李红玫，蒋选利．贵州省草石蚕病害种类调查及其腐烂病化学防治技术的研究［D］．贵阳：贵州大学，2010.
[13] 何冬云，王生荣．玛曲草原土壤真菌分离鉴定及多样性研究［D］．兰州：甘肃农业大学，2010.
[14]］赵杰，孔凡玉．山东省烟草镰刀菌根腐病病原及生物学特性的研究［D］．北京：中国农业科学院，2013.
[15] Pietro A D，Madrid M P，Caracuel Z，et al. *Fusarium oxysporum*：exploring the molecular arsenal of a vascular wilt fungus［J］．Molecular Plant Pathology，2003，4（5）：315－325.

河南棉区落叶型黄萎病菌分布及致病力分化研究*

汪　敏**，赵　杨，丁胜利，李洪连
（河南农业大学植物保护学院，郑州　450002）

摘　要：棉花黄萎病在我国黄河流域、长江流域及西北内陆棉区发生危害严重。近年来，河南省多地出现强致病力落叶型棉花黄萎病菌，但落叶型黄萎病菌在河南棉区系统分布及致病力分化方面的研究较少。本研究从 2014—2015 年，从河南不同棉区采集分离棉花黄萎病菌 38 株，通过形态学方法和分子生物学手段鉴定分离到的棉花黄萎病菌均为大丽轮枝菌（*V. dahliae*）。交配型基因鉴定结果表明棉花黄萎病菌交配型基因均为 *MAT*1 -2 -1，不含有交配型基因 *MAT*1 -1 -1。通过黄萎病菌小种特异性引物 VdAve1F/VdAve1R 与 VdR2F/VdR2R 鉴定黄萎病菌的生理小种，结果表明棉花黄萎病菌生理小种均为小种 2。通过黄萎病菌的落叶型（D）/非落叶型（ND）特异性引物 INTD2F/INTD2R 与 INTND2F/INTND2R 扩增，结果表明棉花黄萎病菌中落叶型菌株有 36 株，占 94.7%；非落叶型菌株有 2 株，占 5.3%，说明落叶型黄萎病菌在河南所有棉区均有分布，且成为优势菌株。通过分生孢子蘸根接种法测定 36 株落叶型棉花黄萎病菌在抗病品种中棉所 41 和感病品种冀棉 11 上的致病力，结果表明河南落叶型棉花黄萎病菌菌株之间致病力存在极显著差异。河南棉区落叶型黄萎病菌可分为强致病力类型和中等致病力类型。其中，强致病力类型 30 株，中等致病力类型 6 株，所占比例分别为 83.33% 和 16.67%。此外，河南棉区不同地区间落叶型棉花黄萎病菌致病力无显著性差异。本研究结果表明河南棉区棉花黄萎病菌以强致病力落叶型菌株为主。

关键词：棉花黄萎病菌；大丽轮枝菌；落叶型菌系；非落叶型菌系；致病力分化

---

* 基金项目：农业部公益性行业科技专项（201503109），河南省棉花产业技术体系建设专项资金资助（S2013 -07 -G04）。

** 作者简介：汪敏，河南新县人，博士，副教授，硕士生导师，主要从事植物真菌病害研究；E - mail：wangm05@sina.com

唐山地区柴胡根腐病病原菌分离鉴定及生物学特性研究*

姜　峰[1**]，马艳芝[1]，客绍英[1]，孙英杰[2]
（1. 唐山师范学院生命科学系，唐山　063000；
2. 中国环境管理干部学院生态学系，秦皇岛　066102）

摘　要：根腐病是影响根茎类中药材产量和质量的主要因素之一。本研究采集河北唐山地区发病的柴胡植株，分离其病原菌，从病原菌的培养性状和形态学特征、内转录区（ITS）系列分析等方面对病原菌进行了鉴定；用回接法检测病原菌的致病性。同时对致病菌的最适生长温度和pH值、孢子最适萌发温度和pH值和孢子临界致死条件进行了研究。结果表明：唐山地区柴胡根腐病致病菌为腐皮镰刀菌（*Fusarium solani*）。致病菌生长的最适温度为25℃，最适生长pH值6～7。孢子在25～30℃时萌发率最高，最适萌发条件为pH值6～8、相对湿度100%。致病菌孢子在25℃水滴中2h开始萌发，9h后萌发率达到百分之百，孢子的临界致死条件为50℃，10min。

关键词：柴胡；根腐病；致病菌鉴定；生物学特性

* 基金项目：国家科技部子课题项目（2011BAI07B05－4）；唐山市科技局项目（15130265a）；河北省科技厅项目（15456420）。

** 作者简介：姜峰，博士，副教授，主要从事药用植物土传病害生物防治研究；E－mail：foodman307@163.com

菌核在稻曲病菌生活史中作用的研究*

范林林，雍明丽，刘亦佳，胡东维**
（浙江大学生物技术研究所水稻生物学国家重点实验室，杭州　310058）

稻曲病是由子囊真菌 *Villosiclava virens* 侵染水稻花丝引起的一种水稻穗部病害。病菌侵染形成的稻曲球上即可产生一层厚厚的厚垣孢子，有些稻曲球表面还会形成一至数个不形状规则的片状菌核。菌核在越冬后可萌发进行有性生殖产生子囊孢子。二者均有作为初侵染源的潜力。自然环境条件下谁在稻曲病菌生活史上发挥了主要作用是稻曲病防治的重要基础和前提。但是近 30 年来随着我国杂交稻超级稻等高产大穗型品种的大范围推广，稻曲病发生普遍加重，发生范围不断向我国南方地区扩展，并在长江中下游地区形成了稻曲病的重灾区。因此，了解稻曲病菌的生活史以及菌核和厚垣孢子在其生活史中的作用对于揭示稻曲病菌的成灾规律及制定田间防治策略具有重要意义。

本课题组通过多年对稻曲病菌的研究，获得以下主要结果：

（1）过去大量文献认为稻曲病菌菌核主要形成于北方稻区。但我们多年的调查发现，在长江中下游地区也普遍存在且数量巨大。在浙江省甬优系列杂交稻种植地区，病穗率高，穗均病粒数多，在严重发病且晚熟的地块，菌核数量可最高达到每亩 15 万个。菌核一般出现在水稻收获前期比较容易观察，因此，临近收获前调查比较可靠。

（2）综合分析浙江省象山县多年气温变化与菌核形成的关系后发现，稻曲球形成前期相对低温的年份菌核较多，播种期偏晚导致水稻花期偏晚的水稻上稻曲球和菌核数量均较多。2015 年秋季长期低温，导致产生菌核的稻曲球比率大幅度提高，最高达到 31%。

（3）室内菌核萌发试验表明，新鲜的稻曲病菌菌核需要经历 1 ~ 2 个月休眠方可萌发。菌核可在光照或黑暗条件下萌发，但只有在光照条件下可形成子实体和产生子囊孢子。黑暗条件下只形成菌落不形成子实体，但菌落可产生分生孢子。菌核黑暗条件下萌发 1 ~ 2 个月后恢复光照可促进菌丝团凝结成为子实体。在稻曲病菌子实体成熟前期，子囊腔分泌类似蜜露的液滴沉积并粘结在子实体表面，阻止了子囊孢子的扩散。子实体在发育过程中受损后，可再次分支并形成新子座。整个菌核萌发可经历 2 ~ 4 个月。每个菌核平均产生 21×10^6 个子囊孢子。

（4）在浙江省象山县自然环境条件下，分布在地表的菌核有 3.1% 可成功越冬。该比率比日本和我国北方等地报道的要低的多，估计是浙江省冬季气温相对较高，土壤微生物活跃的结果。通过田间孢子扑捉试验，发现在在水稻主要生育期内均可扑捉到子囊孢子。同时春季在田间发现了产生子实体的菌核。由此推断，稻曲病菌菌核在其生活史中可能具有重要作用。

（5）人工接种的水稻植株在稻曲球早期进行夜间 15℃ 的低温处理，连续 3 天即可有效诱导菌核的产生。该处理菌核率达到 16.2% 以上，远远高于田间自然形成的菌核率。稻曲球发育中晚期低温诱导的效率极低或无效。目前我们正在进行菌核形成的转录组学分析。

（6）对厚垣孢子田间和室内种子带菌的越冬试验表明，黑色厚垣孢子为休眠孢子，部分能够安全越冬并在来年陆续萌发，但数量极低。

---

* 基金项目：国家自然科学基金（31271999）和国家公益性行业（农业）科研专项（200903039-5）。

** 通讯作者：胡东维，主要从事水稻和小麦真菌病害研究；E-mail：hudw@ zju. edu. cn

就目前证据而言，菌核在稻曲病菌生活史中发挥了更为重要的作用。此外可以推断，杂交稻较长的生育期使水稻发育后期处于相对较低的温度，可能是稻曲病菌能够在长江中下游地区顺利完成生活史并不断加重发生的重要原因。稻曲病的防控策略应更加关注菌核及其越冬和萌发过程。

Biotrophy Lifestyle is Revealed in The Early Stage of Infection Process by GFP Labeled Strain of *Botryosphaeria dothidea* on Fruits of Apple and Pear

GU Xue-ying [1,2], WANG Hong-kai [1], LIN Fu-cheng [1], GUO Qing-yuan [2]

(1. *Biotechnology Institute*, *Zhejiang University*, *Hangzhou* 310058, *China*;
2. *Agricultural College*, *Xinjiang University*, *Urumuqi* 830052, *China*)

Abstract: *Botryosphaeria dothidea* is a destructive pathogen of apple and pear. To investigate the infection process, binary vectors containing various lengths of H3 promoters and TEF promoters fused with GFP and hygromycin B gene cassettes were constructed respectively. These vectors were integrated into genomic DNA of *Botryosphaeria dothidea* with high transformation frequency by ATMT method. Transformants showed strong expression of GFP and hygromycin B genes in cells. Pathogenicity test was performed using a GFP labeled strain of *B. dothidea* on fruits of apple and pear. Results showed that the leading hyphae of the pathogen extend alone the cell wall, the cells of host is still living at the early stage of infection process. This is the first report that a biotrophic stage exists at the early period of infection process by *B. dothidea*, the detailed study of mechanisms of transition of biotrophy to necrotrophy is needed in the future.

Key words: *Botryosphaeria dothidea*; Transformation; GFP; Biotrophy; Fruit

Stachyose is a Preferential Carbon Source Utilized by the Rice False Smut pathogen, *Ustilaginoidea virens*[*]

WANG Yu-qiu[**], LI Guo-bang, GONG Zhi-you,
LI Yan, HUANG Fu, FAN Jing, WANG Wen-ming[***]
(*Rice Research Institute, Sichuan Agricultural University at Wenjiang, Chengdu* 611130, *China*)

Abstract: *Ustilaginoidea virens* (Cooke) Tak. is the causal pathogen of rice false smut (RFS) disease. RFS is an emerging panicle disease, threatening rice production worldwide. However, the biology and pathogenicity of *U. virens* are still not well-understood. In this study, we found that *U. virens* preferentially utilize stachyose over sucrose that is previously reported to be the best carbon source for *U. virens*. Stachyose could promote conidia germination, hyphae elongation and mycelium growth of *U. virens*. Transcriptome and qRT-PCR analyses showed that genes involved in transporting carbohydrate, amino acid, inorganic ion and secondary metabolites were mostly up-regulated by stachyose than by sucrose. Moreover, *Uv8b*_ 6977 encoding a putative major facilitator superfamily (MFS) transporter was highly and specifically induced by stachyose, indicating that it may be a stachyose transporter. Expression levels of several stachyose-inducible MFS genes were also highly induced in *U. virens*-infected rice spikelets especially when early false smut balls appeared, implicating that these genes may play important roles in the formation of false smut balls. Collectively, this work identifies stachyose as a preferential carbon source for *U. virens*, and provides gene candidates to be investigated for clarifying sugar utilization mechanism of *U. virens*.

Key words: Major facilitator superfamily; oligosaccharide; rice false smut; stachyose; *Villosiclava virens*

---

* Foundation item: National Natural Science Foundation of China (31501598)

** First author: WANG Yu-qiu, Master student, specialized in rice false smut disease; E-mail: Yuqiuwang929@ gmail. com

*** Corresponding authors: WANG Wen-ming; E-mail: j316wenmingwang @ 163. com); FANG Jing; E-mail: fanjing7758 @ 126. com

The Autophagy-related gene *BcATG*1 is Involved in Fungal Development and Pathogenesis in *Botrytis cinerea*

REN Wei-chao, ZHANG Zhi-hui, SHAO Wen-yong, YANG Ya-lan, ZHOU Ming-guo, CHEN Chang-jun*

(*College of Plant Protection*, *Nanjing Agricultural University*, *Nanjing*, 210095, *China*)

Abstract: Autophagy, a ubiquitous intracellular degradation process, is conserved from yeast to human. It serves as a major survival function during nutrient depletion stress and is crucial for proper growth and differentiation. In this study, we characterized an atg1 orthologue Bcatg1 in the necrotrophic plant pathogen *Botrytis cinerea*. Quantitative real-time polymerase chain reaction (qRT-PCR) assays showed that the expression of *BcATG*1 was upregulated under carbon or nitrogen starvation conditions. *BcATG*1 can functionally restore the survival defects of the yeast *ATG*1 mutant during nitrogen starvation. Deletion of *BcATG*1 (ΔBcatg1) inhibited autophagosome accumulation in the vacuoles of nitrogen-starved cells. ΔBcatg1 was dramatically impaired in vegetative growth, conidiation and sclerotial formation. Additionally, most conidia of ΔBcatg1 lost the capacity to form the infection-structure appressorium and failed to penetrate onion epidermis. Pathogenicity assays showed that the virulence of ΔBcatg1 on different host plant tissues was drastically impaired, which was consistent with its disability to form appressorium. Moreover, lipid droplets accumulation was significantly reduced in the conidia of ΔBcatg1 but glycerol contents was increased. All the defects of ΔBcatg1 were complemented by reintroducing an intact copy of the wild-type *BcATG*1 into the mutant. These results indicate that *BcATG*1 plays a critical role in numerous developmental processes and is essential to pathogenesis of *B. cinerea*.

Key words: Autophagy; *Botrytis cinerea*; *BcATG*1; Development; Pathogenesis

* Corresponding author: CHEN Chang-jun; E-mail: changjun-chen@ njau. edu. cn

OsmiR169a Negatively Regulates Rice Immunity Against *M. oryzae* by Targeting OsNF-YA genes*

ZHAO Sheng-li**, LI Jin-lu, YANG Nan, XIAO Zhi-yuan, FAN Jing, HUANG Fu, LI yan, WANG Wen-ming***

(*Rice Research Institute, Sichuan Agricultural University at Wenjiang, Chengdu* 611130)

Abstract: MicroRNAs (miRNAs) are kinds of conserved small RNAs existed in both prokaryote and eukaryote. A growing number of data demonstrate that miRNAs are involved in rice immunity against pathogen invasion in plants. Rice is the most important crop supporting food for most of the world' s population, and rice blast is the most serious disease caused by fungal pathogen *Magnaporthe oryzae*. The role of miRNAs in regulating rice resistance against *M. oryzae* were explored in recent years. In a previous study, we picked out more than 30 miRNAs that involved in rice immunity against the blast fungus via deep sequencing of small RNA libraries from susceptible and resistant lines in normal conditions and upon *M. oryzae* infection. Here we reported that miR169a negatively regulates rice immunity against *M. oryzae* by targeting *Nuclear transcription factor Y subunit A* (*OsNF-YA*) genes. Overexpression of miR169a in the susceptible accession TaiPei309 (TP309) resulted in enhanced susceptibility as indicated by increased fungal growth, decreased hydrogen peroxide accumulation at the infection site, and down-regulated expression of defense-related genes. The known target genes of miR169a all belong to NF-YA family. Real-time reverse transcription-polymerase chain reaction assay showed that the expression levels of six target genes were decreased significantly in transgenic lines over-expressing miR169a. Then, miR169a and its eYFP-labeled target genes were transient? co-expressed in *Nicotiana benthamiana*. Western blotting and Laser? Confocal Scanning? microscopy revealed that the protein levels of all target genes of miR169a were decreased significantly. In addition, a time course assay demonstrated that the transcription of miR169 target genes were reduced in susceptible rice line but increased in resistant line upon *M. oryzae* infection. Taken together, our data indicate that miR169a negatively regulates rice immunity against *M.* oryzae by targeting NF-YA genes. [1]

Key words: MicroRNA; *Magnaporthe oryzae*; rice; miR169a; OsNF-YA.

* Funding: National Natural Science Foundation of China (31430072 and 31471761)

** First author: Sheng-Li Zhao, Male, Graduate student; E-mail: 2471248808@ qq. com

*** Corresponding authors: Wen-Ming Wang, E-mail: j316wenmingwang@ 163. com); Yan Li, E-mail: jiazaihy@ 163. com

OsmiRNA444b Negatively Regulates Defense Against *Magnaporthe oryzae* in Rice*

XIAO Zhi-yuan**, WANG Qing-xia, ZHAO Sheng-li, WANG He,
LI Jin-lu, HUANG Fu, FAN Jing, LI Yan, WANG Wen-ming***
(*Rice Research Institute, Sichuan Agricultural University at Wenjiang, Chengdu* 611130)

Abstract: In recent years, research has shown that microRNAs involved in regulating the resistance to rice blast fungus *Magnaporthe oryzae* in rice, but research focusing on a single miRNA is rarely reported. OsmiR444b accumulative level changes after the blast infection, but its regulatory role in rice resistance to rice blast and signaling pathway is unclear. In this study, we provide data to show that OsmiR444b negatively regulates rice defense against *M. oryzae* in rice. First, transgenic rice plants over-expressing OsmiR444b led to enhanced susceptibility as indicated by increased fungal mass, more and bigger disease lesions on the inoculated leaves, and reduced expression level of defense-related genes. After incubating the rice blast fungus, disease phenotypes showed that the number of lesions was more and the size of the lesions was bigger on transgenic lines than those in control plants. Second, qRT-PCR from a time course samples demonstrated that the expression levels of defense-related genes in transgenic lines were significantly reduced than those in the control plants. Third, after DAB staining (DAB, 3, 3′diaminobenzidine-HCl, 5 mg/mL in ddH2O, pH 3.8) demonstrated that t H2O2 accumulation in transgenic lines was less than that in the control plants. Forth, Trypan Blue staining indicated that the mycelia in transgenic leaves were magnificently more than that in the control plants. Taken together, our data indicate that OsmiR444b negatively regulates rice defense against the blast fungus.

Key words: Rice; MicroRNA; *Magnaporthe oryzae*; miR444b; resistance; susceptibility.

* Funding: National Natural Science Foundation of China (31430072 and 31471761)

** First author: Zhi-Yuan Xiao, Male, Graduate student; E-mail: zhiyuan_ 666888@163.com

*** Corresponding authors: Wen-Ming Wang, E-mail: j316wenmingwang@163.com); Yan Li, E-mail: jiazaihy@163.com

Identification of a novel phenamacril-resistance-related gene by the cDNA-RAPD method in *Fusarium asiaticum*

REN Wei-chao, ZHAO Hu, SHAO Wen-yong, MA Wei-wei, WANG Jian-xin, ZHOU Ming-guo, CHEN Chang-jun*

(*Key Laboratory of Monitoring and Management of Crop Diseases and Pest Insects, Ministry of Education, College of Plant Protection, Nanjing Agricultural University, Nanjing* 210095, *China*)

Abstract: *Fusarium asiaticum*, a dominant pathogen of Fusarium head blight (FHB) in East Asia, causes huge economic losses. Phenamacril, a novel cyanoacrylate fungicide, has been increasingly applied to control FHB in China, especially where resistance of *F. asiaticum* against carbendazim is severe. It is important to clarify the resistance-related mechanisms of *F. asiaticum* to phenamacril so as to avoid control failures, and to sustain the usefulness of the new product. A novel phenamacril-resistance-related gene *Famfs*1 was obtained by employing the cDNA random amplified polymorphic DNA (cDNA-RAPD) technique, and was validated by genetic and biochemical assays. Compared with the corresponding progenitors, deletion of *Famfs*1 in phenamacril-sensitive or highly phenamacril-resistant strains caused a significant decrease in effective concentrations inhibiting radial growth by 50% (EC_{50} value). Additionally, the biological fitness parameters (including mycelial growth under different stresses, conidiation, perithecium formation and virulence) of the deletion mutants attenuated significantly. *Famfs*1 not only was involved in the resistance of *F. asiaticum* to phenamacril but also played an important role in adaptation of *F. asiaticum* to the environment. Moreover, our data suggest that the cDNA-RAPD method can be a candidate technique to clone resistance-related genes in fungi. 2015 Society of Chemical Industry

Key words: *Fusarium asiaticum*; phenamacril; resistance; cDNA-RAPD

* Correspondence: E-mail: changjun-chen@ njau. edu. cn

核盘菌 *Ss-FoxE2* 基因互作蛋白筛选的研究[*]

陈　亮[**]，刘言志，张祥辉，潘洪玉[***]

（吉林大学植物科学学院，长春　130062）

摘　要：核盘菌（*Sclerotinia sclerotiorum*（*Lib.*）de Bary）是一种寄主范围非常广泛的病原真菌，可引起 400 多种植物病害，造成巨大的经济损失。在病害循环中，适宜条件下菌核萌发形成子囊盘，成熟子囊盘可释放大量子囊孢子，子囊孢子随风散落在寄主植物上萌发并侵染，造成菌核病害的流行。本实验室在核盘菌 T-DNA 突变体库中筛选到一类含有叉头框（Forkhead-box，Fox）结构域的转录因子。这类转录因子与很多生物过程的调控相关，包括胚胎发育、细胞周期调控、形态特征、细胞分化、代谢等。通过基因敲除其中一个 Fox 类转录因子 Ss-FoxE2，并对敲除突变体表型分析发现，Ss-FoxE2 影响子囊盘的发育。

为了寻找与 *Ss-FoxE2* 基因互作的蛋白，明确其调控途径。本研究采用酵母单杂交的方法在核盘菌 cDNA 文库中筛选其互作蛋白。将 *Ss-FoxE2* 上游 1200bp 包含启动子的序列构建到酵母单杂交诱饵载体上，得到 pHIS2-*Ss-FoxE2* 诱饵载体。通过 LiAc 法将诱饵载体和公司构建的文库质粒依次转化到酵母中，在营养缺陷培养基 SD/Trp/Leu/His 上筛选到 7 个与 *Ss-FoxE2* 互作的蛋白。其中包括泛素结合酶、赖氨酸 2，3-氨基变位相关蛋白、磷脂酶、PHD 锌指蛋白及 3 个未知功能的蛋白。在此基础上，本研究将进一步验证 *Ss-FoxE2* 与这些蛋白的互作，揭示蛋白互作的机制。本研究的结果为深入研究核盘菌转录因子 Ss-FoxE2 调控子囊盘发育途径奠定了基础。

关键词：核盘菌；子囊盘；酵母单杂交；Ss-FoxE2

---

* 基金项目：国家自然科学基金（31471730，31271991）

** 第一作者：陈亮，硕士研究生，植物病理专业；E-mail：chenliang14@ mails. jlu. edu. cn

*** 通讯作者：潘洪玉，教授，主要从事植物病原真菌分子生物学与抗病基因工程研究；E-mail：panhongyu@ jlu. edu. cn

核盘菌 GATA 类转录因子功能分析*

刘 玲**，王翘楚，刘金亮，潘洪玉***

（吉林大学植物科学学院，长春 130062）

摘 要：核盘菌（*Sclerotinia sclerotiorum*（*lib.*）de Bary）是一种寄主范围广泛、危害极其严重的腐生型植物病原真菌。在病害循环中，适宜条件下菌核萌发形成子囊盘，成熟子囊盘可释放大量子囊孢子，子囊孢子随风散落在寄主植物上萌发并侵染，引起植物菌核病并造成病害流行，严重影响作物产量和品质。GATA 转录因子家族是一类能识别（W）GATA（R）基序并与之结合的转录调节因子，是真核生物中重要的转录调控因子，广泛存在于植物、动物和真菌中，可调控多种细胞功能，包括氮代谢、有性发育，分生孢子产生等。已有研究表明，在皮炎芽生菌（*Blastomyces dermatitidis*）、水稻恶苗病菌（*Fusarium fujikuroi*）和构巢曲霉（*Aspergillus nidulans*）中，GATA 转录因子在形态发育的转变过程、氮代谢、分生孢子产生等方面发挥调控作用。本研究采用 qRT-PCR 的方法对核盘菌中 6 个含有 ZnF-GATA 结构域的基因在其不同生长发育时期的表达量进行分析，结果显示：*SsGATA*-1、*SsGATA*-2 与 *SsGATA*-3 在核盘菌的子囊盘中上调表达，*SsGATA*-4、*SsGATA*-5 与 *SsGATA*-6 在不同发育阶段的表达量没有显著差异。为了进一步解析 *SsGATA*-1、*SsGATA*-2 与 *SsGATA*-3 这 3 个转录因子的功能，本研究利用同源重组的方法获得其敲除突变体，表型分析结果显示，在核盘菌●SsGATA-1 敲除突变体中，菌核与子囊盘形成缺陷且致病力下降；●*SsGATA*-2 敲除突变体产生微小的菌核，并且菌核数量与野生型对照相比明显增多；●*SsGATA*-3 敲除突变体气生菌丝量增多。3 个敲除突变体的菌丝生长速率都明显降低。本研究结果表明 *SsGATA*-1、*SsGATA*-2 与 *SsGATA*-3 基因在核盘菌的生长发育及致病过程中发挥重要作用，本研究结果为后续全面系统的研究 GATA 类转录因子在核盘菌中的作用机制奠定基础。

关键词：GATA 转录因子；核盘菌；基因敲除；功能分析

---

* 基金项目：国家自然科学基金（No. 31471730，No. 31271991）

** 第一作者：刘玲，博士研究生，研究方向为植物病原真菌分子生物学与抗病基因工程；E-mail：liuling14 @ mails. jlu. edu. cn

*** 通讯作者：潘洪玉，教授，主要从事植物病原真菌分子生物学与抗病基因工程研究；E-mail：panhongyu@ jlu. edu. cn

核盘菌 *SsMCM*1 基因功能及其互作蛋白的研究*

刘晓丽**，刘金亮，张祥辉，刘言志，张艳华***，潘洪玉***
（吉林大学植物科学学院，长春 130062）

摘　要：核盘菌（*Sclerotinia sclerotiorum*（lib.）de Bary）是一种寄主范围广泛、危害极其严重的腐生型植物病原真菌。由核盘菌引起的植物菌核病是世界性分布的重要病害，主要危害油菜、向日葵、大豆等油料作物和莴苣、胡萝卜等蔬菜作物，尤其在东北地区大豆菌核病发病严重，已成为影响大豆等作物产量和品质的最主要障碍因素之一。MADS-box 基因家族是真核生物中重要的转录调控因子，存在于植物、昆虫、线虫、真菌、低等脊椎动物及哺乳动物中，可调控多种细胞功能，包括初级代谢、细胞周期、细胞识别等。已有研究表明，在酵母菌（*Saccharomyces. cerevisiae*）、轮枝镰刀菌（*Fusarium. verticillioides*）、稻瘟菌（*Magnaporthe. oryzae*）等真菌中，MADS-box 在生长、代谢、生殖、致病力等方面发挥多效调控作用。

本课题组克隆了核盘菌 MADS-box 基因 *SsMCM*1，该基因与酿酒酵母、稻瘟菌等真菌中 MADS-box 基因家族中的 Mcm1 高度同源。为研究核盘菌 *SsMCM*1 基因功能，构建了 RNA 沉默载体 pS1-*SsMCM*1 和 pSD-*SsMCM*1，利用原生质体转化的方法，获得 *SsMCM*1 基因沉默转化菌株。对沉默菌株的生物学功能进行研究，结果表明，SsMCM1 参与核盘菌菌丝营养生长、致病力等。为了深入研究 *SsMCM*1 的功能，阐明与 SsMCM1 互作的蛋白质及调控的靶基因，明确其转录调控网络和调控机制，本课题组在前期工作基础上构建核盘菌 cDNA 文库及含有 *SsMCM*1 基因的重组质粒 pGBKT7-*SsMCM*1，并将其转入酵母细胞中。随后对其进行自激活检测，发现其无自激活活性，可进行酵母双杂交实验。利用 GAL4 酵母双杂交系统，将核盘菌 cDNA pGADT7 融合表达文库中的基因转入含有重组质粒的酵母菌株，并筛选出 6 个与 SsMCM1 转录调控因子互作的蛋白质，并构建了筛选蛋白的沉默和敲除载体，为深入研究筛选蛋白的功能及 SsMCM1 转录调控因子的调控网络奠定基础。

关键词：核盘菌；酵母双杂系统；转录调控因子；MADS-box

---

* 基金项目：国家自然科学基金（No. 31101394，No. 31271991）
** 第一作者：刘晓丽，硕士研究生，植物保护专业，吉林大学植物科学学院
*** 通讯作者：张艳华，教授；E-mail：yh_ zhang@ jlu. edu. cn
潘洪玉，教授；E-mail：panhongyu@ jlu. edu. cn

核盘菌转录因子 SsFox-E2 调控基因鉴定的初步研究*

孙 瑞**，程海龙，张艳华，潘洪玉***
（吉林大学植物科学学院，长春 130062）

摘 要：核盘菌（*Sclerotinia sclerotiorium*（lib.）de Bary）属于子囊菌门核盘菌属。其寄主范围十分广泛，由该菌引起的作物菌核病对我国的农业造成严重经济损失。因此，为有效解决菌核病的防治问题，从分子水平研究核盘菌的生长发育是有效的手段。研究表明，Forkhead 家族蛋白不仅能作为典型的转录因子通过招募共激活因子等调节基因转录，有些还能直接同凝聚染色质结合参与其重构，协同其他转录因子参与转录调节。Forkhead 家族蛋白在细胞发育、细胞周期调控、生物老化和免疫调节等多种生物学过程中发挥作用。通过本实验室之前的研究表明，核盘菌中存在多个 Forkhead 家族转录因子，其中 SsFox-E2 参与调控核盘菌的有性生殖，对子囊盘的发育产生影响。因此，实验室将通过染色质免疫共沉淀（ChIP）等手段明确该转录因子所调控的下游启动子区域序列以及分析其所调控的靶基因。本研究通过 RT-PCR 扩增获得 *SsFoxE2* 基因的完整开放阅读框序列，构建其原核表达载体，利用表达出来的融合蛋白为抗原制备多克隆抗体，获得了大量具有可以与 SsFoxE2 特异性结合且高效价的抗体，为后续的 ChIP 实验奠定了基础。经典的染色质免疫沉淀技术多以哺乳动物细胞或酵母菌为基础建立，核盘菌为丝状真菌，对影响 ChIP 结果准确性的因素不明确，甲醛交联时间和染色体的破碎程度对结果的稳定性和可重复性最重要，因此我们对 ChIP 实验进行优化。优化结果为：甲醛交联时间 15min，将材料用液氮冷冻磨碎，提取染色质，以 10% 功率，2s ON，2s OFF 条件下超声 6-8min 可得到有效的超声片段。在此基础上，本研究将对核盘菌转录因子 SsFox-E2 调控基因进行鉴定与关联分析。上述结果有助于 ChIP 实验的顺利开展，为深入研究 SsFoxE2 转录因子的调控网络奠定基础。

关键词：核盘菌；转录因子；Forkhead 家族；染色质免疫共沉淀

---

* 基金项目：国家自然科学基金（No. 31471730，No. 31271991）

** 第一作者：孙瑞，硕士研究生，研究方向为植物病原真菌分子生物学与抗病基因工程；E-mail：sunrui14@mails.jlu.edu.cn

*** 通讯作者：潘洪玉，教授，主要从事植物病原真菌分子生物学与抗病基因工程研究；E-mail：panhongyu@jlu.edu.cn

第二部分　卵　菌

苹果疫腐病侵染发病条件研究及防治药剂筛选*

刘　芳**，李保华***
（青岛农业大学农学与植物保护学院/山东省植物病虫害综合防控重点实验室，青岛　266109）

苹果疫腐病由恶疫霉菌［*Phytophthora cactorum*（Leb. et Cohn.）Schrot.］侵染所致，在中国各苹果主产区主要为害果实，导致果实腐烂，严重影响产量，降低果园经济效益。目前，国内外对苹果疫腐病的流行条件和预测方法缺乏系统研究，病害始终得不到有效的控制。为了明确苹果疫腐病的侵染时期、侵染条件及预测预报方法，为病害的防治提供依据，本试验在人工控制条件下，测试了温度、湿度对苹果疫腐病菌游动孢子释放、萌发、侵染和病害潜育期的影响。结果表明，疫腐病菌游动孢子囊在0～20℃下均能萌发，并释放出游动孢子，其中，10℃下游动孢子的释放数量最多。在不同温度下处理15min后，孢子囊悬浮液中就能检查到游动孢子，4h后游动孢子的释放量达到高峰，孢子悬浮液中游动孢子的浓度达22.37×10^4个/mL；4h后，随处理时间延长，孢子悬浮液中游动孢子的浓度逐渐降低。离体孢子萌发试验表明，疫腐病菌游动孢子的萌发温度范围为7～34℃，最适温度为23.67℃。游动孢子的萌发需要自由水条件，其他湿度条件下萌发率很低，几乎检测不到萌发的孢子。苹果疫腐病菌游动孢子的侵染温度范围为7.6～33.5℃，最适温度为23.28℃，在此温度下游动孢子完成侵染过程并导致果实发病所需的露时最短。15℃，20℃和25℃下，用游动孢子悬浮液接种的苹果果实，不需要保湿，病菌便可完成侵染过程，并导致果实发病。据此推测，在自然条件下，游动孢子只要随雨水传播到达果实表面，就能够完成全部侵染过程，导致果实发病。果面结露时间越长，病菌的侵染量越大。10℃，15℃，20℃，25℃和30℃下，病菌的侵染量随果面结露时间的变化动态可用5个逻辑斯蒂模型描述。接种疫腐病菌的果实在7.8～32.6℃下可发病，最适发病温度为25.35℃，在此温度下，病害的潜育期最短，仅为3.6d，果实的发病率和病害严重度最高。本研究结果可为苹果疫腐病的侵染预测和病害防治提供依据。

为了筛选防治苹果疫腐病的有效杀菌剂，采用先施药后接种和先接种后施药的方法，在室内离体富士苹果果实上测试了13种杀菌剂保护果实免受疫腐病菌侵染的保护效果和抑制病菌在果实内生长扩展的内吸治疗效果。结果表明，所测试的13种杀菌剂都能有效保护果实防止疫腐病菌侵染，保护效果达100%；其中，氟菌·霜霉威、噁霜·锰锌、烯酰吗啉、氰霜唑、吡唑醚菌酯、双炔酰菌胺和霜脲·锰锌7种杀菌剂的保护效果可维持10d以上，其余6种药剂的保护效果可维持5d以上；当疫腐病菌侵入果实后，13种杀菌剂都不能有效抑制病菌的生长扩展，防止果实发病，没有内吸治疗效果。因此，所测试的13种杀菌剂在病菌侵染之前喷施都能有效阻止病菌侵染，持效期不短于5d，在病菌侵染之后施用则都不能有效抑制侵染病菌扩展致病。

* 基金项目：国家苹果产业技术体系（CARS－28）
** 第一作者：刘芳，山东烟台人，硕士研究生，研究方向为植物病害流行学；E-mail：woshixiaofang22@126.com
*** 通讯作者：李保华，教授，主要从事植物病害流行和果树病害研究；E-mail：baohuali@qau.edu.cn

The Sensitivity Detection of *Phytophthora infestans* to Dimethomorph in Yunnan Province*

XI Jing**, CHEN Feng-ping, ZHAN Jia-sui***

(*Fujian Key Laboratory of Plant Virology*, *Institute of Plant Virology*, *Fujian Agriculture and Forestry University*, *Fuzhou* 350002, *China*)

Abstract: Potato late blight caused by *Phytophthora infestans* is a heavy disease in potato production in Yunnan province. Dimethomorph, inhibiting the synthesis of cell wall, was effective on potato late blight control. However, the sensitivities of *Phytophthora infestans* isolates from potato in Yunnan province to dimethomorph, were determined in this study. The total of *P. infestans* isolates was 173 isolates sampled from 2012 to 2015 in Yunnan province. The results indicated that the EC_{50} values for 2012, 2013, 2014 and 2015 population were ranged from 0.011 to 0.359, 0.373, 0.308, 0.186μg/mL, respectively; with the average EC_{50} value of 0.158, 0.147, 0.157 and 0.148μg/mL. The frequency distribution of the EC_{50} for four populations was all unimodal indicating no resistance subpopulation occurred. According to analysis of ANOVA results showed no significant difference in the EC_{50} among populations. Therefore, our findings suggested that all isolates sampled from 2012 to 2015 in Yunnan province were sensitive. Dimethomorph will be still effective for controlling late blight disease of potato in Yunnan province.

Key words: *Phytophthora infestans*; Population; Dimethomorph; EC_{50}; Controlling

* Funding: China Modern Agricultural Industry and Technology System (Grant No. CARS－10)

** First author: XI Jing, male, master, major in molecular plant pathology; E-mail: 843112687@qq.com

*** Corresponding author: ZHAN Jia-sui, professor, research interests for population genetics; E-mail: Jiasui.zhan@fafu.edu.cn

低温诱导大豆疫霉游动释放的基因表达谱研究*

侯巨梅**，刘　震，崔　佳，曲建楠，左豫虎，刘　铜
（黑龙江八一农垦大学，大庆　163319）

摘　要：由大豆疫霉菌（*Phytophthora sojae* Kaufmann & Gerdemann）引起的大豆疫霉根腐病是大豆的毁灭性病害，给我国的大豆生产带来严重的经济损失。大豆疫霉菌的游动孢子是在作物生长季节中病害的主要传播方式，可以在水中进行游动，通过雨水传播，接近寄主植物萌发形成芽管而形成对植物侵染。大豆疫霉游动孢子释放机理成为研究该病害发生规律和防治的关键因素。尽管目前发现低温可以促进大豆疫霉游动孢子的释放，但是对于低温诱导游动孢子释放的分子机理迄今未见研究报道。本研究以未经低温处理的游动孢子囊为对照组，以经低温处理的游动孢子囊为处理组，构建了基因表达谱。以 FDR < 0.01 且差异倍数 FC（Fold Change）≥2 作为筛选标准，结果经过低温处理的样品有 38 基因上调和 137 基因下调。对差异表达基因进行 Blaxp 注释，差异基因主要包括谷氧还蛋白，糖苷水解酶，锚定重复蛋白，TBP 相关蛋白，乙酰转移酶，RNA 指导的 DNA 内切酶，Flocculin，GTPase 激活蛋白，内切 1，3 - 葡萄糖酶，钙调蛋白，磷酯酶 D 和磷酸 - 2 - 3 - 双脱氧乙醛酶蛋白，因此，推测低温诱导游动孢子释放主要涉及钙信号和磷酯酸信号通路。该结果研究将有助于揭示低温促进大豆疫霉游动孢子释放的分子机理，为防治大豆疫霉提供理论基础。

关键词：大豆疫霉；低温；游动孢子；释放；钙调蛋白

* 基金项目：国家自然科学基金（31301611）

** 第一作者：侯巨梅，副研究员，主要从事植物病理学研究工作；E-mail：amyliutong@163.com

The Adaptation of Ultraviolet Irradiation in the Irish Great Famine Pathogen *Phytophthora infestans*[*]

WU E-jiao[1**], YANG Li-na[1], XIE Ye-kun[1], WANG Xue-song[1], ZHAN Jia-sui[1,2***]
(1. *Fujian Key Laboratory of Plant Virology*, *Institute of Plant Virology*, *Fujian Agriculture and Forestry University*, *Fuzhou*, *Fujian* 350002, *China*;
2. *Key Lab of Biopesticide and Chemical Biology*, *Ministry of Education*, *Fujian Agriculture and Forestry University*, *Fuzhou* 350002, *China*)

Abstract: The amount and distribution of genetic variation in *Phytophthora infestans* populations is one of the key factors influencing the occurrence and epidemic of potato late blight. Ozone layer deletion, ultraviolet irradiation is expected to increase in coming decades, accompanied by an increased occurrence of extreme temperature events. Such global trends are likely to have various major impacts on human society through their influence on natural ecosystems, food production and biotic interactions, including diseases. Here, we present a novel analysis of UV irradiation responses of *P. infestans* to changing environment using common garden experiment method. We found that the pathogen grew slower with the increased of stronger UV irradiation except the isolates sampled from Yunnan. Population differentiation for UV adaptation (Q_{ST}) was 0.94, which was significantly higher than the population differentiation for neutral SSR markers ($F_{ST}=0.12$), illustrating that adaptation to UV irradiation in *P. infestans* was under diversifying selection according to local environments and natural selection favoring for specific genotype and phenotype. Local UV irradiation can lead to local adaptation. We also found the aggressiveness of *P. infestans* was significantly reduced after 180s UV treatments except isolates collected from Yunnan.

Key words: *Phytophthora infestans*; UV irradiation; Natural selection; Genetic diversity

---

* The project was supported by the Modern Agricultural Industry and Technology System (No. CARS－10)

** First author: WU E-jiao, PhD student, major in plant pathology; E-mail: wej2012fafu@163.com

*** Corresponding author: ZHAN Jia-sui, mainly engaged in population genetics; E-mail: Jiasui.zhan@fafu.edu.cn

大蒜/辣椒间作控制辣椒疫病的化学生态学机理*

刘屹湘**，廖静静，朱书生***
（云南农业大学国家生物多样性控制病虫害工程中心，昆明　650201）

摘　要：近年来国内外对大蒜及其抑菌活性物质的研究日益增多，大量研究结果表明大蒜能够有效抑制动物体内的病原菌，同时在田间大蒜也被广泛用于间作和轮作来控制土传病害的发生和传播。大蒜素及其降解产物是大蒜具备广谱抑菌能力的主要活性物质，但是目前相关研究集中在大蒜植株浸出液中大蒜素等物质离体环境下对病原微生物的影响，而关于间作中大蒜根系和其他作物病原菌的相互作用研究较少。

针对这一问题，本研究中设计温室试验和室内试验，温室试验一方面通过设计不同比例的大蒜/辣椒间作观察间作的控病效果，另一方面通过活性炭添加和大蒜移除的方法观察大蒜根系分泌物在间作控病中的作用；室内试验一方面通过根系—微生物互作装置观察辣椒疫霉菌在大蒜根际的行为和活动，另一方面则设计不同浓度梯度的大蒜挥发物、浸提液和根系分泌物，观察其对辣椒疫霉菌菌丝生长的影响、孢子囊释放、游动孢子游动和休止、孢子萌发的影响，并通过 GC-MS 和 HPLC 进一步探索抑菌活性物质。

大田试验结果表明，大蒜和辣椒间作能够显著抑制辣椒疫霉病的发生和扩展。其中，大蒜挥发物和浸提液都能够显著抑制辣椒疫霉菌的生长和传播，有助于抑制辣椒疫霉菌在地上部的传播和侵染；另外，大蒜的根系能够吸引辣椒疫霉菌的游动孢子并使其迅速休止，抑制休止孢的萌发并最终导致裂解，从而干扰辣椒疫霉菌在土壤中的发生和扩散。进一步研究结果表明，根系分泌物是大蒜根系抑制辣椒疫病的主要途径，根系分泌物中的苯并噻唑和大蒜素降解产物都对辣椒疫霉菌表现出显著的抑制效果。

总之，大蒜的挥发物、浸提物和根系分泌物都能够干扰非寄主病原菌辣椒疫霉菌的侵染过程，进而实现田间辣椒疫病的生态管理和控制。

关键词：根系—微生物互作；间作；辣椒疫霉菌；根系分泌物；含硫化合物

---

* 基金项目：国家自然科学基金（项目批准号：31260447）
** 第一作者：刘屹湘，湖南长沙人，博士，植物营养学；E-mail：lyxcm@126.com
*** 通讯作者：朱书生，教授；E-mail：shushengzhu79@126.com

辣椒疫霉游动孢子阶段与菌丝阶段差异蛋白组分析[*]

宋维珮[**]，庞智黎[***]，刘西莉[***]
（中国农业大学植物病理学系，北京　100193）

摘　要：辣椒疫霉（*Phytophthora capsici* Leonian）是一种世界性分布的重要病原卵菌，可以侵染茄科、豆科和葫芦科等多种农作物。每年由辣椒疫霉引起的病害可造成严重的经济损失。辣椒疫霉生活史包括有性生殖和无性繁殖，其中，无性繁殖在一个成长季的多次病害循环中发挥着非常重要作用。

本研究采用同位素标记相对和绝对定量（iTRAQ）方法，进行了辣椒疫霉游动孢子阶段和菌丝阶段蛋白组分析，共鉴定到 2 258个蛋白。对获得的蛋白进行基因功能注释，结果表明这些蛋白主要参与的生物过程包括细胞组分生物合成、组装、定位、对环境刺激响应以及代谢过程。本研究丰富了卵菌蛋白数据库，为同类其他卵菌研究提供了参考。

同时，本研究分析了辣椒疫霉游动孢子阶段和菌丝阶段差异表达蛋白组，结果表明：368 个蛋白在这两个阶段具有显著的表达量差异，其中 150 个蛋白在游动孢子阶段富集，218 个蛋白在菌丝阶段富集。进一步的生物信息学分析表明，这些差异表达蛋白主要涉及细胞壁生物合成、氨基酸代谢、能量产生等重要生命活动。该研究结果为新型杀菌剂设计提供了潜在的分子靶标。

关键词：辣椒疫霉；游动孢子；菌丝；蛋白组

---

* 基金项目：国家自然科学基金（编号：31272061）

** 第一作者：宋维珮，在读硕士研究生；E-mail：jessysong@163.com

*** 通讯作者：庞智黎，博士；E-mail：pzl19870611@126.com

刘西莉，教授；主要从事杀菌剂药理学及病原菌抗药性研究；E-mail：seedling@cau.edu.cn

Genetic Diversity of ATP Synthase F_0 Subunit 6 (*ATP6*) in the *Phytophthora infestans* from Potato*

ZHANG Jia-feng**, ZHU Wen, ZHAN Jia-sui***

(*Fujian Key Lab of Plant Virology*, *Institute of Plant Virology*, *Fujian Agriculture and Forestry University*, *Fuzhou* 350002, *China*)

Abstract: The genetic diversity of ATP synthase F_0 subunit 6 (*ATP6*) in *Phytophthora infestans* was analyzed by sequencing 139 isolates sampled from potato from seven groups of six parts of China. Result show that *ATP6* in *Phytophthora infestans* population had high haplotype diversity ($H_d = 0.538 > 0.5$) but low nucleotide diversity ($P_i = 0.0009 < 0.005$). It is meant that bottlenecks associated with rapid population growth and accumulation of mutations. Most genetic variation in the gene occurred within population, and pairwise population subdivision was at low to moderate level. The haplotype network analysis showed that *ATP6* haplotype were consisted of seven groups. Neutrality tests and mismatch distribution analyses suggested that population expansion might occur in the gene.

Key words: *Phytophthora infestans*; ATP synthase F_0 subunit 6 (*ATP6*); Genetic diversity; Population expansion

---

* Funding: The project was supported by the Modern Agricultural Industry and Technology System (No. CARS－10)

** First author: ZHANG Jiafeng, male, master, major in molecular plant pathology; E-mail: 1079045694@qq.com

*** Corresponding author: ZHAN Jiasui, mainly engaged in population genetics; E-mail: Jiasui.zhan@fafu.edu.cn

Screening for RNA Virus in the Plant Pathogenic Oomycete *Phytophthora infestans*[*]

ZHAN Fang-fang[1**], ZHU Wen[1], ZHAO Zhuo-qun[1], ZHAN Jiasui[1,2***]

(1. *Fujian Key Laboratory of Plant Virology*, *Institute of Plant Virology*, *Fujian Agriculture and Forestry University*, *Fuzhou* 350002, *China*; 2. *Key Lab of Biopesticide and Chemical Biology*, *Ministry of Education*, *Fujian Agriculture and Forestry University*, *Fuzhou* 350002, *China*)

Abstract: In oomycete *Phytophthora infestans*, the causal agent of late blight disease of potato and tomato, four double-stranded RNAs (dsRNAs), namely *P. infestans* RNA virus 1 to 4 (PiRV－1 to PiRV－4) have been reported by sequence analysis and biological properties. In our study, we used genomic approaches to survey the RNA virueses in the *P. infestans* collections from China. *P. infestans* isolates maintained on rye agar slopes were refreshed on rye agar plates for mycelium production. Total RNA was extracted from mycelium of *P. infestans* using RNAiso Plus (TaKaRa) according to manufacturer's instructions and then was subjected to reverse transcription PCR (RT-PCR) assays. PiRV－4 specific primers for RT-PCR assays were based on the sequences downloaded from GenBank and targeted the entire open reading frame of this virus. RT-PCR with the specific primers resulted in the amplification of an expected band of ～2.9 kb from eleven isolates sampled form Shanxi province, China, indicating PiRV－4 infection of *P. infestans*. The purified amplicons were cloned and sequenced. BLAST analysis indicated that the eleven amplicons shared 98 to 99% nucleotide identity with the complete or partial genome sequence of PiRV－4 deposited in the GenBank database (GenBank Accession No. JN400241, JN40042, and JN400243). Further work is needed to investigate the interaction of the virus with its *P. infestans* host, such as its impacts on the growth, sporangia production, and pathogenicity of *P. infestans*.

Key words: *Phytophthora infestans*; Screening; RNA virus

---

* Funding: The project was supported by the Modern Agricultural Industry and Technology System (No. CARS－10)

** First author: ZHAN Fang-fang, research assistant, major in plant pathology; E-mail: zhanfangfang@ fafu. edu. cn

*** Corresponding author: ZHAN Jia-sui, mainly engaged in population genetics; E-mail: Jiasui. zhan@ fafu. edu. cn

大豆疫霉氮素营养吸收在其致病过程中的功能研究

王荣波，张 雄，沈丹宇，许 静，刘 虹，于 佳，窦道龙*

（南京农业大学植物保护学院，南京 210095）

摘 要：在植物与病原微生物的互作过程中，微生物要建立成功的寄生关系，不但要克服寄主的防御反应，还要能有效的从寄主体内吸取自身所需要的营养物质，其中，氮素营养是其从寄主摄取的主要养分之一。通过生物信息学分析，在大豆疫霉中分别鉴定了 8 个铵盐转运子、16 个硝酸盐/肽转运子和 78 个氨基酸转运子，并且与其他所检测物种进行比较分析，发现这些转运子的数目都是显著多于其他卵菌和真菌，表明转运子基因在大豆疫霉中经历了显著地扩张。其中，Rh 型铵盐转运子在卵菌中是保守且特异的，与此相反的是，寡肽转运子在植物和真菌中普遍存在但是在卵菌中没有鉴定到其同源基因，这表明卵菌对氮素的吸收与植物和真菌存在差异。通过分析大豆疫霉在含有不同氮源的合成培养基上生长状态，发现大豆疫霉可以在以不同氮素形式为单一氮源的培养基上生长，并且不同氮素对其生长的影响是有差异的，例如，半胱氨酸显著抑制了其生长，而硝酸盐、甲硫氨酸、谷氨酸、天冬酰胺等则有利于其生长，说明大豆疫霉可以利用不同形式的氮源，但是，对不同氮源的利用存在一定的偏好性。大豆疫霉的转录组数据显示大部分氮素转运子基因在其侵染过程中是上调表达的，并且在 RT-PCR 中得到了验证，表明氮素转运子在大豆疫霉侵染过程中发挥一定的功能。进一步研究发现大豆疫霉侵染的叶片中游离氨基酸的含量是增加的，其中谷氨酸最为显著。基于基因的特异性和受寄主诱导表达的特性，选择铵盐转运子 PsRhs 和氨基酸转运子 PsCAT3 通过 PEG 介导的沉默进一步研究其在致病中的功能，结果发现分别沉默这两个基因后大豆疫霉的致病力明显下降。这些结果表明，大豆疫霉可以利用广泛的氮源，并且在侵染过程中可能会通过干扰寄主的代谢来增加侵染组织中氨基酸的含量，同时氮素转运子在其致病中可能扮演着重要的角色。

关键词：大豆疫霉；铵盐转运子；氨基酸转运子；氮素利用；致病性

* 通讯作者：窦道龙；E-mail：ddou@ njau. edu. cn

Expansion of β-glucosidase Genes was Associated with the Evolution of Pathogenic Types in Filamentous Pathogens

ZHANG Xiong, DOU Dao-long*

(*Department of Plant Pathology*, *Nanjing Agricultural University*, 1 *Weigang Road*, *Nanjing* 210095, *China*)

Abstract: The genetic basis of pathogen adaptation to hosts is a critical issue of molecular evolution. β-glucosidases (BGLUs; EC 3. 2. 1. 21) could hydrolyze cellobiose to two molecules of glucose monomers. Although several studies have recognized that *BGLU* genes are essential for infection, but the knowledge of their pathogenic evolutionary history is unclear. In this study, we identified all *BGLU* genes in 15 fully sequenced oomycetes and fungus genomes. Comparative analysis showed a significant expansion of *BGLU* genes in oomycetes relative to fungus, also an extensively expansion in necrotrophic relative to biotrophic. Our study not only reveals a strong positive correlation between the copy number of *BGLU* genes and the phylum types (oomycetes and fungus), but also the nutritional types (biotrophic, necrotrophic and hemibiotrophic), suggesting that *BGLU* gene expansion has facilitated the evolution of pathogenic types. Further analysis found that BGLU genes could be classified into two groups (*BGLU-A*and*BGLU-B*). *BGLU-B* group genes accounted for the expansion in *Phytophthora* and were under much more relaxed selective pressure. In addition, our results suggest that expression divergence increased with duplication age and transcription factor binding sites (TFBSs) divergence also increased with duplication age in *Phytophthora sojae*. Importantly, expression divergence increased with TFBSs divergence, suggesting that TFBSs divergence has contributed to the retention of duplicated *BGLU* genes.

* Corresponding author: DOU Dao-long, E-mail: ddou@ njau. edu. cn

A Puf RNA-binding Protein Encoding Gene PlM90 Regulates the Sexual and Asexual Life Stages of the litchi Downy Blight Pathogen Peronophythora litchii*

JIANG Li-qun[1]**, YE Wen-wu[2]**, SITU Jun-jian[1], CHEN Yu-bin[1], YANG Xin-yu[2], KONG Guang-hui[2], LIU Ya-ya[1], RUNYANGA J. Tinashe[1], XI Ping-gen[1], JIANG Zi-de[1]***, WANG Yuanchao[2]***

(1. *Guangdong Province Key Laboratory of Microbial Signals and Disease Control*, *South China Agricultural University*, *Guangzhou* 510642, *China*; 2. *Department of Plant Pathology*, *Nanjing Agricultural University*, *Nanjing* 210095, *China*)

Abstract: Sexual and asexual reproduction are two key processes in the pathogenic cycle of many filamentous pathogens. For oomycete plant pathogens, the survival and dispersing are mainly due to the thick wall of oospores for physical protection and energy storage, and the large number of zoospores which have two flagella and motility in water, respectively. However in *Peronophythora litchii*, the causal pathogen for the litchi downy blight disease, functional study with critical regulator(s) of sexual or asexual differentiation has not been carried out due to unavailability of the genetic manipulation system. In this study, we cloned a gene named *PlM*90 from *P. litchii*, which encodes a putative Puf RNA-binding protein. We found that *PlM*90 was highly expressed in stages of sporangia, zoospores and cysts, but relatively lower during cysts germination and plant infection. By polyethylene glycol (PEG)-mediated protoplast transformation, we generated three *PlM*90-silencing transformants and found a delay of zoospores release from sporangia and zoospores encystment, and a severely impaired ability in oospores production, but the pathogenicity of P. litchii was not affected by *PlM*90-silencing. Therefore we conclude that *PlM*90 specifically regulates the sexual and asexual differentiation of *P. litchii*.

Key words: Oomycete; *Peronophythora litchi*; Sexual and asexual reproductions; Zoospores; Oospores

* Funding: China Agriculture Research System (CARS-33-07) and the Natural Science Foundation of Guangdong Province, China (2016A030313408)

** First authors: JIANG Li-qun, Ph. D. candidate, E-mail: 526508682@ qq. com; YE Wen wu, E-mail: yeww@ njau. edu. cn

*** Corresponding authors: JIANG Zi-de; E-mail: zdjiang@ scau. edu. cn

WANG Yuan-chao; E-mail: ycwang@ njau. edu. cn

Intrinsic Disorder is a Common Structural Characteristic of RxLR Effector Proteins in Oomycetes

SHEN Dan-yu, LI Qing-ling, ZHANG Mei-xiang, DOU Dao-long*
(*Department of Plant Pathology*, *Nanjing Agricultural University*, *Nanjing* 210095, *China*)

Abstract: Intrinsic disorder is very common in nature and has never been investigated in oomycete RxLR effectors. We used PONDR VL-XT tool to analyze the abundance of intrinsic disorder in oomycete RxLR effectors. This analysis showed that abundant disorder content was found in RxLR proteins, largely surpassed that in another two effector families. These RxLR proteins exhibited unusual amino acid compositional profile and were enriched in disorder-promoting residues. Furthermore, the distribution of disordered residues presented personal preference that RxLR-dEER regions were enriched in disordered residues. In contrast, the disorder content was depleted in the C-terminal regions, especially for W/Y/L motifs. We also found that around 42% of RxLR proteins were predicted to contain at least one α-MoRF, and most α-MoRFs were located in the C-terminal regions. We postulated that the high abundant disorder content might be relative to effector translocation and manipulation of plant responses. Overall, these results demonstrate that intrinsic disorder is a common characteristic of RxLR proteins, which extends our understanding of RxLR effectors in protein structures, and opens up new directions to explore novel mechanisms of oomycete RxLR effectors.

Key words: Intrinsic disorder; RxLR effector; Oomycete

* Corresponding author: DOU Dao-long, E-mail: ddou@ njau. edu. cn

Phytophthora sojae Avirulence Effector PsAvr3c Target Soybean Serine and Arginine Rich Proteins SRKPs, a Novel Component of Alternative Splicing Complex, to Regulate Plant Immunity

HUANG Jie[1], KONG Guang-hui[1], YAN Ting-xiu[1], HANG Yu[1], ZHANG Ying[1], QIU Min[1], GUO Bao-dian[1], KONG Liang[1], JING Mao-feng[1], Mark Gijzen[2], WANG Yuan-chao[1], DONG Suo-meng[1]

(1. *Department of Plant Pathology, Nanjing Agricultural University, Nanjing, China*;
2. *Southern crop protection and food research center, Agricultural and Agri-Food Canada, London, Canada*)

Abstract: Avirulence (AVR) effectors, recognized by host resistant protein, are a group of key players in plant-microbe interactions. However, how AVR effectors manipulate plant immunity remains poorly characterized. The AVR effector PsAvr3c from soybean (*Glycine max*) root rot pathogen *Phytophthora sojae* is recognized by the soybean Rps3c receptor in a nuclear dependent manner. PsAvr3c also functions as a virulence protein to enhance *Phytophthora* colonization. Recently, we identified two soybean serine/arginine rich proteins GmSRKP1 and GmSRKP2 as interactors of PsAvr3c. We demonstrated that PsAvr3c physically interacts with GmSRKP1/2 and stabilizes GmSRKPs proteins *in vivo*. The GmSRKPs are conserved proteins in plants without any predicted functional domains other than three serine/arginine/lysine rich repeat regions and a nuclear localization signal. Transient gene silencing and over-expression of *GmSRKP*1/2 in soybean hairy roots indicates that GmSRKP1/2 are susceptibility factors. Furthermore, biochemical assays suggest that GmSRKPs possess RNA binding activity *in vitro*, and GmSRKPs also associated with some known alternative splicing factors. In summary, these findings provide evidence that a pathogen AVR effector targets a novel component of alternative splicing complex to subvert plant immunity.

A *Phytophthora* RxLR Effector Manipulates Host Immunity by Regulating SAGA-mediated Histone Acetylation Modification

KONG Liang, QIU Xu-fang, KANG Jian-gang, WANG Yang, QIU Min, LIN Ya-chun, KONG Guang-hui, HUANG Jie, WANG Yan, YE Wen-wu, DONG Suo-meng, WANG Yuan-chao
(*Plant Pathology Department*, *Nanjing Agricultural University*, *Nanjing*, *China* 210095)

Abstract: Plant pathogens secrete a large number of effector proteins to subvert host cell immunity. However, knowledge of the mechanism on how effector proteins manipulate host immunity remain largely unknown. Epigenetic modifications, such as histone modification, serve as important strategy of host innate immunity against various pathogens. Nevertheless, pathogens secretes effector proteins to suppress host histone modification pathways. Here, we shows that PsAvh23, a virulence-essential effector from the soybean stem and root rot pathogen *Phytophthora sojae*, interacts with plant transcriptional coactivator protein ADA2b (alteration/deficiency in activation), a subunit of SAGA complex. The SAGA complex recruits the catalytic subunit Gcn5 (general control nonderepressible 5) via ADA2b to regulate histone acetylation modification. We found that PsAvh23 competes with soybean Gcn5a (Gcn5 subunit of *G. max*) and depletes it out of ADA2b-Gcn5 subcomplex of SAGA complex, resulting in a significant reduction of histone acetylation and H3K9ac *in vitro* and *in vivo*. In addition, we demonstrated that the PsAvh23 down-regulated a subsets of defense-related genesand thisrequires SAGA-mediated histone acetylation. Our findings reveal a novel mechanism that plant pathogens can employ effector proteins to regulate host epigenetic modifications, such as histone acetylation modification, as a general virulence strategy to counteract host defenseat an epigenetic level.

Key words: Effector proteins; Epigenetic modifications; SAGA complex; Host defense

A Nucleus-localized Effector from *Phytophthora sojae* Recruits a N-acetyltransferase Into Nucleus to Suppress Plant Immunity

LI Hai-yang, WANG Hao-nan, JING Mao-feng, KONG Liang, GUO Bao-dian, WANG Yang, LIN Ya-chun, CHEN Han, YANG Bo, KONG Guang-hui, ZHAO Yao, MA Zhen-chuan, WANGYan, YE Wen-wu, DONG Suo-meng, WANG Yuan-chao

(*Department of Plant Pathology*, *Nanjing Agricultural University*, *Nanjing* 210095, *China*)

Abstract: Filamentous fungi and oomycete pathogens secrete many intracellular effectors to manipulate host immunity during infection. Identification of plant targets of these effectors will help to uncover the mechanisms on how effectors suppress PAMP-triggered immunity (PTI) and effector-triggered immunity (ETI) in plants. Previously we have determined that PsAvh52, an effector secreted from soybean root rot pathogen *Phytophthora sojae*, can suppress the cell death induced by both the effectors (Avh238 or Avh241) and PAMPs (INF1 and XEG1) in *Nicotiana benthamiana*. Here, we demonstrated that PsAvh52 interacts with a soybean histone acetyltransferase (GmHAT1), a key factor manipulating epigenetic modifications. GmHAT1 localizes in plant cell cytoplasm, but it was translocalized from the cytoplasm to nucleus when co-expressed with PsAvh52. The nucleus-localized GmHAT1 regulates the expression of plant defense-related genes and attenuates plant resistance by acetylating histones. Taken together, these results indicate that PsAvh52 manipulates epigenetic modifications to enhance *Phytophthora sojae* colonization in soybean.

Key words: *Phytophthora sojae*; effector; plant defense; epigenetic modifications

A Coin with Double sides? *Phytophthora* Essential Effector Avh238 has Dual Functions in Cell Death Activation Andplant Immunity Suppression

YANG Bo[1], WANG Qun-qing[2], JING Mao-feng[1], GUO Bao-dian[1], WANG Hao-nan[1], LI Hai-yang[1], WANG Yang[1], YE Wen-wu[1], WANG Yan[1], DONG Suo-meng[1], WANG Yuan-chao[1]

(1. *Department of Plant Pathology*, *Nanjing Agricultural University*, *Nanjing* 210095, *China*; 2. *Department of Botany and Plant Pathology*, *Oregon State University*, *Corvallis*, *Oregon* 97331)

Abstract: *Phytophthora* pathogens secrete effectors to manipulate host innate immunity and thereby to facilitate infection. Avh238 is anRxLR effector that not only contributes to the virulence of *Phytophthora sojae* but also triggers cell death in various plant species. To date, the detailed molecular basis of Avh238 functions islargely unknown. Here, we showed that transient expression of Avh238 in *Nicotiana benthamiana* revealed its nucleus and cytoplasm localization. Avh238-triggered cell death requires the nuclear localization, but suppression of INF1-triggered plant cell death requires the cytoplasmic localization. Furthermore, we performed a natural variation investigation together with a large scale mutagenesis assay, and identified the key residues that are required for both functions. Taken together, our results demonstrate that the cell death-inducing and cell-death suppression activity of Avh238 are uncoupled. This work uncovers that a conserved *Phytophthora* RxLR effector has evolved to escape the host recognition but retained the host immunity-suppressing activity to enhance virulence *in planta*.

Cloning and Functional Analysis of Succinate Dehydrogenase Gene *PsSDHA* in *Phytophthora sojae* *

PAN Yue-min**, YE Tao**, ZHANG Jin-yuan,
ZHENG Ting, LIU Ten-fei, GAO Zhi-mou***
(*Department of Plant Pathology*, *College of Plant Protection*,
Anhui Agricultural University, *Hefei* 230036)

Abstract: Succinate dehydrogenase (SDH) is one of the key enzymes of the tricarboxylic acid cycle (TCA cycle) and the only multi-subunit one embedded in the membrane. To research the roles of the *SDHA* gene in *Phytophthora sojae*, we first cloned the conservative *PsSDHA* gene to construct the *PsSDHA* silenced expression vector pHAM34-*PsSDHA*, and then utilized PEG to mediate the *P. sojae* protoplast transformation experiment. Through transformation screening, we obtained the silenced mutants A1 and A3 identified by PCR and RT-qPCR, which have good suppression effects. Further study showed that the silenced mutant strains were shorter and more bifurcated; the growth of the silenced mutants was clearly inhibited in 10% V8 culture medium containing 0.7mol/L sodium chloride (NaCl), 0.01% sodium dodecyl sulfate (SDS), 3mmol/L hydrogen peroxide (H_2O_2) and 200ng/mL Congo Red. The pathogenicity of the silenced mutants was significantly reduced compared with the wild-type strain and the control strain. Through studying this gene, we can better understand the position and function of SDH in *P. sojae*, and we can provide theoretical basis for further research.

Key words: *Phytophthora sojae*; Succinate dehydrogenase; RNAi; Phenotypic analysis

---

* Funding: This work was supported by the Commonwealth Specialized Research Fund of China Agriculture (Grant No. 201303018) and the Anhui Provincial Natural Science Foundation (Grant No. 1408085MC56).

** First author: PAN Yue-min and YE Tao contributed equally to the paper

*** Corresponding author: GAO Zhi-mou; E-mail: gaozhimou@126.com

Comparative Transcriptome Analysis Between *Phytophthora infestans*-resistant and-sensitive Tomato*

CUI Jun**, JIANG Ning, LUAN Yu-shi***

(*School of Life science and Biotechnology, Dalian University of Technology*)

Phytophthora infestans is a biotrophic pathogen, causal agent of tomato late blight (LB). The oomycetes, which have a worldwide occurrence and have long been prevalent in Korea, China, the USA, and others, have caused serious economic loss for field-grown tomatoes, and therefore is regarded as a major threat to tomato production. A wild tomato *S. pimpinellifolium* L3708, a resistant line, has been identified to be highly resistant to a wide range of *P. infestans* isolates overcoming *Ph*-1, *Ph*2 and *Ph*-3. Another cultivated tomato, *S. lycopersicum* Zaofen No. 2 was breeded by the Institute of Vegetables and Flowers, Chinese Academy of Agricultural Sciences, Beijing, China. Tomato Zaofen No. 2 is a susceptible accession to a variety of pathogens including *P. infestans*. In this study, we preformed comparative transcriptome profiling study on the differential expressed genes and lncRNAs (DEGs and DELs) between*S. pimpinellifolium* L3708 and *S. lycopersicum* Zaofen No. 2. More than 80 000 000 raw sequence reads were generated from two samples using the high-throughput sequencing. After expression levels (FPKM) for each gene and lncRNAs were calculated, a total of 1 037 genes and 688 lncRNAs showed significantly differential expression in leaves of Sp samples compared with Slz sample, including 463 up-regulated genes and 277 up-regulated lncRNAs. It was found that a total of 128 DEGs might be regulated 127 DELs by co-expressed analysis. GO analysis indicated that oxidoreductase activity, which might play an important role in regulating ROS to reduce damage to plant cells in plant-pathogen interactions, was highly represented, including 17 genes. In this GO term, 6 members of glutaredoxin (GRX) gene family was found, which were regulated 7 DELs. The results of RT-qPCR showed that the expression levels ofthese 6 GRXs and 7 DELs were consistent with them through RNA-Seq. These discoveries will provide us useful information to explain tomato resistance mechanisms against LB.

---

* This work was supported by grants from the National Natural Science Foundation of China (Nos. 31272167, 31471880 and 61472061)

** First author: CUI Jun, Ph. D candidate; E-mail: cuijun@ mail. dlut. edu. cn

*** Corresponding author: LUAN Yu-shi, Professor; E-mail: ysluan@ dlut. edu. cn

Pathogen Identification of Sisal Zebra Disease of China

ZHENG Jin-long[1], YI Ke-xian[1,2], XI Jin-gen[1], GAO Jian-ming[2], ZHANG Shi-qing[2], CHEN He-long[2], HE Chun-ping[1], WU Wei-huai[1], ZHENG Xiao-lan[1], LIANG Yan-qiong[1]

(1. *Environment and Plant Protection Institute*, *Chinese Academy of Tropical Agricultural Sciences*; *Key Laboratory of Integrated Pest Management on Tropical Crops*, *Ministry of Agriculture*, *China*, *Haikou*, *Hainan* 571101; 2. *Institute of Tropical Bioscience and Biotechnology*, *Chinese Academy of Tropical Agricultural Sciences*; *Key Laboratory of Tropical Crop Biotechnology*, *Ministry of Agriculture*, *China*, *Haikou*, *Hainan* 571101)

Abstract: Sisal zebra disease is a devastating oomycetes disease, one of most serious diseases of sisal hemp. There are 12 samples of sisal zebra disease collected from major production areas of Hainan, Guangdong and Guangxi, from which genomic DNA of its pathogens are extracted, and the rDNA-ITS sequence amplification is conducted. The result of sequence comparison is consistent with the morphological identification results. This suggests that all these tested strains of pathogens are Phytophthora nicotianae Breda. Results of phylogenetic tree which is constructed under N-j method show that, 12 tested strains can be divided into two groups and they are of certain geographical differentiation. The above research will lay foundations for future researches in aspects of occurrence regularity, control methods and mechanisms of resistance of agave zebra disease and provide them with theoretical basis.

Key words: Sisal hemp; Zebra disease; RDNA-ITS; Sequence analysis; Pathogen identification

The Biological Functions of Nudix Effector Proteins in Potato Late Blight Famine Pathogen *Phytophthora infestans*

YAN Ting-xiu[1], HUANG Jie[1], LIN Long[1], KONG Guang-hui[1], MA Zhen-chuan[1], YE Wen-wu[1], MA Hong-yu[1], TIAN Zhen-dong[3], ZHENG Xiao-bo[1,2], WANG Yuan-chao[1,2], DONG Suo-meng[1,2*]

(1. *College of Plant Protection*, *Nanjing Agricultural University*, *Nanjing* 210095, *China*; 2. *The Key Laboratory of Integrated Management of Crop Diseases and Pests*, *Ministry of Education*, *Nanjing* 210095, *China*; 3. *College of horticulture*, *Huazhong Agricultural University*, *Wuhan*, *Hubei* 430070, *China*.)

Abstract: Fungal and oomycete plant pathogens secrete effector proteins into host cells to promote infection. Previously, we identified an oomycete effector protein PsAvr3b encodes a Nudix protein with phosphohydrolase activity. However, questions like the in planta biological functions and the evolution of the Nudix effectors remain largely unknown. Here, we took Nudix effectors from potato late blight famine pathogen *Phytophthora infestans* as model to address these questions. We had cloned three Nudix effectors named *NUD*1, *NUD*2 and *NUD*3 from *P. infetans* cDNA, finding that unlike most of other nudix effectors, each of *P. infetans* Nudix effector gene have two introns. Interestingly, NUD3 undergoes alternative splicing with two spliced forms. Furthermore, we found that three Nudix effectors have distinct functions in suppressing hypersensitive response triggered by *Phytophthora* elicitors INF1, NLP and PiXEG1. Transient expressing of Nudix effectors in *N. benthamiana* leaves show that they mainly localized in cell cytoplasm, as well as cell membrane. Infection assay demonstrated that the only NUD3 effector increased susceptibility to *P. infestans*. Mutation of the key residues in Nudix motif abolished both HR suppression and susceptibility functions. We are now screening substrates of three Nudix effectors using HPLC. Our data suggesting that three Nudix effectors may potentially perform distinct biological virulent functions during infection.

* Corresponding author: DONG Suo-meng; E-mail: smdong@ njau. edu. cn

Identification of a Novel Cysteine Rich Effector SCR2 from Plant Oomycete Pathogens

WANG Shuai-shuai[1], XING Rong-kang[1], HU ANGJie[1], YE Wen-wu[1,4], Liliana Cano[2], WU Chih-hang[2], ZHANG Zheng-guang[1], WANG Yuan-chao[1,4], Vivianne Vleeshouwers[3], Sophien Kamoun[2], DONG Suo-meng[1,4 *], ZHENG Xiao-bo[1,4]

(1. *College of Plant Protection*, *Nanjing Agricultural University*, *Nanjing* 210095, *China*; 2. *The Sainsbury Laboratory*, *Norwich Research Park*, *Norwich NR*4 7*UH*, *United Kindom*; 3. *Wageningen UR Plant Breeding*, *Wageningen University and Research Centre*, *Droevendaalsesteeg* 1, *Wageningen* 6708 *PB*, *The Netherlands*; 4. *The Key Laboratory of Integrated Management of Crop Diseases and Pests*, *Ministry of Education*, *Nanjing* 210095, *China*)

Abstract: Oomycete, a group of filamentous plant pathogens, is responsible for most notorious crop diseases such as potato late blight that threaten the global food security. It has been known that many small cysteine rich (SCR) effectorgenes are highly inducedduring oomycete infection, however, the molecular mode of actions of SCR effectors remain largely unknown. A preliminary bioinformatics analysis combined with high-throughput*in planta* screening assay resulted in the identification of a *Phytophthora infestans* cysteine rich effector *SCR*2, which could trigger significant reactive oxygen burst, defence related gene induction and hypersensitive response in *Nicotiana benthamiana*. Unlike many other SCR effectors, *SCR*2 orthologs are wildly presented in oomycete plant pathogens, including species from *Phytophthora*, *Hyaloperonospora*, *Pythium* and *Plasmopara*. Virus induced gene silencing (VIGS) found that *SCR*2 induced host immunity required immune hub regulator BAK1 and akinase MAPKKKξ. Furthermore, chemical inhibition assay suggested that Serine proteases are also involved in *SCR*2 perception. In summary, we identified a novel SCR effector with priming host immunity function. The biochemical functions of *SCR*2 and the mechanisms of recognition deserve further investigations.

* Corresponding author: DONG Suo-meng; E-mail: smdong@ njau. edu. cn

第三部分 病 毒

Beyond the Suppressor Function, γb Protein Directly Participates in *Barley Stripe Mosaic Virus* Replication by Interacting with Replicase αa at Chloroplast*

ZHANG Kun**, ZHANG Yong-liang, YANG Meng, LIU Song-yu, LI Zheng-gang, WANG Xian-bing, HAN Cheng-gui, YU Jia-lin, LI Da-wei***

(*State Key Laboratory of Agro-Biotechnology*, *College of Biological Sciences*, *China Agricultural University*, *Beijing* 100193, *P. R. China*)

Abstract: Viral suppressor of RNA silencing (VSR) were commonly utilized by various viruses to counter host defense. However, whether other functions beyond the suppression of antiviral silencing existed in VSR were largely unexplored. In this study, a novel function of Barley stripe mosaic virus (BSMV) γb is described, subcellular localization analysis showed that γb protein could be recruited to the chloroplast and formed granules in the context of viral infection, additional evidence verified that BSMV replication occurs in the chloroplast. Further studies demonstrated that enhanced association of γb with the chloroplast required its direct interaction with the viral replication protein αa. Mutation analysis revealed that γb is dispensable for the initiation of viral replication and has little effect on the production of viral minus-strand RNA. However, accumulation of viral plus-strand RNA was differentially affected by mutating the γb. Intriguingly, disruption of γb suppressor activity could only moderately affected the viral (+) RNA accumulation, whereas mutant BSMV lacking either the αa-interacting domain or the single-stranded RNA-binding activity within the γb protein exhibited a dramatic reduction in viral (+) RNA level. Our results, for the first time, provide evidence that VSR could directly participate in the viral replication beyond their roles in suppressing antiviral immunity, and would broaden our understanding of the multifunctional roles played by the suppressor proteins in viral infections.

Key words: Barley stripe mosaic virus; γb protein; Chloroplast; Replication; αa replicase protein; Interaction

---

* This work was supported by the National Natural Science Foundation of China (31270184 and 31570143)

** These authors contributed equally to this work

*** Corresponding author: LI Da-wei; E-mail: Lidw@ cau. edu. cn

Dissecting the Role of Hsp70 in Beet Black Scorch Virus Infection*

WANG Xiaoling**, CAO Xiu-ling, LIU Min, ZHANG Rui-qi, ZHANG Xin, WANG Xian-bing, LI Da-wei, ZHANG Yong-liang***

(*State Key Laboratory of Agro-Biotechnology, College of Biological Sciences, China Agricultural University, Beijing* 100193, *P. R. China*)

Abstract: Many RNA viruses subvert large number of host proteins to complete the viral life cycle, such as genomic RNA replication and viral movement, encapsidation and to interfere with host antiviral responses. In order to find host proteins involved in Beet black scorch virus (BBSV) infection, P23, the auxiliary replication protein, and the coat protein (CP) were fused with a FLAG tag, respectively, followed by agro-infiltration assays. Based upon an affinity purification approach and liquid chromatography-tandem mass spectrometry (LC-MS/MS), we identified heat shock protein70 (Hsp70) as a component existing in both complexes that was co-purified with P23 and CP. In this work, we confirmed that Hsp70 interact with both CP and p23 by Co-IP, pull-down, and BiFC, respectively, whereas no interaction was detected between p23 and CP. Furthermore, dual-color trimolecular fluorescence complementation reveals the formation of ternary p23-Hsp70-CP complexes *in vivo*, which localized to vesicles derived from endoplasmic reticulum. Using qRT-PCR, we revealed that both mRNA and protein level of Hsp70 transcripts increased during BBSV infection. In addition, VIGS or inhibitor down-regulation of Hsp70, leads to the decrement of BBSV genomic RNA and coat protein accumulation. *Agrobacterium*-mediated expression of p23 leads to hsp70 mRNA up-regulated whereas CP leads to opposite directions. BBSV CP, which acts in RNA encapsidation and long distance movement, also represses viral RNA replication in a dose-dependent manner. Altogether, in order to achieve persistent infection, CP interact with both P23 and Hsp70 to balance replication and encapsidation via redistributed to ER-associated vesicles. Hence, our findings unveiled Hsp70 spatio-temporally regulates the infection of BBSV in *N. benthamiana via* interacting with different viral components at diverse viral infection stages.

Key words: Beet black scorch virus; P23; CP; Hsp70; Interaction; Infection

---

* This work was supported by the National Natural Science Foundation of China (Grant No. 31470253)

** These authors contributed equally to this work

*** Corresponding author: ZHANG Yong-Liang, Associate Professor; E-mail: cauzhangyl@ cau. edu. cn

Soil Transmission of Cucumber Green Mottle Mosaic Virus Associated with Plant Debris*

LIANG Chao-qiong[2**], Abie Xiao-bing[1,2], LIU Hua-wei[3], MENG Yan[1,2,4], LUO Lai-xin[1,2], LI Jian-qiang[1,2***]

(1. *Department of Plant Pathology, China Agricultural University/Key laboratory of Plant Pathology, Ministry of Agriculture, Beijing* 100193, *P. R. China*; 2. *Beijing Engineering Research Center of Seed and Plant Health/Beijing Key Laboratory of Seed Disease Testing and Control, Beijing* 100193, *P. R. China*; 3. *Molecular Plant Pathology, United States Department of Agriculture, Agricultural Research Service, Beltsville, Maryland*, 20705, *USA*; 4. *Department of Plant Pathology and Microbiology, Iowa State University, Ames* 50011, *USA*)

Abstract: Cucumber green mottle mosaic virus (CGMMV), which belongs to the genus *Tobamovirus*, is a major pathogen of cucurbits. The typical symptoms of CGMMV disease include mosaic pattern on infected leaves and fruit distortions. It is well known that CGMMV could be transmitted by direct mechanical contact with the infected source plant or other contaminated materials, pruning and irrigation, seed, pollen and phytoparasitic weed dodder. However, soil transmission of CGMMV is still a heated debate and few studies support it. The goal of this study is to provide the evidence for soil transmission of CGMMV by RT-PCR (reverse transcription-polymerase chain reaction). The results indicated that tested leaf samples, collected from cucumber plants grown in contaminated soils containing CGMMV-infected cucumber debris, are confirmed CGMMV-positive. Meanwhile, when contaminated soil did not contain CGMMV-infected cucumber debris, tested leaf samples may produce inconclusive or only weakly CGMMV-positive. The study showed that CGMMV can be transmitted by soil containing CGMMV-infected cucumber debris. It is proposed that soil is one of the primary infection sources for spreading of CGMMV. The results have meaningful epidemiological implications for the management of CGMMV disease, particularly regarding the role of soil as CGMMV vector.

Key words: Soil transmission; Cucumber green mottle mosaic virus; Plant debris

---

* Funding: National Natural Science Foundation of China (NSFC) (Project No. 31371910)

** First author: LIANG Chao-qiong, PhD student, mainly focused on the study of plant virus and host interactions; E-mail: lcq19880305@126.com

*** Corresponding author: LI Jian-qiang, research field focused on Seed Pathology and Fungicide Pharmacology; E-mail: lijq231@cau.edu.cn

Identification of Cucumber MicroRNA Targets Responding to Infection of Cucumber Green Mottle Mosaic Virus*

LIANG Chao-qiong[2**], LIU Hua-wei[3], MENG Yan[1,2,4], A BIE Xiao-bing[1,2], LUO Lai-xin[1,2], LI Jian-qiang[1,2***]

(1. *Department of Plant Pathology, China Agricultural University/Key laboratory of Plant Pathology, Ministry of Agriculture, Beijing* 100193, *P. R. China*; 2. *Beijing Engineering Research Center of Seed and Plant Health/Beijing Key Laboratory of Seed Disease Testing and Control, Beijing* 100193, *P. R. China*; 3. *Molecular Plant Pathology, United States Department of Agriculture, Agricultural Research Service, Beltsville, Maryland* 20705, *USA*; 4. *Department of Plant Pathology and Microbiology, Iowa State University, Ames* 50011, *USA*)

Abstract: Cucumber green mottle mosaic virus (CGMMV) is one of the *Tobamovirus*, which has spread fast in China, and has a narrow host range confined to the *Cucurbitaceae*. It causes severe mosaic symptoms with discoloration and deformation on plants. MicroRNAs (miRNAs) are endogenous small RNAs playing an important regulatory function in plant development, stress responses and host-pathogen interaction. MicroRNAs regulate target gene expression by mediating target gene cleavage or inhibition of translation at transcriptional and post-transcriptional levels in higher plants. Many cucumber (*Cucumis sativus*) miRNAs have been identified, but the miRNA targets from CGMMV-infected cucumber is still little known. In this study, we validated a number of miRNA target genes in CGMMV-infected cucumber that have been identified in previous studies. Degradome sequencing for CGMMV-infected cucumber leaf samples produces 10 974 567 raw reads and 4 265 600 of them are unique. Thereinto, 2 080 311 unique transcript mapped reads could be matched with cucumber mRNAs. Then, 25 681 transcripts could be covered cucumber mRNAs with sequence information of cucumber miRNAs using BLASTN program. Finally, 3046 target transcripts for 47 cucumber miRNAs families are experimentally verified by degradome sequencing. The results show that miR159, miR166, miR2673 and csa-miRn6 – 3p negatively regulate expression of their target mRNAs through guiding corresponding target mRNA cleavage, and cleave their target mRNAs mainly at the tenth nucleotide of 5′-end of miRNA. Our results confirm the importance of miRNAs in cucumber developmental processes and interactions between cucumber and CGMMV.

Key words: MicroRNA; Targets; Cucumber; Cucumber green mottle mosaic virus; Degradome sequencing

---

* Funding: National Natural Science Foundation of China (NSFC) (Project No. 31371910)

** First author: LIANG Chao – qiong, PhD student, mainly focused on the study of plant virus and host interactions; E-mail: lcq19880305@126.com

*** Corresponding author: LI Jian-qiang, research field focused on Seed Pathology and Fungicide Pharmacology; E-mail: lijq231@cau.edu.cn

Integrative Analysis Elucidates the Response of Susceptible Rice Plants to Rice Stripe Virus (RSV)*

YANG Jian, ZHANG Fen, LI Jing, CHEN Jin-Ping**, ZHANG Heng-mu**

(State Key Laboratory Breeding Base for Zhejiang Sustainable Pest and Disease Control; Key Laboratory of Plant Protection and Biotechnology, Ministry of Agriculture, China; Zhejiang Provincial Key Laboratory of Plant Virology, Institute of Virology and Biotechnology, Zhejiang Academy of Agricultural Sciences, Hangzhou 310021, China)

Abstract: Rice is not only a staple cereal but also a model monocotyledonous plant, for which much genomic and related expression sequence tag (EST) information is available. Its production is threatened by at least 15 virus or virus-like diseases (Hibino, et al., 1996), one of the most serious of which is rice stripe disease. This is caused by Rice stripe virus (RSV) which is transmitted by the small brown planthopper, *Laodelphax striatellus* Fallén, in a persistent and transovarial propagative manner and which can infect many different species of plants in the family Gramineae. Many host factors are known to be involved in the response of plants to viruses, which could be important for developing new strategies for disease control. Nevertheless, our knowledge of the plant response to RSV infection is still very limited. To investigate how rice responds to RSV infection, here we integrated miRNA expression with parallel mRNA transcription profiling by deep sequencing in susceptible rice plants. A total of 570 miRNAs were identified of which 69 miRNAs (56 up-regulated and 13 down-regulated) were significantly modified by RSV infection. Digital gene expression (DGE) analysis showed that 1 274 mRNAs (431 up-regulated and 843 down-regulated genes) were differentially expressed by RSV infection. The differential expression of selected miRNAs and mRNAs was confirmed by qRT-PCR. Gene ontology (GO) and pathway enrichment analysis showed that a complex set of miRNA and mRNA networks were selectively regulated by RSV infection. In particular, 63 differentially expressed miRNAs were found to be significantly and negatively correlated with 160 target mRNAs. Interestingly, 22 up-regulated miRNAs were negatively correlated with 24 down-regulated mRNAs encoding disease resistance-related proteins, indicating that the host defense responses were selectively suppressed by RSV infection. The suppression of both osa-miR1 423 – 5p-and osa-miR1 870 – 5p-mediated resistance pathways was further confirmed by qRT-PCR. Chloroplast functions were also targeted by RSV, especially the zeaxanthin cycle, which would affect the stability of thylakoid membranes and the biosynthesis of ABA (Havaux, 1998; Finkelstein et al., 2013). All these modifications may contribute to viral symptom development and provide new insights into the pathogenicity mechanisms of RSV.

* Funding: National Science and Technology Support Program (2012BAD19B03) and the State Basic Research Program of China (2014CB160309)

** Corresponding authors: CHEN Jin-Ping; E-mail: jpchen2001@126.com
ZHANG Heng-mu; E-mail: zhhengmu@tsinghua.org.cn

Interaction of HSP20 With a Viral RdRp Changes Its sub-Cellular Localization and Distribution Pattern in Plants*

LI Jing, XIANG Cong-ying, YANG Jian, ZHANG Heng-mu**, CHEN Jian-ping**

(State Key Laboratory Breeding Base for Zhejiang Sustainable Pest and Disease Control; Key Laboratory of Plant Protection and Biotechnology, Ministry of Agriculture, China; Zhejiang Provincial Key Laboratory of Plant Virology, Institute of Virology and Biotechnology, Zhejiang Academy of Agricultural Sciences, Hangzhou 310021, China)

Abstract: Plant heat shock proteins (HSPs) are stimulated in response to a wide array of stress conditions and perform a fundamental role in protecting plants against abiotic stresses. Generally, HSPs function as molecular chaperones, facilitating the native folding of proteins and preventing irreversible aggregation of denatured proteins during stress. HSPs can be classified into five major categories based on molecular mass and sequence homology: HSPp100/ClpB, HSP90, 70 kDa heat shock protein (HSP70/DnaK), chaperonin (HSP60/GroEL), and small heat shock protein (sHSP), which is one of the most abundant and complex groups and is characterized by a conserved α-crystallin domain (ACD) of 80 ~ 100 amino acids in the C-terminal region (Haslbeck et al., 2005). There have been few reports that sHSPs-perform a fundamental role in protecting cells against a wide array of stresses but their biological function during viral infection remains unknown. Rice stripe disease is one of most devastating viral diseases of rice in East Asia (Hibino, 1996). Infected plants often have chlorotic stripes or mottling and necrotic streaks in the newly expanded leaves and growth is stunted. The causal agent, Rice stripe virus (RSV), is one of best-studied rice viruses and is the type member of the genus Tenuivirus. Interestingly, the expression of a rice small heat shock protein 20 (OsHSP20) and its homolog of *N. benthamiana* (NbHSP20) were found to be significantly regulated by the viral infection, which were further validated by digital gene expression profiling and/or qRT-PCR database, suggesting that both might be involved in the host response to RSV infection. To investigate the relationship between the sHSPs and RSV infection, both OsHSP20 and NbHSP20 were then fused with eGFP at their C terminus and introduced into N. benthamiana epidermal cells. In confocal microscopy, both formed numerous granules with a variety of sizes *in vivo* and moved at different speeds in the cytoplasm of non-infected cells; However, almost no GFP granules were detectable in the cytoplasm of RSV-infected cells, suggesting the interaction between the HSP20 and viral proteins. To determine the protein interaction, both sHSPs were used as baits in yeast two-hybrid (YTH) assays to screen an RSV cDNA library and were found to inter-

* Funding: National Science and Technology Support Program (2012BAD19B03) and the Zhejiang Provincial Foundation for Natural Science (LQ14C140003)

** Corresponding authors: ZHANG Heng-mu; E-mail: zhhengmu@tsinghua.org.cn
CHEN Jin-Ping; E-mail: jpchen2001@126.com

act with the viral RNA-dependent RNA polymerase (RdRp) of RSV. Interactions were validated by pull-down and BiFC assays. Further analysis showed that the N-terminus (residues 1 ~ 296) of the RdRp was crucial for the interaction between the HSP20s and viral RdRp and responsible for the alteration of the sub-cellular localization and distribution pattern of HSP20s in protoplasts of rice and epidermal cells of *N. benthamiana*. This is the first report that a plant virus or a viral protein alters the expression pattern or sub-cellular distribution of sHSPs.

三种马铃薯Y病毒属病毒通用型单克隆抗体的鉴定

王永志*，李小宇[1]，张春雨[1]，李启云[1]**，宋景荣[2]，尤　晴[1]，张淋淋[1]
（1. 吉林省农业科学院植物保护研究所，农业部东北作物有害生物综合治理重点实验室，吉林省农业微生物重点实验室，公主岭　136100；
2. 呼伦贝尔市农业科学研究所，扎兰屯　162652）

摘　要：鉴定和分析马铃薯Y病毒属的三种病毒病，即大豆花叶病毒（Soybean mosaic virus，SMV）、马铃薯Y病毒（Potato virus Y，PVY）、三叶草黄脉病毒（Clover yellow vein virus，CLYVV）的通用型单克隆抗体。提取PVY感染的马铃薯叶片和SMV感染的大豆叶片总RNA，RT-PCR克隆PVY和SMV的Coat protein（CP）基因，利用已有的CLYVV CP基因，构建原核表达载体pET28－PVY CP、pET28－SMV CP和pET28－CLYVV CP。诱导、表达和纯化PYV、SMV和CLYVV的CP蛋白。采用ELISA方法，用9株实验室制备的单克隆抗体对PVY、SMV、CLYVV的病毒悬液和表达并纯化的重组CP蛋白进行检测分析。成功表达并纯化了PYV、SMV和CLYVV CP蛋白。ELISA检测表明，9株单克隆抗体均识别SMV病毒，2D3、4F9、4G12和6E7 4株单克隆抗体能够识别PVY病毒，4F9和4G12能够识别CLYVV病毒，但PVY、CLYVV病毒与抗体的亲和力低于SMV病毒；对于重组表达的衣壳蛋白：SMV、CLYVV与2D3、4F9、4G12和6E7四株抗体反应强烈，PVY与4F9和4G12两株抗体反应稍强。综上，本研究鉴定出能识别PVY、SMV、CLYVV的通用单克隆抗体4F9和4G12，为马铃薯Y病毒病毒的检测研究提供了一种广谱性的抗体材料。

关键词：马铃薯Y病毒；大豆花叶病毒；三叶草黄脉病毒；CP蛋白；抗体鉴定

* 第一作者：王永志，从事分子病毒学研究；E-mail：yzwang@126.com
** 通讯作者：李启云，从事植物保护和分子抗病育种研究；E-mail：qyli1225@126.com

不同区域野生大豆 SMV 的检测及遗传结构分析

李小宇*，王永志，张春雨，李启云**，尤 晴，张淋淋
（吉林省农业科学院植物保护研究所，农业部东北作物有害生物
综合治理重点实验室，吉林省农业微生物重点实验室，公主岭 136100）

摘 要：为调查研究野生大豆资源中大豆花叶病毒（Soybean mosaic virus，SMV）感染情况和遗传结构，本研究通过建立 SMV 通用型检测方法，对来源于我国和朝鲜的共 14 个地区的 3 524份野生大豆样品进行 ELISA 分析，通过对病毒分离物 NIb、CP 基因的克隆、测序和序列进化分析，初步阐明野生大豆花叶病毒感染状况和遗传结构。黑龙江省野生大豆 SMV 阳性率比其他省份较低，且阳性率呈现由北向南逐渐升高的趋势，其他省份（吉林、辽宁、朝鲜、）野生大豆 SMV 阳性率均达到60%以上，说明野生大豆的总体带毒率较高。野生大豆 SMV 带毒率呈现出随纬度降低而升高的趋势，这可能是由于气候、传播宿主的变化引起，需要进一步的深入研究加以阐明。从系统进化数据上看，野生大豆 SMV 具有明显的地域差异，在野生大豆资源的异位保护中，要注意预防病毒的交叉侵染。本研究发现野生大豆中广泛存在 SMV 隐症感染。这一现象对抗性鉴定和毒株分离提出新的要求。

关键词：野生大豆；大豆花叶病毒；遗传结构

* 第一作者：李小宇，分子生态学研究；E-mail：lxyzsx@163. com
** 通讯作者：李启云，从事植物保护和分子抗病育种研究；E-mail：qyli1225@126. com

Simultaneous silencing of Two Target Genes Using Virus-Induced Gene Silencing Technology*

ZHU Feng**, ZHOU Yang-kai

(*College of Horticulture and Plant Protection, Yangzhou University, Yangzhou, Jiangsu 225009, China*)

Abstract: Virus-induced gene silencing (VIGS) is an effective strategy for rapid gene function analysis during plant growth and development. Over the last decade, VIGS has been successfully used to knock down the expression of a single gene in various plant species. However, simultaneous silencing of two target genes using VIGS in plants has been rarely reported. It is well established that the NAC (NAM, ATAF, and CUC) transcription factor and salicylic acid (SA) signal transduction pathway play essential roles in response to biotic stresses. Therefore, in this report, we performed VIGS to silence simultaneously the salicylic acid-binding protein 2 (*NbSABP2*) and *NbNAC1* in *Nicotiana benthamiana* to investigate the gene silencing efficiency of simultaneous silencing of two genes. Overlap extension PCR analysis showed that the combination of *NbSABP2* and *NbNAC1* was successfully amplified. Bacteria liquid PCR confirmed that the combination of *NbSABP2* and *NbNAC1* was successfully inserted into the TRV vector. Quantitative real-time PCR results showed that the expression of *NbSABP2* and *NbNAC1* were significantly reduced in 12 days post silenced plants after TRV infiltration compared with the control (TRV: 00) plants. Overall, our results suggest that VIGS can be used to silence simultaneously two target genes only one time infiltration.

* Funding: This work was supported by the National Natural Science Foundation of China (31500209), Natural Science Foundation of the Higher Education Institutions of Jiangsu Province of China (15KJB210007) and Natural Science Foundation of Yangzhou (YZ2015106)

** Corresponding author: Zhu Feng; E-mail: zhufeng@ yzu. edu. cn

通过 RNA-Seq 解析南方水稻矮缩病毒侵染介体昆虫培养细胞后不同时间的转录水平变化*

吴　维[1**]，江朝阳[1]，韩　玉[1]，陈红燕[1]，贾东升[1]，王海峰[2]，魏太云[***]
（1. 福建农林大学植物病毒研究所，福建省植物病毒学重点实验室，福建　350002；
2. 福建农林大学植物保护学院，福建　350002）

摘　要：南方水稻黑条矮缩病毒（Southern rice black-streaked dwarf virus，SRBSDV）由介体白背飞虱（*Sogatella furcifera*，Horváth）以持久增殖型方式传播，其引起的南方水稻黑条矮缩病是水稻生产上一种重要的病毒病害。为了研究 SRBSDV 侵染介体昆虫后，介体响应病毒侵染以及 SRBSDV 的复制增殖机理，我们根据病毒侵染不同阶段利用 Illumina 平台测定了 SRBSDV 在白背飞虱培养细胞中初侵染（2 h）、复制增殖（48 h）、以及病毒完全侵染时（120 h）的转录组。对转录组测序结果进行组装后共得到70 712条 Unigene 序列，通过 GO 聚类和 KEGG Pathway 代谢通路分析发现与生物代谢相关的基因最多。在对四组样品进行差异分析后，共得到 1 485个显著差异表达基因，其中病毒侵染后 2，48，120h 的上调表达基因分别为 306、323 和 359 个，下调表达基因分别为 428、370 和 486 个。对差异基因进行 GO 和 KEGG 聚类分析后结果表明：介体昆虫培养细胞对 SRBSDV 侵染不同时间的响应存在差异，在病毒通过内吞侵入时，会导致培养细胞的细胞膜发生一系列的改变，此时差异基因主要集中在细胞过程相关通路，如细胞连接、膜相关蛋白、吞噬等；在病毒大量复制时，其需要借助细胞内各种原料，导致与代谢相关通路基因大量上调；当病毒在细胞内大量积累时，会导致细胞的死亡，此时与细胞程序性死亡相关通路基因大量上调表达，且免疫通路相关基因在病毒侵染后期表达最高，初侵染时次之，病毒在单个细胞中复制时最低，表明病毒侵染不同时期的介体培养细胞对病毒亲和性存在差异，病毒初侵染引起免疫通路相关基因的差异表达要高于在单个细胞中复制时的差异表达，且 SRBSDV 的侵染对白背飞虱培养细胞是有害的，病毒的大量积累会导致细胞死亡。

关键词：南方水稻黑条矮缩病毒；培养细胞；白背飞虱；转录组

---

* 基金项目：国家自然科学基金项目（31571979）

** 第一作者：吴维，助理实验员；E-mail：wuwei_ 19861115@163. com
*** 通讯作者：魏太云，研究员；E-mail：weitaiyun@163. com

大豆花叶病毒东北 3 号株系全基因组感染性克隆的构建及生物学特性分析*

张春雨[1]，李小宇[1]，张淋淋[2]，王永志[1**]，李启云[1***]
（1. 吉林省农业科学院植物保护研究所，东北作物有害生物综合治理重点实验室，
吉林省农业微生物重点实验室，公主岭　136100；2. 东北农业大学
植物保护学院，哈尔滨　150030）

摘　要：大豆花叶病毒（Soybean mosaic virus，SMV）病是在世界范围内广泛分的大豆病害之一，不仅造成大豆产量降低，还严重影响大豆的品质。SMV 基因组是单链正义 RNA，通过构建感染性克隆有助于对其进行分子水平上的操作与研究。

本实验采用分段扩增的方法对大豆花叶病毒东北 3 号株系（SMV－3）进行了全基因组克隆，获得了 SMV－3 的基因组序列；采用同源重组法获得了 SMV－3 的全长 cDNA 克隆；将其放入 35S 启动子下游，利用基因枪介导，成功感染大豆品种 Williams，证明 SMV－3 感染性克隆构建成功。

进一步对 SMV－3 感染性克隆进行了生物学特性分析，对感病大豆的根、茎、叶、花、青豆、豆荚、种子进行间接 ELISA 检测，结果所示，上部、中上部叶片、花中病毒含量较高，且较稳定，茎、根中虽然也含有病毒，但较叶片中病毒含量稍低。

接种不同抗性品种后进行症状观察和间接 ELISA 分析，结果表明，SMV－3 感染大豆 Williams、Lee－68、Essex、L29，不感染 V94、PI88788、PI96983、L78－379，说明抗性基因 *Rsv*1 和 *Rsv*4 能够抑制感染性克隆 SMV－3 的侵染，而抗性基因 *Rsv*3 对 SMV－3 无抗性。

---

* 基金项目：吉林省科技厅国际合作项目：以大豆花叶病毒为载体的大豆生物反应器研究（20150414044GH）

** 第一作者：王永志，副研究员，博士，研究方向：分子病毒学；E-mail：yzwang@126. com
*** 通讯作者：李启云，研究员，博士，研究方向：植物保护；E-mail：qyli1225@126. com

水稻瘤矮病毒在介体电光叶蝉内的经卵传播机制*

廖珍凤**，毛倩卓，陆承聪，魏太云***
（福建农林大学植物病毒研究所，福建省植物病毒学重点实验室，福建　350002）

摘　要：水稻瘤矮病毒（Rice gall dwarf virus，RGDV）引起的水稻瘤矮病在我国广东、海南省稻区流行。RGDV 主要由介体电光叶蝉以持久增殖型方式进行传播，并可垂直传播。但关于 RGDV 在介体电光叶蝉中是否能经卵传播还存在争议，有关 RGDV 在介体电光叶蝉卵巢内的侵染过程还未有相关报道。本研究通过免疫荧光标记和电镜观察发现 RGDV 首先聚集在卵巢管端丝，由生殖区侵入并大量增殖，并随卵巢的发育逐渐侵染整个卵巢。在此过程中，RGDV 利用病毒编码的非结构蛋白 Pns11 形成的管状结构运输病毒粒体穿过滤泡细胞及微绒毛区进入卵黄区，最终经卵传播给后代。此外，通过显微注射将体外合成的 RGDV S11 基因双链 RNA（dsPns11）导入带毒若虫体内，干扰 RGDV Pns11 蛋白的表达，待昆虫羽化后进行单雌单雄交配实验。对所产的卵进行带毒率检测，结果发现注射 dsPns11 干扰带毒雌虫与健康雄虫进行交配子代的带毒率仅为 5.5%，显著低于带毒雌虫与健康雄虫，及注射 dsGFP 的带毒雌虫与健康雄虫交配子代带毒率（21.3%）。该结果进一步验证了 Pns11 是 RGDV 突破电光叶蝉经卵传播屏障的运输工具。由此，本研究证明了 RGDV 能在电光叶蝉中经卵传播，且阐明了 RGDV 在介体卵巢内的侵染过程，并提出了 RGDV 借助 Pns11 穿过卵巢微绒毛结构入卵的新模型。研究结果为进一步阐明病毒在介体昆虫经卵传播的机制提供了依据，也为开拓新的病害防治途径奠定了基础。

关键词：水稻瘤矮病毒；电光叶蝉；Pns11；经卵传播

* 基金项目：国家自然科学基金项目（31571979）
** 第一作者：廖珍凤，硕士研究生，植物病理学专业；E-mail：995238051@qq.com
*** 通讯作者：魏太云，研究员；E-mail：weitaiyun@163.com

利用叶蝉细胞瞬时表达系统研究水稻瘤矮病毒在介体细胞内的增殖机制*

王海涛**，张晓峰**，谢云杰，王　娟，陈红燕，魏太云***
（福建农林大学植物病毒研究所，福建省植物病毒学重点实验室，福建　350002）

摘　要：很多动植物病毒以昆虫作为媒介进行传播，由此引发的病毒病害严重影响人畜健康和粮食生产安全。由于昆虫介体本身体积小，寿命短，各种器官组织非常精细复杂，而且很多病毒基因组结构复杂，目前尚不易得到侵染性克隆，不能利用反向遗传学的手段进行研究。因此，昆虫细胞系结合相应的瞬时表达载体系统，是研究无侵染性克隆病毒在细胞内增殖必不可少的技术手段。目前，叶蝉细胞系广泛地应用于植物呼肠孤病毒的研究中，然而缺少叶蝉细胞特异的瞬时表达载体，导致无法更深入地探索水稻病毒在叶蝉细胞内的增殖机制。本研究通过定位和克隆黑尾叶蝉 actin 基因的 promoter 序列，首次开发建立了能在叶蝉细胞中高效瞬时表达的载体系统。通过优化转染条件，该系统目前可以转染飞虱、叶蝉以及 sf9 等昆虫细胞并表达多种外源蛋白。同时利用该系统验证水稻瘤矮病毒 RGDV 非结构蛋白 Pns7、Pns9 和 Pns12 的细胞定位以及它们与主要的结构蛋白 P3、P5、P8 之间的互作情况，进一步完善了 RGDV 病毒原质（Viroplasm）形成的分子机制以及其招募结构蛋白的机理。该系统扩宽了昆虫细胞系的平台优势，为研究水稻病毒和昆虫互作机制提供了有力的工具。

关键词：昆虫培养细胞；瞬时表达载体；水稻瘤矮病毒；病毒原质

---

* 基金项目：国家自然科学基金项目（31571979）

** 第一作者：王海涛，博士研究生，植物病理学专业；E-mail：wanghaitao9105@163.com
张晓峰，讲师，主要从事分子植物病毒学；E-mail：zhangxiaofeng911@163.com

*** 通讯作者：魏太云，研究员；E-mail：weitaiyun@163.com

水稻矮缩病毒非结构蛋白 Pns10 的介体叶蝉原肌球调节蛋白的互作*

张玲华**，陈　倩，魏太云***
（福建农林大学植物病毒研究所，福建　350002）

摘　要：水稻矮缩病毒（Rice dwarf virus，RDV）主要由介体黑尾叶蝉以持久增殖型的方式传播。在叶蝉体内，病毒利用自身编码的非结构蛋白 Pns10 形成的小管通道，采取“小管运输病毒”的策略进行安全扩散，以抵御叶蝉的各种免疫攻击。但是单靠 Pns10 小管蛋白不足以使病毒在昆虫体内的扩散变得畅通无阻，还需要依靠昆虫蛋白参与此过程。

鉴于此，本研究利用核蛋白互作的 Clontech 酵母双杂交系统筛选黑尾叶蝉 cDNA 库，验证阳性克隆后，筛选到与 RDV Pns10 互作的候选互作蛋白之一：原肌球调节蛋白（tropomodulin，Tomd）。在本氏烟共定位表达系统中，Tomd 能与 RDV Pns10 共定位；利用双分子免疫荧光互补技术，证明 Tomd 可与 Pns10 互作。利用 sf9 杆状病毒表达系统，发现 Tomd 能与 RDV Pns10 共定位。制备 Tomd 的抗体，免疫标记带毒培养细胞和虫体，结果表明 Tomd 也可与 RDV Pns10 共定位。并且在带毒的培养细胞和黑尾叶蝉内，Tomd 的相对表达量能够随着病毒和 Pns10 相对表达量的提高而上升；在培养细胞和黑尾叶蝉内干扰 Tomd 的表达，病毒和 Pns10 的相对表达量降低，因此认为 Tomd 对 Pns10 行使功能有正调控作用。

由于 Tomd 作为肌动蛋白（actin）慢生长端的唯一盖帽蛋白，对稳定肌动蛋白具有重要作用，而 Pns10 形成的包裹病毒的小管可沿着 actin 纤维丝进行病毒的运输。推测 Pns10 通过与 Tomd 的互作，抑制了 Tmod 的加帽功能，使得 actin 延伸失控。无限增长的 actin 纤维丝为 Pns10 小管伸入邻近细胞，或突破介体组织或膜屏障提供动力，实现基于 actin 的小管运动性。

关键词：水稻矮缩病毒；Pns10；黑尾叶蝉；互作；原肌球调节蛋白

---

* 基金项目：国家自然科学基金基金项目（31130044；31300136）
** 第一作者：张玲华，硕士研究生，植物病理专业；E-mail：313285111@ qq. com
*** 通讯作者：魏太云，研究员；E-mail：weitaiyun@ 163. com

黑尾叶蝉内共生菌 *Nasuia* 与 RDV 的经卵传播有关*

李曼曼**，贾东升，魏太云***

（福建农林大学植物病毒研究所，福建省植物病毒学重点实验室，福建 350002）

摘　要：水稻矮缩病由水稻矮缩病毒（Rice dwarf virus，RDV）引起，主要由介体黑尾叶蝉（*Nephotettix cincticeps*）以持久增殖型方式传播的，并能经卵传播到介体后代。前期研究已明确 RDV 是从介体叶蝉卵巢卵柄上皮鞘侵入卵母细胞，其在经卵传播过程中可直接附着在经卵垂直传播的初生共生菌外周，表明共生菌在 RDV 经卵传播过程中发挥一定的作用。叶蝉体内协同生长的共生菌包括 *Sulcia* 和 *Nasuia*，其中本实验室已证明 RDV 通过与 *Sulcia* 的直接互作介导 RDV 突破经卵传播屏障侵入卵母细胞，而共生菌 *Nasuia* 是否也与 RDV 的经卵传播尚不清楚。本实验首先提取无毒和带毒黑尾叶蝉的总 RNA，经 RT-qPCR 检测发现带毒黑尾叶蝉体内 *Nasuia* 的含量比无毒黑尾叶蝉高；随后收集携带 RDV 黑尾叶蝉所产的卵，并用 Cells-to-CT™ 试剂盒提取单个卵的总 RNA，RT-qPCR 的检测结果表明带毒卵内 RDV 与 *Nasuia* 的相对表达量存在正相关关系。根据文献报道已知黑尾叶蝉菌胞内可特异表达肽聚糖识别蛋白相关的基因 *NcPrp*，干扰该蛋白的表达对共生菌 *Nasuia* 有抑制作用，但对共生菌 *Sulcia* 没有影响。因此，本研究体外合成 *NcPrp* 基因的 dsRNA，采用 RNA 干扰技术抑制黑尾叶蝉体内 *Nasuia* 的含量，RT-qPCR 检测结果表明，注射 dsPrp 后，与对照 dsGFP 相比，携带 RDV 的黑尾叶蝉成虫体内的共生菌 *Nasuia* 相对表达量降低，产卵量和后代卵的带毒率均有降低。由此，我们认为共生菌 *Nasuia* 有利于 RDV 在黑尾叶蝉体内的经卵传播，其具体机制有待进一步深入研究。

关键词：水稻矮缩病毒；黑尾叶蝉；共生菌 *Nasuia*；RNA 干扰

---

* 基金项目：国家自然科学基金项目（31130044，31400136）

** 第一作者：李曼曼，硕士研究生，植物病理学专业；E-mail：pyqflmm@163.com

*** 通讯作者：魏太云，研究员；E-mail：weitaiyun@163.com

苹果褪绿叶斑病毒互作寄主因子的鉴定*

王亚迪**，李　楠，吕运霞，王树桐，胡同乐，王亚南***，曹克强***
（河北农业大学植物保护学院，保定　071001）

摘　要：苹果褪绿叶斑病毒（Apple chlorotic leaf spot virus，ACLSV）是苹果上危害最大的一类潜隐性病毒，广泛分布于世界各地。该病毒单独侵染或与其他病毒混合侵染严重影响果树产量及果实品质，造成严重的经济损失。本研究在利用酵母双杂交技术筛选能与 ACLSV CP、MP 互作的苏俄苹果寄主因子的基础上，通过生物信息学分析，从中选取可能具有重要功能的寄主因子，利用酵母双杂交、Pull-down 技术进行进一步互作验证。主要研究结果如下：

（1）构建带有完整寄主基因 *Malus*4 – 1、*Malus*15 – 1、*Malus*40 – 5、*Malus*72 – 3、*Malus*85 – 1、*Malus*86 – 1 的酵母双杂交表达载体。利用共转化法将诱饵载体 pGBKT7 – CP、pGBKT7 – MP 与带有上述 6 种寄主基因的酵母表达载体转化至酵母菌株 Y2H gold 中，将共转化产物涂布于四缺（SD/-Ade/-His/-Leu/-Trp/X-α-Gal）培养基上，观察酵母的生长及显色情况，验证寄主功能基因与寄主基因之间的互作。结果表明，寄主因子 *Malus*15 – 1、*Malus*40 – 5 在四缺平板上菌落呈明显蓝色，证明 CP、MP 与寄主因子 *Malus*15 – 1、*Malus*40 – 5 互作。

（2）构建原核表达载体 pET28a-CP、pET28a-MP、pET28a – 15 – 1 和 pET28a – 40 – 5，然后将重组载体转化大肠杆菌 Rosetta（DE3）感受态细胞，并诱导表达目的蛋白。采用 Pull-down 方法，将洗脱产物经 SDS-PAGE、Western blot 检测蛋白间的互作。结果表明，以 CP 为诱饵捕获 *Malus*15 – 1 蛋白，洗脱产物中有 21.5 kDa、16 kDa 两条明显条带；以 MP 蛋白为诱饵捕获 *Malus*15 – 1 蛋白，洗脱产物中有 50.8 kDa、16 kDa 两条条带，证明 CP、MP 与 *Malus*15 – 1 之间存在直接相互作用。同样分别以 CP、MP 为诱饵捕获 *Malus*40 – 5 蛋白，洗脱产物中仅有一条条带，证明 CP、MP 与 *Malus*40 – 5 蛋白之间无相互作用。*Malus*15 – 1 为苏俄苹果 PR10 基因。

研究结果将为揭示寄主基因在 ACLSV 侵染过程中具有的生物学功能及两蛋白互作的意义奠定基础。

关键词：苹果褪绿叶斑病毒；苏俄苹果；酵母双杂交；Pull-down；蛋白互作

---

* 基金项目：国家苹果现代产业技术体系（CARS – 28）；国家自然科学基金资助项目（31201487）；“河北省青年拔尖人才计划”资助

** 第一作者：王亚迪，在读硕士研究生，研究方向：植物病害流行与综合防治；E-mail：2760522298@qq.com

*** 通讯作者：王亚南，副教授，博士，从事植物病毒学研究；E-mail：wyn3215347@163.com
曹克强，博士，教授，从事植物病害流行与综合防治研究；E-mail：ckq@hebau.edu.cn

苹果褪绿叶斑病毒植株体内分布*

李　楠**，王亚迪，吕运霞，丁　丽，王树桐，胡同乐，王亚南***，曹克强***
（河北农业大学植物保护学院，保定　071001）

摘　要：苹果褪绿叶斑病毒（Apple chlorotic leaf spot virus，ACLSV）是世界性分布最广的一类苹果潜隐性病毒，该病毒会导致树体衰弱、产生慢性病害，严重影响果树的生长发育。繁育无毒苗木是目前苹果病毒病害最为有效的防治措施。但 ACLSV 在常规品种上不表现明显症状，加之树体大、病毒含量受气候因素影响，在不同生育期病毒在树体内分布不均，影响了病毒的检测效率。因此，了解病毒在植株体内的分布规律是准确诊断的前提，也是果树无毒栽培的基本保障。本研究利用实时荧光定量 RT-PCR 技术，对苹果不同生长发育时期、不同组织部位 ACLSV 进行测定，揭示病毒分布规律，为 ACLSV 的准确诊断及无毒苗木生产奠定基础。取得的具体研究结果如下。

（1）ACLSV 实时荧光定量 RT-PCR 检测体系的优化：参考王鹏等（2013）的引物，对检测体系进行了优化。测定了内参基因和病毒基因引物特异性，确定病毒特异性引物最佳引物浓度为 10μmol/L、最佳退火温度为 57℃，优化后的检测体系灵敏度为 10^{-6} 模板稀释液，并且验证了该方法在田间检测中的可靠性。

（2）ACLSV 在苹果植株体内的分布：采用实时荧光定量 RT-PCR 技术，对 2015 年 9 月中旬至 2016 年 4 月中旬苹果树体内 ACLSV 基因相对表达量进行了测定。从生长发育时期、树体方位和组织部位 3 个角度对病毒含量变化进行了分析，结果表明：①从生长发育时期角度进行分析，在苹果一年生枝条树皮组织中，以 9 月份 ACLSV RNA 表达量为参照，病毒含量的整体趋势是从 9 月中旬到 10 月中旬大幅度升高，10 月中旬达到这 8 个月份中的最高峰，此后病毒含量开始下降，3 月起呈上升趋势。②从树体方位角度分析，在苹果一年生枝条树皮组织中，以东向 ACLSV RNA 的表达量为参照，ACLSV 在 4 个方位上的病毒含量差异比较显著，位于南向的病毒含量在所有采集时期中均高于其他方向，而西向与北向的差异较小且大部分时期都低于东向。③从组织部位角度分析，在苹果的盛花期，以幼叶中 ACLSV RNA 的表达量为参照，ACLSV 在花中的含量最高，幼叶最低，ACLSV 含量由高到低的顺序依次为：花、两年生枝条树皮组织、芽、一年生枝条树皮组织和幼叶。

关键词：苹果褪绿叶斑病毒；分布规律；RT-PCR

* 基金项目：国家苹果现代产业技术体系（CARS－28）；国家自然科学基金资助项目（31201487）；“河北省青年拔尖人才计划”资助

** 第一作者：李楠，在读硕士研究生，研究方向：植物病害流行与综合防治；E－mail：875726290@ qq. com
*** 通讯作者：王亚南，副教授，博士，从事植物病毒学研究；E-mail：wyn3215347@ 163. com
曹克强，博士，教授，从事植物病害流行与综合防治研究；E-mail：ckq@ hebau. edu. cn

水稻瘤矮病毒利用非结构蛋白 Pns11 突破介体电光叶蝉唾液腺释放屏障*

毛倩卓**，廖珍凤，吴　维，魏太云***
（福建农林大学植物病毒研究所，福建　350002）

摘　要：大量的植物病毒利用介体昆虫以持久性行的方式进行传播，病毒在昆虫内侵染循环，最后到达唾液腺，进而随唾液一起在昆虫再次取食的过程中传播出去。昆虫的唾液腺细胞中分布着大量表面折叠的内腔结构并与唾管连通，形成内腔的基底质膜，成为病毒从唾液腺细胞释放到唾液中的最后一道膜屏障。水稻瘤矮病毒（Rice gall dwarf virus，RGDV）是呼肠孤病毒科（Reoviridae）植物呼肠孤病毒属（phytoreovirus）的成员，由介体电光叶蝉以持久性增值型的方式进行传播。本研究利用电镜观察了 RGDV 在电光叶蝉唾液腺中的分布，发现病毒会利用电子致密的纤维丝状和囊泡状结构以类似内吞的形式进入内腔释放到唾液腺中。具体表现为病毒首先附着在内腔的表面，进而向内腔中凹陷并进一步形成包裹着多个病毒粒体的小泡进入到内腔中。通过免疫荧光标记技术和免疫电镜技术发现，这种电子致密的纤维丝状或者囊泡状结构由病毒编码的 Pns11 蛋白组成，而且 Pns11 与内腔基底质膜的肌动蛋白（actin）、病毒粒体分别共定位。酵母双杂交试验也证实 Pns11 可以与 actin 互作。表明 Pns11 在病毒释放到内腔的过程中起着介导作用。此外，在收集到的带毒电光叶蝉唾液中也检测到大量的 Pns11 蛋白。当利用 RNAi 技术抑制 Pns11 在唾液腺的表达，发现病毒不能释放到内腔中，且传毒率下降。综合上述结果，本研究推测 Pns11 在 RGDV 突破介体昆虫唾液腺屏障的过程中起着关键的作用。

关键词：水稻瘤矮病毒；电光叶蝉；唾液腺屏障；RNA 干扰

* 基金项目：国家自然科学基金项目（31571979，31401712）
** 第一作者：毛倩卓，在职博士，植物病理学专业；E-mail：hermione9@163.com
*** 通讯作者：魏太云，研究员；E-mail：weitaiyun@163.com

甘蔗斑袖蜡蝉（*Proutista moesta* Westwood）研究初报*

唐庆华**，朱　辉，宋薇薇，余凤玉，牛晓庆，覃伟权***
（中国热带农业科学院椰子研究所，文昌　571339）

摘　要：槟榔黄化病是一种毁灭性病害，该病仅在印度和中国（海南省）发生，给两国槟榔产业造成了巨大的经济损失。印度学者报道该国槟榔黄化病由甘蔗斑袖蜡蝉（*Proutista moesta* Westwood），该虫还可传播非致死性病害椰子（根）枯萎病（现更名为椰子卡拉拉枯萎病，Kerala wilt disease of coconut palms）。在海南，笔者于 2015 年 1 月 15 日在文昌发现了 *P. moesta*，随后对该虫进行了一年多的调查研究。初步调查结果显示，该虫的寄主有槟榔、油棕、椰子、玲珑椰子、矮琼棕、散尾葵、狐尾椰子和甘蔗，嗜食椰子和油棕。调查结果还显示在椰子苗圃中该虫 7—10 月种群数量最多，12 月至翌年 3 月数量最少。*P. moesta* 多在 16:00 ~ 18:00 进行交配，交配时间为 10 ~ 20min。由于印度和中国槟榔黄化病的病原植原体分属于 16SrXI 组和 16SrI 组，因此，该虫是否为中国槟榔黄化病的媒介昆虫尚需进一步验证。但是，*P. moesta* 接种实验面临种种考验：田间种群数量少，生活史较长，虫源尚无法得到保障；槟榔植株高大，田间实验存在一定危险性和难度；通常染病槟榔植株进入结果期后才开始慢慢表现黄化症状，如果用幼苗进行接种然后再移栽到田间继续观察则需要数年时间才能完成媒介昆虫性质验证；*P. moesta* 以及寄主槟榔、气候条件等也会影响接种实验。

关键词：槟榔黄化病；媒介昆虫；*Proutista moesta*

---

* 基金项目：海南省重大科技项目（ZDZX2013008、ZDZX2013019）；海南省重点项目（314144）

** 第一作者：唐庆华，博士，助理研究员，研究方向为病原细菌 – 植物互作功能基因组学及植原体病害综合防治；E-mail：tchuna129@163.com

*** 通讯作者：覃伟权，研究员；E-mail：QWQ268@163.com

酵母双杂交筛选甜菜坏死黄脉病毒 RNA2 编码蛋白的寄主互作因子*

侯丽敏**，万 琪，姜 宁，张永亮，韩成贵，王 颖***
（中国农业大学，植物病理学农业部重点实验室，北京 100193）

摘 要：甜菜丛根病是由甜菜坏死黄脉病毒（Beet necrotic yellow vein virus，BNYVV）所引起的一种“土传”病害，直接影响甜菜产量和含糖量，危害严重，难以治理。而 RNA2 是 BNYVV 编码蛋白最多的 RNA，一共编码 6 个蛋白，分别与病毒的包装、运动、菌传、转录后沉默抑制有关，是侵染寄主必不可少的 RNA。因此可以通过酵母双杂交系统研究这些蛋白与烟草的互作关系，筛选到与之有关的基因，有助于进一步阐明 BNYVV 的致病机制及病毒与寄主植物相互作用的机理，为寻找防治 BNYVV 策略提供理论依据。

本研究将 RNA2 编码的蛋白 CP、p42、p14、54ku 克隆到酵母双杂交诱饵载体 pGBKT7 上，首先鉴定这四个蛋白对酵母菌 Gold 均未表现出毒性和自激活活性，之后分别将诱饵质粒 pGBKT7-CP、pGBKT7-p42、pGBKT7-p14、pGBKT7－54 与烟草 cDNA 文库（购于 Clontech 公司）进行酵母结合，在四缺营养缺陷型培养基和 X-α-Gal 的条件下筛选可能与诱饵蛋白互作的阳性克隆，提取阳性克隆质粒，通过 PCR 鉴定阳性克隆并送测序。CP、p14 和 p42 均没有筛到可能互作蛋白，其中 CP、p14 可以形成聚集体，可能自身互作较强，导致与其他蛋白的互作检测不到；p42 定位于胼胝质和胞间连丝，不能定位和转运到核内，可能不适合用酵母双杂交系统筛选互作蛋白。以含有 54ku 的 pGBKT7－54 为诱饵，获取了一些阳性酵母菌落，在 NCBI 数据库中对测序结果 Blast 比对，并将比对信息进行分类合并及功能分析。根据测序和比对的结果，初步筛选到几个寄主因子，主要是与植物次生代谢相关的蛋白质。目前对这些寄主因子的互作验证及其功能的进一步研究正在进行中。

关键词：酵母双杂交；BNYVV；54ku；寄主因子；互作

致谢：感谢于嘉林教授、李大伟教授和王献兵教授对本研究的建议。

* 基金项目：国家自然科学基金项目（31401708）和高校基本业务费（2016QC014）资助
** 第一作者：侯丽敏，硕士生，主要从事甜菜病毒的研究；E-mail：houlimin@ cau. edu. cn
*** 通讯作者：王颖，副教授，主要从事植物病毒学研究；E-mail：yingwang@ cau. edu. cn

芸薹黄化病毒三种基因型的生物学特性比较研究*

张晓艳**，王　颖，张宗英，李大伟，于嘉林，韩成贵***
（中国农业大学，植物病理学农业部重点实验室与农业生物
技术国家重点实验室，北京　100193）

摘　要：芸薹黄化病毒（Brassica yellows virus，BrYV）是一种侵染十字花科作物的马铃薯卷叶病毒属的病毒，该病毒在我国的发生分布比较广泛，自然条件下由蚜虫以持久循回非增殖的方式传播。病毒基因组 5′端的序列差异比较大，根据其 5′端序列特点可分为 A、B、C 三种基因型。本实验室已获得 BrYV 三种基因型的侵染性 cDNA 克隆，能够通过农杆菌浸润接种的方法系统侵染本生烟。

马铃薯卷叶病毒属多个成员 P0 被鉴定具有 RNA 沉默抑制活性，由于各个 P0 蛋白之间的氨基酸序列差异比较大，表现出来的抑制子活性强弱也存在差异。本实验室前期的研究工作中发现 $P0^{BrA}$是一个很强的基因沉默抑制子，不仅能够引起局部沉默，在接种 GFP 转基因本生烟 16c 上能够抑制由单链 GFP 所诱发的系统沉默。BrYV 三种基因型 P0 的氨基酸序列同源性为 86.7% ~ 90.8%。通过将各 P0 与单链 GFP 共注射实验表明，BrYV 三种基因型 P0 都具有很强的抑制局部沉默和抑制系统沉默的能力。P0 除了具有抑制 RNA 沉默能力以外还能够在本生烟叶片上引起细胞坏死反应。通过在本生烟叶片上瞬时表达 P0 比较由其产生的坏死症状发现，$P0^{BrB}$和 $P0^{BrC}$比 $P0^{BrA}$要延迟坏死症状的出现。并且发现将 $P0^{BrA}$第 228 位的氨基酸 F 突变为 L 后不影响其抑制子功能，但是能够延迟坏死症状的出现。

马铃薯卷叶病毒（PLRV）能够与豌豆耳突病毒 2（PEMV 2）发生协生互作，我们的研究结果表明 BrYV 三种基因型分别与 PEMV 2 复合侵染本生烟同样能够发生协生互作，将 BrYV 和 PEMV 2 混合接种本生烟 2 周以后能够在上部叶片观察到叶片扭曲和坏死的症状，检测结果表明在 PEMV 2 的辅助下 BrYV 的积累量增加。

通过比较发现 BrYV 三种基因型 P0 都是很强的基因沉默抑制子，但是，在本生烟上产生的坏死症状存在差异，并且找到一些影响症状的关键氨基酸位点，为进一步研究病毒致病性的分子机理提供了基础。

关键词：芸薹黄化病毒；基因沉默

致谢：感谢王献兵教授和张永亮副教授对本研究工作的建议。

* 基金项目：国家自然科学基金项目（31371909、31071663）和 111 引智计划资助
** 第一作者：张晓艳，博士生，主要从事植物病毒的研究；E-mail：xiaoyan433@cau.edu.cn
*** 通讯作者：韩成贵，教授，主要从事植物病毒学与抗病毒基因工程；E-mail：hanchenggui@cau.edu.cn

芸薹黄化病毒运动蛋白的原核表达纯化及抗血清制备*

赵航海**，张晓艳，王　颖，张宗英，李大伟，于嘉林，韩成贵***
（中国农业大学，植物病理学农业部重点实验室与农业生物
技术国家重点实验室，北京　100193）

摘　要：芸薹黄化病毒（Brassica yellows virus，BrYV）属于黄症病毒科（*Luteoviridae*）马铃薯卷叶病毒属（*polerovirus*），该属病毒在全球范围内广泛分布，能够侵染十字花科、茄科、藜科和禾本科等多科植物，造成较为严重的损失。病毒通过蚜虫以持久循回非增殖方式进行传播，属于韧皮部局限病毒，但与一些病毒复合侵染时可以突破韧皮部局限。BrYV 是向海英博士等在对我国各省市的 9 种十字花科作物进行病毒检测和全序列分析后发现的一种在我国分布广泛的马铃薯卷叶病毒属新病毒，并且根据 BrYV 的序列分析结果，将其分为三种基因型，分别命名为 BrYV-A，BrYV-B 和 BrYV-C，这三种基因型的 3′端都非常保守，5′端序列差异较大。

BrYV 的基因组为正单链 RNA。基因组 RNA 含有 7 个 ORF，其中的 ORF4 通过渗漏扫描表达，编码运动蛋白（movement protein，MP），该蛋白应与病毒胞间运动相关。目前的研究表明，有一些病毒的 MP 蛋白与病毒的致病性及症状表现密切相关，明确其在病毒致病机制中所承担的角色，对深入研究 BrYV 对于十字花科植物的致病性和症状形成机理有着重要意义。

为了研究 BrYV 的 MP 蛋白，我们将 BrYV-A 的 MP 克隆到 pDB-His-MBP 表达载体上，通过转化大肠杆菌 Rosetta（DH3）菌株，利用 0.1mM IPTG 在 18℃条件下诱导表达 12h，通过超声破碎，离心后将上清经过 Ni^{2+} 亲和层析柱纯化得到目的蛋白。

为了检测 BrYV 的 MP 蛋白的表达水平，我们还委托中科院遗传发育研究所制备了该蛋白的多克隆抗体，将纯化得到的总量约 3mg 的 BrYV MP 蛋白作为抗原，免疫健康大白兔获得抗血清。之后利用该抗血清对不同样品进行了 Western blotting 检测，发现抗血清可以有效检测到在本生烟中瞬时表达的 BrYV MP 蛋白，在接种 BrYV 病毒的本生烟接种叶和上位叶上也能检测到 MP 蛋白的表达，并且抗血清的特异性比较好。

BrYV 的 MP 蛋白抗血清的制备为芸薹黄化病毒的检测及其运动蛋白的功能研究打下了材料基础。

关键词：芸薹黄化病毒；运动蛋白；原核表达；抗血清制备

致谢：感谢王献兵教授和张永亮副教授对本研究工作的建议。

---

* 基金项目：国家自然科学基金项目（31371909）资助

** 第一作者：赵航海，硕士生，主要从事植物病毒与寄主的分子互作研究；E-mail：hanghaizhao@cau.edu.cn

*** 通讯作者：韩成贵，教授，主要从事植物病毒学与抗病毒基因工程；E-mail：hanchenggui@cau.edu.cn

苹果锈果类病毒传播途径探究*

吕运霞**，杨金凤，李　楠，王亚迪，王树桐，胡同乐，王亚南***，曹克强***
（河北农业大学植物保护学院，保定　071001）

摘　要：苹果锈果病又名花脸病或裂果病，是危害苹果较为严重的非潜隐性病毒之一，也是我国苹果的重要检疫性对象。苹果锈果病的病原为苹果锈果类病毒（Apple scar skin viroid，ASSVd），属于马铃薯纺锤块茎类病毒科（Pospivioidae）、马铃薯纺锤块茎类病毒属（Pospiviroid），主要侵染苹果和梨。有研究表明ASSVd可通过嫁接、修剪工具传播，种子及根接能否传播存在分歧（Desvignes，1999；Hyun，2006；郭超，2014），至今未见详细报道。

为明确ASSVd传播方式，本研究以带毒组培苗和无毒组培苗为试材，人工模拟田间不同传播方式，包括汁液沾染、修剪工具修剪、根系搭接，组织培养60d后运用RT-PCR技术检测植株心叶带毒情况，结果表明：组培条件下，组培苗伤口沾染带毒汁液后，随着带毒汁液沾染时间的延长组培苗带毒几率增加。沾染10s，扩繁无毒组培苗带毒率几乎为0；沾染2min，扩繁无毒组培苗带毒率为6.7%；沾染5min，带毒率增至为9.7%。组培剪沾染带毒汁液后立即进行组培苗的扩繁，使组培苗带毒率为20%；组培剪剪取带毒苗后立即进行无毒组培苗的扩繁，使组培苗带毒率增至16.7%。无毒组培苗培养基中滴加带毒汁液，培养后带毒率为33.3%；带毒、无毒组培苗共培养后，无毒组培苗带毒率为30.0%。因此推测，ASSVd可通过汁液沾染、修剪工具、带毒土壤及根系搭接传播，其中，根系传染风险最高，但需要进行进一步的田间试验进行验证。

关键词：苹果锈果类病毒；传播途径；RT-PCR

---

* 基金项目：国家苹果现代产业技术体系（CARS-28）；河北省高等学校科学技术研究项目（YQ2014023）；河北省青年拔尖人才计划

** 第一作者：吕运霞，在读硕士研究生，研究方向：植物病害流行与综合防治；E-mail：1390741310@qq.com

*** 通讯作者：王亚南，副教授，博士，从事植物病毒学研究；E-mail：wyn3215347@163.com

曹克强，博士，教授，从事植物病害流行与综合防治研究；E-mail：ckq@hebau.edu.cn

豇豆花叶病毒属病毒广谱 RT-PCR 检测方法的建立*

廖富荣[1]**，林武镇[1,2]，方志鹏[1]，沈建国[3]，陈　青[1]，陈红运[1]
（1. 厦门出入境检验检疫局检验检疫技术中心，海沧　361026；2. 福建农林大学，福州　350002；3. 福建出入境检验检疫局检验检疫技术中心，福州　350001）

摘　要：根据 RNA2 的外壳蛋白基因保守区域设计简并引物（ComoV－2－F2/ComoV－2－R1），建立了豇豆花叶病毒属（Comovirus）的 RT-PCR 检测方法。利用该方法成功用于扩增 9 种豇豆花叶病毒属病毒（APMoV、BBTMV、BBSV、BPMV、CPMV、CPSMV、RCMV、RaMV 和 SqMV）的 14 个分离物，而不能从其他豇豆花叶病毒亚科病毒和健康寄主植物中扩增到预期大小条带，显示出良好的广谱性和特异性。通过在简并引物的 5′端引入测序引物（ComoV－2－F2－M4/ComoV－2－R1－M1），成功进行 PCR 产物的直接测序，并且不会影响其通用性（图 2B）。按 1∶10 比例在 PCR 扩增中同时加入简并引物（ComoV－2－F2－M4/ComoV－2－R1－M1）和通用测序引物（M4/M1），结果表明，其灵敏度可以提高 1 000倍。另外，在简并引物的 5′端引入富含 AT 的非互补序列，在并未改变其通用性情况下，其灵敏度可以提高 100 倍。序列测定分析表明，测定的病毒序列与相应病毒均具有最高的序列同源性，可用于病毒种类的快速鉴定。该方法可用于豇豆花叶病毒属的广谱检测及病毒种类的快速鉴定，而且有利于新病毒种类的发现。

关键词：豇豆花中病毒；RT－PCR；病毒鉴定

* 基金项目：福建省自然科学基金项目（2015J01148）；国家质检总局科技项目（2015IK190）

** 第一作者：廖富荣，高级农艺师，主要从事植物病原鉴定及检测方法研究；E-mail：LFR005@163.com

叶蝉共生菌 *Sulcia* 介导水稻矮缩病毒经卵传播*

贾东升**，毛倩卓，陈 勇，刘宇艳，王海涛，吴 维，陈 倩，陈红燕，魏太云***
（福建农林大学植物病毒研究所，福建省植物病毒学重点实验室，福建 350002）

摘 要：水稻矮缩病毒（Rice dwarf virus，RDV）属于呼肠孤病毒科（Reoviridae），植物呼肠孤病毒属（Phytoreovirus）成员，主要由介体黑尾叶蝉以持久增殖型方式传播，并经卵传播。但 RDV 的经卵传播机制尚不清楚。本研究通过免疫荧光标记对 RDV 侵入卵巢的过程进行了系统的分析，发现病毒首先侵染卵柄，随后扩散到卵柄上皮鞘，然后侵入卵母细胞，因此推测病毒是通过卵柄上皮鞘侵入卵母细胞。根据已有的文献报道叶蝉内共生菌 *Sulcia* 也经卵垂直传播，本研究对 RDV 和共生菌 *Sulcia* 在卵巢的侵染进行共标记，发现共生菌 *Sulcia* 未侵入卵柄上皮鞘时，RDV 仅侵染卵柄部位，当 *Sulcia* 侵入卵柄上皮鞘时 RDV 也侵入卵柄上皮鞘，且 RDV 粒体分布在共生菌表面。同时通过电镜在卵柄上皮鞘和卵母细胞内均观察到 RDV 吸附或嵌入共生菌 *Sulcia* 表面。此外，通过 RT-qPCR 对带毒和无毒黑尾叶蝉体内 Sulcia 的含量进行比较，发现带毒叶蝉体内的共生菌含量比无毒黑尾叶蝉高，且带毒卵内的共生菌与 RDV 含量呈正相关。由此我们初步认为共生菌 *Sulcia* 侵入卵的过程中介导 RDV 侵入卵，并推测两者间存在互作而导致 RDV 吸附在共生菌表面。为此，本研究通过酵母双杂交和 GST pull-down 技术证明共生菌 *Sulcia* 外膜蛋白（OMP）的 BSA 结构域与 RDV 外壳蛋白 P2 的 N 端（1－688aa）存在特异性互作。当用提取的 RDV 粒体与卵巢组织在体外孵育，发现 RDV 可与卵巢卵柄上皮鞘部位的共生菌相结合，且原核表达的 P2 蛋白 N 端多肽也可与卵柄上皮鞘的共生菌在体外相结合，而通过注射法将 *Sulcia* 外膜蛋白 OMP 的多克隆抗体注入叶蝉腹部封闭共生菌，导致 RDV 无法侵入卵柄上皮鞘。该研究进一步表明 RDV 的 P2N 与共生菌外膜蛋白间存在直接的互作。综合以上结果，本研究首次明确叶蝉共生菌 *Sulcia* 通过外膜蛋白与 RDV 粒体表面 P2 蛋白间的直接互作介导 RDV 突破经卵传播屏障侵入叶蝉卵柄上皮鞘，并侵入卵母细胞。

关键词：水稻矮缩病毒；黑尾叶蝉；共生菌 *Sulcia*；经卵传播

---

* 基金项目：国家自然科学基金项目（31571979）
** 第一作者：贾东升，讲师，从事水稻病毒与介体昆虫互作机制研究；E-mail：jiadongsheng2004@163.com
*** 通讯作者：魏太云，研究员；E-mail：weitaiyun@163.com

水稻矮缩病毒诱导黑尾叶蝉细胞凋亡的机制研究*

陈　倩**，郑立敏，王海涛，张玲华，魏太云***
（福建农林大学植物病毒研究所，福建省植物病毒学重点实验室，福建　350002）

摘　要：许多持久增殖型的植物病毒在介体昆虫内高效增殖的同时，对介体昆虫的生长发育也会产生不利的影响。早期研究发现水稻矮缩病毒（Rice dwarf virus，RDV）侵染可降低黑尾叶蝉若虫的存活率和成虫寿命，同时昆虫内部器官会出现一定程度的细胞病变。最近有报道发现水稻齿叶矮缩病毒（Rice ragged stunt virus，RRSV）可在介体褐飞虱唾液腺通过诱导细胞凋亡，促进病毒释放至植物内，这为我们研究 RDV 对介体昆虫的伤害机制提供了启示。

本研究用高浓度的 RDV 提纯病毒侵染黑尾叶蝉培养细胞，在病毒侵染 3 天后，出现细胞凋亡显著的形态学特征，CCK－8 法显示带毒细胞活力显著降低，表明 RDV 对黑尾叶蝉培养细胞有一定的伤害性。用可鉴定细胞凋亡晚期细胞核 DNA 断裂特征的 TUNEL 凋亡原位检测法，观察到多于对照的凋亡小体，证明 RDV 侵染可诱导黑尾叶蝉细胞凋亡。JC－1 流式细胞术发现有 20.87% 细胞因线粒体跨膜电位下降，发生不可逆转的凋亡，推测 RDV 引发的细胞凋亡依赖于内源线粒体通路。RT－qPCR 结果显示，RDV 的侵染特异性激活线粒体通路的细胞凋亡的起始蛋白 caspase 2，并且病毒增殖动态与 caspase 2、抑制凋亡蛋白 IAP1 相对表达水平趋势基本一致，表明 RDV 的侵染诱导了依赖于内源线粒体通路的细胞凋亡。对 caspase 2 和 IAP1 进行基因沉默，结果显示病毒积累量显著低于对照，说明病毒的增殖与凋亡趋势呈正相关，推测 RDV 引发的细胞凋亡有利于病毒侵染。

本研究利用黑尾叶蝉培养细胞体系，应用细胞凋亡验证的经典方法，证明 RDV 的侵染可以诱导细胞凋亡，首次从细胞凋亡角度研究 RDV 对黑尾叶蝉生长发育的不利影响的机制。研究结果为此类持久性增殖型的植物病毒与介体昆虫的共进化研究提供借鉴。

关键词：水稻矮缩病毒；黑尾叶蝉；细胞凋亡；培养细胞

---

* 基金项目：国家自然科学基金项目（31130044；31300136）
** 第一作者：陈倩，助理研究员；E-mail：chenqian_07@126.com
*** 通讯作者：魏太云，研究员；E-mail：weitaiyun@163.com

南方水稻黑条矮缩病毒非结构蛋白 P6 是病毒增殖的关键因子*

韩 玉**，贾东升，毛倩卓，陈 倩，陈红燕，魏太云***
（福建农林大学植物病毒研究所，福建省植物病毒学重点实验室，福州 350002）

摘 要：南方水稻黑条矮缩病毒（Southern rice black-streaked dwarf virus，SRBSDV）引起的南方水稻黑条矮缩病给我国农业生产造成巨大的损失。SRBSDV 编码的非结构蛋白 P5－1、P6 和 P9－1 是病毒侵染形成病毒原质（viroplasm）的组分，参与病毒的复制和装配。本实验通过首先免疫荧光标记技术在介体培养细胞中观察 SRBSDV 侵染后 P5－1、P6 和 P9－1 表达的先后顺序，结果发现 P6 在 SRBSDV 侵染白背飞虱细胞时最早表达并不断增大，随后 P5－1 和 P9－1 蛋白表达。随后利用 RNAi 技术分别干扰 P5－1、P6 和 P9－1 蛋白在白背飞虱培养细胞中的表达，均导致 SRBSDV 的侵染率显著降低。抑制 P5－1 蛋白表达后，P5－1 和 P9－1 的表达量明显降低，但 P6 的表达量和对照组相当；抑制 P6 蛋白表达后，P5－1、P6 和 P9－1 的表达量都显著降低；抑制 P9－1 蛋白表达后，除了自身蛋白的表达量降低外，另外两个蛋白的表达水平不受影响。为了进一步验证 P6 蛋白的功能，通过微针注射法将源于 P6、P5－1 和 P9－1 基因的 dsRNA 分别导入饲毒 1 d 的白背飞虱体内，以 dsGFP 处理为对照，注射后 5 d 后对白背飞虱消化道进行免疫荧光标记。共聚焦显微镜观察发现，与对照组相比，分别抑制病毒原质三个组分蛋白的表达均可阻碍病毒在昆虫体内的增殖，且病毒被局限在初侵染的上皮细胞内。其中 dsP6 处理组带毒率显著下降，且 P6 蛋白被抑制后 P5－1 和 P9－1 蛋白均不表达；而抑制 P5－1 或 P9－1 蛋白的表达后，昆虫的带毒率不受影响，且在初侵染上皮细胞内仍可观察到 P6 蛋白的表达。该研究结果表明 P6 蛋白是病毒原质形成中的关键因子，最早表达并参与激发或调控病毒原质其他组分蛋白的表达，抑制 P6 蛋白的表达则完全阻碍病毒的增殖。该研究结果同时表明 P6 蛋白是阻断病毒增殖最理想的靶标。

关键词：南方水稻黑条矮缩病毒；白背飞虱；病毒原质；复制；关键因子

---

* 基金项目：国家自然科学基金项目（31300136，31400136，31401712）
** 第一作者：韩玉，硕士研究生，植物病理学专业；E-mail：1337465791@ qq. com
*** 通讯作者：魏太云，研究员；E-mail：weitaiyun@ 163. com

首次在中草药竹叶子上检测到黄瓜花叶病毒*

张　旺[1**]，申　杰[1,2]，孙现超[1***]
（1. 西南大学植物保护学院，分子植物病理学实验室，北碚　400716；
2. 重庆市药物种植研究所，南川　408435）

摘　要：重庆是全国重要的中药材产地之一，大面积的山区生长着数千种野生和人工培植的中药材，在全国产量最大的有黄连、五倍子、金银花、厚朴、黄柏、杜仲、元胡等。竹叶子是竹叶子属（*Streptolirion*）是鸭跖草科下的单种属植物，仅含竹叶子（*S. volubile*）一种。中国有竹叶子（*S. volubile* Edgew.）1 种，产西南部、中部、西北至东北部。其性味，甘；平。全草入药，具有清热利尿的作用。2015 年 4 月到 2015 年 9 月对重庆南川、涪陵地种植的具有疑似病毒病症状的苍术、紫苏、千金藤、竹叶子等中草药植物调查及采样。通过斑点酶联免疫吸附法（DOT-ELISA）检测番茄斑萎病毒（Tomato spotted wilt virus，TSWV）、马铃薯 Y 病毒（Potato virus Y，PVY）、芜菁花叶病毒（Tunip mosaic virus，TuMV）、黄瓜花叶病毒（Cucumber mosaic virus，CMV）、烟草花叶病毒（Tobacco mosaic virus，TMV）5 种病毒病的发生情况，并进一步对 CMV 检出样品采用反转录 PCR（RT-PCR）方法复检，最后将检出样品通过摩擦接种法回接本氏烟并进行 DOT-ELISA 和 RT-PCR 验证，结果表明，竹叶子受到黄瓜花叶病毒（CMV）的侵染，其余样品的病原目前尚未确定。据了解，本研究为国内首次报道 CMV 能够侵竹叶子，进一步明确了黄瓜花叶病毒的寄主范围。

关键词：竹叶子；CMV；DOT-ELISA；RT-PCR

---

* 基金项目：国家自然科学基金（30900937），中央高校基本科研业务费专项资金（XDJK2016A009，2362015xk04）中国烟草总公司四川省公司科技项目（201202007），中国烟草总公司湖南省公司科技项目（14－16ZDAa02）

** 第一作者：张旺，硕士研究生，从事植物病理学研究

*** 通讯作者：孙现超，博士，研究员；E－mail：sunxianchao@163.com

小麦黄花叶病毒（WYMV）两个山东分离物的全基因组序列及进化分析*

耿国伟**，于成明，王德亚，顾　珂，李向东，原雪峰***
（山东农业大学植物保护学院植物病理系，山东省农业微生物重点实验室，泰安　271018）

摘　要：小麦黄花叶病毒（Wheat yellow mosaic virus，WYMV）是小麦黄花叶病的主要病原之一，在我国分布广泛，对小麦生长、发育构成严重危害。WYMV 粒子呈弯曲线状，长度分为 300nm 和 600nm 两种。WYMV 为二分体基因组，由两条线性的单链正义 RNA 组成。RNA 1 全长约 7.6kb，RNA 2 全长约 3.6kb，分别编码一个 270kDa 和 100kDa 的多聚蛋白。WYMV RNA 的 5′端有共价结合的 VPg，3′端有一个 poly（A）尾。

利用 5′RACE 结合一步法 RT-PCR 分别从山东泰安与临沂冬小麦上克隆了 WYMV 的全基因组序列（TADWK 和 LYJN）。这两个分离物基因组间核苷酸一致性分别为 97.21%（RNA1）和 95.12%（RNA2）。通过对目前已报道的共 14 个 WYMV 分离物的基因组不同部分的分析，表明 5′UTR 是 WYMV 基因组变化幅度最大的区域，而编码区（ORF）及 3′UTR 的序列一致性较高且变动幅度小。另外，发现 LYJN RNA2 的 5′UTR 与已报道的所有分离物 RNA2 的核苷酸一致率仅为 90% 左右。综合分析 RNA1 和 RNA2 的系统发生树，表明 WYMV 各基因组片段呈单独进化特征，不同分离物间存在 RNA 重排。RNA 重组分析显示在 LYJN RNA1 和 TADWK RNA2 发现了 RNA 重组。此研究说明 WYMV 在山东地区存在分离物分化现象，并且表明 WYMV 的 5′UTR 是基因组中的突变热点。5′UTR 的高频率突变可能导致潜在的 IRES（核糖体内部进入位点）活性变化，从而导致 WYMV 不同分离物的蛋白表达差异并影响致病力。

从山东临沂的 4 个小麦品种（LM4、AK58、TS23、LY502）克隆的 WYMV 为同一分离物，序列同源性大于 99%；但是就其田间发病情况来看 LM4、AK58、TS23 这 3 个品种较之 LY502 要严重的多，并且 LY502 要比其他 3 个品种发病晚。不同小麦品种对 WYMV 的抗病性差异可能与翻译起始因子（eIF4E）的变异有关。

关键词：小麦黄花叶病毒；系统发生树；RNA 重组；进化

* 基金项目：国家自然科学基金资助项目（31370179）；山东省自然科学基金资助项目（ZR2013CM015）
** 第一作者：耿国伟，硕士研究生，主要从事分子植物病毒学研究；E-mail：guowgeng@163.com
*** 通讯作者：原雪峰，教授，博士生导师，主要从事分子植物病毒学研究；E-mail：snowpeak77@163.com

烟草丛顶病毒不同分离物中 p35 蛋白的差异表达机制*

王德亚**，于成明，刘珊珊，逯晓明，原雪峰***
（山东农业大学植物保护学院植物病理系，山东省农业微生物重点实验室，泰安　271018）

摘　要：烟草丛顶病毒（Tobacco bushy top virus，TBTV）属于番茄丛矮病毒科幽影病毒属。TBTV 基因组由一条正义单链 RNA（+ssRNA）组成，编码 4 个 ORF，5′端缺乏帽子结构（m7GPPPN），3′末端也不带 poly（A）尾。p35（ORF1）蛋白是通过不依赖帽子翻译机制而表达，核心调控元件为类 BTE 元件（最早在 BYDV 中发现）。通过萤火虫荧光素酶（Fluc）载体和 TBTV 全长 RNA 的体外翻译分析，表明其核心组件为可能结合 eIF4E 的 SL-I 以及与 5′UTR 形成远距离 RNA-RNA 互作的 SL-III B；并且其结构稳定性对于 p35 的不依赖帽子翻译也至关重要。

TBTV 不同分离物的体外翻译实验表明，p35 蛋白的表达呈现约 3 倍的差异。通过嵌合病毒的翻译分析，定位了造成 p35 蛋白表达差异的调控区域：RI 区（基因组 5′端的 500nt）和 RV 区［3′端的 3 138nt 到 3 885nt 区间［包含类 BTE 元件］。进一步的突变实验表明，RV 区对于 p35 差异表达的调控并不是由于类 BTE 元件内的核苷酸突变；而发生在 RV 区中 BTE 元件区域之外的核苷酸突变也仅造成 p35 的细微变化。而通过 RNA 结构预测，在 p35 表达量低的 TBTV 分离物和嵌合病毒中均发现具有典型 BTE 结构的比例很低；而 p35 表达量高的 TBTV 分离物则拥有高比例的典型 BTE 结构特征。RI 区对于 p35 蛋白的差异表达的调控机理正在研究中。因此，TBTV 不同分离物的 p35 差异表达的内在机制部分源自于 BTE 区域的结构进化。

关键词：烟草丛顶病毒；不依赖帽子翻译；类 BTE 元件；RNA 结构

* 基金项目：国家自然科学基金资助项目（31370179）；山东省自然科学基金资助项目（ZR2013CM015）
** 第一作者：王德亚，博士研究生，主要从事分子植物病毒学研究；E-mail：wangdeyasdny@163. com
*** 通讯作者：原雪峰，教授，博士生导师，主要从事分子植物病毒学研究；E-mail：snowpeak77@163. com

Long-distance RNA-RNA Interaction may be Associated with -1 Programmed Ribosome Frameshift in Tobacco bushy top virus*

YU Cheng-ming**, WANG De-ya, GENG Guo-wei, WANG Guo-lu, LU Xiao-ming, YUAN Xue-feng***

(Department of Plant Pathology, College of Plant Protection, Shandong Agricultural University, Daizong Road No. 61, Taian 271018, P. R. China)

Abstract: Tobacco bushy top virus (TBTV) has a (+) ssRNA genome with 4152 nucleotides. The viral genome without 5′-cap and 3′-poly (A) encoded four putative open reading frames (ORFs). The RNA-dependent RNA polymerase (RdRp) is expressed via -1 programmed ribosome frameshift of p35 with molecular weight of p98. Mutation analysis have proved that the heptanucleotide sequence for -1 type frameshift of RdRp is ^{946}GGAUUUU in TBTV, and the optimal distance between ^{946}GGAUUUU and downstream potential RNA structure was the range of 6- 9 nt. We also identified that the 3′ terminal 200 nt could form RNA-RNA interaction with potential RNA structure located downstream of ^{946}GGAUUUU through electrophoretic mobility shift assay (EMSA). We further mapped this RNA-RNA interaction occurring between positions of 925 ~ 1 120 and 4 003 ~ 4 152 in TBTV. Through In-line probing experiment, we analyzed the structure of 925 ~ 1 120 and 4 003 ~ 4 152 as well as their structural alteration caused by the RNA-RNA interaction. Finally, we identified the sites of the long-distance RNA-RNA interaction locating at positions of 1 023 ~ 1 025 (AGU) and 4 137 ~ 4 139 (ACU). The function of this long-distance RNA-RNA interaction will be identified for -1 type frameshift.

Key words: Tobacco bushy top virus; Ribosomal frameshift; RNA-RNA interaction

---

* 基金项目：国家自然科学基金资助项目（31370179）；山东省自然科学基金资助项目（ZR2013CM015）

** 第一作者：于成明，博士研究生，主要从事分子植物病毒学研究；E-mail：ycm2006. apple@ 163. com

*** 通讯作者：原雪峰，教授，博士生导师，主要从事分子植物病毒学研究；E-mail：snowpeak77@ 163. com

广东省番木瓜畸形花叶病毒的发现与鉴定*

吴自林**，李华平***
（华南农业大学农学院，广州　510642）

摘　要：2016 年 5 月从广州南沙区采集了 18 个番木瓜感病样品，运用 RT-PCR 技术，检测到 13 个样品为番木瓜畸形花叶病毒（Papaya leaf-distortion mosaic virus，PLDMV）。这是首次在中国广东省田间发现番木瓜感染了 PLDMV。继续对该批样品的病毒进行扩增，通过测序获得了这 13 个 PLDMV 的全基因组序列，通过 NCBI 比对，找到了 5 个已报道的 PLDMV 全基因组参考序列。系统发育分析表明，PLDMV 各分离物分为两个大簇，南沙的这 13 个 PLDMV 分离物与海南的 2 个分离物分在同一簇内，中国台湾和日本的 3 个分离物则分在另一个簇里。由序列相似性分析结果可知，南沙这 13 个 PLDMV 分离物间的氨基酸序列相似性在 99.4% ~100%，与海南的 2 个分离物间的氨基酸序列相似性在 99.1% ~99.4%，与中国台湾和日本的 3 个分离物间的氨基酸序列相似性在 95.3% ~96.0%。以上结果显示了各分离物之间具有一定的地理差异性。经调查得知，该批发病植株多是在 2015 年 2 月以后种植的，在此之前并未在广东省发现过 PLDMV。而目前在中国只有海南省和台湾省有 PLDMV 的报道，由此我们推测该病原物可能是经番木瓜幼苗传入广东省。本实验室培育的商业化转基因番木瓜“华农 1 号”，只能对番木瓜环斑病毒有很好的防治效果，PLDMV 的发生和流行，可能对番木瓜的生产造成较大的影响，因此在农业生产中应对 PLDMV 引起必要的重视。

关键词：番木瓜；番木瓜畸形花叶病毒；鉴定；序列分析

---

* 基金项目：农业部农产品质量安全监管专项经费

** 第一作者：吴自林，博士研究生，植物病理学；E-mail：wuzilin2008@126. com

*** 通讯作者：李华平，教授；E-mail：huaping@scau. edu. cn

Disruption of a Conserved Stem-loop Structure Located Upstream of Pseudoknot domain in Tobacco mosaic virus Enhances Viral RNAs Accumulation

GUO Song[1], WONG Sek-Man[1,2,3]

(1. *Department of Biological Sciences*, *National University of Singapore*, *Republic of Singapore*; 2. *Temasek Life Sciences Laboratory*, *Singapore*, *Republic of Singapore*; 3. *National University of Singapore Research Institute in Suzhou*, *Jiangsu*, *PRC*)

Abstract: A conserved stem-loop structure was validated to be present in Tobacco mosaic virus (TMV) genome. It was formed with 24 nucleotides from partial TMV coat protein (CP) and partial 3′ UTR sequences. RNA structural prediction showed that this stem-loop structure was conserved among most tobamoviruses. Both of the disrupted stem-loop structure and sequence deletion mutants of TMV demonstrated a rapid replication and a higher viral RNA accumulation in *Nicotiana benthamiana* protoplasts. The TMV mutant with complete and incomplete mirrored stem-loop structure showed about 4 and 30 times higher viral RNA accumulation level than that of the wild-type, respectively. The complete mirrored stem-loop TMV mutant expressed similar CP level as that of TMV. The incomplete mirrored stem-loop mutant showed about 40% higher CP expression than TMV. These results suggest that the stem-loop structure of TMV plays a *cis*-regulating role in virus replication.

Cloning, Prokaryotic Expression and Monoclonal Antibody Preparation of Sonchus yellow net virus *P* gene*

DENG Jie, QU Si-yi, SHEN Yang-yang, ZHU Xiao-ling, SHI Man-ling**

(*College of Life and Environmental Sciences*, *Hangzhou Normal University*, *Hangzhou* 310036, *China*)

Abstract: Sonchus yellow net virus (SYNV) belongs to the *Nucleorhabdovirus* genus of *Rhabdoviridae* family. The genome of SYNV contains six genes whose order is 3″-N-P-sc4-M-G-L-5″. P protein interacts with the N protein to regulate transitions between transcription and replication. The *P* gene of SYNV was amplified by RT-PCR, and PCR products were transformed into *E. coli* DH5a. The sequencing results showed that SYNV *P* gene was composed of about 1038 nucleotides encoding a polypeptide of 346 amino acids, and it had 96% ~99% nucleotide sequence identities with those of the other 4 reported SYNV isolates.

The cloned SYNV *P* gene was inserted into prokaryotic expression vector Pet-30a to construct recombinant expression vector Pet-30a-p, which was transformed into *E. coli* BL21 (DE3). After induced by IPTG, an about 42 kDa fusion protein was obtained and purified through Ni^{2+}-NTA affinity column. The purified recombinant P was used to immunize BALB/c mice for producing MAbs. Two monoclonal antibodies named 3C7 and 4D1 were obtained. The result of western blot showed that two MAbs could react specifically with recombinant P protein of SYNV. When coating antigen concentration of 3ug/mL, the ELISA titres of ascitic fluids of MAbs from cell lines 3C7 and 4D1 were 1 : 64 000 and 1 : 128 000. An indirect enzyme-linked immunosorbant assay (id-ELISA) method based on MAb from cell lines 3C7 was set up for P protein detection, and id-ELISA optimal working dilution was antigen P dilution of 1 : 4 000 and monoclonal antibody dilution of 1 : 8 000. An id-ELISA based on MAb from cell lines 4D1 was set up for P protein detection, and id-ELISA optimal working dilution was antigen P dilution of 1 : 4 000, monoclonal antibody dilution of 1 : 16 000 . MAb from cell lines 4D1 could successfully detect 0. 188μg/mL purified P protein or SYNV virus in plant sap at 1 : 320 dilution. MAbs from cell lines 4D1 was capable of testing the P protein expressed in tobacco plants through infiltration inoculation by the eukaryotic expression vector of plant. The experimental results indicated two MAbs were applicable to specifically detect not only P protein but also SYNV virus.

Key words: Sonchus yellow net virus; P protein; Cloning; prokaryotic expression; monoclonal antibody.

* Founding: National Natural Science Foundation of China (31371916)

** Corresponding author: SHI Man-ling; E-mail: manling. shi@ hznu. edu. cn

CTV 对寄主植物营养与蚜虫适合度影响的研究*

关桂静**，王洪苏，刘金香***
（西南大学柑桔研究所，重庆　400712）

摘　要：由柑橘衰退病毒（Citrus tristezavirus，CTV）引起的柑橘衰退病，是世界上最具经济重要性的柑橘病害之一，而褐色橘蚜（*Toxoptera citricida*）是其主要的传播媒介。植物病毒的媒介传播是一个复杂的生物过程，涉及寄主植物、媒介昆虫和植物病原。病毒侵染会对媒介昆虫及植物产生直接或间接的影响。本研究以褐色橘蚜、CTV、西蒙斯甜橙为研究对象，测定了健康及感染 CTV 植株中氨基酸含量、碳氮化合物比例变化及褐色橘蚜取食健康、感染 CTV 植株对其寄主选择行为及其在寄主植物上的生长发育、种群增长、繁殖力及寿命等生物学特性的影响，并通过刺吸电位图谱（electrical penetration graph，EPG）分析橘蚜取食健康、感染 CTV 植株的取食行为差异以及携带病毒和不带毒的褐色橘蚜体内过氧化物酶（SOD）、过氧化氢酶（CAT）、多酚氧化酶（PPO）等保护酶和乙酰胆碱酯酶（AchE）、谷胱甘肽 S-转移酶（GST）、羧酸酯酶（CaE）等解毒酶活性的变化，并测定了橘蚜取食健康和感病植物后其内源性茉莉酸、水杨酸含量及与茉莉酸、水杨酸途径相关的基因表达。结果发现褐色橘蚜更喜欢在感染 CTV 的西蒙斯上取食且西蒙斯是否感染 CTV 对橘蚜的生长发育、繁殖力、寿命、种群增长并无显著性差异；其次，携带病毒的褐色橘蚜体内保护酶系和解毒酶系活性均显著性升高。其中，CTV 侵染与否对西蒙斯的氨基酸含量、碳氮化合物比例的影响和橘蚜取食健康、感病植物后其内源性茉莉酸、水杨酸含量及与茉莉酸、水杨酸途径相关的基因表达正在进行中。研究结果为进一步明确褐色橘蚜与 CTV 之间的互作、揭示褐色橘蚜的适应性及传毒机制等补充了资料。

关键词：柑橘衰退病毒；褐色橘蚜；刺吸电位图谱；适合度；取食行为

---

* 基金项目：重庆市基础与前沿研究计划项目（cstc2015jcyjBX0043），中央高校基本科研业务费专项（XDJK2014C131）
** 第一作者：关桂静，硕士研究生，研究方向：分子微生物学；E-mail：guanshero@163.com
*** 通讯作者：刘金香，副研究员，研究方向：分子植物病理学；E-mail：Ljinxiang@126.com

Genomic Variability and Molecular Evolution of Asian Isolates of Sugarcane streak mosaic virus

LIANG Shan-shan[1,2], ALABI Olufemi J. [3], DMAJ Mona B. [3], FU Wei-lin[1], SUN Sheng-ren[1], FU Hua-ying[1], CHEN Ru-kai[1], MIRKOV T. Erik [3], GAO San-ji[1]

Abstract: Sugarcane streak mosaic virus (SCSMV), an economically important causal agent of mosaic disease of sugarcane, is a member of the newly created genus Poacevirus in the familyPotyviridae. In this study, we report the molecular characterization of three new SCSMV isolates from China (YN-YZ211 and HN-YZ49) and Myanmar (MYA-Formosa) and their genetic variation and phylogenetic relationship to SCSMV isolates from Asia and the type members of the family Potyviridae. The complete genome of each of the three isolates was determined to be 9 781 nucleotides (nt) in size, excluding the 30 poly (A) tail. Phylogenetic analysis of the complete polyprotein amino acid (aa) sequences (3 130 aa) revealed that all SCSMV isolates clustered into a phylogroupspecific to the genus Poacevirus and formed two distinct clades designated as group I and group II. Isolates YNYZ211, HN-YZ49 and MYA-Formosa clustered into group I, sharing 96.8% ~99.5% and 98.9% ~99.6% nt (at the complete genomic level) and aa (at the polyprotein level) identity, respectively, among themselves and 81.2% ~ 98.8% and 94.0% ~99.6% nt (at the complete genomic level) and aa (at the polyprotein level) identity, respectively, with the corresponding sequences of seven Asian SCSMV isolates. Population genetic analysis revealed greater between-group (0.190 ± 0.004) than within-group (group I = 0.025 ± 0.001 and group II = 0.071 ± 0.003) evolutionary divergence values, further supporting the results of the phylogenetic analysis. Further analysis indicated that natural selection might have contributed to the evolution of isolates belonging to the two identified SCSMV clades, with infrequent genetic exchanges occurring between them overtime. These findings provide a comprehensive analysis ofthe population genetic structure and driving forces for the evolution of SCSMV with implications for global exchange of sugarcane germplasm.

Molecular Characterization of Two Divergent Variants of Sugarcane bacilliform viruses Infecting Sugarcane in China

SUN Sheng-ren[1], DAMAJ Mona B.[2], ALABI Olufemi J.[2], WU Xiao-bin[1], MIRKOV T. Erik[2], FU Hua-ying[1], CHEN Ru-kai[1], GAO San-ji[1] *

(1. *Key Laboratory of Sugarcane Biology and Genetic Breeding*, *Ministry of Agriculture*, *Fujian Agriculture and Forestry University*, *Fuzhou*, *Fujian* 350002, *China*; 2. *Department of Plant Pathology and Microbiology*, *Texas A&M AgriLife Research and Extension Center*, *Weslaco*, *Texas* 78596, *USA*)

Abstract: Sugarcane bacilliform viruses (SCBV; genus *Badnavirus*, family *Caulimoviridae*) are considered economically important pathogens of sugarcane, limiting the exchange of its germplasm worldwide. Similar to other badnaviruses, SCBV are genetically diverse and highly complex, hence the need for a thorough analysis of its genetic structure and diversity. In the present study, we report the molecular characterization of two new SCBV isolates (SCBV-CHN1 and SCBV-CHN2) from China and their genetic variation and phylogenetic relationship with 10 global SCBV isolates. The complete genomes of SCBV-CHN1 and SCBV-CHN2 were determined to be 7 764 and 7 629 nucleotides (nt) in size, respectively. Both isolates displayed a typical *Badnavirus* genome organization with three open reading frames (ORFs) but differed in their predicted sizes and putative scanning model for P2 (ORF2-encoded) and P3 (ORF3-encoded) protein translation. Phylogenetic analysis revealed the segregation of complete genomes, individual ORFs, and the RT/RNase H region of the 12 SCBV isolates (two from this study and 10 from the GenBank) into nine phylogroups. SCBV-CHN1 was more closely aligned with SCBV-BB in the SCBV-H clade with both isolates sharing 83% nt sequence identity and low genetic distance (0. 19), based on analysis of the RT/RNase H region. In contrast, SCBV-CHN2 clustered with SCBV-BT into a distinct clade designated as SCBV-G. Recombination analysis identified SCBV-CHN1 and SCBV-CHN2 as putative new recombinant variants arising from putative inter- and intra-specific recombination events. Our findings provide evidence for the presence of distinct genotypic variants of SCBV affecting sugarcane in China and contribute to the understanding of genetic diversity and evolution of sugarcane badnaviruses.

Key words: *Badnavirus*; Genetic variant; Recombination; Sequence analysis; Sugarcane; Sugarcane bacilliform viruses

* Corresponding author: GAO San-ji; E-mail: gaosanji@yahoo.com

Prevalence and RT/RNAse H Genealogy of Isolates of Sugarcane bacilliform viruses from China

WU Xiao-bin[1], ALABI Olufemi J.[2], DAMAJ Mona B.[2], SUN Sheng-ren[1], MIRKOV T. Erik[2], FU Hua-ying[1], CHEN Ru-kain[1,3], GAO San-ji[1,3,*]

(1. *Key Laboratory of Sugarcane Biology and Genetic Breeding, Ministry of Agriculture, Fujian Agriculture and Forestry University, Fuzhou, Fujian* 350002, *China*; 2. *Department of Plant Pathology and Microbiology, Texas A&M AgriLife Research and Extension Center, Weslaco, Texas* 78596, *USA*; 3. *Guangxi Collaborative Innovation Center of Sugarcane Industry, Guangxi University, Nanning, Guangxi* 530004, *China*)

Abstract: Sugarcane bacilliform viruses (SCBV; genus *Badnavirus*; family *Caulimoviridae*) are a threat to the global exchange of sugarcane germplasm. A total of 280 sugarcane leaf tissue samples collected from six provinces in China, 25 from three states in the USA and five from Queensland, Australia were tested for the presence of SCBV by polymerase chain reaction (PCR) using newly designed degenerate primers targeting a 720 base pair (bp) fragment of the reverse transcriptase/ribonuclease H (RT/RNAse H) genomic region. PCR amplified fragments from 94 SCBV-positive samples were then cloned, sequenced and analyzed for their genetic diversity. The results revealed considerable haplotype diversity within individual SCBV isolates. Recombination analyses showed weak signatures of recombination among some of the SCBV sequences. Phylogenetic analysis revealed the segregation of global SCBV isolates into three major monophyletic clades encompassing 18 subgroups, including five previously undescribed subgroups named as SCBV-N to -R. Population genetic analysis data indicated that relatively low levels of genetic exchange have occurred between SCBV populations from different sugarcane-producing regions of the world. Together with the new set of degenerate SCBV-specific primers designed in this study, our results will advance the understanding of SCBV population structure in a semi-perennial host plant and aid the screening of global sugarcane germplasm to minimize the spread of genetic variants of the virus via contaminated plant materials.

* Corresponding authors: CHEN Ru-Kai and GAO San-Ji at Fujian Agriculture and Forestry University, Fuzhou, Fujian, China. E-mail addresses: wuxiaobin@ 163. com (X. -B. Wu), alabi@ tamu. edu (O. J. Alabi), mbdamaj@ ag. tamu. edu (M. B. Damaj), e-mirkov@ tamu. edu (T. E. Mirkov), ssr03@ 163. com (S. -R. Sun), mddzyfhy@ 163. com (H. -Y. Fu), chenrk0805@ 126. com (R. -K. Chen), gaosanji@ yahoo. com (S. -J. Gao)

病毒对梨离体植株生根及细胞分裂素氧化酶基因表达的影响*

陈　婕，王国平，洪　霓**

（华中农业大学植物科学技术学院，农业微生物学国家重点实验室，武汉　430070）

摘　要：苹果茎沟病毒（Apple stem grooving virus，ASGV）和苹果茎痘病毒（Apple stem pitting virus，ASPV）是我国栽培梨上普遍携带的两种病毒，茎尖培养是获得无毒梨种质的主要方式，而提高离体植株的生根效率是繁育无病毒种质的一个重要环节。为明确病毒对梨离体植株生根效率的影响，本研究选取带 ASGV 的沙梨金水 2 号、带有 ASGV 和 ASPV 的西洋梨红贝雷莎及康弗伦斯 3 个品种的离体植株为试验材料，以 3 个品种的无病毒离体植株为对照，将带病毒与无病毒离体植株同时置于 MS 生根培养基培养，于培养 5 ~ 40d 每 5d 测量一次生根数量、根长及愈伤大小。结果显示，病毒对不同种质的梨离体植株生根效率的影响程度存在差异，其中金水 2 号带病毒植株的生根率、根数、根长和愈伤直径均显著低于无病毒植株的相应数值；而康弗伦斯和红贝雷莎的带病毒离体植株的生根率、根数、根长和愈伤大小均无差异，结果表明，病毒对金水 2 号离体植株的生根有明显的抑制作用，对康弗伦斯和红贝雷莎生根无明显影响。采用 qRT-PCR 技术测定了细胞分裂素氧化酶（CKX）基因在带病毒与无病毒梨离体植株的顶端、茎基部和根部相对表达量，结果显示，在生根 20d 时，金水 2 号的 *CKX* 基因表达量较带病毒植株上调 30 倍以上；康弗伦斯无病毒植株 3 个部位的 *CKX* 基因表达量较带病毒植株分别上调 1.1 ~ 1.7 倍；而“红贝雷莎”无病毒与带病毒植株的 *CKX* 基因表达量无明显差异，本研究结果初步表明病毒对离体植株生根效率的影响与植株体内 *CKX* 基因表达量有一定的相关性。

关键词：苹果茎沟病毒；苹果茎痘病毒；离体培养；细胞分裂；基因表达

* 基金项目：国家农业部公益性行业计划（201203076-03）；梨现代农业技术产业体系（nycytx-29-08）

** 通讯作者：洪霓；E-mail：whni@ mail. hzau. edu. cn

柑橘黄化脉明病毒虫媒初探*

刘翠花**，周　彦，周常勇***

（中国农业科学院柑桔研究所，西南大学柑桔研究所，
国家柑桔工程技术研究中心，重庆　400712）

摘　要：柑橘黄化脉明病毒（Citrus yellow vein clearing virus，CYVCV）引起的柑桔黄化脉明病是国内外新近发生的一种柑桔病害。该病最早发现于巴基斯坦，随后印度、土耳其等地也有报道。近年来该病在我国云南、四川等地也有分布。CYVCV 可侵染大多数柑橘属植物，其中以对柠檬和酸橙的为害最为严重，造成春梢、秋梢新发叶片侧脉脉明，后期叶片皱缩，畸形，叶背面侧脉附近呈水渍状等症状。老叶症状依然可见。CYVCV 也可危害香橼、莱檬、甜橙等柑橘类型，造成潜症带毒。此外，豇豆、菜豆、藜麦是 CYVCV 的草本寄主。CYVCV 可通过嫁接传播，也可经绣线菊蚜、豆蚜从柠檬传至豆科作物以及在豆科作物之间传播。尚不清楚 CYVCV 依靠哪种媒介昆虫在柑橘间传播。

本研究从毒源植株上采集褐色橘蚜、橘二叉蚜、黄蜘蛛、黑刺粉虱、绣线菊蚜，抽提其核酸后进行 RT-PCR 检测，发现褐色橘蚜可携带 CYVCV。将脱毒后的褐色橘蚜饲喂于毒源甜橙植株 24h，然后用毛笔将成虫挑至健康的甜橙幼苗上取食 24h，每株放置 50 头蚜虫，然后用毒死蜱将蚜虫杀死。重复 20 次。将传毒后的甜橙幼苗放置在一个干净的网罩中，生长温度控制在 25℃左右，并科学地进行水肥管理。3 个月后对传毒后的甜橙幼苗进行 RT-PCR 检测，结果显示均未检测到 CYVCV。初步表明褐色橘蚜不能在甜橙之间传播 CYVCV。

本研究首次对甜橙间 CYVCV 的传播媒介进行了探索，对柑橘黄化脉明病毒病的防控有一定的指导意义。

关键词：柑橘黄化脉明病毒；传毒媒介

---

* 基金项目：两江学者计划；中央高校基本科研专项（XDJK2015A009）；重庆基础及前沿研究项目（cstc2015cyjBX0043）

** 第一作者：刘翠花，博士研究生，西南大学柑橘研究所

*** 通讯作者：周常勇，研究员，主要从事柑橘病毒类病害研究和无病毒繁育体系建设；E-mail：zhoucy@ swu. edu. cn

柑橘黄脉病毒侵染对尤力克柠檬叶绿素代谢的影响*

金　鑫**，张艳慧，唐　萌，周　彦***
（西南大学柑桔研究所，中国农业科学院柑桔研究所，重庆　400712）

摘　要：柑橘黄脉病毒（Citrus yellow vein clearing virus，CYVCV）引起的柑橘黄脉病是一种新近发生的柑橘病害。柠檬感染 CYVCV 后的症状表现为春稍、秋稍新发叶片侧脉脉明及侧脉附近黄化，后期叶片反卷、皱缩、畸形。以健康和感染 CYVCV 的尤力克柠檬叶片为材料，对叶绿素及合成中间物质含量进行测定，结果显示感病柠檬叶绿素 a、叶绿素 b 和总叶绿素含量以及合成中间物质中胆色素原（PBG）、尿卟啉原Ⅲ（Urogen Ⅲ）、粪卟啉原Ⅲ（CoprogenⅢ）、原卟啉Ⅸ（Proto Ⅸ）、Mg-原卟啉Ⅸ（Mg-Proto Ⅸ）和原叶绿素酸酯（Pchlide）含量均显著降低，但 δ-氨基酮戊酸（ALA）积累。运用实时荧光定量 PCR 分析叶绿素代谢途径关键基因相对表达，结果显示感病柠檬合成途径中编码谷氨酰-tRNA 还原酶的 *HEMA* 基因表达上调了 5 倍，其余基因表达量均有所降低，从而导致 ALA 积累。分解途径中编码叶绿素酶的 *CLH* 基因表达量上调了 45 倍。结果表明，尤力克柠檬感染 CYVCV 后叶绿素含量减少是合成和分解途径共同作用的结果。

关键词：CYVCV；尤力克柠檬；叶绿素代谢

---

* 基金项目：重庆市两江学者项目；中央高校基本科研业务费（XDJK2015A009）；重庆市基础与前沿研究重点项目（CSTC2015JCYJBX0043）

** 第一作者：金鑫，2014 级硕士研究生，专业：植物病理学；E-mail：jinxin773741127@ sina. com
*** 通讯作者：周彦，研究员，研究方向：分子植物病理学；E-mail：zhouyan@ cric. cn

柑橘黄化脉明病毒在尤力克柠檬叶片叶柄中的免疫酶标定位*

邓雨青**，李 平，马丹丹，李中安***
（西南大学柑桔研究所，中国农业科学院柑桔研究所，重庆 400712）

摘 要：柑橘黄化脉明病毒（Citrus yellow vein clearing virus，CYVCV）引起的柑橘黄脉病是一种新近发生的柑橘病害，主要造成柠檬和酸橙叶片黄化、扭曲、明脉，偶尔伴有环斑和叶脉坏死。以健康和感染 CYVCV 的尤力克柠檬叶片、叶柄为材料，利用直接组织点免疫技术（Direct tissue blot immunoassay，DTBIA）和荧光免疫技术对尤力克柠檬叶片、叶柄中的病毒进行免疫定位。DTBIA 试验结果表明，CYVCV 主要分布于叶柄叶肉细胞以及维管束周围。利用荧光免疫进一步研究发现：在健康以及感病的叶片中都观察到阳性反应，表明该方法不适宜尤力克柠檬叶片的检测。而对叶脉中的 CYVCV 定位时，发现感病柠檬出现阳性反应，健康柠檬无阳性反应，且 CYVCV 主要分布在韧皮部、分泌腔周围以及维管束周围的薄壁细胞，与 DTBIA 结果一致，表明该方法可用于检测尤力克柠檬叶脉组织。

关键词：柑橘黄化脉明病毒；免疫标记；定位

* 基金项目：重庆市基础与前沿研究计划重点项目（CSTC2015JCYJBX0043）；中央高校基本科研业务费（XDJK2015A009）
** 第一作者：邓雨青，硕士研究生，专业：植物病理学；E-mail：dengyuqing828@163. com
*** 通讯作者：李中安，研究员，研究方向：分子植物病理学、细胞生物学；E-mail：zhongan@ cric. cn

柑橘脉突病毒实时荧光定量 RT-PCR 检测体系的建立与应用*

王艳娇[1**]，崔甜甜[1]，黄爱军[2]，陈洪明[1]，李中安[1]，周常勇[1]，宋 震[1***]

（1. 西南大学柑桔研究所，中国农业科学院柑桔研究所，重庆 400712；
2. 赣南师范大学国家脐橙工程研究中心，江西 341000）

摘 要：柑橘脉突病毒（Citrus vein enation virus，CVEV）可引起柑橘脉突病及木瘤病。目前，该病毒在多个国家均有发生，包括中国、日本、印度、秘鲁、南非、西班牙、美国和土耳其。为了快速、灵敏地检测 CVEV，通过设计特异性引物（EVqF4/ EVqR4），优化反应条件建立了 CVEV 的实时荧光定量 RT-PCR 检测体系。该方法特异性良好；检测灵敏度比常规 RT-PCR 高 100 倍；标准曲线循环阈值与模板浓度呈良好的线性关系，相关系数为 0.992，扩增效率 101.8%；检测组内和组间变异系数均小于 2.85%，重复性较好。利用所建立的实时荧光定量方法对柑橘植株进行检测发现，代酸橙植株和象杂柑植株中 CVEV 分布不均匀，其中，根部病毒滴度最高分别为 81 261拷贝/500 ng RNA 和 22 660拷贝/500 ng RNA，茎及叶部病毒滴度相对较低。

关键词：柑橘脉突病毒；实时荧光定量 RT-PCR；应用

---

* 基金项目：农业部公益性行业（农业）科研专项经费项目（201203076）；重庆市基础及前沿研究项目（CSTC2014jcyjA80033）；重庆市两江学者计划项目

** 第一作者：王艳娇，硕士研究生；E-mail：wangyanjiao91@126.com

*** 通讯作者：宋震，副研究员，主要从事柑橘病毒类病害研究；E-mail：songzhen@cric.cn

柑橘衰退病毒弱毒株系与强毒株系互作研究*

刘 勇[1,3]，王国平[1,2]，洪 霓[1,2**]

（1. 华中农业大学农业微生物国家重点实验室，武汉 430070；2. 华中农业大学植物科技学院，湖北省植物病害检测与安全控制重点实验室， 武汉 430070；3. 湖北省农业科学院果树茶叶研究所，武汉 430064）

摘　要： 由柑橘衰退病毒（Citrus tristeza virus，CTV）引起的柑橘衰退病是柑橘上发生最为普遍且危害严重的病毒病，该病毒广泛分布于世界各柑橘产地。CTV 存在明显的致病性分化，CTV 强毒株系在敏感的柑橘植株上诱导产生木质部茎陷点症状，严重时引起树体快速衰退或死亡；CTV 弱毒株系在寄主植物上一般不产生明显的症状。本研究选取甜橙、莱檬和 HB 柚为寄主植物，采用 CTV 弱致病力株 N4 和强致病力株 N21 交互接种，同时以单一接种这 2 个株系的植株为对照，进行生物学和病毒含量变化的分析。发现接种约 18 个月后 N21 在这 3 中寄主植物上均可引起新叶褪绿或黄化及新梢的生长受到抑制，且 N4 免疫接种未能减轻 N21 引起症状表现。为明确 CTV 强毒株 N21 在弱毒株 N4 免疫接种柑橘植株中的相对含量变化，选取 N4 和 N21 感染的 3 种寄主各选取 20 株，分别交互接种 N21 和 N4，以 30d 为间隔，采用强弱毒株鉴别引物进行 RT-PCR 分析。结果显示，CTV 弱毒株 N4 的检出率随强毒株 N21 接种周期的延长而逐渐下降，在交互接种的 HB 柚和莱檬植株中，N4 含量于 N21 接种 60d 后逐渐降低，在甜橙植株中，N4 含量于 N21 接种 90 d 后逐步降低，至接种 150d 后，N4 检出率分别降至 45%、45% 和 75%；而 N21 含量变化则呈现相反的趋势，N21 在交互接种的莱檬植株中增殖相对较快，60dpi 检出率均达 100%。在 90dpi，N21 在交互接种的 HB 柚中的检出率达 100%。而在甜橙植株中，N21 增殖相对较慢，90d 时检出率为 70%，至 120dpi 检出率达 100%。因此 CTV 弱毒株 N4 对强毒株 N21 的抑制作用与其寄主存在一定的关系。为确认 RT-PCR 检测阴性的植株中 CTV 弱毒株 N4 是否存在，于接种 210d 后进一步采用半巢式 RT-PCR 方法对部分检测结果为阴性的植株进行了检测，结果显示这些植株中都能扩增到 CTV 弱毒株的特异性目标片段，表明 CTV 弱毒株系在交互接种植株中随接种时间延长而含量降低，但并未完全降解。此外，RT-PCR 结果显示，先接种强毒株系 N21 对后期接种的 N4 增殖有明显的抑制作用，在单一接种 N4 的梾檬植株中于第 30d 即可扩增出弱毒株的特异片段；而在接 N21 + N4 植株中，至 6 ~ 18 个月时仅部分植株扩增出弱毒株的特异片段，这些结果表明，在 N4 和 N21 混合接种的莱檬植株中，强毒株 N21 的复制能力明显高于弱毒株或对弱毒株 N4 的复制有一定的抑制作用。

关键词： 柑橘衰退病毒；株系；互作

* 基金项目：国家自然科学基金资助项目（No. 31272145；30871684）

** 通讯作者：洪霓，教授，研究方向为果树病毒；E-mail：whni@ mail. hzau. edu. cn

瑞昌山药病毒病病原鉴定

贺　哲*，黄　婷，秦双林，崔朝宇，蒋军喜**
（江西农业大学农学院，南昌　330045）

摘　要：山药是江西省瑞昌市重要的经济作物之一。近年来，山药病毒病在瑞昌山药种植区发生严重，其症状表现为叶片出现花叶、斑驳，明脉，影响山药正常的光合作用，对瑞昌山药的品质和产量造成很大影响。为了探明瑞昌山药病毒病病原，以便给当地山药病毒病发生规律研究和有效防治提供理论依据，采用分子生物学技术对其病原种类进行了鉴定。从瑞昌市高丰、范镇、桂林及白杨 4 个乡镇的山药种植区中采集获得 18 个感病叶片样本，提取样本中的总 RNA，利用 RT-PCR 技术，通过马铃薯 Y 病毒科通用引物，对病毒序列进行 PCR 扩增，对扩增产物进行割胶回收、序列测定和序列分析。结果表明，这 18 个分离物与日本山药花叶病毒均具有最高的序列同源性（86% ~87%），并且在构建的系统发育树上与日本山药花叶病毒共处于一个分支上；根据马铃薯 Y 病毒属 CP 同源性 80% 的种类划分标准，认为瑞昌山药病毒病病原为日本山药花叶病毒（Japanese yam mosaic virus）。18 个分离物 CP 氨基酸序列之间同源性为 92.9% ~ 99.5%，同源性大小与病害发生地及其与症状类型之间不存在明显的相关性。

关键词：山药病毒病；分子鉴定；RT-PCR；日本山药花叶病毒

* 第一作者：贺哲，硕士生，主要从事分子植物病理学研究；E-mail：hezhe1993@163.com
** 通讯作者：蒋军喜，教授，博士，主要从事植物病害综合治理研究；E-mail：jxau2011@126.com

Characterization of a New Badnavirus from *Wisteria sinensis*

LI Yong-qiang[1], DENG Cong-liang[2], ZHOU Qi[2]

(1. *College of Plant Science and Technology*, *Beijing University of Agriculture*, *Beijing* 102206;
2. *Plant Laboratory of Beijing Entry-Exit Inspection and Quarantine Bureau*, *Beijing* 100026)

Abstract: *Wisteria sinensis*, also known as Chinese wisteria, is highly valued for the foliage and the blue or white pea-like flowers adorning walls and pergolas as ornamental plants. As other flowering plants, *W. sinensis* were prone to be affected by different biotic agents especially wisteria mosaic disease caused by *Wisteria vein mosaic virus* and *Cucumber mosaic virus* which reduced their quality and ornamental value. In June of 2015, a *Wisteria sinensis* plant with mosaic and crinkle symptom in the leaves was observed in Beijing. To identify the causal agent (s), small RNA deep sequencing was conducted with the symptomatic leaves. A previously undescribed badnavirus, tentatively named wisteria bacilliform virus (WBV) was identified together with WVMV. The complete genome sequence of WBV was subsequently cloned with PCR and the circular double-stranded DNA genome of this virus consisted of 7 362bp in size with four open reading frames (ORFs 1 to 4) on the plus strand. Sequence analysis showed this virus shared highest (69%) nt sequence identity with *Pagoda yellow mosaic associated virus* (PYMAV). In the RT-RNase H region of the ORF3 encoded polyprotein, this virus shared 74%, the highest nt sequence identity with PYMAV. Phylogenetic analysis provided further evidence that the virus identified in this study is probably a member of a new species in the genus Badnavirus.

Key words: *Wisteria sinensis*; Deep sequencing; Badnavirus

胶体金免疫层析试纸条检测香蕉束顶病毒方法的建立*

刘　娟**，饶雪琴，李华平***
（华南农业大学农学院，广州　510642）

摘　要：香蕉是热带和亚热带地区重要的经济作物，在世界水果贸易和消费中占有极其重要的位置，中国是世界第三大香蕉生产国。而香蕉束顶病毒（Banana bunchy top virus，BBTV）引致的香蕉束顶病毒病是世界香蕉产业生产中的主要限制因子之一，目前还没有切实有效的防控方法，其中种植无病种苗是当前防控该病的主要措施之一，而无病种苗的生产有耐于快速、高效和简便的检测方法。本研究在免疫渗透技术的基础上，以胶体金作为示踪标记物、硝酸纤维膜为介质，通过毛细管作用使样品溶液在介质上运动，待测物同层析材料上的受体发生特异性亲和反应。通过探索香蕉束顶病毒单克隆抗体及AP标记羊抗鼠IgG最佳浓度和探针制备的最适胶体金pH值及抗体浓度等条件，建立了胶体金免疫层析试纸条技术，并对该技术的特异性、灵敏度、稳定性和重复性效果等进行了评价。结果表明，探针制备的最适胶体金pH值为6.0，BBTV抗体与胶体结合的最佳浓度为12μg/mL，所建立的胶体金免疫层析试纸条法仅对携带BBTV的香蕉样品才呈阳性反应。这表明所建立的BBTV胶体金免疫层析试纸条法可以快速、有效、方便、可靠的检测香蕉植株是否带有BBTV病毒。本研究对于香蕉束顶病毒病的防治具有重要的意义。

关键词：香蕉；香蕉束顶病毒；胶体金免疫层析试纸条；检测

---

* 基金项目：公益性（农业）行业科技专项（201203076－07）
** 第一作者：刘娟，硕士研究生，植物病理学；E-mail：1296757041@qq.com
*** 通讯作者：李华平，教授；E-mail：huaping@scau.edu.cn

小 RNA 测序结合 RT-PCR 鉴定一种侵染梨的负义单链 RNA 病毒*

刘华珍，王国平，洪　霓**
（华中农业大学植物科学技术学院，农业微生物学国家重点实验室，武汉　430070）

摘　要：RNA 沉默是植物抗病毒机制之一，小 RNA（sRNA）测序及现代生物信息学技术的发展为病毒鉴定提供了一种新手段。本研究采用 sRNA 测序技术对一株表现疑似病毒病的梨树叶片进行分析，采用 Velvet 软件对 sRNA 进行拼接，然后对获得的 contigs 在 NCBI 数据库进行 BLASTn 和 BLASTX 搜索，鉴定出 5 个 contigs 与欧洲山梣环斑病毒属（*Emaravirus*）的部分病毒的 RNA1，RNA3 和 RNA7 编码蛋白存在较低的相似性。根据这些 contigs 序列设计引物进行 RT-PCR 扩增，获得了这些 RNA 的部分片段序列，序列比对结果显示，所获片段的核苷酸和推测编码蛋白氨基酸序列与欧洲山梣环斑病毒属病毒基因组及编码蛋白存在一定的相似性（20% ~ 34%），其中，RNA1 编码依赖于 RNA 的 RNA 复制酶（RdRp），具有该属病毒的 RdRp 保守基序，初步判断该病毒为一种新的欧洲山梣环斑病毒属病毒。此外，采用该属病毒的保守引物进行 RT-PCR 扩增，获得了 RNA2 的部分序列和 RNA5 全长序列。欧洲山梣环斑病毒属病毒为一类侵染植物的负义单链 RNA 病毒，本研究首次从梨树上鉴定到类似该属的病毒，并发现该病毒感染与梨树病害症状表现有关，研究结果为加深对梨树病毒病的认识提供了新的信息。

关键词：RNA 测序；RT-PCR；负义单链 RNA 病毒

---

* 基金项目：国家农业部公益性行业计划（201203076-03）；梨现代农业技术产业体系（nycytx-29-08）
** 通讯作者：洪霓；E-mail：whni@ mail. hzau. edu. cn

SRBSDV 侵染及温度胁迫对传毒介体白背飞虱代谢组的影响

冯文地*，钟 婷，周国辉**
（华南农业大学植物病毒研究室，广州 510642）

摘 要： 南方水稻黑条矮缩病毒（Southern rice black-streaked dwarf virus，SRBSDV）经迁飞性昆虫白背飞虱（*Sogatella furcifera*）以循回增繁持久型方式传播。白背飞虱不能耐受35℃以上高温和4℃以下低温，因此病害的发生与温度密切相关。前期研究发现，带毒白背飞虱对高温的耐受性显著提高，对低温的耐受性显著降低。为了探析 SRBSDV—白背飞虱—温度三者之间的互作机制，本研究运用 GC-MS（Gas chromatography-mass spectrometry）技术对经过高温（36℃）、低温（5℃）胁迫处理4h 后的无毒及饲毒白背飞虱进行代谢组测定。结果表明，适温（25℃）下，白背飞虱应答 SRBSDV 侵染的差异代谢物有 16 种显著上调，22 种显著下调，以糖和多元醇为主；应答低温和高温胁迫的差异代谢物分别有 16 种和 20 种显著上调，22 种和 42 种显著下调，以氨基酸类、三羧酸循环有关代谢物居多。应答病毒侵染-低温复合胁迫的差异代谢物有 8 种显著上调，12 种显著下调，下调的代谢物主要为三羧酸循环有关及脂肪酸类物质。应答病毒侵染-高温复合胁迫的差异代谢物有 29 种显著上调，29 种显著下调，上调物质中主要是糖和多元醇类，下调代谢物则以三羧酸循环有关产物、脂肪酸及氨基酸类为主。分析认为，白背飞虱应对病毒侵染和高、低温胁迫分别采取了不同的代谢通路，但两者存在显著的交叉影响，SRBSDV 通过改变白背飞虱氨基酸代谢、糖代谢及脂类代谢等生理活动从而影响后者对高、低温的耐受性。

关键词： 南方水稻黑条矮缩病毒；白背飞虱；代谢组；耐温性

---

* 第一作者：冯文地，硕士研究生，从事植物病毒研究；E-mail：768334676@ qq. com
** 通讯作者：周国辉，教授，从事植物病毒及病毒病害研究；E-mail：ghzhou@ scau. edu. cn

水稻橙叶植原体基因组序列测定及分析

朱英芝[*]，何园歌，周国辉[**]

（华南农业大学农学院，广州　510642）

摘　要：水稻橙叶植原体（*Candidatus* Rice Orange Leaf Phytoplasma）由电光叶蝉和黑尾叶蝉传播侵染水稻韧皮部细胞，引起水稻橙叶病。该病害20世纪80年代末至90年代初曾在我国华南局部地区流行成灾，近年再度暴发并呈扩大蔓延趋势。为了深入研究病菌致病机理，本研究以感病水稻叶鞘组织总DNA构建文库，通过Illumina HiSeq 2000平台对文库进行测序，获得21 751 038个Raw Reads；采用基因拼接软件CLC Genomic workbench（CLC-bio，Demmark）对获得的序列进行denovo拼接获得总长为285 321 174bp的74 342条Contigs；利用本地化BLAST比对已报道植原体参考基因组，筛选出总长563 399bp与植原体匹配的60个contings；采用PCR扩增获得各Contig之间的gap序列，最终获得近全长水稻橙叶植原体基因组序列。所得序列总长606 884bp，GC含量为28.2%，编码661个蛋白，包含35个tRNA和2个rRNA操纵子。同时获得一个长度为4 197bp的完整质粒序列。为下一步研究提供了基础信息。

关键词：水稻橙叶病；植原体；基因组序列

* 第一作者：朱英芝，博士后，从事植物病理学研究；E-mail：zhuyingzhi376@163.com

** 通讯作者：周国辉，教授，从事植物病毒及病毒病害研究；E-mail：ghzhou@scau.edu.cn

与南方水稻黑条矮缩病毒 P5-1 互作的白背飞虱蛋白分析

马思琦*，涂　智，周国辉**

（华南农业大学农学院，广州　510642）

摘　要：南方水稻黑条矮缩病毒（Southern rice black-streaked dwarf virus，SRBSDV）经迁飞性昆虫白背飞虱（*Sogatella furcifera*）以循回增繁持久性方式传播。SRBSDV 侵入白背飞虱体内后，需借助介体体内相关蛋白形成病毒基质，进而得以复制和增殖。前人研究表明，SRBSDV 编码的非结构蛋白 P5-1 是病毒基质的重要组份。为探索虫体内与病毒 P5-1 互作的蛋白组份及其功能，本研究以 P5-1 为诱饵，通过酵母双杂交技术筛选白背飞虱 cDNA 文库，得到了 2 个与 P5-1 互作的蛋白，分别为一个核糖体蛋白组分（ribosomal protein）和一个烯醇酶（enolase）。PCR 扩增获取这两个互作蛋白的基因全长序列，经酵母双杂交和双分子荧光互补实验，证实了它们与 P5-1 的互作。进一步地，通过 RT-qPCR 分析这两个蛋白基因在无毒和带毒白背飞虱体内表达水平，结果显示，核糖体蛋白亚基在带毒虫体内的表达量相比于无毒虫有显著提高，而烯醇酶基因在带毒虫和无毒虫体内的表达量无显著差异。该研究为深入解析介体因子在病毒基质形成中的作用奠定了基础。

关键词：南方水稻黑条矮缩病毒；白背飞虱；病毒—介体互作；病毒基质

* 第一作者：马思琦，硕士研究生，从事植物病毒研究；E-mail：1340084034@ qq. com

** 通讯作者：周国辉，教授，从事植物病毒及病毒病害研究；E-mail：ghzhou@ scau. edu. cn

南方水稻黑条矮缩病毒 P9-1 病毒质蛋白与单链 RNA 结合特性研究

吴鉴艳，陶小荣
（南京农业大学植物保护学院分子植物病毒实验室，南京　210095）

摘　要：南方水稻黑条矮缩病毒（Southern rice black streaked dwarf virus，SRBSDV）是呼肠孤病毒科（Reoviridae）斐济病毒属（*Fijivirus*）2 组的一个新种。由该病毒基因组 S9 编码的 P9-1 是个病毒质蛋白，本文采用琼脂糖凝胶迁移阻滞试验（Electrophoretic mobility shift assay，EMSA）在体外初步分析 SRBSDV P9-1 蛋白结合单链 RNA 的生化特性。非变性蛋白胶及凝胶阻滞分析结果表明 SRBSDV P9-1 蛋白在自然状态下形成多种多聚体类型，其中，八体构型蛋白是蛋白存在及蛋白结合单链 RNA 的主要形式；丙氨基酸替换实验结果表明 N 端带正电荷氨基酸 R26、R39、K43、K44 是蛋白结合单链 RNA 的主要位点，并且与 C 端末尾 23 个氨基酸一样都影响蛋白八体的形成；与 SRBSDV 同科同属的水稻黑条矮缩病毒（Rice black streaked dwarf virus，RBSDV）编码的 P9-1 蛋白同样是一个病毒质蛋白，两者都包含 347 个氨基酸，最大的差异存在于 131 ~ 160 位残基序列，氨基酸交换实验表明 SRBSDV P9-1 的 131 ~ 160 位氨基酸残基并不能拯救 RBSDV P9-1 形成更多的八体蛋白，反而减弱了 RBSDV P9-1 结合单链 RNA 的活性；这些结果表明仅有 RNA 结合位点但不具备蛋白八体构型将不能有效结合单链 RNA，而只有当蛋白构象与结合位点同时存在并相互匹配时，蛋白结合 RNA 的活性才达到最大。

关键词：病毒质蛋白；八体构型；RNA 结合位点；蛋白构象；RNA 结合活性

番茄免疫蛋白受体Sw-5b的亚细胞定位与抗性功能研究

陈小姣，陈虹宇，陶小荣
（南京农业大学植物保护学院，南京　210095）

摘　要：番茄斑萎病毒属病毒是一类对农业生产造成严重危害的植物多分体负义链RNA病毒，而Sw-5b是番茄上进化出的一个针对番茄斑萎病毒属病毒具有广谱抗性的免疫受体蛋白。近年来的研究表明免疫受体蛋白的亚细胞定位在抗病信号通路中发挥着重要作用，番茄Sw-5b是一个含有额外的N端结构域（NTD）的CC-NB-LRR类型的免疫受体蛋白，该蛋白的亚细胞定位以及在抗病中的作用还并不清楚。本研究将Sw-5b与YFP融合后发现YFP-Sw-5b既定位在细胞质同时也定位在细胞核，融合的YFP-Sw-5b依然对番茄斑萎病毒具有抗性。在YFP-Sw-5b的N端融合核输出信号（NES）将Sw-5b拉到细胞质，瞬时表达发现细胞质定位导致细胞死亡活性显著增强，但是转基因NES-YFP-Sw-5b植株接毒测试表明细胞质定位的Sw-5b丧失了对TSWV的抗性；在YFP-Sw-5b的N端融合核定位信号（NLS）将Sw-5b拉到细胞核内，发现其诱导的细胞死亡明显减弱，但是却足以抵抗病毒，表明细胞核定位对于Sw-5b的抗性具有重要作用。将Sw-5b的激发子NSm融合NES，发现定位在细胞质的NSm能够诱导细胞死亡，而将NSm融合NLS定位至细胞核则不再诱导HR，表明Sw-5b与NSm的识别发生在细胞质。此外，我们发现Sw-5b，将Sw-5b P-loop保守基序突变Sw-5b不再能够被激活，突变体YFP-Sw-5b也不再能够诱导细胞死亡，但是该突变体依然可以定位于细胞核，同样将*SGT*1沉默，YFP-Sw-5b也不能诱导细胞死亡，但在*SGT*1沉默植株中，YFP-Sw-5b依然能够进核，表明该抗性蛋白有没有激活都可以定位于细胞核。我们进一步发现Sw-5b的细胞核定位信号是由额外的NTD结构域决定的，并且发现NTD结构域的进核对于Sw-5b的抗性是必需的。

关键词：免疫受体；细胞定位；抗性

Applications of Next Generation Sequencing in Plant Virology*

CAO Meng-ji[1**], ZHOU Chang-yong[1], LI Ru-hui[2], DING Shou-wei[3]

(1. *National Citrus Engineering Research Center*, *Citrus Research Institute*, *Southwest University/Chinese Academy of Agricultural Sciences*, *Chongqing* 400712, *China*; 2. *USDA-ARS*, *National Germplasm Resources Laboratory*, *Beltsville*, *MD* 20705, *USA*; 3. *Department of Plant Pathology and Microbiology*, *University of California*, *Riverside*, *CA* 92521, *USA*)

Abstract: Nextgeneration sequencing (NGS) technologies provide a high throughput, efficient andfast DNAsequencing platform compared to the standard and traditional technologies. The technologies have been rapidly applied to several areas of plant virology including virus/viroid genome sequencing, discovery and detection, ecology, replication and antiviral mechanism since 2009. We report here the use of small interfering RNAs (siRNAs) and Hiseq RNA sequencing for the study of virus-host interaction and virus detection/diagnosis. Analysis of the siRNA sequencing data revealed the presence of a genetically distinct class of virus-activated siRNAs (vasiRNAs) in *Arabidopsis thaliana*. We propose that antiviral RNAi activates broad-spectrum antiviral activity via widespread silencing of host genes directed by vasiRNAs in addition to specific antiviral defense by viral siRNAs. Both siRNA and Hiseq RNA sequences were used to identify known and unknown pathogens from citrus, fruit trees and sweet potato. The results showed that fast, accurate, and full indexing and identification of the pathogens were achieved by both methods. It is expected that NGS will play veryprominent roles in fundamental and applied research areas of plant virology.

Key words: Next generation sequencing; Virus-activated siRNAs; Virus identification

---

* Funding: National Natural Science Foundation of China (No. 31501611); Fundamental Research Funds for the Central Universities (No. XDJK2016B021 and 20700907); Agro－scientific Research in the Public Interest (No. 201203076)

** Corresponding author: CAO Meng-ji, research field focused on the study of plant virus and host interactions; E-mail: caomengji@cric. cn

Development of RT-LAMP Assay for the Detection of Maize Chlorotic Mottle Virus in Maize

JIAO Zhi-yuan, CHEN Ling, XIA Zi-hao, ZHAO Zhen-xing,
LI Ming-jun, ZHOU Tao, FAN Zai-feng*

(*Department of Plant Pathology*, *China Agricultural University*, *Beijing* 100193, *China*)

Abstract: Maize (*Zea mays* L.) has been widely cultivated in the world, but unfortunately virus diseases of maize severely threat maize production. Maize chlorotic mottle virus (MCMV) is one of the most important virus pathogensin corn which can induce symptoms such as mosaic, stunting and leaf necrosis. Whatis worse, MCMV couldcause corn lethal necrosis (CLN) when co-infect with a potyvirus. As the only established member of the genus *Machlomovirus* in the family *Tombusviridae*, MCMV can be transmitted through infected maize seeds, beetles and mechanical inoculation leading to high risk of the transmission of MCMV. Therefore, it is crucial to develop a rapid and reliable method to detect MCMV.

In this study, we established a rapid and effective method to detect MCMV using reverse transcription loop-mediated isothermal amplification (RT-LAMP). The results showed that the reaction mixtures kept at 63℃ for 60 min were optimal and the sensitivity of the RT-LAMP was 10 to 100 fold than conventional PCR method for detecting MCMV. No cross-reactivity was detected with other most common and important viral pathogens infecting maize in China. In addition, RT-LAMP products could be stained with SYBR Green I in-tube, which facilitated detection of MVMC through observing fluorescentrather than ethidium staining following gel electrophoresis. The method was verified by testing field samples collected from Yunnan province and showed high reliability and sensitivity. Taken together, we established a rapid, reliable and visible method for MCMV detection in maize.

Key words: Maize chlorotic mottle virus; RT-LAMP; Detection method

* Corresponding author: FAN Zai-feng; E-mail: fanzaifeng@ 126. com

利用酵母双杂交系统研究桑脉带相关病毒核外壳蛋白自身相互作用

张 璐[1]，朱丽玲[1]，潘瑞兰[2]，陈保善[2,3*]，蒙姣荣[1,3*]

（1. 广西大学农学院，南宁 530005；2. 广西大学生命科学与技术学院，南宁 530005；
3. 亚热带农业生物资源保护与利用国家重点实验室，南宁 530005）

摘 要： 桑脉带相关病毒（Mulberry vein banding-associated virus，MuVBaV）属布尼安病毒科番茄斑萎病毒属，是广西桑树病毒病的主要病原。为研究其核外壳蛋白（nucleocapsid protein，N）的自身相互作用，本研究利用分离泛素酵母双杂交膜系统构建了 N 蛋白的诱饵表达质粒和猎物表达质粒并转化酵母菌株 NMY51。诱饵表达质粒 pBT3-SUC-MuVBaV-N 或 pDHB1-MuVBaV-N 和阳性对照质粒 pOst1-NubI 共转化的酵母菌在筛选性营养缺陷型培养基上 SD-trp-leu-his-ade（SD-AHLW）均能正常生长，而与阴性对照质粒 pPR3-N 共转化的酵母在营养缺陷型 SD-AHLW 平板上不能正常生长，表明 N 蛋白诱饵质粒的表达产物没有自我激活活性，对酵母细胞也没有毒性。N 蛋白诱饵表达质粒及其猎物表达质粒共转化的酵母也可以在 SD-AHLW 培养基上正常生长，说明 N 蛋白在酵母细胞中可发生自我相互作用。本研究结果为今后研究 MuVBaV N 蛋白的功能以及通过筛选桑 cDNA 文库获得互作寄主因子奠定了良好的基础。

关键词： 桑脉带病毒；核外壳蛋白；自身相互作用；分离泛素酵母双杂交膜系统

* 通讯作者：陈保善；蒙姣荣

A Novel Mycoreovirus from *Sclerotinia sclerotiorum* Reveals Cross-family Horizontal Gene Transfer and Evolution of Diverse Viral Lineages*

LIU Li-jiang[1,2], GAO Li-xia[1,2], CHENG Jia-sen[1], FU Yan-ping[1], LIU Hui-quan[3], JIANG Dao-hong[1], XIE Jia-tao[1]**

(1. *State Key Laboratory of Agricultural Microbiology*, *The Provincial Key Lab of Plant Pathology of Hubei Province*, *College of Plant Science and Technology*, *Huazhong Agricultural University*, *Wuhan* 430070, *People's Republic of China*; 2. *Key Laboratory of Biology and Genetic Improvement of Oil Crops*, *Ministry of Agriculture of People's Republic of China*, *Oil Crops Research Institute*, *Chinese Academy of Agricultural Sciences*, *Wuhan* 430062, *China*; 3. *NWAFU-PU Joint Research Center*, *State Key Laboratory of Crop Stress Biology for Arid Areas*, *College of Plant Protection*, *Northwest A&F University*, *Yangling* 712100, *Shanxi Province*, *People's Republic of China*.)

Abstract: As the largest dsRNA virus family, the evolutionary relationship of reoviruses remains still unclear for the lack of enough representative members and evolutionary clues. Here, we performed the molecular cloning and complete genome of a novel mycoreovirus *Sclerotinia sclerotiorum* Reovirus 1 (SsReV1) which had a genome of 28 055 base pairs (bp) and 11 dsRNA segments. In combination with unique molecular features, virions shape and composition, and phylogenetic analysis, SsReV1 was distinct from all known reoviruses and may represent a new taxonomic unit in the *Reoviridae* family. More importantly, two conserved domains, double-stranded RNA binding motif (dsRBM, pfam 00035) and reovirus sigma C capsid protein (Reo_ σC, pfam04582), were identified in the genome of SsReV1, that were widespread in diverse virus lineages in the further analysis. Sequence comparison and phylogenetic analysis revealed that multiple cross-family horizontal gene transfer (HGT) events may occur among reoviruses, dsDNA viruses, single-stranded RNA viruses and even cellular organisms. Interestingly, the dsRBM of SsReV1 shared a closest phylogenetic relationship with those of two human proteins (Q7Z6F6/P19525, E7EVJ4) with a strong support. These results indicate that SsReV1 is a new taxonomic representative in the *Reoviridae* family and reoviruses may have a large-scale gene-communications with other virus lineages more common than previously recognized, providing new insights into the diversity, global ecology and evolution of reoviruses and segmented dsRNA viruses.

Key words: *Sclerotinia sclerotiorum*; Mycoreovirus; Horizontal gene transfer

---

* Funds: Special Fund for Agro-scientific Research in the Public Interest (201103016), the Key Project of the Chinese Ministry of Education (313024), the Fundamental Research Funds for the Central Universities (2662015PY107), and the China Agriculture Research System (CARS-13)

** Corresponding author: XIE Jia-tao; E-mail: jiataoxie@mail. hzau. edu. cn

苹果茎痘病毒 CP 及 TGB 序列扩增及基因变异分析*

龚卓群，谢吉鹏，陈冉冉，范在丰，国立耘，周　涛**
（中国农业大学植物病理学系，农业部植物病理学重点实验室，北京　100193）

摘　要： 苹果茎痘病毒（Apple stem pitting virus，ASPV）是一种危害苹果和梨等重要果树的重要潜隐病毒。ASPV 在全世界苹果产区均有分布，在我国发生率高，危害较严重。为了解 ASPV 的发生和基因组重要编码区的变异情况，利用自行设计的引物从陕西咸阳样品、实验室保存的组培苗以及丽嘎品种幼苗中分别扩增得到 ASPV 的 1 191nt 的外壳蛋白（coat protein，CP）及 1 155nt 的三基因盒（the triple gene block，TGB）序列，分别记为 ShXXY-S121-CP、ShXXY-S121-TGB；HB-TGB；LG-CP。将所得序列在 NCBI 中进行 BLAST，与 GenBank 中收录的 ASPV 的 CP、TGB 序列在核苷酸水平上一致性分别高达 88%、97%，在氨基酸水平上一致性分别高达 93%、99%。用 DNAMAN 将所得序列与 NCBI 中登录号为 NC_ 003462、LM999967、FN433599、KJ522472 以及实验室之前得到的新疆阿克苏地区苹果上的分离物 ASPV-AKS 进行比对。结果表明 ShXXY-S121-CP、LG-CP 与中国苹果分离物 KJ522472、印度苹果分离物 FN433599 氨基酸相似度较高，分别为 92.81%、91.41%；ShXXY-S121-TGB、HB-TGB 与中国苹果分离物 KJ522472、印度苹果分离物 FN433599 氨基酸相似度较高，分别为 86.65%、85.69%。通过比对发现，ASPV 各分离物 CP 序列较 TGB 序列保守，且近 3′端 600nt 相似度高，5？端 500nt 变异性较大。就 ASPV 的 TGB 而言，ORF2 保守性较高，ORF3、ORF4 变异性大。

关键词： 苹果茎痘病毒；外壳蛋白；三基因盒；基因变异

---

* 基金项目：国家现代苹果产业技术体系 nycytx－08－04－02。

** 通讯作者：周涛，副教授，主要从事植物病毒学研究；E－mail：taozhoucau@ cau. edu. cn

我国部分地区苹果样品苹果锈果类病毒的检测*

陈冉冉，谢吉鹏，龚卓群，国立耘，范在丰，周　涛**
（中国农业大学植物病理学系，农业部植物病理学重点实验室，北京　100193）

摘　要：我国是苹果生产第一大国，苹果是我国栽培面积最大的一类果树。苹果病毒病是一种系统性侵染病害，在我国发生普遍，已成为限制我国苹果优质高产的关键因素。苹果锈果类病毒（Apple scar skin viroid，ASSVd）属于马铃薯纺锤块茎类病毒科（Pospiviroidae）。苹果树一旦感染 ASSVd 后再枝干和叶片上不表现症状，但通常在结果后表现果实表面花脸、锈果等症状，显著降低食用价值和经济价值，造成严重经济损失。利用特异扩增 ASSVd 全长基因组的特异性引物对采集自山东，陕西和黑龙江的 18 份苹果枝条样品，山东和北京市场表现花脸症状的 7 份苹果果实样品，以及山西表现花叶症状的 4 份苹果叶片样品进行 RT-PCR 检测。结果表明上述样品中 ASSVd 的阳性检出率为 72.41%，正在进行序列克隆和分析。结合近些年来苹果锈果病、花脸病在我国一些地区发生的情况，苹果育苗中应对采穗母本树进行危险性病毒和类病毒的检测，降低危险性病毒病害的发生率，促进我国苹果生产健康高效发展。

关键词：苹果锈果类病毒；病毒检测；RT-PCR

* 基金项目：国家现代苹果产业技术体系 nycytx-08-04-02。

** 通讯作者：周涛，副教授，主要从事植物病毒学研究；E-mail：taozhoucau@ cau. edu. cn

A New Potyvirus Identified in Phragmites plants by Small RNA Deep Sequencing

YUAN Wen, DU Kai-tong, FAN Zai-feng, ZHOU Tao*

(*Department of Plant Pathology*, *China Agriculture University*, *Beijing* 100193)

Abstract: In July 2015, Phragmites plants showing virus-like symptoms such as systemic streaked chlorosis and necrosis of leaves were observed in an apple garden in Tianshui city, Gansu province (China). Total RNA was isolated from diseased leaves and subjected to small RNA library preparation. Deep sequencing of small RNA and bioinformatics analysis indicated a potential potyviral pathogen. The existence of potyvirus in the Phragmites samples was confirmed by RT-PCR with degenerate primers. The primary nucleotide sequence of the potyvirus was obtained by three contigs assembling from small RNAs. Overlapped contigs were confirmed by RT-PCR with specific primers designed on either side of the overlapped region. The near full-genome sequence of this potyvirus is 9 319 nucleotides (nt) in length and contains a large open reading frame encoding a polyprotein of 3 014 amino acids. According to the molecular demarcation criterion for potyviruses, the potyvirus detected in Phragmites plants represents a distinct member of the genus Potyvirus since it shared coat protein amino acid sequence identities of less than 76%.

Key words: Potyvirus; genome sequence; contig; coat protein

* Corresponding author: ZHOU Tao; E－mail: taozhoucau@ cau. edu. cn

甘蔗种苗传播病害病原检测与分子鉴定*

李文凤**，王晓燕，黄应昆***，单红丽，张荣跃，尹　炯，罗志明
（云南省农业科学院甘蔗研究所，云南省甘蔗遗传改良重点实验室，开远　661699）

摘　要：我国蔗区（尤其云南）生态多样化，甘蔗病害病原种类复杂多样，多种病原复合侵染，尤其种苗传播病害病原具有潜育期和隐蔽性，传统方法难以诊断。精准有效地对甘蔗种苗传播病害病原进行诊断检测，明确监测病害致病病原是科学有效防控甘蔗种苗传播病害的基础和关键。本研究针对甘蔗种苗传播病害诊断检测基础薄弱、主要病害病原种类及株系（小种）不明等关键问题，以严重为害我国甘蔗生产的黑穗病、宿根矮化病、病毒病、白叶病等种苗传播病害为对象，系统建立了甘蔗黑穗病、宿根矮化病、白条病、赤条病、花叶病（SCMV、SrMV、SCSMV）、黄叶病、杆状病毒病、斐济病和白叶病 9 种种苗传播病害 11 种病原分子快速检测技术，为甘蔗种苗传播病害精准有效诊断、脱毒种苗检测及引种检疫提供了关键技术支撑。通过从不同生态蔗区广泛采集有代表性的甘蔗黑穗病、宿根矮化病、病毒病、白叶病等病害样品，系统分离鉴定明确病害病原种类及主要株系（小种）。结果表明：①云南和广西蔗区甘蔗黑穗病病原为甘蔗鞭黑粉菌（*Ustilago scitaminea* Sydow），病菌存在致病性和生理小种分化，云南蔗区存在生理小种 1、小种 2 及新小种 3。②21 个蔗区 21 批 1270 个样品宿根矮化病检测、测序分析，致病菌为 *Leifsonia xyli* subsp. *xyli*，其核苷酸序列完全一致大小为 438bp（GenBank 登录号 JX424816、JX424817），与 GenBank 中巴西、澳大利亚（登录号 AE016822、AF034641）RSD 致病菌相似性为 100%；与美国路州（AF056003）的有 1 个碱基错配和 1 个碱基插入，相似性为 99.54%。③引起甘蔗花叶病病原有甘蔗花叶病毒（Sugarcane mosaic virus，SCMV）、高粱花叶病毒（Sorghum mosaic virus，SrMV）、甘蔗条纹花叶病毒（Sugarcane streak mosaic virus，SCSMV）3 种，SCSMV 阳性检出率 100%、SrMV 阳性检出率 27.27%、SCMV 阳性检出率 1.3%，SCSMV 为最主要病原（扩展蔓延十分迅速、致病性强），SrMV 为次要病原，且存在 2 种病毒复合侵染。83 份 SCSMV 和 34 份 SrMV 克隆及测序分析，SCSMV 序列分为 3 个大类群，中国分离物聚为 1 个类群，印度分离物聚为 2 个类群，中国分离物除 JX467699 外全聚在一起。SrMV 形成 2 个组：Ⅰ和Ⅱ组，组间又分为 2 个亚组。不同来源 SrMV 在系统树中交叉存在，云南分离物在各个分支中普遍存在，表现出很高的遗传多样性。④甘蔗杆状病毒病病原为甘蔗杆状病毒（Sugarcane Bacilliform virus，SCBV），其核苷酸序列大小为 589bp，与 SCBV-Australia 核苷酸和氨基酸序列相似性为 74.0% 和 84.1%；同 SCBV-Morocco 核苷酸和氨基酸序列相似性为 67.1% 和 66.7%，可见，SCBV 不同分离物间基因组序列存在高度变异特性。⑤89 份黄叶病毒阳性样品 RT-PCR-RFLP 分析结果表明，云南蔗区检测的甘蔗黄叶病毒（Sugarcane yellow leaf virus，SCYLV）全部为 BRA 基因型。⑥甘蔗白叶病病原为 SCWL 植原体（Sugarcane white leaf phytoplasma），其核苷酸序列大小为 210bp，与 GenBank 中公布的 SCWL 植原体基因组序列同源性在 99.05% ~100%。系统进化分析，SCWL 序列分为 3 个类群，云南保山分离物聚为 1 个类群，与越南、缅

---

* 基金项目：现代农业产业技术体系建设专项资金资助（CARS－20－2－2）；云南省现代农业产业技术体系建设专项资金资助；云南省“人才培养”项目（2008PY087）

** 第一作者：李文凤，云南石屏人，研究员，主要从事甘蔗病害研究；E-mail：ynlwf@163.com

*** 通讯作者：黄应昆，研究员，从事甘蔗病害防控研究；E-mail：huangyk64@163.com

甸、泰国聚在一起；云南临沧分离物聚为1个类群，与泰国、印度、日本聚在一起。研究结果丰富了种苗传播病害相关理论和技术基础，为监测预警和科学有效防控甘蔗种苗传播病害奠定了重要基础。

关键词：甘蔗；种传病害；病原检测；分子鉴定

甜橙和柚类中柑橘衰退病毒 *p*25 种群的分子变异*

王亚飞[1**]，刘慧芳[1]，周　彦[2]，王雪峰[2]，孙现超[1]，周常勇[2***]，青　玲[1,2***]

（1. 西南大学植物保护学院，植物病害生物学重庆市高校级重点实验室，重庆　400716；
2. 中国农科院柑橘研究所，重庆　400712）

摘　要：柑橘衰退病毒（Citrus tristeza virus，CTV）是长线形病毒属（*Closterovirus*）正义单链 RNA 病毒，由其引起的柑橘衰退病是柑橘上具有经济重要性的病害，在世界各柑橘产区传播为害。此前研究表明 CTV *p*25 基因片段编码的蛋白是其 RNA 沉默抑制子之一，然而有关 CTV 该抑制子片段遗传变异的情况尚不清楚。明确不同寄主中 CTV 强弱毒株系 *p*25 基因片段的变异情况，对于了解 CTV 致病机理以及防控柑橘衰退病都有积极意义。本研究以本实验室保存的甜橙强毒株系 TR-L514 和弱毒株系 CT31、PeraIAC，以及取自柚类的强毒株系 CT3、CT22、CT23 和弱毒株系 CT9，通过分子克隆及序列测定建立 CTV *p*25 种群，进而对其遗传多样性和变异水平进行分析。

我们对 7 个种群共 177 条长度为 672bp 的序列比对分析发现，在所有的种群中我们共检测到 17 种不同的单倍型，单个种群内部通常有一种或更多的单倍型出现。CT31 种群碱基突变频率最低，为 1.72×10^{-4}；PeraIAC 种群碱基突变频率最高，为 4.17×10^{-4}。甜橙中的 CTV 强毒株种群突变克隆百分比为 17.4%，碱基突变频率为 2.59×10^{-4}，弱毒株种群突变克隆百分比为 17.6%，碱基突变频率为 2.92×10^{-4}；柚类中的 CTV 强毒株种群突变克隆百分比为 11.8%，碱基突变频率为 2.94×10^{-4}，弱毒株种群突变克隆百分比为 14.8%，碱基突变频率为 2.76×10^{-4}。在不同寄主 CTV 不同株系中的 *p*25 种群的突变克隆百分比及碱基突变频率均比较接近。在所检测到的 34 个碱基突变中，碱基替代突变有 18 个，为主要的碱基突变类型，而碱基插入突变有 7 个，碱基缺失突变有 9 个。在 18 个碱基替代突变中，寄主为甜橙的 CTV *p*25 种群检测到 2 个 A→G 碱基转换突变，而在寄主为柚类的 CTV *p*25 种群没有检测到，其他类型的碱基转换突变在甜橙和柚类 CTV *p*25 种群均有检测到。甜橙中发现的 7 个碱基替代突变有 6 个是碱基转换，占比高达 85.7%，柚类中发现的 11 个碱基替代突变有 7 个是碱基转换，占比 63.6%。

关键词：柑橘衰退病毒；*p*25 基因；种群；分子变异

---

* 基金项目：国家公益性行业（农业）科研专项（201203076－04）
** 第一作者：王亚飞，西南大学植物保护学院 2015 级博士研究生
*** 通讯作者：周常勇，博导，研究员，主要从事分子植物病毒学研究；E-mail：zhoucy@ cric. cn
青玲，四川会理人，博导，教授，主要从事分子植物病毒学研究；E-mail：qling@ swu. edu. cn

云南赛葵黄脉病毒的群体遗传分析*

任江平**，荆陈沉，余化斌，吴舒劼，王锦锋，吴根土，青　玲***
（西南大学植物保护学院，植物病害生物学重庆市高校级重点实验室，重庆　400716）

云南赛葵黄脉病毒（Malvastrum yellow vein Yunnan virus，MaYVYV）为双生病毒科（Geminiviridae）菜豆金色花叶病毒属（*Begomovirus*）的一个伴随 β 卫星分子的单组份双生病毒，于2005 年在云南地区的田间杂草赛葵上分离得到，主要由媒介昆虫烟粉虱传播。本实验室在四川部分地区发现该病毒常与其他双生病毒复合侵染赛葵、锦葵、胜红蓟等杂草，为了明确 MaYVYV 的遗传变异情况以及追踪病毒的起源进化，本研究将实验室获得的 4 条 MaYVYV 分离物全基因组序列和 GenBank 已登录的 10 条 MaYVYV 分离物全基因组序列进行了群体遗传分析。序列比对结果表明，分离自四川和广西的 4 个 MaYVYV 分离物全基因组序列与已登录的 MaYVYV 分离物序列相似性均大于99%。基于 MaYVYV 基因组变异分析表明，MaYVYV 在进化上比较保守，在全基因组上的变异分布呈不均衡性。MaYVYV 的群体进化分析表明，*AC*1 和 *AV*1 这两个重要的病毒基因功能编码区的核苷酸多样性较低，说明 *AC*1 和 *AV*1 区较为保守，而 *AV*1 与 *AV*2 的重叠区域及 *AC*2 区的变异较大。同时，MaYVYV 分离物的群体呈现明显的扩张趋势。系统发育树显示这 14 个 MaYVYV 分离物在进化树上可以分为 3 个组群，分别代表 3 个不同的地区，说明 MaYVYV 的进化与地理来源具有相关性。

关键词：云南赛葵黄脉病毒；单组份双生病毒；群体变异

---

* 基金项目：教育部新世纪优秀人才支持计划项目（NCET－12－0931）

** 第一作者：任江平，西南大学植物保护学院 2017 届硕士研究生

*** 通讯作者：青玲，四川会理人，博导，教授，主要从事分子植物病毒学研究；E-mail：qling@ swu. edu. cn

Identification of Diverse Mycoviruses Through Metatranscriptomics Characterization of the Viromes of Wheat Fusarium Head Blight pathogens

ZHANG Zhong-mei [1,2], PENG Yun-liang [1], JIANG Dao-hong [2]

(1. *Institute of Plant Protection, Sichuan Academy of Agricultural Sciences/Key Laboratory of Integrated Pest Management on Crops in Southwest China, Ministry of Agriculture,* *Chengdu* 610066; 2. *College of Plant Sciences & Technology of Huazhong Agricultural University, Wuhan* 430070)

Abstract: To characterize the viromes of wheat Fusarium Head Blight pathogens, a high-throughput sequencing-based metatranscriptomic approach was used to detect viral sequences. Total RNA from mycelia of 150 *Fusarium spp.* strains isolated from diseased wheat grain from Sichuan province in 2015 was sequenced. Sequence data were assembled *de novo*, and contigs with predicted amino acid sequence similarities to viruses in the non-redundant protein database were selected. The analysis identified 55 genome segments representing at least 14 mycoviruses, and 11 of them were previously undescribed. The novel mycovirus genomes showed similarity to 5 distinct lineages including *Mononegavirales*, *ourmiavirus*, *Narnaviridae*, *Gamaflexiviridae* and *Partitiviridae*, respectively.

Key words: *Fusarium* spp.; metatranscriptome; mycovirus

通过宏转录组测序鉴定多种小麦赤霉病菌真菌病毒*

张重梅[1,2]**，彭云良[1]，姜道宏[2]***
（1. 四川省农业科学院植物保护研究所，西南农作物有害生物治理重点实验室，成都 610066；
2. 华中农业大学植物科学技术学院，武汉 430070）

摘 要：为了研究小麦赤霉病菌（*Fusarium* spp.）的病毒组特性，对2015年150个四川 *Fusarium* spp. 菌株进行了宏转录组测序。经 de novo 组装和 contig 筛选，鉴定出代表至少14种真菌病毒的基因组的55个基因组片段，其中，11种病毒在镰刀菌中首次报道。新发现的镰刀菌真菌病毒分别与分子负链 RNA 病毒目 Mononegavirales、欧尔密病毒属 *ourmiavirus*、裸露病毒科 Narnaviridae、伽马线形病毒科 Gamaflexiviridae 和双分病毒科（Partitiviridae）在内的5个不同的 lineages 具相似性。

关键词：小麦赤霉病菌；宏转录组；真菌病毒。

* 基金项目：国家科技支撑计划（2012BAD19B04）和四川省财政专项（2011JYGC06－021）

** 第一作者：张重梅，湖北鹤峰人，实习研究员，华中农业大学在读博士，从事植物病理学研究；E-mail: zhongmeizhang12@163.com

*** 通讯作者：姜道宏，博士，教授，从事植物病理学研究；E-mail：daohongjiang@mail.hzau.edu.cn

Transcriptomic Changes in *Nicotiana benthamiana* Plants Inoculated with the Wild Type or an attenuated Mutant of Tobacco Vein Banding Mosaic Virus*

GENG Chao[1,2], WANG Hong-yan[2], LIU Jin[1], YAN Zhi-yong[1,2], TIAN Yan-ping[1,2], YUAN Xue-feng [1,2], GAO Rui[1], LI Xiang-dong[1,2]**

(1. *Laboratory of Plant Virology*, *Department of Plant Pathology*, *College of Plant Protection*, *Shandong Agricultural University*, *Tai'an*, *Shandong* 271018, *China*; 2. *Shandong Provincial Key laboratory for Agricultural Microbiology*, *Tai'an*, *Shandong* 271018, *China*)

Tobacco vein banding mosaic virus (TVBMV) is a potyvirus mainly infecting solanaceous crops. Helper component proteinase (HCpro) is an RNA silencing suppressor and the virulence determinant of potyviruses including TVBMV. The mutations of D198 to K and IQN motif to DEN in HCpro eliminated its RNA silencing suppression activity and attenuated the symptoms of TVBMV in *Nicotiana benthamiana* plants. Here, we used RNA-seq analysis to compare differential genes expression between the wild type (T-WT) TVBMV and an artificially attenuated mutant (T-HCm) carrying both mutations mentioned above at 1, 2 and 10 days post agroinfiltration (dpai). At 1 and 2 dpai, the *N. benthamiana* genes related to ribosome synthesis were up-regulated, whereas those related to lipid biosynthetic/metabolic process and responses to extracellular/ external stimuli were down-regulated in the inoculated leaves inoculated with T-WT or T-HCm. At 10 dpai, T-WT infection resulted in the repression of photosynthesis-related genes, which associated with the chlorosis symptom. T-WT and T-HCm differentially regulated RNA silencing pathway, suggesting the role of RNA silencing suppressor HCpro in virus pathogenesis. The salicylic acid and ethylene signaling pathway were induced but jasmonic acid signaling pathway were repressed after T-WT infection. The infection of T-WT and T-HCm differentially regulate the genes involved in the auxin signaling transduction, which associated with the stunting symptom caused by TVBMV. These results illustrate the dynamic nature of TVBMV- *N. benthamiana* interaction at the transcriptomic level.

* Funding: This work was supported by grants from the National Natural Science Foundation of China (NSFC; 31571984, 31501612)

** Corresponding author: LI Xiang-dong; E-mail: xdongli@ sdau. edu. cn

芜菁花叶病毒编码蛋白与拟南芥 SWEET 家族蛋白的互作研究*

王 艳[1**]，祝富祥[1]，孙 颖[1]，潘洪玉[1]，李向东[2]，刘金亮[1***]
（1. 吉林大学植物科学学院，长春 130062；2. 山东农业大学植物保护学院，泰安 271018）

摘 要：芜菁花叶病毒（*Turnip mosaic virus*，TuMV）是马铃薯 Y 病毒科（*Potyviridae*）马铃薯 Y 病毒属（*Potyvirus*）的重要成员。该病毒具有广泛的寄主范围，可侵染至少 156 属的 300 多种植物，尤其对十字花科蔬菜造成严重威胁，同时也是传播最广、破坏性最强的侵染芸薹属植物的病毒，给农业生产带来巨大损失。研究 TuMV 病毒与寄主在蛋白水平上的互作，不仅能够深入了解病毒与寄主互作机理，而且深化对病毒致病机理与寄主抗病机制的了解，为抗病种质资源的选育提供理论依据。本研究利用 TuMV JCR06 分离物 *p*3 基因（GenBank 登录号：KP165425）构建诱饵载体，通过酵母双杂交融合的方法，从拟南芥酵母双杂交 cDNA 文库中筛选到一个与 P3 蛋白互作的糖转运蛋白（AtSWEET1），并用共转化和双分子荧光互补（BiFC）证据进一步验证 P3 蛋白与 AtSWEET1 蛋白互作。亚细胞定位分析结果显示 P3 与 AtSWEET1 蛋白均定位于本生烟表皮细胞的细胞膜上。为了明确 P3 蛋白与 TuMV 互作的重要氨基酸，进而解析 P3 蛋白在病毒与寄主互作中的作用机制。本研究分别针对位于 P3 蛋白的 62、67、123、188、299、300 位的 6 个氨基酸位点及终止 *PIPO* 基因表达的 *p*3 基因的 555 位碱基等 7 个氨基酸位点，构建诱饵载体突变体，结果显示，除 299 位氨基酸突变后不与拟南芥 AtSWEET1 蛋白互作外，其余 6 个位点突变后的 P3 蛋白均与 AtSWEET1 蛋白存在互作，推测 P3 蛋白 299 位氨基酸可能是参与病毒与寄主互作的关键氨基酸。为了验证 P3 蛋白是否与拟南芥中除 AtSWEET1 之外其他 16 个 SWEET 家族成员互作，通过酵母双杂交方法初步证明 P3 蛋白与拟南芥 AtSWEET4、AtSWEET15 蛋白之间存在互作。利用酵母双杂交和 BiFC 方法证明，在 TuMV 编码的蛋白中，除了 P3 蛋白，还有 HC-Pro、VPg 和 NIa-Pro 与 AtSWEET1 蛋白发生互作。本研究对分析 TuMV 病毒与寄主 SWEET 蛋白互作的分子机制和丰富对马铃薯 Y 病毒属病毒与寄主互作机理奠定了理论基础。

关键词：芜菁花叶病毒；P3；SWEET 蛋白；蛋白互作

---

* 基金项目：国家自然科学基金项目（31201485，31271991）

** 第一作者：王艳，硕士研究生，主要从事分子植物病毒学的研究；E-mail：w1178014711@163.com

*** 通讯作者：刘金亮，吉林大学植物科学学院；E-mail：jlliu@jlu.edu.cn

第四部分　细　菌

A New Gene ACH51_ 14495 Deficiency of *Ralstonia solanacearum* YC45 Affects its Colony Morphology and Hypersensitive Reaction*

SHE Xiao-man[1,2]**, HE Zi-fu[1]***

(1. *Plant Protection Research Institute*, *Guangdong Academy of Agricultural Sciences*, *Guangzhou*, *China*; 2. *Guangdong Provincial Key Laboratory of High Technology for Plant Protection*, *Guangzhou*, *China*)

Abstract: *Ralstonia solanacearum* strainYC45 isolated from *Rhizoma Kaempferiae*. The genome of YC45 was completely sequenced (Genbank NO. CP011997, NO. CP011998). A new gene ACH51_14495which has no homologous sequence with other organisms was found. ACH51_14495 was predicted to be a candidate of T3SS effectors by T3SS effector prediction website. In order to investigate the function of the gene ACH51_14495 in strain YC45, ACH51_14495mutant of YC45 was constructed. Accordingto the upstream and downstream sequence of YC45 gene ACH51_14495, the PCR primers were designed to amplify the 500bp-fragments of upstream and downstream of ACH51_14495 gene, respectively. The fragments and gentamycin gene were cloned into suicide vector pK18mobsacB, resulting in recombinant plasmid pK18-ACH51_14495-Gm. The plasmid was then introduced into *R. solanecarum* strain YC45. The ACH51_14495mutant, named YC45 – Δ14495, was generated by homologous recombination and selected by three-steps method. The mutant was identified by PCR. After incubation at 30℃ for 2 days on tetrazolium chloride (TZC) medium, the mutant showed small, crimson colonies, while its wild-type strain exhibited large, irregular round, fluidal and white colonies with pink center. The results of hypersensitive reaction test showed that the mutant could not cause necrosis, but the wild-type strain YC45 might produce necrosis on five-leaf stage tobacco after 24h post inoculation . Thus, the gene ACH51_14495 deficiency obviously affected bacterial colony morphology and hypersensitive reaction of YC45, and the gene was related with the pathogenicity of YC45.

Key words: *Ralstonia solanacearum*; Mutant; Hypersensitive reaction

* 基金项目：广东省科技计划项目（2014B070706017；2015A020209057）；广东省农业科学院院长基金项目（201513）

** 第一作者：佘小漫，副研究员，硕士，研究方面植物细菌；E-mail：lizer126@ 126. com

*** 通讯作者：何自福，研究员；E-mail：hezf@ gdppri. com

GGDEF 结构域蛋白影响蜡样芽胞杆菌 905 生物膜的形成*

杨　旸**，段雍明，崔　实，李　燕，王　琦***
（中国农业大学植物病理学系，农业部植物病理学重点开放实验室，北京　100193）

摘　要： 芽胞杆菌（*Bacillus* spp.）在自然环境中广泛存在，具有适应性广、作用机制多样、抗逆性强、环境友好等优点，被认为是一种重要的生防因子。生防芽胞杆菌在自然环境中表达防病相关性状时往往受到到各种因素的影响，因此，研究影响生防性状表达的调控机制是发挥生防菌防病效果的重要理论基础。蜡样芽胞杆菌 905 是本课题组从小麦根部上分离获得的一株植物根际促生细菌，温室和大田应用结果显示，其可以在小麦根部大量定殖，促进小麦生长。

C-di-GMP 是在细菌中广泛存在的第二信使，影响细菌的生物膜、运动性、毒性等生物学功能并且在不同的细菌中差异较大。生物信息学分析发现 c-di-GMP 的合成酶具有 GGDEF 结构域。本研究利用同源重组原理，将蜡样芽胞杆菌 905 基因组中的 B. cereus_ 905_ mapped_ 4203 基因进行了敲除突变。该基因编码 217 个氨基酸，包括 2 个跨膜结构域和一个 GGDEF 结构域。该基因在蜡样芽胞杆菌、苏云金芽胞杆菌、炭疽芽胞杆菌、韦氏芽胞杆菌和蕈状芽胞杆菌中氨基酸序列同源性都在 80% 以上，在其他芽胞杆菌中却没有同源基因。有趣的是，在肺炎链球菌中也有该同源蛋白。研究发现，该敲除突变体的生长情况与野生型菌株没有明显差异，但是其附着在底部的生物膜（submerged biofilm）形成显著减弱，而互补菌株的生物膜表型接近野生型。文献报道 GGDEF 结构域与 c-di-GMP 的合成相关而部分退化的 GGDEF 结构域可能会作为 c-di-GMP 的受体调控下游基因。本研究中的 B. cereus_ 905_ mapped_ 4203 基因保守结构域 GGDEF 的部分位点已经改变，推测其可能作为 c-di-GMP 的合成酶或受体发挥作用，有待进一步实验证明。

生防菌发挥生防作用的前提条件是可以与寄主植物形成良好的互作关系，尤其是可以稳定定殖于植物根部。已有研究证明生物膜的形成对细菌的稳定定殖发挥极其重要的作用。目前，在很多革兰氏阴性菌中都已证明 c-di-GMP 与生物膜的形成有关，但是在芽胞杆菌中尚无有效证据证明 c-di-GMP 与生物膜形成的相关性。本课题组已有研究表明蜡样芽胞杆菌 905 形成的 submerged biofilm 主要成分为胞外 DNA 和蛋白，本文研究的 GGDEF 结构域蛋白具体影响 submerged biofilm 形成的哪一部分还有待研究。本研究丰富了蜡样芽胞杆菌生物膜的调控网络，为芽胞杆菌定殖机制的研究开辟了新的思路。

关键词： 生物防治；蜡样芽胞杆菌；GGDEF 结构域；生物膜

---

* 基金项目：中国农业大学基本科研业务专项资金项目

** 第一作者：杨旸，博士研究生，中国农业大学植物病理学系；E-mail：katherine523@163.com

*** 通讯作者：王琦，教授，主要从事植物病害生物防治与微生态方面研究；E-mail：wangqi@cau.edu.cn

云南元江蔗区首次检测发现由 *Acidovorax avenae* subsp. *avenae* 引起的甘蔗赤条病*

单红丽**，李文凤，黄应昆***，王晓燕，张荣跃，罗志明，尹　炯
（云南省农业科学院甘蔗研究所，云南省甘蔗遗传改良重点实验室，开远　661699）

摘　要：甘蔗赤条病（Sugarcane red stripe disease）是由细菌引起的一种世界性重要甘蔗病害之一。2015 年笔者在对国家甘蔗体系云南元江综合试验站甘蔗新品种（系）进行病虫害调查时，观察发现疑似甘蔗赤条病症状植株。本研究采用 PCR 方法，首次在云南元江蔗区的甘蔗上检测到引起甘蔗赤条病的病原菌 *Acidovorax avenae* subsp. *avenae*，证实 *Acidovorax avenae* subsp. *avenae* 已传入云南。将检测阳性的 8 份典型甘蔗赤条病样品 PCR 产物进行测序分析，8 条序列大小均为 1 495bp（Genbank 登录号：KU948657 – KU948661），BLAST 检索结果表明所得序列是导致甘蔗赤条病的 *Acidovorax avenae* subsp. *avenae* 16S rDNA 序列，其与 GenBank 中登录的 *Acidovorax avenae* subsp. *avenae* 16S rDNA 序列（Genbank 登录号：CP002521）同源性在 99% 以上，并在系统发育树中处于同一分支。田间调查结果显示：云蔗 93 – 194 高度感病，病株率为 8% ~ 30%，严重田块高达 80%；福农 38 号、福农 39 号、桂糖 29 号、桂糖 31 号、粤糖 60、粤糖 55 号、云蔗 05 – 51、柳城 05 – 136、ROC22 在田间表现抗病。甘蔗赤条病是一种极其危险的细菌性病害，建议有关部分立即采取相应的预防与管理措施，防止其扩散蔓延，确保我国甘蔗安全生产和蔗糖产业可持续发展。

关键词：云南蔗区；*Acidovorax avenae* subsp. *avenae*；PCR 检测；甘蔗赤条病

---

* 基金项目：现代农业产业技术体系建设专项资金资助（CARS – 20 – 2 – 2）；云南省现代农业产业技术体系建设专项资金资助

** 第一作者：单红丽，云南保山人，助理研究员，主要从事甘蔗病害研究；E-mail：shhldlw@163.com

*** 通讯作者：黄应昆，研究员，从事甘蔗病害防控研究；E-mail：huangyk64@163.com

Effects of Amino Acids and α-keto Acids on Diffusible Signal Factors production in *Xanthomonas campestris* pv. *campestris*

Abdelgader Diab, ZHOU Lian, WANG Xing-yu, CAO Xue-qiang, HE Ya-wen
(*State Key Laboratory for Microbial Metabolism, School of Life Sciences and Biotechnology, Shanghai Jiao Tong University, Shanghai* 200240, *China*)

Members of the diffusible signal factor (DSF) family are a novel class of quorum sensing (QS) signals and responsible for the regulation of extracellular degradative enzyme and extracellular polysaccharide biosynthesis in diverse Gram-negative bacteria. Previous study has identified that *Xanthomonas campestris* pv. *campestris* (*Xcc*), the causal agent of black rot in crucifers, produces four DSF-family signals (DSF, BDSF, CDSF and IDSF) during cell culture. Although the structures of these four DSF-family signals have been characterized, many questions in their biosynthesis remain to be addressed, especially the metabolic origins of the DSF-family signals in *Xcc* and the role of amino acids in the whole synthetic pathway. To investigate whether amino acids influence the production of DSF-family signals, a total of 21 amino acids was added separately to the Δ*rpfC* culture in the medium XYS. DSF-family signals were extracted and measured by HPLC. Our results showed that individual addition of 15 amino acids significantly increased BDSF production. Addition of 3 branched-chain amino acids significantly increased DSF or IDSF production. We also found that addition of three branched-chain α-keto acids significantly increased DSF-family signals production. The molecular mechanisms underlying the roles of these compounds in DSF-family signals bioysnthesis will be further explored in the future.

Characterization of Phosphate-Solubilizing Bacteria isolated From Agricultural Soil*

ZHAO Wei-song, GUO Qing-gang, WANG Pei-pei,
LI She-zeng, LU Xiu-yun, MA Ping**
(*Institute of Plant Protection, Hebei Academy of Agricultural and Forestry Sciences, Integrated Pest Management Center of Hebei Province, Baoding* 071000, *China*)

Abstract: Phosphorus is one of the major macro-nutrients for plant growth, and phosphorus security is emerging as one of the greatest global sustainability challenges of the 21st century. Phosphorus availability is limited in soils because of its fixation as insoluble phosphates of iron, aluminum or calcium and all these are unavailable to plants. Phosphate-solubilizing microorganisms can convert these insoluble phosphates into soluble forms. Five bacterial strains were isolated and shown different degrees of phosphate solubilizing activity (88.69 to 231.71μg/mL). Strain 105 – 10, an excellent phosphate-solubilizing bacteria isolated from agricultural soil, was systematic studied for its inorganic phosphate-solubilizing capability *in vitro*. The strain 105 – 10 was incubated in the National Botanical Research Institutes Phosphate (NBRIP) medium in 250 mL Erlenmeyer flasks, supplemented with tricalcium phosphate [$Ca_3(PO_4)_2$]. The concentration of soluble phosphate were determined periodically during 6 d incubation. Strain 105 – 10 could grow and demonstrate a better phosphate-solubilizing activity in NBRIP medium with pH values between 5.0 and 9.0 (94.49 ~ 144.99μg/mL). In addition, the suitable phosphate-solubilizing conditions for strain 105 – 10 were under 40℃, 180 rpm, with 1% ~5% volume of bacteria inoculated. Meanwhile, the solubilization activity of the strain was associated with a decrease in the pH and the release of organic acids in culture. The results of present study indicated that strain 105 – 10 can be used as a strong phosphate solubilizer in agricultural environments.

Key words: Phosphate-solubilizing bacteria; Soluble phosphate; Phosphate-solubilizing activity

* 第一作者：赵卫松，博士，助理研究员，主要从事环境微生物资源开发与利用；E-mail：zhaoweisong1985@163.com
** 通讯作者：马平，博士，研究员，主要从事植物病害生物防治研究；E-mail：pingma88@126.com

Functional Characterization of the *pilO* gene of *Acidovorax citrulli*

ZHANG Ying[a], XIONG Xi[a], XU Yu-bin[a], GUO Meng-lin[a], PENG Feng[a], WANG Gang[a,b]

(*Henan University a. College of Life Science/b. Institute of Bioengineering*, *Kaifeng* 475004, *China*)

Bacterial fruit blotch of melons caused by *Acidovorax citrulli* is a world-wide disease of cucurbit production. The biogenesis and function of bacterial type-IV pili is controlled by more than 10 genes of which is *pilO*. Specific disruption of the *pilO* gene from *A. citrulli* strain AAC－3 was achieved through homologous recombination. Transmission electron microscope observation showed that the ΔpilO mutant strain could not form type-IV pili. The ΔpilO mutant strain growth is basically not affected, but the motility ability decreased significantly. Disruption of *pilO* in *A. citrulli* caused a 80% reduction in biofilm formation and a 55% decrease in disease efficacy. These results suggested that *pilO* may be one of the key genes involved in type-IV pili conformation and *pilO* plays an important role in biofilm formation and pathogenicity of *A. citrulli*.

水稻白叶枯菌 DSF 家族群体感应信号天然降解的机制及其生物学功能

王杏雨，周　莲，何亚文*
（上海交通大学生命科学技术学院，微生物代谢国家重点实验室，上海　200240）

摘　要：群体感应（Quorum sensing，QS）是微生物之间相互交流的一种重要联络方式，细菌通过监测环境中群体感应信号分子浓度来感知细菌密度从而调控基因表达。近期研究结果发现，细菌中同样存在着天然的群体信号降解机制，在适当条件下引导细菌退出群体感应。

水稻白叶枯菌（*Xanthomonas oryzae* pv. *oryzae*，*Xoo*）能侵染水稻，引发水稻白叶枯病。DSF 家族（Diffusible signal factor family）信号分子介导的 QS 系统在 *Xoo* 侵染水稻过程中起着关键作用。DSF 家族信号分子的生物合成与信号传导是由 *rpfABCDEFG* 基因簇负责完成。在野油菜黄单胞菌中，我们发现 *rpfB* 基因具有脂酰辅酶 A 连接酶编活性，能够降解 DSF 信号。本研究进一步证实：在 *Xoo* 菌株 PXO99A 中，RpfB 同样参与 DSF 家族信号分子的降解。敲除 *rpfB* 基因，能显著提高 DSF 和 BDSF 的产量，在此基础上过表达 *rpfB*，所得菌株不产 DSF 或 BDSF；*rpfB* 的表达具有细胞密度依赖性，并受到 RpfC/G 双组份系统及全局性转录因子 Clp 负调控，这些结果均与野油菜黄单胞菌中的结果一致。此外，我们还发现了 RpfB 在 *Xoo* 中还有一些独特功能。首先，*rpfB* 敲除株虽产生大量 DSF 信号分子，但其胞外酶和胞外多糖的产生量及其对水稻的致病性都有所降低。其次，在 *rpfC* 突变体中敲除 *rpfB* 可以进一步提高 DSF 产生量，却部分削弱了菌黄素的产生。我们推测这些差异性可能与 *Xoo* 本身对碳源的利用及其在侵染过程中与植物之间的互作存在相关性。综上所述，RpfB 依赖的 DSF 群体感应信号降解机制广泛存在于黄单胞菌中，但其调控的生物学功能和调控方式具有一定的菌株特异性。

关键词：水稻白叶枯菌；群体感应；天然解除

* 通讯作者：何亚文；E-mail：yuaiji4ever@163. com

Resuscitation and Pathogenicity Test of the Viable But Nonculturable Cells of *Acidovorax citrulli* *

KAN Yu-min, JIANG Na, HAN Si-ning, LUO Lai-xin**, LI Jian-qiang

(*Department of Plant Pathology, China Agricultural University/Beijing Engineering Research Center of Seed and Plant Health* (*BERC-SPH*) / *Beijing Key Laboratory of Seed Disease Testing and Control* (*BKL-SDTC*). *Beijing* 100193, *P. R. China*)

Abstract: Bacterial fruit blotch (BFB) is a devastating disease of cucurbits and has caused significant economic losses of cucurbits industry worldwide. The causal agent, *Acidovorax citrulli* (Ac), is a gram-negative bacterium that infects a wide range of cucurbits including watermelon and melon. Infected seeds are the most important source of primary inoculums and copper compounds are the most widely-used bactericide to control BFB. Our previous experiments confirmed that Ac (strain AAC00 – 1) could enter into VBNC state by the induction of cupric ion, which may increase probability of false negative results for seed health testing by bio-PCR, a method based on the bacterial culturability. In this research, we demonstrated that VBNC state of Ac induced by different concentration of $CuSO_4$ for different time (50μM $CuSO_4$ for 3h, 10μmol/L $CuSO_4$ for 5 d and 5μmol/L $CuSO_4$ for 15 d) could be resuscitated by some appropriate conditions. The experiment used LB broth, cell-free suspension (CFS) and different concentration of EDTA (the ratio to Cu^{2+} was 1 : 1, 1.5 : 1 and 2 : 1) to resuscitate the VBNC cells, the results indicated that after treated by different time, culturable Ac could be detected in all tested samples. Adding EDTA (the ratio to Cu^{2+} was 1.5 : 1) was the most efficient method to resuscitate Ac from VBNC state. The average frequency in two biological replications of resuscitation in 50μmol/L $CuSO_4$ for 3 h, 10μmol/L $CuSO_4$ for 5 d and 5μmol/L $CuSO_4$ for 15 d was 83.33%, 83.33% and 100%, respectively. The quantity of resuscitated culturable cells in the microcosms was 4.80×10^6 CFU/mL, 1.68×10^6 CFU/mL and 6.80×10^5 CFU/mL, respectively. All the resuscitated and re-culturable cells of *A. citrulli* recovered pathogenicity on watermelon seedlings by inoculation. However, the VBNC cells of *A. citrulli* could not colonize and infect the watermelon plants in present results.

Key words: *Acidorvorax citrulli*; VBNC; Resuscitation; Pathogenicity

---

* 基金项目：本研究得到公益性行业（农业）科研专项（201003066）资助

** Corresponding author: LUO Lai-xin; E-mail: luolaixin@ cau. edu. cn

茉莉酸介导 *N* - 酰基高丝氨酸内酯对植物软腐病的抗性调控

赵　芊[1,2]，靳晓扬[1]，刘　方[1,2]，贾振华[1,2]，宋水山[1,2*]
（1. 河北省科学院生物研究所，石家庄　050051；
2. 河北省农作物主要病害微生物控制工程技术研究中心，石家庄　050051）

摘　要： *N* - 酰基高丝氨酸内酯（N-acyl-homoserine lactones，AHLs）是革兰氏阴性细菌分泌的一种用于协调菌群活动的群体感应（Quorum Sensing，QS）信号分子，也是介导植物 - 细菌跨界信息交流的重要信号分子。分析比较了多种 AHLs（C6 - HSL、C8 - HSL、3OC6 - HSL、3OC8 - HSL）在拟南芥对软腐病的抗性激发中的作用，发现 3OC8 - HSL 在处理 48h 后可以显著降低软腐欧文氏菌 *Erwinia carotovora* subsp *Carotovora*（ECC）在拟南芥中的定殖，降低软腐病的发病率；采用高效液相色谱 - 质谱联用法分析发现 3OC8 - HSL 能够显著诱导拟南芥体内茉莉酸（JA）含量的升高，至处理 48h 达到最高值；之后接种病原菌 ECC，发现 ECC 本身也能刺激拟南芥 JA 含量的升高，但 3OC8 - HSL 预处理能够更强烈的激发拟南芥 JA 的合成；qRT-PCR 结果显示 JA 合成的关键基因 AOS、AOC、LOX2 受 3OC8 - HSL 的诱导上调表达，JA 通路关键基因 PDF1. 2、VSP1、MYC2 也受 3OC8 - HSL 的诱导上调表达；JA 通路突变体 *coi*1 和 *jar*1 无论是在 ECC 的菌落定殖和发病率方面还是在 JA 通路关键基因的表达方面均不感应 3OC8 - HSL 的调控。以上结果表明 JA 介导 3OC8 - HSL 对植物软腐病的抗性调控，这为揭示植物—细菌跨界信息交流的分子机制提供有力证据，对提高农作物抗病能力具有重要的指导意义。

关键词： *N* - 酰基高丝氨酸内酯；茉莉酸；软腐病；植物 - 细菌跨界通讯

---

* 通讯作者：宋水山

The Signal Peptide-like Segment Affects HpaXm Transport, but without the Pathogenicity of *Xanthomonas citri* subsp. *malvacearum* *

LIU Yue**, MIAO Wei-guo***, ZHENG Fu-cong***

(*Hainan Key Laboratory for Sustainable Utilization of Tropical Bioresource/ College of Environment and Plant Protection, Hainan University, Haikou* 570228, *China*)

HpaXm, encoded by the *hpaXm* gene of *Xanthomonas citri* subsp. *malvacearum* (*Xcm*), is a novel harpin protein described from cotton leaf blight bacteria. We predicted that the N-terminal leader peptide (1 – MNSLNTQIGANSSFL – 15) was a putative signal peptide of hpaXm, and determined here whether this segment are related to HpaXm transport or the pathogenicity of the pathogen *Xcm*. First, transgenic tobacco lines expressing the full-length *hpaXm* and the signal peptide-like segment-deleted mutant *hpaXmΔLP* were developed using transformation mediated by *Agrobacterium tumefaciens*. Then the extracted soluble-protein activity was determined. Additionally the defensive responses and resistance to tobacco mosaic virus (TMV) of transgenic tobacco were identified. Finally, differences in the association sites between the endogenously expressed *hpaXm* and *hpaXmΔLP* in transgenic tobacco tissues were demonstrated. Molecular identification showed that the target genes had already integrated into the recipient genomes of transgenic tobacco respectively and were expressed normally. Soluble proteins extracted from plants transformed with *hpaXm* and *hpaXmΔLP* were bio-active. Defensive responses of *hpaXm* and *hpaXmΔLP* as micro-HR were observed on transgenic tobacco leaves. Disease resistance bioassays showed that tobacco plants transformed with *hpaXm* had slightly enhanced resistance to TMV compared to those transformed with *hpaXmΔLP*. Immune colloidal-gold detection showed that the *hpaXm* expressed in transgenic tobaccos was transferred to the plasma membrane and the cell wall of tobacco, but in the deletion mutant *hpaXmΔLP* expressed in transgenic tobacco could not cross the membrane to reach the cell wall. In summary, the N-terminal signal peptide-like fragment (1 ~ 45bp) in HpaXm sequence was not necessary for endogenous expression and bioactivity of HpaXm in transgenic tobacco and the pathogenicity of *Xanthomonas citri* subsp. *malvacearum*. But in the deletion mutant it prevented *hpaXmΔLP* being transferred to the cell wall, thus slightly reducing its resistance to TMV.

Key words: HpaXm; Suspected signal peptide; HR reaction; Defensive response; Immune colloidal gold position

---

* National Natural Science Foundation of China (31360029, 31160359), National agricultural industrial technology system (CARS – 34 – GW8); Doctoral Fund of Ministry of Education (20124601110004)

** First author: LIU Yue, PhD candidate, was engaged in the study of pathology; E-mail: heizi2327@ 126. com

*** Corresponding authors: MIAO Weiguo; E-mail: weiguomiao1105@ 126. com

ZHENG Fucong; zhengfucong@ 126. com

中华常春藤细菌性叶斑病病原菌的鉴定*

张小芳**，付丽娜，李　雪，李兴明，姬广海***
（云南农业大学，农业生物多样性与病虫害控制教育部重点实验室，昆明　650201）

摘　要：2015年春季在昆明市区绿化带种植的中华常春藤上发现一种由细菌侵染而引起的病害，称为中华常春藤细菌性叶斑病。通过发病症状、菌落形态观察、致病性测定、Biolog分析，16S rDNA序列和核糖体DNA内转录间隔区（Internal Transcribed Spacer，ITS）序列分析比较，对昆明地区的常春藤叶斑病病原菌及系统进化关系进行研究。分离病原菌接种中华常春藤叶片完成科赫法则验证，发病初期在叶片表面形成带有黄色叶晕的不规则褐色斑点，后期叶片边缘形成倒V字型坏死和起皱。将菌株CCT1和CCT6测序结果与现有的黄单胞菌菌株的16S rDNA序列和核糖体DNA的ITS序列构建进化树，结果均显示病原菌与野油菜黄单胞菌的序列相似度最大，属于同一支。研究确定该病原菌为野油菜黄单胞菌（*Xanthomonas campestris*）。这是中国首次报道由*X. campestris*引起的中华常春藤叶斑病。

关键词：中华常春藤；细菌性叶斑病；野油菜黄单胞菌；鉴定

---

* 基金项目：国家自然基金（31360002，31460458）；农业部公益性行业专项（201303015）；省重点新产品计划（2014BB005）；云南高校创新团队（云教科〔2014〕22号）

** 第一作者：张小芳，硕士研究生，从事植物细菌病害的研究；E-mail：384952241@qq.com

*** 通信作者：姬广海，教授，从事植物病理学研究；E-mail：550356818@qq.com

Establishment of a Loop-Mediated Isothermal Amplification Assay for Rapid Detection of *Xanthomonas oryzae* pv. *oryzicola* *

ZHANG Xiao-fang, FU Lina, LIU Ya-ting,
ZHANG Hai-yan, LI Xue, JI Guang-hai**
(1. *College of Plant Protection*, *Yunnan Agricultural University*, *Kunming* 650201, *China*; 2. *College of Agronomy And Xbiotechnology*, *Yunnan Agricultural Universitykunming* 650201, *China*)

Abstract: A visual loop-mediated isothermal amplification detection of *Xanthomonas oryzae* pv. *oryzicola*, the causal agent of bacterial leaf streak disease was developed with SYBR Green I as a fluorescence indicator in this study. Four special Loop-mediated isothermal amplification primers and two loop primers were designed based on putative glycosyltransferase gene loci which is unique for *X. oryzae pv. oryzicola*. Loop-mediated isothermal amplification assay efficiently amplified the target element in less than 60 min at 65℃. LAMP assay was evaluated for specificity and sensitivity. A positive colour (Fluorescent Green) was only observed in the presence of Xooc by addition of SYBR Green I to amplification, whereas none of other closely related *Xanthomonas* showed a colour change. The lower limit of detection of this LAMP assay was about 100 fg/μL DNA and 3×10^3 CFU/mL Bacterium Suspension, which was 100 times more sensitive than PCR method. Establishment of LAMP provides a new alternative method for on-site quarantining and large-scale monitoring of this pathogen.

Key words: *Xanthomonas oryzae* pv. *oryzicola*; Loop-mediated isothermal amplification; Detection

* Funding: This Manuscript Has Been Supported By The National Natural Foundation Of China (31460458, 31360002), The R&D Special Fund For Public Welfare Industry (Agriculture) Of China (201303015) And Program For Innovative Research Team (In Science And Technology) In University Of Yunnan Province [(2014) 22]

** Corresponding Author: JI Guang-hai; E-mail: 550356818@ qq. com

枯草芽胞杆菌 NCD－2 菌株的 GFP 标记及其生物膜形成和定殖特性检测*

董丽红[1,2**]，郭庆港[2]，李社增[2]，鹿秀云[2]，
王培培[2]，张晓云[2]，赵卫松[2]，马　平[***]
（1. 河北农业大学植物保护学院，保定　071000；2. 河北省农林科学院植物保护研究所，河北省农业有害生物综合防治工程技术研究中心，农业部华北北部作物有害生物综合治理重点实验室，保定　071000）

摘　要： 枯草芽胞杆菌防治作物土传病害的机制包括竞争作用、颉颃作用、诱导抗性等，而枯草芽胞杆菌在植物根际有效的定殖是其发挥生防作用机制，实现其生防效果的前提。研究证明，枯草芽胞杆菌的根际定殖能力与其生物膜形成能力有关。枯草芽胞杆菌 NCD－2 菌株能有效防治作物黄萎病及中药材根腐病。为明确 NCD－2 菌株在棉花根部的定殖规律，本试验构建了可在 NCD－2 菌株中表达的绿色荧光（GFP）质粒 PMC123。通过电击转化的方法将质粒 PMC123 导入到 NCD－2 菌株中，荧光显微镜观察发现，GFP 标记的 NCD－2 菌株表现很强的荧光信号。比较了野生型 NCD－2 菌株与 GFP 标记的菌株在 LB 培养基中的生长速率，结果表明，GFP 标记的菌株与野生型菌株生长速率基本一致。在 MSgg 培养基中比较了野生型 NCD－2 菌株与 GFP 标记菌株的生物膜形成能力，结果发现，野生型 NCD－2 菌株与 GFP 标记菌株均能形成稳定的生物膜，通过激光共聚焦显微镜可以观察到 GFP 标记的 NCD－2 菌株具有立体网状结构的生物膜形成过程。将 GFP 标记的 NCD－2 菌株接种到棉花幼苗根部，取棉根的不同部位以及根部的横切面和纵切面进行激光共聚焦观察，结果发现，NCD－2 菌株不仅可在棉花根表定殖也可定殖到棉花根内部。在根表定殖时，NCD－2 菌株首先在根表有一附着点，然后成堆出现，形成生物膜并沿根系细胞间隙有规律的排列。NCD－2 菌株在棉花主根的定殖能力比在侧根的定殖能力强。NCD－2 菌株在棉花根尖的定殖能力较差而在棉花根部的分生区、伸长区和成熟区的定殖能力较好。NCD－2 菌株在棉花根内部定殖时既可以在细胞间隙部位穿过根表进入根组织内部，主要定殖在维管柱细胞中。NCD－2 菌株也可通过伤口进入细胞组织内部，并通过维管柱细胞向上传导。以上研究结果表明枯草芽胞杆菌 NCD－2 菌株能够有规律的定殖到棉花根部，为进一步研究枯草芽胞杆菌的生防机制提供科学依据。

关键词： NCD－2 菌株；GFP；生物膜；定殖能力

---

* 基金项目：国家自然科学基金（31272085）；国家自然科学基金（31572051）
** 第一作者：董丽红，在读博士，研究方向为植物病害生物防治与分子植物病理学；E-mail：xingzhe56@126. com
*** 通讯作者：马平，博士，研究员，主要从事植物病害生物防治研究；E-mail：pingma88@126. com

棉花根系分泌物对枯草芽胞杆菌 NCD－2 菌株趋化性的影响*

郭庆港[2]**，董丽红[1,2]，李社增[2]，鹿秀云[2]，王培培[2]，张晓云[2]，赵卫松[2]，马　平[2]***

（1. 河北农业大学植物保护学院，保定　071000；2. 河北省农林科学院植物保护研究所，河北省农业有害生物综合防治工程技术研究中心，农业部华北北部作物有害生物综合治理重点实验室，保定　071000）

摘　要：植物根系分泌物是植物根系在生长过程中释放到介质中的全部有机物质，主要包括一些氨基酸、有机酸、糖类、多酚类等低分子量的化合物。棉花根系分泌物中的有机物质，不仅为根际微生物的繁殖提供营养物质，同时还可作为信号分子调控根际微生物的群体结构。趋化性是细菌识别环境中的化学因素并朝着其喜好的条件移动的结果，趋化作用是微生物在植物根部定殖的前提。为明确棉花根系分泌物对枯草芽胞杆菌 NCD－2 趋化性的影响，本研究首先比较了 NCD－2 菌株在 5 个棉花品种（冀棉 11、中植棉 2 号、中棉所 41、鲁棉 29 和海岛棉 Pima 90）根际的定殖能力，结果表明，NCD－2 菌株在 Pima 90 的根际定殖能力最强，播种 35d 后在根际的群体数量达到 7.63×10^5 CFU/g 根重，其次是冀棉 11，出苗 35d 后根际群体数量达到 4.73×10^5 CFU/g 根重。NCD－2 菌株在而在中植棉 2 号的根际定殖能力最弱，播种 35d 后根际 NCD－2 菌株的群体数量为 6.51×10^4 CFU/g 根重。利用再循环水培系统收集不同棉花品种的根系分泌物，利用高效液相色谱（HPLC）分析了不同棉花品种的根系分泌物中氨基酸的成分及含量，结果发现在所测试的 5 个棉花品种中，氨基酸的种类基本相同，而分泌物中氨基酸的相对含量存在差异，精氨酸、酪氨酸、缬氨酸、赖氨酸、异亮氨酸、苯丙氨酸在冀棉 11 的根系分泌物中相对含量较高。苏氨酸、脯氨酸、半胱氨酸在 Pima 90 的根系分泌物中相对含量较高。而丙氨酸和亮氨酸在鲁棉 29 的根系分泌物中相对含量较高。甲硫氨酸在中植棉 2 号的根系分泌物中相对含量较高。NCD－2 菌株对不同品种的棉花根系分泌物的趋化性测定结果表明，NCD－2 菌株对冀棉 11 根系分泌物的趋化作用最强，趋化距离达到 2.10 cm，NCD－2 菌株对中棉所 41 根系分泌物的趋化作用最弱，趋化距离达到 0.04 cm。进一步测定发现，NCD－2 菌株对 3 种有机酸中苹果酸的趋化作用最强，趋化距离达到 3.67 cm；NCD－2 菌株对 15 种氨基酸中的精氨酸、丙氨酸和赖氨酸趋化作用明显，趋化距离分别为 3.20、3.00 和 3.00 cm；同时发现该菌株对甘氨酸和色氨酸有负趋化性。NCD－2 菌株对 7 种糖分的趋化性相对较弱。RT-qPCR 检测冀棉 11 的根系分泌物对 NCD－2 菌株趋化性相关基因 *cheA* 和 *cheD* 表达量的影响，结果表明 NCD－2 菌株与冀棉 11 的根系分泌物互作 30min 后 *cheA* 基因上调。本试验为进一步明确有益微生物与植物互作机制提供科学依据。

关键词：根系分泌物；NCD－2 菌株；趋化性；RT-qPCR

---

* 基金项目：国家自然科学基金（31272085）；国家自然科学基金（31572051）

** 第一作者：郭庆港，博士，副研究员，主要从事植物病害生物防治研究；E-mail：gqg77@163.com

*** 通讯作者：马平，博士，研究员，主要从事植物病害生物防治研究；E-mail：pingma88@126.com

不同温度和烟草品种抗性对烟草青枯病潜育期的影响*

曾乙心[1**]，何永宏[2]，赵明丽[2]，刘月静[3]，陈兴全[2]，
匡希茜[2]，谢　勇[2***]，秦西云[4***]
（1. 云南农业大学烟草学院，昆明　650201；2. 云南农业大学植物保护学院，
昆明　650201；3. 云南省烟草公司文山州公司，文山　663000；
4. 云南省烟草农业科学研究院，昆明　650106）

摘　要：为了探索烟草青枯病在云南不同烟区的发生流行规律，本研究在准确测定青枯病菌菌悬液浓度的基础上，采用温室盆栽试验筛选出最佳接种方法，再通过人工接种方法最终确定烟草青枯病的侵染阈值和潜育期。研究结果表明，用菌悬液蘸根是烟草青枯病菌的最佳接种方法；烟草青枯病的侵染阈值为 3×10^{2} CFU/g；烟草青枯病潜育期受温度影响大，表现出潜育期随温度的增加而缩短，在最适发病区间内，烟草青枯病平均潜育期为 24 天；潜育期与烟草品种抗性相关，抗性越强，潜育期越长，反之越短。

关键词：烟草青枯病；潜育期；侵染阈值；病害流行

* 基金项目：中国烟草总公司云南省公司资助项目（2015YN07）；云南省高校科技创新团队支持计划（云教科〔2014〕22 号）
** 第一作者：曾乙心，贵州贵阳人，在读硕士研究生，主要从事烟草病害研究
*** 通讯作者：秦西云，研究员，主要从事烟草病虫害防控技术研究
谢勇，共同通讯作者

十字花科黑腐病菌转录调控因子 HpaR1 调控 *gum*B 表达的研究

苏辉昭[1]，吴　柳[1]，祁艳华[1]，刘国芳[1]，陆光涛[1,2*]，唐纪良[1,2*]

（1. 广西大学生命科学与技术学院，南宁　530005；

2. 亚热带农业生物资源保护与利用国家重点实验室，南宁　530005）

摘　要：实验室前期研究发现，十字花科黑腐病菌中的一个 GntR 家族转录调控因子 HpaR1 正调控与过敏反应及致病力相关基因的表达，并且对自身基因的表达起负调控作用。本论文研究了 HpaR1 对黄原胶产量相关基因的调节机制。十字花科黑腐病菌中的黄原胶是由 *gum* 基因簇基因编码的蛋白合成的，此基因簇主要的启动子为 gumB 启动子。HpaR1 突变后，gumB 的转录水平显著降低，凝胶阻滞研究发现 HpaR1 能够直接结合在 gumB 启动子区。研究还发现，HpaR1 和 RNA 聚合酶分别结合在 gumB 启动子区的 -21 到 +10，-42 到 +28 之间。进一步研究发现，HpaR1 能够增强 RNA 聚合酶与 gumB 启动子区的结合能力，进而提高 gumB 的转录水平。这些结果证明，HpaR1 对 gumB 转录调控的寄主与目前研究发现的 MerR 家族转录激活因子的调控机制相似但又有不同。

关键词：转录调控因子；HpaR1；*gum*B；调控机理

* 通讯作者：陆光涛；E-mail：lugt@ gxu. edu. cn

唐纪良；E-mail：jltang@ gxu. edu. cn

玉米青枯病主要分离物对玉米苗期的影响*

李石初**，唐照磊，杜　青，农　倩，磨　康
（广西农科院玉米研究所，国家玉米改良中心广西分中心，南宁　530006）

摘　要：通过试验研究，弄清玉米青枯病主要分离物对玉米苗期生长的影响，明确玉米青枯病的主要致病菌，为防控该病害提供理论依据。用玉米青枯病主要分离物，禾谷镰刀菌、串珠镰刀菌、禾生腐霉菌和肿囊腐霉菌进行盆栽接种试验，玉米播种30d后进行测量玉米叶片数量、叶片重量、植株高度及根系长度，对结果进行统计分析。禾生腐霉菌和肿囊腐霉菌对玉米苗期的影响起主要作用。腐霉菌是玉米青枯病的主要病原菌，在玉米大田生产上防控玉米青枯病要针对腐霉菌。

关键词：根系；叶片；高度；分离物；玉米青枯病

* 基金项目：国家现代农业产业技术体系广西玉米创新团队建设项目（nycytxgxcxtd－07）；广西农业科学院基本科研业务专项基金资助项目（桂农科2013YZ21、桂农科2014YZ19、2015YT29）

** 第一作者：李石初，副研究员，从事玉米病虫害发生流行规律、防控技术及种质资源抗性鉴定研究；E-mail：shichuli@aliyun.com

CatB is Critical in Total Catalase Activities and Confers Sensitivity to Phenazine-1-carboxylic Acid both in *Xanthomonas oryzae* pv. *oryzae* and *Xanthomonas oryzae* pv. *oryzicola*

PAN Xia-yan[1], XU Shu, WU Jian, DUAN Ya-bing, ZHOU Ming-guo*

(*College of Plant Protection*, *Nanjing Agricultural University*, *Key Laboratory of Pesticide*, *Jiangsu Province*, *Nanjing* 210095, *China*)

Abstract: Rice bacterial leaf blight and rice bacterial leaf streak are two major diseases of rice, which caused by *Xanthomonas oryzae* pv. *oryzae* (*Xoo*) and *Xanthomonas oryzae* pv. *oryzicola* (*Xoc*) respectively. Phenazine-1-carboxylic acid (PCA) is a natural-product isolated from *Pseudomonas* spp. to control many important rice diseases in China. In previous studies, we found that PCA disturbed redox balance, which resulted in accumulation of reactive oxygen species (ROS) in target *Xoo*. In this study, we found PCA significantly up-regulated the transcript levels of *catB* and *katE* (catalase), and PCA-sensitivity was reduced when adding exogenous catalase in both *Xoo* and *Xoc*. Furthermore, *catB* deletion mutants of *Xoo* and *Xoc* decreased total catalase activities dramatically and became more sensitive to PCA. Interestingly, deleted *catB* in two strains led to up-regulation of *katE*, but this up-regulation did not reduce PCA-sensitivity significantly. In contrast, deletion mutants of *srpA* and *katG* (catalase) caused little PCA-sensitivity changes. Taken together, we systematically investigated all members of catalase in *Xoo* and *Xoc* under PCA-induced stress and demonstrated *catB* was a crucial member of catalase in total catalase activities, which could protect *Xoo* and *Xoc* against PCA-induced stress.

* Corresponding author: ZHOU Ming-guo; E-mail: mgzhou@ njau. edu. cn

黄单胞菌属非编码 RNA 的研究进展[*]

严玉萍[**]，钟 晰，王雪峰[***]
（西南大学柑桔研究所，中国农业科学院柑桔研究所，重庆 400712）

摘 要：黄单胞菌属（*Xanthomonas*）是一类能引起多种单子叶和双子叶植物感病的革兰氏阴性细菌，严重危害水稻、甘蓝、番茄、柑橘等多种经济作物，其入侵和增殖依赖于三型分泌系统（Type III Secretion System，T3SS）和其他毒素因子。细菌非编码 RNA 能通过与靶 mRNA 互作，在转录后水平调控基因的表达，或直接与蛋白互作，影响细胞的各种生理功能。主要介绍了细菌非编码 RNA 的分类、及其对细菌蛋白调节、生长代谢、基因转录以及毒力调控等方面的研究进展，并重点对黄单胞菌属中少数的非编码 RNA 及其生物学功能进行综述，以期为黄单胞菌引起的作物病害防控提供新的思路。

关键词：黄单胞菌；非编码 RNA；蛋白调节；生长代谢；毒力调控

---

* 基金项目：中央高校基本科研业务费（XDJK2015A009）

** 第一作者：严玉萍，硕士研究生，研究方向：分子植物病理学；E-mail：838711762@ qq. com

*** 通讯作者：王雪峰，副研究员，研究方向：分子植物病理学；E-mail：wangxuefeng@ cric. cn

利用重测序技术研究水稻白叶枯病菌对噻枯唑的抗性机制

梁晓宇[*]，于晓玥，段亚冰，王建新，周明国[**]
（南京农业大学植物保护学院，南京 210095）

摘　要：水稻白叶枯病是水稻上的重要病害之一。噻枯唑是我国防治水稻白叶枯病最主要的药剂之一，这种药剂在我国的使用历史在三十年以上，开展抗药性机制研究十分必要。水稻白叶枯病菌（*Xanthomonas oryzae* pv. *oryzae*，*Xoo*）2-1-1 是野生型 *Xoo* ZJ173 接种到喷施有噻枯唑的水稻上筛选分离得到的对噻枯唑抗性菌株。但 2-1-1 菌株在离体环境下表现对噻枯唑更加敏感，EC_{50}值为 ZJ173 的一半。噻枯唑可以抑制 ZJ173 的胞外多糖合成，不能抑制抗性菌株 2-1-1 的胞外多糖合成。基于以上表型特征，我们分别对 ZJ173 和 2-1-1 进行基因重测序技术研究，构建 350bp 小片段文库，利用 HiSeq 测序平台进行双末端测序，测序深度 >100X。结果表明两种菌株均与标准菌株 MAFF 311018 亲缘性最高，基因组匹配率为 99.37%。单核苷酸多态性（SNP）分析发现 ZJ173 与 2-1-1 基因组与 MAFF 311018 相比有多处点突变，点突变的研究有利于水稻白叶枯病菌对噻枯唑的抗性机制的进一步了解。

关键词：噻枯唑；水稻白叶枯病菌；抗性机制；重测序技术

* 第一作者：梁晓宇，河北张家口人，博士生，从事植物病害化学防治研究；E-mail：2012202020@ njau. edu. cn
** 通讯作者：周明国，江苏南通人，博士生导师，主要从事植物病害化学防治研究；E-mail：mgzhou@ njau. edu. cn

A Promoter-less Transposon for the Trapping of *rplY*, a Novel Gene Crucial to Virulence in *Pectobacterium carotovorum* subsp. *carotovorum* that is Activated by Leaf Extracts from *Zantedeschia elliotiana*

JIANG Huan[1,2*], JIANG Meng-yi[1*], YANG Liu-ke[1], MA Lin[1], WANG Chun-ting[1], WANG Huan[1], QIAN Gou-liang[1], FAN Jia-qin[1**]

(1. *Department of Plant Pathology*, *Nanjing Agricultural University*, *Nanjing* 210095, *China*; 2. *Center for Biotechnology*, *Chongqing Yudongnan Academy of Agricultural Sciences*, *Chongqing* 408000, *China*)

Abstract: Previous studies have demonstrated the interactions between bacterial pathogens and plant hosts, from the viewpoint both of the plant and bacterium, using many approaches, including proteomic and transcriptomic analyses and genomic and post-genomic studies. However, there are some limitations due to some undetectable proteins using the current tools or incomplete understanding due to miscellaneous post-transcriptional processes. In this study, we constructed a library of approximately 6 000 insertion mutants of the *Pectobacterium carotovorum* subsp. *carotovorum* strain PccS1 using *mariner*, a promoter-less kanamycin resistance transposon, from which approximately 500 insertional mutants were found to have kanamycin resistance dependent on plant extracts from the leaves of *Zantedeschia elliotiana*; only one of these mutants showed seriously attenuated virulence on the hosts and impaired multiplication in the media without the plant extract. The gene in the insertion site of the mutantwas named*rplY*, and it was activated in PccS1 both at the gene transcript and protein expression levels with a significant increase in the promoter strength when the cells encountered the plant extract. It was revealed that deletion of *rplY* reduced the virulence of the pathogen and also decreased the motility and production of extracellular enzymes, such as protease, pectatelyase and cellulase, but did not affect the formation of biofilm, compared with the wild-type strain. The strain of the gene deletion mutant complementary with*rplY* partly restored the virulence, motility and production of the exo-enzymes. Meanwhile, the virulence of the *rplY* gene deletion mutant could be partly restored by the plant extract. The data indicate that the plant extract induces PccS1 virulence and multiplication partly through*rplY* activation, and*rplY* is crucial to virulence, motility and exo-enzyme production in *P. carotovorum*. This work is a valuable complement to the approaches on bacterial-plant interactions and on the identification of novel genes associated with pathogenicity.

Key words: Bacterial-plant interaction; *Pectobacterium carotovorum*; *Mariner* transposon; Plant extract; *Zantedeschia elliotiana*

* First author: The first two authors contributed equally to this research.

** Corresponding author: JIANG Huan, FAN Jia-qin; E-mail: fanjq@ njau. edu. cn

Characterisation of *Pectobacterium carotovorum* Proteins Differentially Expressed During Infection of *Zantedeschia elliotiana in vivo* and *in vitro* Which are Essential for Virulence[*]

WANG Huan[1*], YANG Zhong-ling[1*], DU Shuo[1], ZHANG Pei-fei[1], MA Lin[1], LIAO Yao[1], WANG Yu-jie[1], Ian Toth[2], FAN Jia-qin[1**]

(1. *College of Plant Protection*, *Nanjing Agricultural University*, *Nanjing* 210095, *China*; 2. *Cell and Molecular Sciences*, *James Hutton Institute*, *Dundee DD2 5DA*, *UK*)

Abstract: Identifying phytopathogen proteins that are differentially expressed in the course of establishing an infection is important to better understand the infection process. *In vitro* approaches, using plant extracts added to culture medium, have been used to identify such proteins but the biological relevance of these findings for *in planta* infection are often uncertain until confirmed by *in vivo* studies. Here, we compared proteins of *Pectobacterium carotovorum* subsp. *carotovorum* strain PccS1 differentially expressed in LB medium supplemented with extracts of the ornamental plant *Zantedeschia elliotiana* cultivar 'Black magic' (*in vitro*) and in plant tissues (*in vivo*) by two-dimensional electrophoresis coupled with mass spectrometry. A total of 53 differentially expressed proteins (>1.5 fold) were identified (up-regulated or down-regulated *in vitro*, *in vivo* or both). Proteins that exhibited increased expression *in vivo* but not *in vitro* or in both conditions were identified, and deletions made in a number of genes encoding these proteins, four of which (*clpP*, *mreB*, *flgK*and *eda*) led to loss of virulence on *Z. elliotiana*. While *clpP*, *flgK* and *mreB* have previously been reported as playing a role in virulence on plants, this is the first report of such a role for *eda*, which encodes 2-keto-3-deoxy-6-phosphogluconate (KDPG) aldolase, a key enzyme in Entner-Doudoroff metabolism. The results highlight the value of undertaking *in vivo* as well as *in vitro* approaches for the identification of new bacterial virulence factors.

Key words: Bacterial-plant interactions; Virulence; *Pectobacterium carotovorum*; Plant extract; *Zantedischia elliotiana*; Entner-Doudoroff

---

* The first two authors contributed equally to this research

** Corresponding author: FAN Jia-qin; E-mail: fanjq@njau.edu.cn

Isolation and Identification of Antagonistic Bacteria Against Soil-borne Pathogenic Fungus of Tobacco*

SONG Xi-le**, DING Yue-qi, KANG Ye-bin***

(*College of Forestry, Henan University of Science and Technology, Luoyang*, 471003, *China*)

Abstract: This study is aimed to isolate and screen bacteria strains from tobacco rhizospheric soil, which have antagonistic activities against soil-borne pathogenic fungus of tobacco. Soil dilution plating technique and dual culture technique on agar plate were used for isolate and screen antagonistic bacteria. The antagonistic strains were identified by morphological features observation, physiological and biochemical experiments, and 16S rDNA sequence analysis. 30 rhizospheric soil were collected from tobacco filed in Luoning, Ruyang, Songxian and Yiyang countries in Luoyang, Henan province. 180 bacterial isolates were obtained from 30 rhizosphere soil samples by separation and purification, and 49 strains have antagonistic function to the soil-born pathogenic fungi tested in this trial. Strains TS681、TS682、TS683 were strongly antagonistic to three kinds of pathogenic fungi, and the inhibition zones to *Pythium aphanidermatum*, *Phytophthora parasitica* and *Furasium oxysporum* respectively more than 7. 0 mm, 10. 0 mm and 9. 0mm with the antagonistic rate respectively reaching above 51. 4%, 68. 8% and 60. 0%. All this three strains were identified as *Paenibacillus polymyxa*, and they may have the potential value of development and application for biological controlling of soil-borne fungi diseases of tobacco.

Key words: Tobacco; Soil-born pathogenic fungi; Antagonistic bacteria; *Paenibacillus polymyxa*

---

* 基金项目：河南省烟草公司科技项目（2012M10）；洛阳市烟草公司项目（2014M06）

** 第一作者：宋喜乐，硕士研究生，主要从事植物免疫学研究；E-mail：xilesong@163. com

*** 通讯作者：康业斌，教授，主要从事植物免疫学研究；E-mail：kangyb999@163. com

甘蔗赤条病病原菌分离与鉴定

李晓燕，王锦达，傅华英，陈如凯，高三基*
（福建农林大学国家甘蔗工程技术研究中心，福州）

摘　要：甘蔗赤条病（Red stripe of sugarcane）是一种常见的细菌性病害，由燕麦食酸菌燕麦亚种（*Acidovorax avenae* subsp. *avenae*，*Aaa*）引起。本研究从感病的甘蔗品种福农41号分离纯化获得病原菌株系AAA411，从病原菌菌落及菌体的形态观察、巢式PCR分子检测证实为Aaa病原菌。将分离出来的Aaa菌液（10^6CFU/mL）接种到健康的甘蔗品种福农38号幼苗叶片，接种后3d，可以引起叶片红褐色条斑，6d后发病率趋于稳定，达60%。侵染后的叶片经PCR验证Aaa病原已成功侵染到甘蔗植株。

关键词：甘蔗赤条病；燕麦食酸菌燕麦亚种；病原菌形态；巢式PCR；分子鉴定

* 通讯作者：高三基；E-mail：gaosanji@ yahoo. com

柑橘黄龙病病原－媒介－寄主互作研究进展*

胡 燕**，王雪峰，周常勇***

（西南大学柑桔研究所，中国农业科学院柑桔研究所，
国家柑桔工程技术研究中心，重庆 400712）

摘 要：柑橘黄龙病（Citrus Huanglongbing，HLB）是当前全球柑橘生产上最具毁灭性的柑橘病害。该病害能够侵染几乎所有栽培柑橘品种，柑橘木虱是该病害田间传播的天然媒介，目前该病尚无有效治愈方法。由于病原菌尚不能纯培养，因此，对病原生物学特性知之甚少，极大制约了对该病的基础研究和防控技术研发。近年来，随着基因组学、转录组学和蛋白组学等技术的快速发展，在病原入侵后对寄主的代谢、激素调节的影响及寄主防卫反应、病原在木虱体内传播及其对木虱发育、代谢及倾向性影响，以及病原通过改变寄主植物的挥发性信息物等间接调控媒介昆虫行为等方面取得了显著进展。本文对该病害病原－媒介－寄主互作相关研究进行综述，并对今后相关研究进行了展望，以期对柑橘黄龙病研究有所促进。

关键词：柑橘黄龙病；木虱；互作

* 基金项目：重庆市两江学者计划（2013）；中央高校基本科研业务费创新团队项目（XDJK2014A001）

** 第一作者：胡燕，2014 级硕士研究生，专业：植物病理学；E-mail：15213170817@163.com

*** 通讯作者：周常勇，研究员，主要从事柑橘病毒类病害研究和无病毒繁育体系建设；E-mail：zhoucy@swu.edu.cn

柑橘黄龙病菌非自主转座元件的多态性及其高频转座*

胡　燕**，王雪峰***，周常勇***
（西南大学柑桔研究所，中国农业科学院柑桔研究所，
国家柑桔工程技术研究中心，重庆　400712）

摘　要：最近从柑橘黄龙病菌（*Candidatus* Liberibacter asiaticus）中鉴定了两个微型反向重复转座元件（Miniature inverted-repeat transposable elements，MITEs），即 MCLas-A 和 MCLas-B，又称非自主型转座元件，其中 MCLas-A 因具有跳跃性而被认为是活性的转座元件。这两个非自主转座元件的上游基因均为推断的转座酶基因。本研究旨在分析该上游转座酶基因的序列变异，并探讨这些序列变异与非自主转座元件活性之间的相关性，并通过大规模调查评估田间样品中两个 MITEs 转座频率。设计可扩增推断转座酶基因和下游转座元件的引物，对来自中国、巴西和美国的 43 个代表性黄龙病菌分离物进行 PCR 和序列分析。PCR 扩增可产生 6 种主要的扩增子，进而对所有扩增子类型进行测序发现共有 12 种序列型，其中 3 种序列型（T4、T5-2、T6）为首次报道。在序列型 T5-2 和 T6 中检测到重组事件，且所有巴西样品均含有这两种序列型。值得注意的是，在活性转座元件 MCLas-A 的上游序列中没有发现序列变异和重组现象，表明转座酶基因的保守性可能与非自主元件的转座活性紧密相关。聚类分析结果显示，所有序列可分为包含 5 个亚组的 2 个组群，且部分亚组对应不同的样品来源，尤其是云南瑞丽样品、巴西圣保罗样品和佛罗里达少量样品。为了评估 MITEs 的转座情况，对近期从广东、广西、江西、福建、湖南、贵州、云南收集的 350 个样品和 2011—2012 年提取的 50 个广西样品 DNA 进行扩增，仅在广东和贵州的 10 个样品（2.86%）中检测到了 MCLas-A，其中推断的转座产物在所有近期收集的样品中占主导地位，而 MCLas-A（60%）在前期收集的 50 个样品中占主导地位，表明 MCLas-A 近期可能存在高频转座的现象。

关键词：柑橘黄龙病菌；微型反向重复转座元件；遗传多态性；高频转座

---

* 基金项目：重庆市两江学者计划（2013），中央高校基本科研业务费创新团队项目（XDJK2014A001）
** 第一作者：胡燕，2014 级硕士研究生，专业：植物病理学；E-mail：15213170817@163.com
*** 通讯作者：王雪峰，副研究员，主要从事柑橘病害研究；E-mail：：wangxuefeng@cric.cn
周常勇，研究员，主要从事柑橘病害研究；E-mail：zhoucy@cric.cn

西藏分离的短小芽孢杆菌 GBSW19 生物学特性研究及全基因组分析*

顾　沁，邵贤坤，伍辉军，高学文**

（南京农业大学植物保护学院植物病理学系，农作物生物灾害综合治理教育部重点实验室，南京　210095）

摘　要：芽孢杆菌属（*Bacillus* spp.）是一种革兰氏阳性菌，好氧或兼性厌氧，能产生多类次生代谢产物，具有较好的抑制病原菌和促进植物生长的活性，广泛应用于生物农药领域。前期，本课题组从西藏极端环境中分离到一株短小芽孢杆菌 GBSW19 菌株，该菌株具有低温适生性，并且能高效降解作物秸秆。在本研究中，我们进一步研究了该菌株的生物学特性。大田试验结果表明，GBSW19 能显著提高水稻秸秆的降解速率，并且能提高土壤速效氮、速效磷等元素的含量，这说明该菌株能增加土壤养分，提高土壤肥料。菌株的抗逆性研究结果表明，GBSW19 菌株具有较强的抗氧化胁迫、抗酸碱胁迫和耐低温的能力。我们还对该菌株的进行了全基因组测序分析。结果表明，GBSW19 菌株全基因组大小为 3 675 201bp，共 3 620个基因。通过基因功能分析发现，GBSW19 中存 6 个基因编码蛋白参与酸性 pH 胁迫调控，9 个基因可能编码外泌纤维素酶和半纤维素酶，同时还发现 3 个基因编码蛋白可能参与过氧化氢胁迫调控。本研究表明，GBSW19 菌株具有开发成生物农药的潜力。

关键词：短小芽孢杆菌；降解秸秆；抗逆性

* 基金项目：公益性行业（农业）科研专项（201303015）

** 通讯作者：高学文，博士，教授，从事生物防治与细菌分子生物学研究；E-mail：gaoxw@ njau. edu. cn

柑橘抗/感溃疡病品种中内生细菌多样性分析*

吴思梦**，刘 冰***
（江西农业大学农学院，南昌 330045）

摘 要：溃疡病是柑橘上的一种检疫性病害，对柑橘生产造成严重影响。该病对柑橘品种的选择性很强，以甜橙类受害最为严重。前期试验中发现柑橘抗病品种中的可培养内生细菌相对于感病品种来说种类少、数量多、强颉颃菌株比重大。为进一步探索柑橘内生细菌在寄主品种抗溃疡病中的作用，本研究分别提取温州蜜柑（抗溃疡病品种）和纽荷尔脐橙（感溃疡病品种）叶片与果实的基因组DNA，进行16s rDNA V4高变区扩增后建库测序，将所有样品的Effective Tags聚类为OTUs并注释。分析结果表明，样品的主要细菌类群为泛菌属（*Pantoea*），盐单胞菌属（*Halomonas*），假单胞菌属（*Pseudomonas*），藤黄单胞菌属（*Luteimonas*），拟杆菌属（*Bacteroides*）以及不动杆菌属（*Acinetobacter*）等。其中温州蜜柑叶片和果实中比例最大的细菌类群均为盐单胞菌属（*Halomonas*），但果实中细菌类群更多样化；纽荷尔脐橙叶片中最多的细菌类群也为盐单胞菌属（*Halomonas*），但要比温州蜜柑叶片中的比例小，而果实中存在盐单胞菌属（*Halomonas*）比例较小，其他细菌类群丰富且比例均匀，泛菌属（*Pantoea*）比例稍大。4个样品中内生细菌物种丰富程度依次为：纽荷尔脐橙果实 > 温州蜜柑果实 > 纽荷尔脐橙叶片 > 温州蜜柑叶片。由此可见，温州蜜柑和纽荷尔脐橙叶片、果实中的内生菌菌群存在明显差异但又有共同之处，结合前期试验结果和柑橘品种发病程度，推测柑橘内生细菌可能与寄主抗病能力之间有一定关系，确切关系还需要进一步的深入分析和试验验证。

关键词：柑橘品种；溃疡病；细菌类群

* 基金项目：国家自然科学基金（31460139），江西省科技支撑计划项目（20121BBF60024）
** 第一作者：吴思梦，硕士研究生，主要从事植物病理学研究；E-mail：1367894992@qq.com
*** 通讯作者：刘冰，博士，副教授，主要从事植物病理学方面的教研工作；E-mail：lbzjm0418@126.com

基于锁式探针的密执安棒状杆菌高通量检测方法研究*

李志锋[1]，冯建军[2,3]**，吴绍精[2,3]，王忠文[1]，程颖慧[2,3]
（1. 广西大学农学院，南宁　530004；2. 深圳出入境检验检疫局动植物检验检疫技术中心，深圳　518045；3. 深圳市检验检疫科学研究院，深圳　518010）

摘　要：密执安棒状杆菌（*Clavibacter michiganensis*）为放线菌目、微球菌科、密执安棒状杆菌属，是重要的植物病原细菌。其中，密执安棒状杆菌的5个亚种：*Clavibacter michiganensis* subsp. *nebraskensis*（缩写CMN）、*Clavibacter michiganensis* subsp. *michiganensis*（缩写CMM）、*Clavibacter michiganensis* subsp. *insidiosus*（缩写CMI）、*Clavibacter michiganensis* subsp. *sepedonicus*（缩写CMS）、*Clavibacter michiganensis* subsp. *tessellarius*（缩写CMT），危害性大、寄主分布广，并且5个亚种的形态学特征近似，鉴别很难，其中前4个亚种是我国进境植物检疫性有害生物。因此，密执安棒状杆菌属是目前检验检疫研究的重点。本研究根据ITS和celA基因序列分别设计了CMN-PLP、CMM-PLP、CMI-PLP、CMT-PLP和CMS-PLP等5条锁式探针（已申请国家发明专利），建立五重滚环扩增体系，并利用生物素标记的下游引物进行扩增，将扩增产物与偶联上微球的捕获探针进行特异性杂交，最后利用液相悬浮芯片仪进行检测。结果表明：①20株供试菌株中，仅靶标菌能与相对应的锁式探针特异性结合，并且当混合样品中同时含有4种、3种或者2种靶标菌时，五重检测体系均可同时检测出对应微球显示阳性信号，而近似种或其他属供试菌株检测结果呈阴性；②建立的五重检测体系可以在一个反应管中同时检测五种混合供试靶标菌，也可检测出单一靶标菌DNA，其检测灵敏度阈值为100fg/μL，同时五对锁式探针相互之间无交叉反应；③该五重检测体系三次重复试验的平均荧光值变异系数均小于10%，表明具有良好的重复性。因此，本研究中基于锁式探针的5种密执安棒状杆菌悬浮液相芯片检测技术具有高通量、特异性强和灵敏度高的特点，解决了常规多重检测中加入多对引物造成体系优化的困难，为密执安棒状杆菌的分类鉴定提供了为新的方法和技术支持。

关键词：密执安棒状杆菌；锁式探针；滚环扩增；液相芯片

* 基金项目：深圳市海外高层次人才创新创业专项资金项目（KQC201109050077A）和深圳局科技项目（SZ2011002）

** 通讯作者：冯建军；E-mail：sccfjj@126.com

A Global Transcriptional Regulatormodulates Production of Multiple Virulence Factors in *Dickeyazeae* EC1

ZHOU Jia-nuan, LV Ming-fa, TANG Ying-xin, ZHAO Gang, CHEN Shao-hua, HE Fei, CUI Zi-ning, JIANG Zi-de, CHANG Chang-qing, ZHANG Lian-Hui*

(1. *Guangdong Province Key Laboratory of Microbial Signals and Disease Control*, *Department of Plant Pathology*, *South China Agricultural University*, *Guangzhou* 510642, *China*; 2. *Institute of Molecular and Cell Biology*, 61 *Biopolis Drive*, *Singapore* 138673)

Abstract: Transcriptional regulators play crucial roles in modulation of virulence factor production in pathogens. In this study, we have characterized a SlyA/MarR family transcription factor SlyA in *Dickeya zeae* strain EC1, the causal agent of rice root rot disease. In-frame deletion of *slyA* significantly decreased the production of phytotoxinzeamines, changed the expression patterns of cell wall degrading enzymes, enhanced bacterial motility, reduced biofilm formation, and attenuated the pathogenicityof strain EC1 on host plants. Transcriptome analysis showed that SlyA modulates the expression of about 25% of the genome, including the genes encoding various pathogenesis-related proteins, five types of secretion systems and quorum sensing systems. In addition, based on the transcriptome data, we explored the regulatory pathway of SlyA. The results showed that *in trans* expressionof *slyA* in the AHL QS mutant *expI*recovered the phenotypes of motility and biofilm formation, and *in trans* expressionof a newly identified regulator genein the *slyA* mutant rescued the phenotypes of zeamines and the production of cell wall degrading enzymes. Taken together, the findings from this study unveiled a key transcriptional regulatory network involved in modulation of virulence factor production and overall pathogenicity of *D. zeae* EC1.

Key words: *Dickeya zeae* EC1; Virulence factor; Zeamines

* Corresponding author: ZHANG Lian-hui; E-mail: lhzhang01@ scau. edu. cn

中国烟草青枯菌遗传多样性和致病力分析

黎妍妍[1,2*]，冯　吉[2]，王　林[3]，Tom Hsiang[4]，李锡宏[2]，黄俊斌[1**]

（1. 华中农业大学植物科学技术学院，武汉　430070；2. 湖北省烟草科学研究院，武汉　430030；3. 湖北中烟工业有限责任公司，武汉　430040；4. 加拿大圭尔夫大学环境科学学院，加拿大圭尔夫市　N1G2W1）

摘　要：由青枯劳尔氏菌（*Ralstonia solanacearum*）引起的青枯病是我国烟草（*Nicotiana tabacum*）最严重的土传病害。2012—2014 年，通过在我国五个植烟区进行烟草青枯病分布情况调查并进行样本采集，发现烟草青枯病分布在西南烟区、东南烟区、长江中上游烟区和黄淮烟区，在北方烟区尚未发现烟草青枯病。从这四个发病烟区分离获得 89 个烟草青枯菌菌株，并对其进行了遗传多样性和致病力分析。根据内切葡聚糖酶基因序列，我国烟草青枯菌为演化型 I（亚洲分支），并可划分为 7 个序列变种（13、14、15、17、34、44 和 54）。其中，序列变种 15、17、34 和 44 已有报道；序列变种 13 和 14 在本研究中首次报道可以侵染烟草。同时，也发现了一个新的序列变种 54。不同区域的烟草青枯菌表现出不同的遗传多样性水平：地理分布越往北，烟草青枯菌的遗传多样性越低。通过人工接种 3 个不同抗性的烟草品种测定了 27 个代表性菌株的致病力。基于病害流行下曲线面积（AUDPC），聚类分析可将 27 个菌株划分为不同的致病型（高、中、低），在序列变种和致病型间不存在明显的相关性。

关键词：烟草青枯病；劳尔氏菌；演化型；序列变种

* 第一作者：黎妍妍，博士研究生，主要从事植物细菌病害研究

** 通讯作者：黄俊斌，教授；E-mail：junbinhuang@ mail. hzau. edu. cn

柑橘溃疡病菌 *gpd*1 基因影响病原菌毒性、游动性和生物膜的形成*

葛宗灿**，邹丽芳，蔡璐璐，马文秀，陈晓斌，陈功友***
（上海交通大学农业与生物学院，上海 200240）

摘　要：甘油作为微生物重要的能量来源及抗逆因子，对其生存具有重要作用。*gpd*1 基因编码甘油-3-磷酸脱氢酶，是甘油合成代谢中的关键酶。本研究分析了柑橘溃疡病菌（*Xanthomonas citri* subsp. *citri*，*Xcc*）*gpd*1 基因的生物学功能。对 *Xcc* 强度菌株 049 和弱毒菌株 021 分别进行 *gpd*1 基因的缺失突变，发现 *gpd*1 基因缺失仅影响弱毒菌株 021 在感病柑橘寄主上的毒性，功能互补菌株能够恢复毒性至野生型水平。多种生物学表型分析显示：与野生型菌株相比，021 菌株的 *gpd*1 突变体群体聚集能力和游动性降低；生物膜形成能力提高；胞外多糖的产生没有显著变化。这些结果表明 *gpd*1 基因在弱毒菌株中是重要的毒性因子，其涉及的甘油代谢途径可能在强弱毒菌株对于寄主柑橘的毒性具有不同的作用。

关键词：柑橘溃疡病菌；*gpd*1 基因；毒性；生物膜；细菌游动性

---

* 基金项目：国家自然科学基金资助项目（31371905，32470235）；公益性行业（农业）科研专项（201303015-02）

** 第一作者：葛宗灿，河北保定人，硕士研究生，主要从事分子植物病理学研究；E-mail：gezongcan@163.com
*** 通讯作者：陈功友，教授，主要从事分子植物病理学研究；E-mail：gyouchen@sjtu.edu.cn

野游菜黄单胞菌不同亚型细胞色素 C 在抵御外界 H_2O_2 胁迫中的作用研究

武　健，潘夏艳，徐　曙，罗剑英，段亚冰，周明国*
（南京农业大学，南京　210095）

摘　要：野游菜黄单胞菌 *Xanthomonas campestris* pv. *campestris*（*Xcc*）是油菜黑腐病的致病菌，主要为害油菜的茎、叶和角果。H_2O_2 是一种氧化剂，本身就是植物体产生以抵御外界病原菌侵染的重要物质，同时也是植物细胞抗侵染过程中的重要的信号分子。细胞色素 C 是呼吸链下游负责传递电子的蛋白质，除呼吸链传递电子的作用外，还参与了细胞抗氧化的重要功能，其可以将细胞中的 H_2O_2 通过反应转化为水，降低细胞氧化压水平。我们通过对野游菜黄单胞中不同亚型细胞色素 C 的敲除发现，不同细胞色素 C 在细胞中担负的作用是不同的，只有一种细胞色素 C 在抵御 H_2O_2 的过程中发挥了重要作用，其对 H_2O_2 的敏感性发生显著提高。这对我们研究病原菌侵染植物的过程有重要意义，同时对病害防治也有指导作用。

关键词：野游菜黄单胞菌；细胞色素 C；基因敲除；H_2O_2

* 通讯作者：周明国；E-mail：mgzhou@ njau. edu. cn

一个新机制参与青枯病菌造成的细菌性枯萎病

卢海彬
（西北农林科技大学，杨凌　712100）

摘　要：青枯雷氏菌是一种危害性很大的植物病原菌，它可以侵染多种作物，造成严重的产量损失。然而我们现在对其致病的分子机制知之甚少。体外病原菌侵染实验表明在接种后一周，青枯雷氏菌 GMI1000 可以抑制根的伸长、促进根毛形成和根尖细胞死亡。HrpG 是一个 GMI1000 的关键调节因子，它的突变直接导致 GMI1000 引起上面 3 种根部表型的丧失。HrpB 处于 HrpG 调节子的下游，它直接可以激活 GMI1000 的三型分泌系统，促进病原菌侵染。*hrpb* 不能促进根毛形成和根尖细胞死亡，但是仍然可以抑制根的伸长。进一步的番茄致病性实验表明，*hrpb* 的致病能力要比 *hrpg* 的致病能力强 5 倍左右。我们的实验证明除了三型分泌系统和效应因子以外，一条不依赖于 HrpB 新途径在其感染农作物过程中起着重要作用。

关键词：青枯雷氏菌；*Hrpb*；体外浸染

野油菜黄单胞菌菌黄素的生物合成机制研究

曹雪强*，周　莲，何亚文**
（生命科学技术学院，微生物代谢国家重点实验室，上海交通大学，上海　200240）

摘　要：黄单胞菌属细菌（以下简称黄单胞菌）是重要的植物病原细菌，可以侵染400多种植物，在农业生产中造成严重危害。黄单胞菌合成一种附膜的黄色色素，又称为菌黄素。菌黄素能保护细菌抵抗光伤害，并提高细菌在寄主体表的附生能力。菌黄素的缺失还能影响黄单胞菌系统侵染的能力，因此，菌黄素是黄单胞菌的一类重要致病相关因子。

前期初步结果显示：菌黄素是一类特殊的溴代芳基多烯脂类化合物，不同结构的菌黄素在溴原子数、芳香环甲基取代及烯烃链长方面存在差异。Poplawsky等首先在野油菜黄单胞菌（*Xanthomonas campestris* pv. *campestris*，缩写 *Xcc*）中克隆了负责菌黄素生物合成的基因簇（*pigA-G*），然而该基因簇中基因的功能并未得到深入研究。我们实验室阐明了 *Xcc* 及水稻白叶枯菌（*X. oryzae* pv. *oryzae*，缩写 *Xoo*）中 *pig* 基因簇中基因 *xanB*2 的功能。XanB2 催化分支酸合成3-羟基苯甲酸（3-HBA）和4-羟基苯甲酸（4-HBA），并提出3-HBA可能是菌黄素合成前体的假说。

本研究以 *Xcc* 野生型菌株XC1为研究对象，对 *pig* 基因簇中的基因进行逐一无痕敲除。通过分析每株突变体菌黄素的合成，结合每个基因编码蛋白质结构域的预测，我们推断菌黄素可能通过Ⅱ-型聚酮与Ⅱ-型脂肪酸杂合机制合成。我们从生物信息学、遗传学、生物化学等方面对 *pig* 基因簇中的 *Xcc*4015 及 *Xcc*4012 进行了深入的研究。研究结果显示：敲除 *Xcc*4015 或 *Xcc*4012，导致无菌黄素合成和胞外3-HBA和4-HBA含量的积累。进一步体外酶活测定实验表明：*Xcc*4015 可以激活3HBA和4HBA腺苷酰化，然后连接至 *Xcc*4012（ACP，aryl carrier protein）上的磷酸泛酰巯基。我们推测3HBA-S-ACP或4HBA-S-ACP作为起始单元在其他酶（酮基合酶、酮基还原酶、脱水酶等）的作用下合成芳基多烯链，进一步在修饰酶（卤化酶、糖基转移酶、酰基转移酶等）的作用下合成完整的菌黄素。该研究为以菌黄素为靶标杀菌药物的研发提供了理论依据。

关键词：黄单胞菌；菌黄素；生物合成；基因敲除

* 第一作者：曹雪强，博士研究生；E-mail：xqcao2014@ sjtu. edu. cn
** 通讯作者：何亚文，研究员；E-mail：yawenhe@ sjtu. edu. cn

3 种半选择性培养基对细菌性疮痂病菌的选择性和回收率研究

崔子君[*]，蒋 娜，白凯红，岳珊珊，罗来鑫[**]，李健强
（中国农业大学植物病理学系/种子病害检验与防控北京市重点实验室，北京 100193）

摘 要：细菌性疮痂病（bacterial spot）是番茄和辣椒等茄科作物上的常见病害，4 个种的黄单胞菌（*Xanthomonas euvesicatoria*、*X. vesicatoria*、*X. perforans* 和 *X. gardneri*）均可引起该病害。疮痂病菌在世界大多数国家和地区均有报道，被欧洲及地中海植物保护组织（EPPO）列为 A2 类检疫性有害生物。种子带菌为该病害远距离传播的主要途径，也是田间重要的初侵染来源。因此建立一套简便、准确、灵敏、高效的种子带菌检测技术，对控制该病害的发生与传播具有重要意义。BIO-PCR 检测是种子带菌检测的常用方法，即种子提取液在半选择性培养基上富集培养后用 PCR 的方法检测靶标病原菌。用于疮痂病菌富集培养的半选择性培养基主要有 CKTM、mTMB 及 Xan-D 等。本研究采用平板计数法对不同来源的 8 株疮痂病菌（4 株 *X. e*，1 株 *X. v*，2 株 *X. p*，1 株 *X. g*）在 CKTM、mTMB 及 Xan-D 3 种半选择性培养基上的回收率进行比较研究，同时采用平板涂布法测定 3 种半选择性培养基对疮痂病菌的选择性。结果显示，在 5% 显著水平下，疮痂病菌在 CKTM、mTMB 及 Xan-D 3 种半选择性培养基上的平均回收率分别为 79%、155% 和 114%，在 mTMB 培养基上显著高于 CKTM 和 Xan-D；8 株疮痂病菌在 mTMB 及 Xan-D 培养基上能够产生区别于大多数非靶标菌的特征，而在 CKTM 培养基上特征不明显，因此 mTMB 及 Xan-D 的选择性较好。综合 3 种培养基的回收率与选择性，确定 mTMB 和 Xan-D 为 BIO-PCR 法检测种子携带疮痂病菌的半选择性培养基。

关键词：细菌性疮痂病菌；半选择性培养基；回收率；选择性

---

* 第一作者：崔子君，硕士研究生；E-mail：cuizijunczj@163.com
** 通讯作者：罗来鑫，副教授，主要从事种传细菌病害研究；E-mail：luolaixin@cau.edu.cn

Resuscitation and Pathogenicity Test of the Viable but Nonculturable Cells of *Acidovorax citrulli**

KAN Yu-min, JIANG Na, HAN Si-ning, LUO Lai-xin**, LI Jian-qiang

(*Department of Plant Pathology, China Agricultural University / Beijing Engineering Research Center of Seed and Plant Health (BERC-SPH) / Beijing Key Laboratory of Seed Disease Testing and Control (BKL-SDTC). Beijing* 100193, *P. R. China*)

Abstract: Bacterial fruit blotch (BFB) is a devastating disease of cucurbits and has caused significant economic losses of cucurbits industry worldwide. The causal agent, *Acidovorax citrulli* (Ac), is a gram-negative bacterium that infects a wide range of cucurbits including watermelon and melon. Infected seeds are the most important source of primary inoculums and copper compounds are the most widely-used bactericide to control BFB. Our previous experiments confirmed that Ac (strain AAC00-1) could enter into VBNC state by the induction of cupric ion, which may increase probability of false negative results for seed health testing by bio-PCR, a method based on the bacterial culturability. In this research, we demonstrated that VBNC state of Ac induced by different concentration of CuSO4 for different time (50μmol/L $CuSO_4$ for 3 h, 10μmol/L $CuSO_4$ for 5 d and 5μmol/L $CuSO_4$ for 15 d) could be resuscitated by some appropriate conditions. The experiment used LB broth, cell-free suspension (CFS) and different concentration of EDTA (the ratio to Cu^{2+} was 1 : 1, 1.5 : 1 and 2 : 1) to resuscitate the VBNC cells, the results indicated that after treated by different time, culturable Ac could be detected in all tested samples. Adding EDTA (the ratio to Cu^{2+} was 1.5 : 1) was the most efficient method to resuscitate Ac from VBNC state. The average frequency in two biological replications of resuscitation in 50μmol/L $CuSO_4$ for 3h, 10μmol/L $CuSO_4$ for 5 d and 5μmol/L $CuSO_4$ for 15 d was 83.33%, 83.33% and 100%, respectively. The quantity of resuscitated culturable cells in the microcosms was 4.80×10^{6} CFU/mL, 1.68×10^6 CFU/mL and 6.80×10^5 CFU/mL, respectively. All the resuscitated and re-culturable cells of *A. citrulli* recovered pathogenicity on watermelon seedlings by inoculation. However, the VBNC cells of *A. citrulli* could not colonize and infect the watermelon plants in present results.

Key words: *Acidorvorax citrulli*; VBNC; Resuscitation; Pathogenicity

* Funding: This research was supported by the Special Fund for Agro-scientific Research in the Public Interest (No. 201003066)

** Corresponding author: LUO Lai-xin; E-mail: luolaixin@cau.edu.cn.

第五部分　线　虫

一年生马尾松苗对松材线虫病的抗性评价*

陶　毅**，张玉焕，祝乐天，陈　晨，孙　思，陈芳妮，
贾　宁，吴路平，廖美德，王新荣***
（华南农业大学农学院，广州　510642）

摘　要：松材线虫病是松树的一种毁灭性病害，不同种源的马尾松对松材线虫病抗病性存在差异，培育出高抗的树苗在林业建设上具有积极意义。本研究用“截枝套管法”，于2013年7月评价了同一种源的马尾松1年生树苗对松材线虫病的抗性。共计接种30棵马尾松苗，每棵接种5 000条松材线虫，每天观察接种结果，接种45d后，按抗性分级标准进行分级和分离线虫。结果表明：①同一种源的马尾松1年生树苗抗性存在差异。抗性等级划分为高抗、中抗、低抗、感病、高感五个等级。一年生树苗在接种15d后，发病率为0；16～30d后，发病率是73.3%；31～45d后，发病率达到80.0%。②在接种45d后，按抗性等级每级选出3株树苗进行线虫分离。发现各个抗性级别的马尾松均可以分离出松材线虫。症状越明显的植株，分离出的松材线虫数量越多，当到达3级，及接种枝以下部分枯死，顶端泛黄，松针下垂时，分离出的松材线虫数量达到最大；植株逐渐枯死，分离出的松材线虫数量也随之降低，植株完全枯死，松材线虫数量减少。该结果对指导马尾松造林具有重要的生产实践意义。

关键词：一年生马尾松；松材线虫病；抗性

* 基金项目：重大林业行业公益项目（项目编号：201204501）；国家公派留学基金（2014）

** 第一作者：陶毅，本科生

*** 通讯作者：王新荣，教授，主要从事植物线虫学研究；E-mail：xinrongw@ scau. edu. cn

万寿菊秸秆综合利用途径及其杀线作用的初步研究

徐 返[1*]，曹 睿[1]，陈志星[2]，王海宁[3]，李 凡[1]，王 扬[1**]，谢 勇[1]
（1. 云南农业大学植保学院，昆明 650201；2. 昆明虹之华园艺有限公司，昆明 651700；3. 云南农业大学香料研究所，昆明 650051）

摘 要：万寿菊的杀线活性广受关注，本研究对万寿菊秸秆的综合利用进行了初步尝试。通过生物测定和温室试验，研究了万寿菊秸秆、添加万寿菊秸秆的平菇栽培基质及菊叶对根结线虫的抑杀作用，明确了万寿菊秸秆对根结线虫的防治效果。研究结果表明，万寿菊叶水提取物对南方根结线虫校正死亡率达到100%，与各处理和对照相比差异显著（$P<0.05$），而且，添加万寿菊叶的处理显著提高了平菇栽培基质水提取物的杀线活性（$P<0.05$）；栽培基质经过栽培平菇后获得的菌糠，其杀线活性相对于新鲜栽培基质有了显著提高（$P<0.05$）；盆栽试验显示，添加不同比例的万寿菊秸秆的处理中，9∶1 处理对促进番茄植株株高、植株叶片生长和植株地下部生长效果显著，而较高菊叶含量的处理（1∶1）抑制植株叶片生长，菌糠和菊叶处理中以平菇菌糠：菊叶质量比为 3∶1 的比例防治根结线虫的效果最好，其防治效果与化学药剂噻唑膦在统计学意义上相当。

关键词：万寿菊；秸秆；菌糠；南方根结线虫

* 第一作者：徐返，曹睿
** 通讯作者：王扬，谢勇

土沉香根结线虫的发生与病原鉴定

苏圣淞[1]，周国英，李　河，王　姣，何苑皞，刘君昂
（中南林业科技大学经济林培育与保护教育部重点实验室；
中南林业科技大学森林有害生物防控湖南省重点实验室，长沙　410004）

摘　要：土沉香是我国特有的热带乡土树种，也是我国生产中药沉香的唯一植物资源。本研究对海南省澄迈县土沉香根结线虫病为害与发生情况进行了调查与研究，并采用比较形态学结合rDNA-ITS序列分析的方法对病原进行了分离鉴定，结果表明，土沉香受到根结线虫不同程度的危害，其中以苗木与幼龄林为主，严重苗圃死亡率达30%，幼龄林死亡率达45%，受感染的土沉香表现为植株矮小，失绿黄化，生长缓慢，根部形成大小形状不一的根结，大部分2年生以上的感病土沉香根颈部会形成畸形的瘤块；通过对线虫的rDNA-ITS序列进行PCR扩增测序，Blast比对分析，并构建系统发育树准确鉴定了发生在澄迈县土沉香苗木上的象耳豆根结线虫（*Meloidogyne enterolobii* Yang & Eisenback）。本文较为详细地阐述了土沉香苗木根结线虫的症状，确定了象耳豆根结线虫这一土沉香病原根结线虫，为土沉香根结线虫的防治与进一步研究奠定了基础。

关键词：土沉香；象耳豆根结线虫；rDNA-ITS

大豆胞囊线虫 CLE 多肽激素调控维管束干细胞信号通路介导取食细胞形成的分子机制

郭晓黎

（华中农业大学植物科学技术学院，武汉　430070）

摘　要：植物 CLE（CLV3/EMBRYO SURROUNDIN REGION）多肽激素可以调控分生区干细胞的增殖和分化，在植物生长发育的不同阶段发挥着重要作用。胞囊线虫可以分泌 CLE 多肽效应子通过口针注射到植物细胞，调控线虫取食细胞的形成。以往鉴定的线虫 CLE 效应子与植物 A 型 CLE 多肽的功能相似，拟南芥和大豆 CLE 受体 CLV1、CLV2/CRN 以及 RPK2 在线虫侵染植物中可以识别线虫 CLE 效应子并辅助取食细胞的形成，受体基因沉默可以增加植物对线虫的抗性。最近我们从大豆胞囊线虫中鉴定出一类新的 CLE 多肽，与植物 B 型 CLE 多肽—导管分化抑制因子 TDIF（tracheary element differentiation inhibitory factor）的功能类似，可以抑制维管形成层细胞的分化并促进其增殖。进一步的研究发现 TDIF-TDR-WOX4 介导的维管形成层细胞增殖在线虫寄生过程中起着重要作用。线虫分泌的 A 型和 B 型多肽效应子协同调控细胞分裂进而促进胞囊线虫取食结构的形成。

关键词：大豆胞囊线虫；多肽激素；维管束；取食细胞

辣椒抗南方根结线虫基因同源序列的克隆与分析*

陆秀红[1,2]**，张　雨[3]，李梦桐[3]，黄金玲[1,2]，张　禹[1,2]，刘志明[1,2]***
（1. 广西农业科学院植物保护研究所，南宁　530007；2. 广西作物病虫害生物学重点实验室，南宁　530007；3. 广西大学行建文理学院，南宁　530005）

摘　要：根结线虫（*Meloidogyne* spp.）病是辣椒（*Capsicum annuum* L.）的重要病害之一，选育和推广抗病品种是防治该病最经济、有效、安全的方法，利用基因工程技术获得抗病基因是一条新的抗病育种途径。本试验根据已克隆植物抗病基因的 NBS（Nucleotide binding site）保守区域设计简并引物，对来源于广西的抗南方根结线虫野生辣椒材料基因组进行体外扩增，获得约 500bp 的预期片段；克隆获得 12 条具有完整开放阅读框（ORF）和 NBS-LRR 结构域的序列。聚类分析结果显示，L－6、L－10、L－11、L－16 和 L－18 这 5 条 RGAs（Resistance-gene analogs）与番茄抗胞囊线虫基因 *Hero*、番茄抗丁香假单孢菌基因 *Prf*、番茄抗根结线虫基因 *Mi* 聚为一类，这 3 个基因为 NBS-LRR 基因；同源性分析表明，这 12 条 RGAs 与已知抗病基因相应区域的氨基酸序列的相似性为 21%～78%，其中 L－6 与 Mi 相应区域氨基酸相似性为 78%，表明 L－6 可作为抗根结线虫基因的候选材料。本研究发现的 RGAs 为分子标记辅助育种和功能性抗病相关基因的克隆奠定了基础。

关键词：辣椒；南方根结线虫；抗病基因同源序列

---

* 基金项目：国家自然科学基金（31460465）；广西农科院科技发展基金（2015JZ20，桂农科 2016JZ06）

** 第一作者：陆秀红，广西大新人，副研究员，从事植物线虫学研究；E-mail：lu8348@126.com
*** 通迅作者：刘志明，研究员，从事植物线虫学研究；E-mail：liu0172@126.com

海南黄秋葵象耳豆根结线虫的鉴定*

丁晓帆**，丁佳丽，殷金钰
（海南大学环境与植物保护学院，海口 570228）

摘　要：2015 年 12 月从海南省乐东县采集到黄秋葵［*Abelmoschus esculentus*（L.）Moench］根结线虫病病样，由于黄秋葵产区根结线虫为害严重，因此对该根结线虫群体进行形态特征描述和分子生物学特征分析，旨在为黄秋葵根结线虫种类的准确鉴定及有效防治提供理论依据。运用形态学特征鉴定的方法初步将该线虫群体鉴定为象耳豆根结线虫（*Meloidogyne enterilobii* Yang & Eisenback，1983），其主要形态鉴别特征：雌虫口针基球有纵沟，会阴花纹卵形，具粗糙和平滑的线纹，背弓中等到高，通常圆，侧线不明显，肛阴区通常无线纹；雄虫头冠高，圆或平，头区无环纹，口针粗壮，锥体尖，基杆圆柱形，基球大而圆，与基杆明显分离；二龄幼虫头冠平，头区略与虫体分离，从肛门后尾向后明显变细，尾端常有缢缩，尾端细，钝圆。结合 ITS、IGS 及 mtDNA 序列分析和系统发育树等分子生物学方法，进一步证实了该线虫群体为象耳豆根结线虫，且黄秋葵为首次发现的象耳豆根结线虫新寄主。

关键词：象耳豆根结线虫；黄秋葵；海南省；形态鉴定；分子鉴定

---

* 基金项目：2016 年度海南省自然科学基金项目（20163040）；海南大学教育教学研究课题立项项目（hdjy1605，hdjy1606）
** 第一作者：丁晓帆，广东潮州人，副教授，硕士；从事植物线虫学研究；E-mail：dingxiaofan526@163.com

根瘤菌 Sneb183 诱导大豆抗孢囊线虫根系差异蛋白质组及抗性相关代谢通路分析*

王媛媛[1**]，田　丰[2]，郭春红[2]，朱晓峰[2]，陈立杰[2]，段玉玺[2***]
（1. 沈阳农业大学生物科学技术学院；2. 沈阳农业大学北方线虫研究所，沈阳　110161）

摘　要：大豆孢囊线虫病是为害世界大豆生产最严重的病害之一，并且对大豆生产造成巨大的经济损失。根瘤菌 Sneb183 是能够抑制大豆孢囊线虫病发生的费氏中华根瘤菌（*Sinorhizobium fredii*）。通过裂根法证明了根瘤菌 Sneb183 可以诱导大豆产生对孢囊线虫的系统抗性，抵御大豆孢囊线虫的侵染。利用 iTRAQ 技术鉴定了接种大豆孢囊线虫大豆根系的蛋白，和同时接种根瘤菌和大豆孢囊线虫的大豆根系的蛋白质组，并且比较了两组处理的差异蛋白质组。在 2 个处理中，共鉴定出 456 个差异蛋白质，其中，上调的有 244 个蛋白，下调的有 212 个蛋白。蛋白质组为研究植物—病原物或者植物—微生物的互作提供了很好的观察途径。接种根瘤菌 Sneb183 诱导大豆产生的差异蛋白质充分参与了植物的生物过程、分子功能、细胞组分等。

通过 BLAST2GO 和 KEGG 注释的生物信息学方法，分析接种大豆孢囊线虫和同时接种大豆孢囊线虫和根瘤菌的大豆根系的差异蛋白质组参与的相关代谢通路。注释分析表明，鉴定出的差异蛋白共参与了 118 种代谢途径，其中钙信号途径、类黄酮合成途径、磷酸戊糖途径、谷胱甘肽代谢途径等与植物能量代谢和抗胁迫相关，从而诱导大豆抵御线虫侵染，提高植物抗性。通过分析差异蛋白参与的代谢通路，进一步了解根瘤菌诱导大豆产生系统抗性的机制。

关键词：根瘤菌 Sneb183；诱导；大豆孢囊线虫；蛋白质；代谢通路

---

* 基金项目：农业部公益性行业科研专项（201503114－4）；现代农业产业技术体系建设专项（CARS－04－PS13）
** 第一作者：王媛媛，讲师，从事植物线虫病害防治研究；E-mail：wyuanyuan1225@163.com
*** 通讯作者：段玉玺，教授，博士生导师，主要从事植物线虫学研究；E-mail：duanyx0717@163.com

南方根结线虫颉颃菌株的筛选及对番茄根结虫病的防治*

王　帅**，朱晓峰，王媛媛，段玉玺，陈立杰***
（沈阳农业大学北方线虫研究所/植物保护学院，沈阳　110866）

摘　要：本研究旨在寻找对南方根结线虫有防治效果的微生物资源及相应的防治方法。试验初期以南方根结线虫（*Meloidogyne incongnita*）为靶标，利用贝式皿浸渍法筛选了1 200多份真菌及细菌发酵液，在此基础上，针对南方根结线虫（*M. incongnita*）进行温室盆栽与田间试验。室内离体杀线虫活性试验结果表明，真菌210、细菌246发酵液24h触杀根结线虫校正死亡率分别为74.44%及70.00%，48h真菌210发酵液触杀南方根结线虫（*M. incongnita*）效果达到97.63%。依据形态学观察及ITS序列分析方法鉴定方法，初步鉴定真菌210为曲霉（*Aspergillus* sp.），细菌246为蜡状芽孢杆菌（*Bacillus cereus*）。田间试验设计采用随机区组法对2个活性菌株进行番茄根结线虫病防效试验。试验中菌株发酵液稀释10倍，5.0%阿维菌素稀释1 000倍后灌根处理，设5个处理：真菌210发酵液200mL/棵；阿维菌素200mL/棵；细菌246发酵液200mL/棵；100mL真菌210发酵液+100mL阿维菌素/棵；100mL细菌246发酵液+100mL阿维菌素/棵；另设无菌水对照。每个处理3个小区，每个小区40株。30d、60d后调查番茄长势及根际土壤J_2数量，处理5个月后以根结指数和根际根结线虫J_2增殖抑制率计算防治指标。结果表明真菌210发酵液处理对番茄根结线虫病有较好防效，其抑制率分别为50.7%及42.8%，与5.0%阿维菌素有相似的防治效果；真菌210+阿维菌素处理60d后植株株重增加22.74%，果实重量明显增加。结合田间试验结果，确定真菌210发酵液作为盆栽试验材料，利用区组随机排列对活性较好的微生物菌株210发酵液对番茄南方根结线虫病进行盆栽防效试验，试验中发酵液稀释10倍，1.8%、5.0%阿维菌素稀释1 000倍。试验共5个处理：真菌210发酵液200mL/棵；1.8%阿维菌素200mL/棵；5.0%阿维菌素200mL/棵；真菌210发酵液100mL+1.8%阿维菌素100mL/棵；真菌210发酵液100mL+5.0%阿维菌素100mL/棵；无菌水200mL/棵对照。每个处理10次重复。处理后45d后，以根结指数和根际根结线虫J_2增值抑制率为防治指标。结果显示真菌发酵液对南方根结线虫有很好的防效。

关键词：微生物；发酵液；番茄；南方根结线虫

---

* 基金项目：国家自然科学基金（31471748）

** 第一作者：王帅，研究方向：植物线虫生物防治；E-mail：ahau1060wangshuai@163.com

*** 通讯作者：陈立杰，教授，博士生导师，主要从事病害生物防治和植物线虫学研究；E-mail：chenlijie0210@163.com

巨大芽孢杆菌 Sneb207 诱导大豆抗大豆胞囊线虫生理机理*

周园园**，许俐霞，赵 丹，刘 睿，王媛媛，朱晓峰，
刘晓宇，段玉玺***，陈立杰***
（沈阳农业大学植物保护学院，北方线虫研究所，沈阳 110161）

摘 要：植物经诱导子处理后，激发植物内部的免疫机制，植物会快速的产生防御反应，从而起到防治病害的作用。为明确生防菌巨大芽孢杆菌 Sneb207 诱导大豆产生的抗大豆孢囊线虫的生理机理，本试验将 Sneb207 发酵液包衣感病品种辽豆 15，接种相同数量的大豆孢囊线虫二龄幼虫，并进行盆栽试验，并利用次氯酸钠－酸性品红染色法研究巨大芽孢杆菌 Sneb207 包衣后对大豆根内大豆胞囊线虫 3 号生理小种不同虫龄侵染和发育的影响，并测定接种线虫后大豆的光合性能、叶绿素含量的变化。结果表明，生防菌 Sneb207 处理的根系中，线虫侵染数量降低，且各龄期出现时间均滞后。Sneb207 发酵液包衣接线虫后，大豆净光合速率（Pn）和叶绿素含量均高于未包衣接线虫的处理。生防菌包衣处理后，提高大豆的光合作用，可降低线虫对大豆的破坏，进而影响植物的生长。

关键词：巨大芽孢杆菌；大豆孢囊线虫；光合

---

* 基金项目：农业部公益性行业科研专项（201503114－4）；现代农业产业技术体系建设专项（CARS－04－PS13）
** 第一作者：周园园，博士，主要从事线虫学生物防治研究；E-mail：zhouyuanyuan6616@163.com
*** 通讯作者：段玉玺，教授，博士生导师，主要从事植物线虫学研究；E-mail：duanyuxi6407@163.com
陈立杰，教授，博士生导师，主要从事植物病理学和植物线虫学研究；E-mail：chenlijie0210@163.com

Sneb821 诱导番茄抗根结线虫转录组测序及分析*

赵　丹**，周园园，尤　杨，朱晓峰，王媛媛，段玉玺，陈立杰***
（沈阳农业大学北方线虫研究所，沈阳　110866）

摘　要：根结线虫（*Meloidogyne* spp.）是一种高度专化型的杂食性植物病原线虫，其寄主十分广泛，由于轮作、抗线品种和化学防治等方法依然存在局限性，国际上逐渐采用生物防治的方法来控制病害的严重发生。Sneb821 是由沈阳农业大学北方线虫研究所分离获得并保存，Sneb821 可诱导番茄抗南方根结线虫。为系统研究其诱抗机理，采用新一代高通量测序技术 Illumina HiSeq™2500 对不同处理的番茄样品进行转录组测序和生物信息学分析。共得到 34. 71Gb Clean Data，各样品 Q30 碱基百分比均不小于 93. 57%。接种南方根结线虫 6d 后，Sneb 821 包衣处理与未处理相比，共有 1 997个差异基因，其中上调 847 个，下调 1 150个，以 KEGG 数据库作为参考，依据代谢途径可将 unigenes 定位到 406 个代谢途径分支；Sneb 821 包衣后接种线虫与未处理接种线虫相比，共有 1 932个差异基因，其中上调 962 个，下调 970 个，依据代谢途径可将 unigenes 定位到 332 个代谢途径分支。同时，对差异表达基因进行基因本体注释，蛋白质直系同源簇注释，结果显示，GO 注释和 COG 注释分别将差异表达基因分为了 53 个和 26 个功能类别，共涉及物质及能量代谢、信号传导、转录调控及防卫反应等诸多生理生化过程。其中，参与到激素信号转导途径，过氧化物酶代谢，铁离子运输等过程中的差异基因最多。过滤掉编码的肽链过短（少于 50 个氨基酸残基）或只包含单个外显子的序列，共发掘 553 个新基因。使用 BLAST 软件将发掘的新基因与 NR，Swiss-Pro，GO，COG，KOG，Pfam，KEGG 数据库进行序列比对，获得 377 个新基因的注释信息。这些数据为进一步明确诱抗机制及抗病基因功能验证等研究提供了依据。

关键词：番茄；根结线虫；诱导抗性；转录组测序

---

* 基金项目：国家自然科学基金（31471748）
** 第一作者：赵丹，博士研究生；E-mail：zhaodan_ 1201@163. com
*** 通讯作者：陈立杰，教授，博士生导师；E-mail：chenlijie0210@163. com

Redescription of *Bursaphelenchus Parapinasteri* (Tylenchina: Aphelenchoididae) Isolated from *Pinus thunbergii* in China with a Key to the *Hofmanni*-group*

MARIA Muna-war[1], FANG Yi-wu[1], GU Jian-feng[2], LI Hong-mei[1**]

(1. *Department of Plant Pathology*, *Nanjing Agricultural University*, *Nanjing* 210095, *P. R. China*; 2. *Technical Centre*, *Ningbo Entry-Exit Inspection and Quarantine Bureau*, 9 *Mayuan Road*, *Ningbo* 315012, *Zhejiang*, *P. R. China*)

Abstract: The systematics and identification of *Bursaphelenchus* species are very important due to regulatory and management issues caused by two economically damaging species, *B. cocophilus* and *B. xylophilus*. During routine nematological survey of Changgang Mountain, Zhoushan city, Zhejiang Province, China, *Bursaphelenchus parapinasteri* was detected on dead *Pinus thunbergii*. It is re-described morphologically coupled with new molecular characterisation. Detailed morphology of the spicule, female gonad, hemizonid position, arrangement of male caudal papillae, and female tail terminus shape are documented and illustrated. The ITS-RFLP patterns and the ITS1/2, partial 28S D2-D3 and partial 18S gene sequences were characterised. The phylogenetic analyses revealed that *B. parapinasteri* belongs to the *hofmanni*-group *sensu* Braasch and is close to *B. anamurius*, *B. hofmanni*, *B. mazandaranense*, *B. paracorneolus*, *B. pinasteri* and *B. ulmophilus*. A diagnostic key to species of the *hofmanni*-group is also presented. It is imperative to re-describe and photo-document the species which lack essential morphological and phylogenetical details. This will help to improve its relevance for posterity and to identify species, which are pivotal to our understanding of the deeper levels of phylogenetic relationships, among species of genus *Bursaphelenchus*.

Key words: Molecular; Morphology; Morphometrics; Phylogeny; Taxonomy

* Funding: National Natural Science Foundation of China (Grant No. 31471751) and Ningbo Science and Technology Innovation Team (2015C110018)

** Corresponding author: LI Hong-mei; E-mail: lihm@ njau. edu. cn

Knockdown of Oesophageal Glands Gene by Transgenic Tobacco Plant-mediated RNAi in the Plant Parasitic Nematode *Radopholus similis* *

LI Yu, WANG Ke, LU Qi-sen, Wang Zhen-yue, LI Hong-lian**

(*Department of Plant Pathology*, *Henan Agricultural University*, *Zhengzhou* 450002, *China*)

Abstract: The burrowing nematode *Radopholus similis* is a migratory plant parasitic nematode which severely harms many agricultural and horticultural crops. Cysteine proteinases play key roles in development, invasion, parasitism, pathogenesis and immune evasion in nematodes and many other animal parasites. Nematode cysteine proteinases mainly include cathepsin B-, K-, L-, S-, and Z-like cysteine proteinases, and cathepsin L and cathepsin B have been extensively studied in recent years. However, the functions of cathepsin S gene in *R. similis* (*Rs-cs*-1) and other plant parasitic nematodes have not been reported. Understanding the *Rs-cs*-1 of *R. similis* would allow us to find new targets and approaches for its control. In this study, a 438-bp partial cDNA fragment of *Rs-cs*-1 was chosen as the target for RNAi. The constructed plant RNAi vector pFGC-Rs-cs_2 contained a 438-bp sense and antisense *Rs-cs*-1 cDNA fragment, a CHSA intron and an OCS terminator, and the cDNA fragments as inverted repeats under the control of CaMV35S promoter to produce the hairpin *Rs-cs*-1 dsRNA. The plant RNAi vectors were introduced into *Agrobacterium tumefaciens* (EHA105) via the freeze-thaw method. The independently generated transgenic lines were analyzed by PCR and Southern blot. The transcript abundance of *Rs-cs*-1 dsRNA appeared diversity in different single copy transgenic lines (No. 4 and 13). The *Rs-cs*-1 dsRNA expression level in line No. 4 was 3.07 times significantly ($P < 0.05$) higher than that in the line No. 13 of *Rs-cs*-1 transgenic tobacco plants. Using plant-mediated RNAi, we confirmed that *Rs-cs*-1 expression in nematodes was significantly suppressed in transgenic tobacco plants expressing *Rs-cs*-1 dsRNA. These results provide a scientific basis for further studies of the new methods for controlling plant parasitic nematodes.

---

* Funding: This study was funded by the Special Fund for Agro – scientific Research in the Public Interest (201503112)

** Corresponding author: LI Hong-lian; E-mail: honglianli @ sina. com

河南省小麦根腐线虫不同种群的遗传多样性研究*

逯麒森，李　宇，徐　平，赵　贝，王振跃，李洪连**
（河南农业大学，植物保护学院，郑州　450002）

摘　要：小麦根腐线虫（Wheat root lesion nematode）是一种重要的植物病原线虫，属于短体线虫属（*Pratylenchus* spp.）。小麦被其侵染后地上部的受害症状并不明显，易与营养缺乏及其他根部病害相混淆，因此该线虫的危害性较隐蔽。该线虫寄主范围及分布地区广泛，在亚洲、澳洲、美洲和欧洲等均有发生，严重危害小麦等多种作物的安全生产，引起了世界许多国家和地区的关注与研究。为了明确河南省不同地区小麦根腐线虫的分布情况及其不同种群的遗传多样性，本研究对河南省17个地县56个采样点的小麦根系土壤进行了分离鉴定，通过提取不同种群小麦根腐线虫的单条线虫DNA，利用特异性引物进行了18S rDNA和28S rDNA D2-D3区两种分子标记的PCR扩增，PCR产物纯化回收后进行了克隆测序，测序结果与NCBI中其他小麦根腐线虫的不同种群进行序列比对并构建了系统进化树。主要研究结果如下：

小麦根腐线虫在河南省不同地区普遍发生，大部分地区的采集样品中均分离出了该线虫，呈不均匀分布。利用DNAMAN软件对小麦根腐线虫不同种群18S rDNA的核苷酸序列进行了分析，不同种群之间的同一性为97.1%～100%。根据小麦根腐线虫不同种群和GenBank中其他小麦根腐线虫种群的18s rDNA的核苷酸序列，利用MEGA5软件构建了系统进化树，结果表明供试的小麦根腐线虫所有种群被分为两支，Pra5单独为一支，其他所以种群聚为第二支；在第二支中Pra7和Pra8种群聚在同一亚支，亲缘关系较近。尽管18Sr DNA的序列相对更保守，具有更高的同一性，且不同种群的PCR产物大小均为895bp，但根据该序列构建的进化树与28S rDNA D2-D3区序列构建的进化树的聚类结果相似。分子标记28S rDNA的序列在供试小麦根腐线虫不同种群之间相对保守，具有96.5%～99.7%的同一性。系统进化树结果表明小麦根腐线虫所有种群被分为两支，*P. pseudocoffeae*（KT175531）和*P. pseudocoffeae*（KT175533）聚为一支，其他种群聚为第二支；后者又可分为两个亚支，其中Pra2和Pra7种群聚为第一亚支，其他所有种群聚为第二亚支。从18S rDNA系统进化树中可以看到Pra4和Pra6，Pra3，Pra7和Pra8种群的亲缘关系较近，而在28S rDNA系统进化树中Pra2和Pra7亲缘关系较近。这些结果充分表明供试小麦根腐线虫不同种群之间存在显著的遗传变异和丰富的遗传多样性。

关键词：小麦根府线虫；分离鉴定；PCR扩增；遗传分析

* 基金项目：本研究受到国家公益性行业科研专项（201503112）
** 通讯作者：李洪连，教授，主要从事植物土传病害研究；E-mail：honglianli@sina.com

基于 EST 数据库开发禾谷孢囊线虫的微卫星标记*

牛雯雯**，马居奎，鞠玉亮，王　暄，李红梅***
（南京农业大学，农作物生物灾害综合治理教育部重点实验室，南京　210095）

摘　要：禾谷孢囊线虫（*Heterodera avenae*）于 1989 年首次在我国湖北省天门县发现，迄今已在河南、河北、山东、江苏、安徽等 16 个省（市/自治区）分布危害，给我国小麦生产及粮食安全带来了极大的威胁。近年来针对小麦孢囊线虫的形态学、生物学、抗病品种筛选以及防控技术等研究有较多的报道，然而关于我国禾谷孢囊线虫种群遗传结构和多样性的研究相对滞后。

微卫星，即简单重复序列（Simple sequence repeat，SSR），是一种广泛应用于种群遗传结构和多样性分析的分子标记。本研究基于禾谷孢囊线虫 EST 序列设计了 210 对 SSR 标记引物，通过降落 PCR 和短串联重复（STR）分型检测，共筛选到 13 对多态性较好的 SSR 标记引物。利用禾谷孢囊线虫群体 JSPX 对这 13 对 SSR 引物进行多态性分析，发现 CS-31、CS-60、CS-102、CS-121、CS-137、CS-179、CS-187 和 CS-8384 等 8 对 SSR 标记引物所对应的微卫星位点，无效等位基因频率都很低，低至为 0，多态性信息含量（PIC 值）均大于 0.25，表明它们的多态性较高。此外测试 13 对 SSR 引物的哈德温（Hardy-Weinberg）平衡和连锁遗传，并利用 Sequencial Bonferroni 校正 *P* 值，发现以上 8 对 SSR 引物均未偏离哈德温平衡（$P < 0.05$），也不存在连锁遗传现象（$P < 0.05$）。结果表明，这 8 对 SSR 标记引物具有良好的多态性，可用于分析我国禾谷孢囊线虫不同地理群体的遗传结构和多样性。

关键词：禾谷孢囊线虫；EST；SSR；多态性分析

* 基金项目：国家自然科学基金项目（31471751）和农业部公益性行业科研专项（201503114）
** 第一作者：牛雯雯，山东东营人，硕士研究生，从事植物线虫学研究；E-mail：niuwenwen18@126.com
*** 通讯作者：李红梅，教授，博导，从事植物线虫学研究；E-mail：lihm@njau.edu.cn

松材线虫侵染下马尾松的转录组响应及其生理变化研究*

谢婉凤[1,2]**，黄爱珍[1]，李慧敏[1]，冯丽贞[1]，张飞萍[1]***，郭文硕[1,2]
（1. 福建农林大学林学院，福州　350002；福建农林大学金山学院，福州　350002）

摘　要：松材线虫病是为害马尾松生长的毁灭性病害之一，可导致其枯萎死亡。马尾松的基因表达也会因松材线虫的侵染而发生改变，进而影响生理防御能力。为揭示此变化过程，本研究利用高通量测序技术构建了松材线虫侵染1d，2d，3d的马尾松针叶mRNA表达谱，通过与接种无菌水的对照马尾松样本中的mRNA表达情况进行比较，结果发现，松材线虫侵染抑制了马尾松针叶细胞中植物与病菌互作途径内的钙结合蛋白基因和CC-NBS-LRR基因表达，使得类线虫抗性蛋白（nematode resistance protein-like）基因和抗病基因（disease resistance gene）也下调表达，因而可能阻碍植物体的抗病防御。此外，大部分差异表达基因在苯丙烷代谢、黄酮类物质合成、植物激素信号等途径富集，致使受侵染的马尾松样本中的总酚、总黄酮等次生化合物含量减少、植物激素浓度降低；与此同时，抗氧化酶相关基因也在受侵染的样本中表达下调，抗氧化酶活性也因此而下降。且随着侵染虫量的增加，上述大部分基因的表达丰度则不断下降。由此可见，松材线虫侵染后，马尾松细胞中与信号响应、抗病防御相关的基因表达降低、抗氧化物质含量减少以及抗氧化酶活性下降，致使其防御能力减弱，这也是导致其最终枯萎死亡的重要原因。

关键词：马尾松；松材线虫；mRNA表达谱；抗病基因；次生化合物

---

* 基金项目：福建省财政厅项目（K81139238，K8911010）；福建自然科学基金项目（2013J01074）
** 第一作者：谢婉凤，讲师，主要研究方向：林木病害生理及其分子机制研究
*** 通讯作者：张飞萍，教授，主要从事林木害虫综合治理研究研究；E-mail：fpzhang1@163. com

第六部分
抗病性

Hydrophobin Protein from *Trichoderma harzianum* Induced Maize Resistance to Maize Leaf Spot Pathogen *Curvularia lunata*[*]

YU Chuan-jin[**], LU Zhi-xiang, XIA Hai, GAO Jin-xin, DOU Kai, WANG Meng, LI Ya-qian, CHEN Jie[***]

(*Key Laboratory of Urban Agriculture* (*South*), *Ministry of Agriculture*, *School of Agriculture and Biology*, *Shanghai Jiao Tong University*, *Shanghai* 200240, *P. R. China*; *State Key Laboratory of Microbial Metabolism*, *Shanghai Jiao Tong University*, *Shanghai* 200240, *P. R. China*)

Abstract: Corn *Curvularia* leaf spot disease was a frequently occurring disease to maize leaf, which caused by *Curvularia lunata* leaf spot pathogen, this disease was very common in corn planting area in China. Therefore, the prevention and treatment of the disease was of great significance. Hydrophobins, low molecular mass (about 10 kDa) secreted proteins of fungi, were characterized by moderate to high levels of hydrophobic material (hydrophobic substances) to cover on the surface of the microbial cells and the presence of eight conserved cysteine (Cys) residues and formed 4 disulfide bond. A class II hydrophobin gene *hyd*1 was cloned. A 2kb fragment of the *hyd*1 promoter sequence upstream *hyd*1, *hyd*1 and DsRed were ligated together. The expression cassete P_{hyd1}: DsRed and P_{hyd1}: *hyd*1: DsRed were introduced into the KO 16 (*hyd*1 disruption transformant) and showed that *hyd*1was localized in the plasma membrane by the lazer confocal fluorescense microscope. B73 seeds that treated with the conidia of different *hyd*1 transformants and grew in greenhouse, the 4-leaf stage leaves were inoculated *C. lunata in vitro* and *in vivo*. The results demonstrated that the maize leaf lesion size that treated by KO 16 was larger than WT (wild type), the disease index of the overexpression treatment (OE3 and OE5) maize was lowest. The callose accumulation was increased in the maize leaves infiltrated with in vitro expressed Hyd1 protein comparied with control. It also indicated that *hyd*1 influenced *Trichoderma* colonization in maize roots by the asssy of SEM and qPCR. Meanwhile, Hyd1 protein was detected on the plasma membrane of maize root by the IEM and it was demonstrated that *hyd*1 acted as MAMPs (Microbial-associated molecular patterns) to induce the maize resistance. The yeast bait plasmid pGBKT7-*hyd*1was constructed and screened the maize root cDNA Library and found that *hyd*1 interacted with *ubiquilin*1. We verified the two interaction proteins again by BiFC and pulll down method and also identified *ubiquilin*1 and *hyd*1 interaction motif in yeast cells. At the same time, it was proved that the signal peptide sequence and the cysteine amino acid of Hyd1 were necessary for the interaction between Ubiquilin1-N and Hyd1. The *hyd*1 and *ubiquilin*1 were expressed in *A. thiana* and then promoted the plant resistance synergistically. RNA-seq

---

* Funds: National Natural Science Foundation of China (No. 30971949); China Agriculture Research System Project and Basic Work of Science and Technology Special Project (2014FY120900)

** First author: YU Chuan-jin, PhD, Research: Pant and microbe interaction; E-mail: yuchuanjin1013@163. com

*** Corresponding author: CHEN Jie, Professor. Research interests: plant pathology and biocontrol; E-mail: jiechen59@sjtu. edu. cn analyzed maize leaf global genes expression that regulated by the *hyd*1 at the time of spraying *C. lunata* for 24 h, and the result showed that it probably regulated by BAK1 of BR (brassinosteroids, BR) pathway signaling transduction and it was preliminarily verified by the mutant *BANK*1 –4. Furthermore, it indicated that the genes of potassium channe protein, CaM and WRKY differentiated notably. We also detected the CaM in different maize leaves, the CaM concentration in KO 16 treatment maize leaves was lower than WT, and the OE3 and OE5 was highest of all. The *hyd*1 and *ubiquilin*1 were localized on the plasma membrane and Hyd1 promoted the stability of the Ubiquilin1 in *Nicotiana benthamiana*. Interaction between Hyd1 and Ubiquilin1, perhaps the Ubiquilin1 might interact with plant receptor BAK1 and together with Hyd1 to form a complex, thereby stimulating the relevant resistance of the plant. This study was to lay a foundation about molecular mechanism of *hyd*1 acted as MAMPs interacted with plant.

Key words: *Trichoderma harzianum*; Hydrophobin; *Ubiquilin*1; MAMPs; Protein interaction; Induce systemic resistance

43 份甘蔗创新种质材料抗甘蔗花叶病鉴定评价*

李文凤**，王晓燕，黄应昆***，单红丽，张荣跃，罗志明，尹　炯
（云南省农业科学院甘蔗研究所，云南省甘蔗遗传改良重点实验室，开远　661699）

摘　要：甘蔗花叶病是一种重要的世界性甘蔗病害，近年来已成为中国蔗区发生最普遍。危害最严重的病害之一。前期研究表明引起中国蔗区甘蔗花叶病病原有甘蔗花叶病毒（Sugarcane mosaic virus，SCMV）、高粱花叶病毒（Sorghum mosaic virus，SrMV）和甘蔗线条花叶病毒（Sugarcane streak mosaic virus，SCSMV）3 种，其中，SrMV 和 SCSMV 是最主要致病病原。利用抗病品种是防治该病最经济有效措施，然而目前能同时兼抗几种病毒病原的品种和资源较为缺乏，发掘新抗病种质对选育抗病品种具有重要意义。本研究以甘蔗热带种‘路打士’与滇蔗茅‘云南 95－19’进行远缘杂交获得的 41 份创新种质为材料，采用苗期人工接种和后期 RT-PCR 检测法，对包括亲本在内的 43 份创新种质材料进行了 SrMV、SCSMV 双重抗性鉴定评价。结果表明，43 份创新种质材料中，对 SCSMV 表现 1 级高抗到 3 级中抗的有 22 份，4 级感病到 5 级高感的有 21 份；对 SrMV 表现 1 级高抗到 3 级中抗的有 31 份，4 级感病到 5 级高感的有 12 份。综合分析结果显示，云 09－603、云 09－604、云 09－607、云 09－608、云 09－619、云 09－622、云 09－633、云 09－635、云 09－656、云滇 95－19 10 份材料对 SCSMV 和 SrMV 均表现 1－2 级抗病，占 23.26%，其中，云 09－604、云 09－607、云 09－619、云 09－633、云 09－656、云滇 95－19 6 份材料对两种病毒均表现为 1 级高抗。研究结果明确了各创新种质材料对甘蔗花叶病两种主要致病病原的抗性，筛选出双抗 SCSMV 和 SrMV 的创新种质 10 份，为进一步利用抗病种质开展抗花叶病育种提供了科学依据和抗源材料。

关键词：甘蔗；创新种质；花叶病；抗性鉴定

---

* 基金项目：现代农业产业技术体系建设专项（CARS－20－2－2）；云南省现代农业产业技术体系建设专项

** 第一作者：李文凤，云南石屏人，研究员，主要从事甘蔗病害研究；E-mail：ynlwf@163.com

*** 通讯作者：黄应昆，研究员，从事甘蔗病害防控研究；E-mail：huangyk64@163.com

Pathogenicity Analysis of *Magnaporthe oryzae* Populations of Laos on Monogenic Lines of Rice*

LIU Shu-fang, DONG Li-ying, LI Xun-dong, YANG Qin-zhong**

(*Institute of Agricultural Environment & Resources*, *Yunnan Academy of Agricultural Sciences*, *Kunming* 650205, *China*)

Abstract: Rice blast, caused by *Magnaporthe oryzae* is one of the most destructive disease of rice worldwide, and the use of resistant cultivars is the most economic, effective and environment-friendly measure. Understanding the pathogenicity of *M. oryzae* population to known resistant genes will facilitate the effectively selection of resistant genes in rice breeding programs. In present study, total of 153 *M. oryzae* strains collected from Borikhamxay, Vientiane, Saravane, Savanaket provinces of Laos were pathotyped on 24 monogenic lines (MLs) with the genetic background of Lijiangxingtuanheigu (*Oryza sativa* subsp. *japonica*), which have been developed by IRRI. The results showed that no strains were found to be virulent on MLs carrying *Pi9* and *Pik-h*, respectively, indicating that these two genes could be very important resources for rice breeding to improve modern cultivars for resistance to rice blast. Five genes, i. e. *Pi1*, *Pik-m*, *Pik*, *Pi7* and *Pik-p* with lower susceptible frequency of less than 6% to the *M. oryzae* strains used in this study, could also be useful resources for disease-resistant breeding; The susceptible frequency of *Piz*, *Pita*, *Pi2* and *Pi5* genes to 153 blast strains are 9.8%, 10.46 and 13.1%, respectively, indicating that these genes could be rationally employed in rice breeding. The other genes, such as *Pish*, *Pi12*, *Pita* -2 and *Pi20* etc., could not used as resistant donors alone in rice breeding programs for blast, but could be pyramided with other genes showing the overlapping of resistant spectrum with them into cultivars to broaden their resistant spectrum to blast. In addition, pathotyping of the blast strains on MLs would provide important information for us to select strains virulent to a certain resistant gene to analyze the mechanism of loss function of Avirulent gene in *M. oryzae* corresponding to the resistant gene in rice.

Key words: *Magnaporthe oryzae*; Pathogenicity analysis; Monogenic lines; Rice

---

* Acknowledgements: This research was funded by Yunnan Provincial Science and Technology Department, International Cooperation Project (2014IA009). We thank Dr. Bounneuang Douangboupha, and Mr. Bounpheng Sihomchanh for their help in field survey and collecting samples

** Corresponding author: YANG Qin-zhong, qzhyang@163.com

Identification of a New Gene for Resistance to *Magnaporthe oryzae* in *Oryza glaberrima* *

DONG Li-ying[1], XU Peng[2], LIU Shu-fang[1], LI Xun-dong[1], ZHOU Jia-wu[2], LI Jing[2], DENG Wei[2], TAO Da-yun[2], YANG Qin-zhong[1] **

(1. *Institute of Agricultural Environment & Resources*, *Yunnan Academy of Agricultural Sciences*; 2. *Institute of Food Crops*, *Yunnan Academy of Agricultural Sciences. Kunming* 650205, *P. R. China*)

Abstract: The African cultivated rice (*Oryza glaberrima*), which is one of the two cultivated species of genus *Oryza* and was independently domesticated from the wild progenitor *O. barthii* about 2000 to 3500 years ago, has been cultivated in West Africa. Although their grain yield and enhanced palatability etc. are lower than the Asian cultivated rice, another cultivated species *O. sativa* of genus *Oryza*, *O. glaberrima* possesses many important traits such as weed competitiveness, drought tolerance, resistant to insects and diseases, which could be very useful for improvement of Asian cultivated rice. In previous study, twenty-four accessions of *O. glaberrima* showing highly resistance to rice blast in blast nursery of Yunan, were crossed and further backcrossed with *O. sativa* subsp. *japonica* cultivar Dianjingyou 1 (DJY1) used as recurrent parent to generate advanced backcrossed BC_5F_4 populations, respectively. In order to identify new blast resistance genes in *O. glaberrima*, the introgression lines (ILs) carrying genomic fragments of *O. glaberrima* was inoculated with blast strains to screen resistant ILs. One of the resistant ILs, IL-Og106 was backcrossed with DJY1 to produce BC_6F_1 seeds. The F_2 population consisting 131 BC_6F_2 individuals were inoculated with blast strin 09BS－10－5a. Segregation of resistant (R) and susceptible (S) progenies of 131 BC_6F_2 individuals fitted a 3 : 1 ratio (101 R : 30 S, χ^2 = 0.3078, $P>0.50$). The results suggested that a single dominant gene in IL-Og106 from *O. glaberrima* confers resistance to 09BS－10－5a. In order to map this blast resistant locus onto the chromosome of rice, bulked segregant analysis method was used to analyze the resistant donor, susceptible parent, resistant pool and susceptible pool with SSR markers. As a result, SSR marker RM30 locating on the long arm of chromosome 6 of rice was found to be linked with this gene, at a genetic distance of 5.3 cM. This gene was tentatively designated as *Pi66* (t).

Key words: *Magnaporthe oryzae*; Resistance gene; Identification; *Oryza glaberrima*

* Acknowledgements: This research was funded by Yunnan Provincial Science and Technology Department (2014IA009), and by National Natural Science Foundation of China (31360462)

** Corresponding author: YANG Qin-zhong; E-mail: qzhyang@163.com

Influence of *Glomerella cingulata* Infection on the Antioxidant System of Susceptible and Resistant Apple Cultivars*

ZHANG Ying**, LI Bao-hua, LI Gui-fang, DONG Xiang-li, WANG Cai-xia***

(*College of Agronomy and Plant Protection, Key Lab of Integrated Crop Pest Management of Shandong Province, Qingdao Agricultural University, Qingdao* 266109)

Abstract: Glomerella leaf spot (GLS) caused by *Glomerella cingulata* is a newly emergent disease that results in severe defoliation and fruit spots. Currently, GLS is not effectively controlled except for the traditional spraying fungicides. Through studying the influence of the pathogen infection on the antioxidant system of susceptible and resistant apple cultivars, we expect to provide a better understanding of apple resistance mechanisms against *G. cingulata*, and for the future breeding of apple genotypes resistant to this pathogen. Biochemical analysis revealed that different patterns of Reactive oxygen species (ROS) and antioxidant mechanism exist between the inoculated susceptible and resistant apple cultivars. The ROS burst occurred in the inoculated plants of resistant 'Fuji' but not in those of the susceptible 'Gala' apple leaves. In the inoculated 'Fuji' apple leaves, CAT, POD and SOD activities increased significantly, however, in the inoculated 'Gala' apple leaves, the three enzymatic activities were increased less compared with the control leaves. Correspondingly, the total antioxidant capacity in the inoculated 'Fuji' apple leaves markedly increased and maintained at a higher level than that in non-inoculated leaves. However, the accumulation of total phenols was not observed in the inoculated 'Fuji' apple leaves, while it was a little increase in the inoculated 'Gala' apple leaves at 5 dpi and 7 dpi. The PAL and PPO activities were markedly increased in the inoculated 'Fuji' apple leaves, while they were slightly elevated in inoculated 'Gala' apple leaves compared with the control leaves. Level of chitinase was higher in the inoculated 'Fuji' apple leaves than the 'Gala', while the β-1, 3-glucanase expression was inverse in the inoculated 'Fuji' and 'Gala' apple leaves. This study has improved our understanding of the differential regulatory mechanisms of susceptible and resistant apple cultivars against *G. cingulata* infections. Our study illustrates that the resistance of 'Fuji' apple cultivar to GLS should be closely related to the antioxidant system

Key words: Antioxidant system; Disease resistance; Glomerella leaf spot; ROS

---

* Funding: Chinese Modern Agricultural Industry Technology System (No. CARS-28); National Natural Science Foundation of China (No. 31272001 and No. 31371883); Tai-shan Scholar Construction Foundation of Shandong province and Graduate Student Innovation Program of Qingdao Agricultural University (QYC201509)

** First author: ZHANG Ying, master, Research on fruit tree pathology; E-mail: zying1017@126.com

*** Corresponding author: WANG Cai-xia, professor, Research on fruit tree diseases; E-mail: cxwang@qau.edu.cn

A Conserved *Puccinia striiformis* effector Interacts with Wheat NPR1 and Reduces Induction of *Pathogenesis-related* genes in Response to Pathogens

WANG Xiao-dong[1,2,3*], YANG Bao-ju[1*], LI Kun[1], KANG Zhen-sheng[2], Dario Cantu[4], Jorge Dubcovsky[1,5**]

(1. *Department of Plant Science*, *University of California*, *Davis*, 95616, *USA*; 2. *State Key Laboratory of Crop Stress Biology for Arid Areas and College of Plant Protection*, *Northwest Agriculture and Forestry University*, *Yangling* 712100, *P. R. China*; 3. *Department of Plant Pathology*, *Agriculture University of Hebei*, *Baoding* 071000, *P. R. China*; 4. *Department of Viticulture and Enology*, *University of California*, *Davis* 95616; *USA*. 5. *Howard Hughes Medical Institute* (*HHMI*), *Chevy Chase*, *MD* 20815, *USA*)

Abstract: In Arabidopsis, NPR1 is a key transcriptional co-regulator of systemic acquired resistance. Upon pathogen challenge, NPR1 translocates from the cytoplasm to the nucleus, where it interacts with TGA-bZIP transcription factors to activate the expression of several pathogenesis-related genes. In a screen of a yeast two-hybrid library from wheat leaves infected with *Puccinia striiformis* f. sp. *tritici*, we identified a conserved rust effector that interacts with wheat NPR1 and named it *Puccinia* NPR1 interactor (PNPi). PNPi interacts with the NPR1/NIM1-like domain of NPR1 via its C-terminal domain. Using bimolecular fluorescence complementation assays, we detected the interaction between PNPi and wheat NPR1 in the nucleus of *Nicotiana benthamiana* protoplasts. A yeast three-hybrid assay showed that PNPi interaction with NPR1 competes with the interaction between wheat NPR1 and TGA2. 2. In barley transgenic lines over expressing *PNPi*, we observed reduced induction of multiple *PR* genes in the region adjacent to *Pseudomonas syringae* pv. *tomato* DC3000 infection. Based on these results, we hypothesize that PNPi has a role in manipulating wheat defense response via its interactions with NPR1.

Key words: Wheat; Stripe rust; NPR1; Pathogen effector; Yeast two-hybrid; Bimolecular fluorescence complementation; Transgenic barley

* First author: These authors contributed equally to this work

** Corresponding author: Jorge Dubcovsky; E-mail: jdubcovsky@ ucdavis. edu

Microarray Analysis on Differentially Expressed Genes Associated with Wheat *Lr*39/41 Resistance to *Puccinia Triticina*

WANG Xiao-dong [1], BI Weis-huai[1], GUO Yu-xin[1], LI Xing[1*], LIU Da-qun[1]

(*Department of Plant Pathology*, *Agriculture University of Hebei*, *Baoding* 071000, *P. R. China*)

Abstract: Wheat leaf rust, caused by *Puccinia triticina* (*Pt*), is one of the most severe fungal diseases threatening global wheat production. Utilization of leaf rust resistance (*Lr*) genes is the major solution for this issue. Wheat isogenic lines carrying *Lr*39/41 gene still showed a moderate or higher resistance to most of the leaf rust races detected in China. In our study, a typical hypersensitive response (HR) has been observed by microcopy in *Lr*39/41 isogenic lines inoculated with avirulent *Pt* race from 24 hours post inoculation, indicating a possible NBS-LRR and program cell death-related pathway for the mechanism of *Lr*39/41 gene. Two *Lr*39/41 resistance-associated suppression subtractive hybridization (SSH) libraries have been established. One with 3 456 clones constructed between *Pt*-inoculated and non-inoculated *Lr*39/41 isogenic lines. Another with 2 544 clones constructed between *Pt*-inoculated *Lr*39/41 isogenic lines and Thatcher susceptible lines. Microarray hybridization has been carried out on the established SSH libraries with RNAs extracted from *Pt*-inoculated and non-inoculated *Lr*39/41 isogenic lines and Thatcher susceptible lines, respectively. Differentially expressed genes (DEGs) were analyzed by significance analysis of microarrays (SAM) and a total of 386 clones with significant induction or suppression were further sequenced. Sequences were then annotated with GenBanknr database, URGI wheat genome database and gene ontology (GO) database. In total of 36 *Lr*39/41 resistance-related DEGs were found, many of which were previously reported in plant defense response. Four *Lr*39/41 resistance-related DEGs were also detected as Thatcher susceptible-related DEGs, indicating a potential involvement of basal resistance for these genes. Relative expression levels of 12 selected *Lr*39/41 resistance-associated DEGs during *Pt* infection were further validated by qRT-PCR assay. A brief profile of DEGs associated with *Lr*39/41 resistance had been drafted.

Key words: Wheat leaf rust resistance; *Lr*39/41; Suppression subtractive hybridization; Microarray; qRT-PCR

* Corresponding authors: LI Xing; E-mail: lxkzh@163.com

LIU Da-qun; E-mail: ldq@hebau.edu.cn

TaSYP71, a Qc-SNARE, Contributes to Wheat Resistance Against *Puccinia striiformis* f. sp. *tritici*

LIU Min-jie[1], PENG Yan[1], LI Hua-yi[1], DENG Lin[1],
WANG Xiao-jie[1*], KANG Zhen-sheng[1*]

(*State Key Laboratory of Crop Stress Biology for Arid Areas and College of Plant Protection*, *Northwest A&F University*, *Yangling* 712100, *China*)

Abstract: *N*-ethylmaleimide-sensitive factor attachment protein receptors (SNAREs) are involved in plant resistance; however, the role of SYP71 in the regulation of plant-pathogen interactions is not well known. In this study, we characterized a plant-specific SNARE in wheat, TaSYP71, which contains a Qc-SNARE domain. Three homologues are localized on chromosome 1AL, 1BL and 1DL. Using *Agrobacterium*-mediated transient expression, TaSYP71 was localized to the plasma membrane in *Nicotiana benthamiana*. Quantitative real-time PCR assays revealed that *TaSYP71* homologues was induced by NaCl, H_2O_2 stress and infection by virulent and avirulent *Puccinia striiformis* f. sp. *tritici* (*Pst*) isolates. Heterologous expression of TaSYP71 in *Schizosaccharomyces pombe* elevated tolerance to H_2O_2. Meanwhile, H_2O_2 scavenging gene (*TaCAT*) was down-regulated in *TaSYP71* silenced plants treated by H_2O_2 compared to that in control, which indicated that *TaSYP71* enhanced tolerance to H_2O_2 stress possibly by influencing the expression of *TaCAT* to remove the excessive H_2O_2 accumulation. When *TaSYP71* homologues were all silenced in wheat by the virus-induced gene silencing system, wheat plants were more susceptible to *Pst*, with larger infection area and more haustoria number, but the necrotic area of wheat mesophyll cells were larger, one possible explanation that minor contribution of resistance to *Pst* was insufficient to hinder pathogen extension when *TaSYP71* were silenced, and the necrotic area was enlarged accompanied with the pathogen growth. Of course, later cell death could not be excluded. In addition, the expression of pathogenesis-related genes were down-regulated in *TaSYP71* silenced wheat plants. These results together suggest that TaSYP71 play a positive role in wheat defence against *Pst*.

Key words: *Puccinia striiformis* f. sp. *tritici*; Wheat; SNARE; Virus-induced gene silencing; Resistance

* Corresponding authors: KANG Zhen-sheng; E-mail: kangzs@ nwsuaf. edu. cn
WANG Xiao-jie; E-mail: wangxiaojie@ nwsuaf. edu. cn

TaTypA, a Ribosome-Binding GTPase Protein, Positively Regulates Wheat Resistance To the Stripe Rust Fungus

LIU Peng, Thwin Myo, MA Wei, LAN Ding-yun, QI Tuo, GUO Jia, SONG Ping, GUO Jun, KANG Zhen-sheng

(*State Key Laboratory of Crop Stress Biology for Arid Areas*, *College of Plant Protection*, *Northwest A&F University*, *Yangling* 712100, *China*)

Abstract: Tyrosine phosphorylation protein A (TypA/BipA) belongs to the ribosome-binding GT-Pase superfamily. In many bacterial species, TypA acts as a global stress and virulence regulator and also mediates resistance to the antimicrobial peptide-bactericidal permeability-increasing protein (BPI). However, the function of *TypA* in plants under biotic stresses is not known. In this study, we isolated and functionally characterized a stress-responsive *TypA* gene (*TaTypA*) from wheat, with three copies located on chromosomes 6A, 6B and 6D, respectively. Transient expression assays indicated chloroplast localization of TaTypA. The transcript levels of *TaTypA* were up-regulated in response to treatment with methyl viologen, which induces ROS in chloroplasts through photoreaction, cold stress and infection by an avirulent strain of the stripe rust pathogen. Knock down of the expression of *TaTypA* through virus-induced gene silencing (VIGS) decreased the resistance of wheat to stripe rust accompanied by weakened reactive oxygen species (ROS) accumulation and hypersensitive response (HR), an increase in *TaCAT* and *TaSOD* expression, and an increase in pathogen hyphal growth and branching. Our findings suggest that *TaTypA* contributes to resistance in an ROS-dependent manner.

Key words: TypA; *Puccinia striiformis* f. sp. *tritici*; ROS; Virus-induced gene silencing; *Triticum aestivum*; Tyrosine phosphorylation

Functional Characterization of a CBL-interacting Protein Kinase Gene *TaCIPK*16 in Wheat

QI Tuo, LAN Ding-yun, MA Wei, KANG Zhensheng, GUO Jun
(*State Key Lab of Crop Stress Biology for Arid Areas and College of Plant Protection*, *Northwest A&F University*, *Yangling* 712100, *China*)

Abstract: Calcium is one of the second messengers in plants, which is used for adaptive responses against these stimuli via perceiving the signal from membrane receptors and switching on intracellular signaling cascade. Calcineurin B-like proteins (CBLs) are Ca^{2+} sensor and regulate the activity of CBL-intracting protein kinase (CIPK) in the plant cells. There are many increasing evidence implicating the CBL-CIPK network plays an important role in various environmental stresses but less work reported its roles on biotic stresses. In this study, we functionally characterized a CBL-interacting protein kinase gene (*TaCIPK*16) in wheat in response to *P. striiformis* f. sp. *tritici* (*Pst*) infection. The transcript levels of *TaCIPK*16 were up-regulated in response to avirulent race of *Pst*, whereas transcript levels were almost steady to virulent race of *Pst*. The expression of *TaCIPK*16 was increased after treatment with abscisic acid (ABA), salicylic acid (SA), cold and drought. TaCIPK16-GFP fusion protein is localized throughout the cell, especially in the nucleus and plasma membrane. Yeast two-hybrid assay showed that *TaCIPK*16 strongly interacts with *TaCBL4* and *TaCBL9*. Knock down of the expression of *TaCIPK*16 through virus-induced gene silencing (VIGS) decreased the resistance of wheat to stripe rust accompanied by weakened reactive oxygen species (ROS) accumulation and hypersensitive response (HR), an decrease in *TaPR*1, *TaPR*2 and *TaPR*5 expression and an increase in pathogen hyphal growth and branching. Our findings suggest that *TaCIPK*16 contributes wheat immunity against *Pst*.

Key words: Wheat; Calcineurin B-like proteins (CBLs); CBL-interacting protein kinases (CIPKs); *Puccinia striiformis* f. sp. *tritici*; Virus-induced gene silencing (VIGS)

小麦抗源 Yaco“S”成株期抗条锈病主效 QTL 定位*

李海洋**，吴建辉[2]，穆京妹 1，王琪琳[2]，曾庆东[2]，
黄丽丽[2]，康振生[2]***，韩德俊[1]***
（1. 西北农林科技大学农学院/旱区作物逆境生物学国家重点实验室，杨凌 712100；
2. 西北农林科技大学植物保护学院/旱区作物逆境生物学国家重点实验室，杨凌 712100）

摘　要：小麦条锈病是由小麦条锈菌（*Puccinia striiformis* f. sp. *tritici*）引起的一种气流传播的真菌病害，是全球性的小麦主要病害之一。成株期抗病性具有广谱性和持久性特点，因此发掘新抗源，鉴定持久抗病新基因对我国小麦抗病育种具有重要实践意义。小麦抗源 Yaco“S”，从 2008 年到 2016 年在天水和杨凌都表现出了良好的成株期抗病性。为了定位 Yaco“S”的抗病基因/QTL，以感病品种铭贤 169 为母本，以 Yaco“S”为父本杂交获得 184 个 $F_{2:3}$家系，藉此建立抗感池，分别利用 90K 和 660K 小麦 SNP 芯片，对双亲和抗感池进行扫描，初步确定主效 QTL 在 2B 染色体，以该染色体多态性 SSR 标记构建遗传连锁图谱，并根据相关 SNP 信息开发 KASP 标记加密图谱，并以此对 184 个 $F_{2:3}$群体进行主效 QTL 作图。将 Yaco“S”成株期抗条锈病主效 QTL 定位于小麦 2BS 染色体，两侧最近标记分别为 Xbarc55 和 Xgwm148，并命名为 Qyryac. nwafu-2B，该 QTL 的 LOD 值 5.3~26.6，可解释表型变异 13.5%~33.5%。通过与定位在 2B 染色体上的已有抗条锈基因和 QTL 进行比较，认为该抗病主效 QTL 可能是一个新的 QTL。本研究利用 SNP 标记和 SSR 标记相结合的方法可在 $F_{2:3}$代快速定位基因，Yaco“S”的抗病 QTL 和与它们的连锁的分子标记可以在抗条锈育种起到重要的作用。

关键词：小麦条锈病；QTL 作图；持久抗病性；SNP，KASP 标记

---

* 基金项目：国家基础研究“973”计划（2013CB127700）；国家十二五支撑计划（2012BAD19B04）；国家自然科学基金（31371924）；高等学校学科创新引智计划资助项目（B07049）

** 第一作者：李海洋，硕士研究生，从事小麦抗条锈病遗传育种研究；E-mail：919006151qq. com. com
*** 通讯作者：康振生，教授，主要从事小麦与条锈菌互作研究；E-mail：kangzs@ nwsuaf. edu. cn
韩德俊，教授，主要从事小麦抗条锈病遗传育种研究；E-mail：handj@ nwsuaf. edu. cn

Screening and Identifying Novel Resistance Sources to Stripe Rust of Wheat*

MU Jing-mei[2**], WANG Qi-lin[1,3], WU Jian-hui[1,3], HUANG Li-li[1,3], KANG Zhen-sheng[1,3***], HAN De-jun[1,2***]

(1. *State Key Laboratory of Crop Stress Biology in Arid Areas*, *Northwest A&F University*, *Yangling* 712100, *China*; 2. *College of Agronomy*, *Northwest A&F University*, *Yangling* 712100, *China*; 3. *College of Plant Protection*, *Northwest A&F University*, *Yangling* 712100, *China*)

Abstract: Stripe rust, caused by *Puccinia striiformis* Westend. f. sp. *tritici* Erikss, is one of the most damaging diseases in the world. To identify new sources of resistance and the resistance genes to stripe rust, 120 wheat entries were assessed for stripe rust resistance at adult plant stage in Yangling, Shaanxi province artificially inoculated and in Tianshui, Gansu province naturally infected from 2010 to 2015 in field test. Nine races including current prevalent stripe races (CYR32, CYR33, V26), potential prevalent races (Sull-4, Sull-5, Sull-7) and past prevalent stripe races (CYR23, CYR29, CYR31) were inoculated separately at seedling stage in the greenhouse. Presence or absence of known resistance genes were postulated with molecular marker linked with *Yr*5, *Yr*9, *Yr*10, *Yr*15, *Yr*17, *Yr*18, *Yr*26, combining with resistance spectra. The result showed 58 wheat entries had adult-plant resistance to stripe rust almost all tested years. Among these, 2 entries including Xinan314-1 and Mulan were resistant to all tested races in seedlings. Molecular test implied that Xinan314-1 and Mulan did not include any tested genes, however, due to their wide-spectra resistance, it was speculated that the 2 entries might have novel unknown all-stage resistance genes. Two entries including LinY867 and Bouquet had seedling susceptibility to all tested races, molecular test showed that LinY867 and Bouquet do not include any tested genes. It was speculated that they might have unknown adult-plant resistance genes. Fifty-four entries were resistant to at least one tested race, indicating they might have race-specific resistance. Among these, Bainong160 included *Yr*17, Zhoumai17, Shaanmai536 included *Yr*9, Annong1401 included *Yr*26. Datang991, Qian15, Yeereryan-1-3 include *Yr*9 + *Yr*17, Annong1236, Xinong418 and Jiumai2 included *Yr*9 + *Yr*26, the remaining entries did not include any tested genes, these 47 entries might have unknown genes or several QTLs. It was concluded that adult-plant resistance appears to be the most effective type of resistance protecting wheat from stripe rust currently.

Key words: Stripe rust resistance; Resistance spetra; Molecular marker test

---

* 基金项目：国家十二五支撑计划（2012BAD19B04）；国家自然科学基金（31371924）；国家小麦产业技术体系（CARS-3-1-11）高等学校学科创新引智计划资助项目（B07049）

** 第一作者：MU Jing-mei；E-mail：mu. jing. mei@163. com

*** 通讯作者：康振生；E-mail：kangzs@nwsuaf. edu. cn

韩德俊；E-mail：handj@nwsuaf. edu. cn

52 个甘蔗新品种（系）抗甘蔗褐锈病评价*

李文凤**，张荣跃，黄应昆***，单红丽，王晓燕，尹 炯，罗志明，仓晓燕
（云南省农业科学院甘蔗研究所，云南省甘蔗遗传改良重点实验室，开远 661699）

摘 要：由黑顶柄锈菌（*Puccina. melanocephala* H. Sydow & P. Sydow）引起的甘蔗褐锈病是世界性的甘蔗重要病害之一，目前在我国甘蔗主产区特别是湿热蔗区常年发生流行，造成巨大的经济损失，严重影响着蔗糖产业的可持续发展。利用抗病品种是控制该病害最经济有效的途径。为明确近年来国家甘蔗体系育成的新品种（系）对甘蔗褐锈病的抗性，确定其应用潜力，为合理布局和推广使用这些新品种（系）提供依据。本研究采用已报道的抗褐锈病基因 *Bru*1 稳定的分子标记对52 个甘蔗新品种（系）进行抗褐锈病基因 *Bru*1 的分子检测，同时在云南甘蔗褐锈病高发蔗区德宏、保山 2 个国家甘蔗体系综合试验站连续 3 年对各测试品种进行田间自然抗病性调查分析。结果表明，52 个甘蔗新品种（系）中，29 个检测出含抗褐锈病基因 *Bru*1，23 个未检测到 *Bru*1；检测到 *Bru*1 的 29 个新品种（系）中，27 个连续 3 年田间调查均未发现有甘蔗褐锈病发生，而粤糖 42 号和桂糖 31 号 2 个品种则 3 年中有 1 年中度感染甘蔗褐锈病；23 个未检测出抗褐锈病基因 *Bru*1 的新品种（系）中，16 个连续 3 年田间调查结果甘蔗褐锈病中度至重度发生，而粤甘 34 号、福农 15 号、闽糖 01 - 77、桂糖 32 号、柳城 05 - 136、云蔗 06 - 80、甘蔗 02 - 707 等 7 个品种则未见发生褐锈病，预示这 7 个品种可能含有新抗褐锈病基因源。本研究结果显示，34 个新品种（系）具有抗褐锈病的应用潜力。

关键词：甘蔗；新品种（系）；褐锈病；抗病性

* 基金项目：现代农业产业技术体系建设专项资金资助（CARS - 20 - 2 - 2）；云南省现代农业产业技术体系建设专项资金资助；云南省“人才培养”项目（2008PY087）

** 第一作者：李文凤，云南石屏人，研究员，主要从事甘蔗病害研究；E-mail：ynlwf@ 163. com

*** 通讯作者：黄应昆，研究员，从事甘蔗病害防控研究；E-mail：huangyk64@ 163. com

甘蔗核心种质花叶病发生调查及自然抗病性分析*

李文凤**，单红丽，黄应昆***，王晓燕，张荣跃，尹　炯，罗志明
（云南省农业科学院甘蔗研究所，云南省甘蔗遗传改良重点实验室，开远　661699）

摘　要：甘蔗花叶病是一种重要的世界性甘蔗病害，近年来已成为中国蔗区发生最普遍，危害最严重的病害之一，利用抗病品种是防治该病最经济有效的措施，鉴定和筛选抗病种质资源对选育抗病品种具有重要意义。为明确甘蔗核心种质花叶病田间自然发生情况，分析评价其对花叶病的自然抗病性，确定其应用潜力，为发掘抗花叶病基因资源提供依据。本研究以国家甘蔗种质资源圃种植保存宿根 6 年的 107 份甘蔗核心种质为材料，用 5 级分级调查法（1 级发病率 0.00%，免疫；2 级发病率 0.01% ~10.00%，高抗；3 级发病率 10.01% ~33.00%，中抗；4 级发病率 33.01% ~66.00%，感病；5 级发病率 66.01% ~100%，高感），对各种质材料甘蔗花叶病的发生情况进行系统调查，并利用 RT-PCR 检测及 CP 基因克隆测序方法对其中的 86 份带症种质样品病原进行系统鉴定。结果表明，107 份核心种质中，22 份未发现感染甘蔗花叶病，占 20.56%；其余 85 份样品不同程度感染甘蔗花叶病，占 77.43%，其中，表现 4 ~5 级严重感染的有 33 份，占 30.84%；2 ~3 级发病的有 52 份，占 48.60%。85 份带症样品中检测到 2 种病毒：高粱花叶病毒（Sorghum mosaic virus，SrMV）和甘蔗线条花叶病毒（Sugarcane streak mosaic virus，SCSMV），其中，SrMV 阳性检出率为 44.71%，SCSMV 阳性检出率 70.59%，23 个样品（27.05%）检出含 2 种病毒。从 SCSMV 和 SrMV 检测阳性样品中选择代表性 RT-PCR 产物进行克隆测序，测序结果与 GenBank 上已报道的 SCSMV 和 SrMV CP 基因核苷酸序列同源性分别达 99% 和 94% 以上。研究结果明确了各核心种质对甘蔗花叶病的自然抗性，初步筛选出 22 份对甘蔗花叶病自然免疫的核心种质；揭示了引起国家甘蔗种质资源圃核心种质甘蔗花叶病的致病病原有 2 种：SrMV 和 SCSMV，并存在 2 种病毒复合侵染现象，其中 SCSMV 为最主要致病病原。

关键词：甘蔗；核心种质；花叶病；病原检测；自然抗病性

* 基金项目：现代农业产业技术体系建设专项资金资助（CARS－20－2－2）；云南省现代农业产业技术体系建设专项资金资助

** 第一作者：李文凤，云南石屏人，研究员，主要从事甘蔗病害研究；E-mail：ynlwf@163.com

*** 通讯作者：黄应昆，研究员，从事甘蔗病害防控研究；E-mail：huangyk64@163.com

iTRAQ Quantitative Proteomics Analysis of the Defense Response of Wheat Against *Puccinia striiformis* f. sp. *tritici*

YANG Yu-heng, YU Yang, BI Chao-wei

(*College of Plant Protection*, *Southwest University*, *Chongqing* 400715)

Abstract: Wheat stripe rust, caused by *Puccinia striiformis* f. sp. *tritici* (*Pst*), is considered one of the most aggressive diseases to wheat production. In this study, we used an iTRAQ-based approach for the quantitative proteomic comparison of the incompatible *Pst* race CYR23 in infected and non-infected leaves of the wheat cultivar Suwon11. A total of 3 475 unique proteins were identified from three key stages of interaction (12, 24, and 48 h post-inoculation) and control groups. Quantitative analysis showed that 530 proteins were differentially accumulated by *Pst* infection (fold changes >1.5, $P<0.05$). Among these proteins, 10.54% was classified as involved in the immune system process and stimulus response. Intriguingly, bioinformatics analysis revealed that a set of reactive oxygen species metabolism-related proteins, peptidyl-prolyl cis-trans isomerases (PPIases), RNA-binding proteins (RBPs), and chaperonins was involved in the response to *Pst* infection. Our results were the first to show that PPIases, RBPs, and chaperonins participated in the regulation of the immune response in wheat and even in plants. This study aimed to provide novel routes to reveal wheat gene functionality and better understand the early events in wheat-*Pst* incompatible interactions.

Key words: Quantitative proteomics; Wheat; *Puccinia striiformis* f. sp. *tritici*; Incompatible reaction

Disease Resistance Through Infection Site Production of a Toxic α-SNAP that Impairs Vesicle Trafficking

Andrew Bent

(*Professor*, *Department of Plant Pathology*, *University of Wisconsin-Madison*, *Madison*, *WI*, *USA*)

Abstract: Current paradigms for mechanisms of plant disease resistance emphasize expression of R gene products or MAMP receptors that mediate pathogen detection, causing cell wall reinforcement, production of antimicrobial compounds, and in some cases, hypersensitive response (HR) cell death. Paradigms are less well developed for mechanisms of plant defense against biotrophic pathogens that overcome preformed plant defenses, escape detection by R gene products, and secrete effectors that block other typical plant immune responses. We have identified at least part of the mechanism for soybean*Rhg*1, a complex locus that confers resistance to soybean cyst nematode (SCN). Soybean is one of the world's four largest food crops. SCN is by far the most yield-reducing pathogen of soybean in the U. S. SCN often does not cause obvious above-ground disease symptoms, so many growers do not try enough to fight this disease. SCN is also one of the most damaging pathogens of soybean in China. SCN invade plant roots and establish a feeding site called a "syncitium" near the root vascular cylinder. This feeding site takes over and massively reprograms multiple surrounding root cells. For successful growth and reproduction, SCN must keep plant syncitium cells active and cooperative for at least three weeks. *Rhg*1 is widely used by plant breeders and soybean farmers to control SCN, and *Rhg*1 has been known for many years to function by causing early failure of the developing syncitium. We first reported in 2012 that one of the *Rhg*1 genes causing SCN resistance encodes an unusual alpha-SNAP protein. α-SNAP functions with NSF (an ATPase) to mediate vesicle trafficking by forming the core SNARE recycling machinery. α-SNAPs are conserved across eukaryotes. α-SNAP stimulates SNARE complex disassembly by NSF, which enables future rounds of vesicle fusion. We now have evidence that the soybean *Rhg*1 α-SNAP interacts less well with NSF, impairs exocytosis and other vesicular trafficking, and is cytotoxic at high doses. This toxicity is dose-dependent, and can be prevented by shifting the cellular ratio toward wild-type α-SNAPs. *Rhg*1 α-SNAP levels increase in the SCN-induced syncitium. Hence this very important disease resistance functions by an interesting mechanism: the host carries a toxic variant of a core housekeeping protein, the abundance of the protein is low enough in most tissues to not harm plant yields, but the biotrophic interface established by the pathogen is disrupted by elevated abundance of the protein, which in the case of *Rhg*1 disrupts cellular vesicle trafficking.

Comparative Transcriptome Analysis Between *Phytophthora infestans*-resistant and-sensitive Tomato*

CUI Jun**, JIANG Ning, LUAN Yu-shi***

(*School of Life science and Biotechnology, Dalian University of Technology*)

Abstract: *Phytophthora infestans* is a biotrophic pathogen, causal agent of tomato late blight (LB). The oomycetes, which have a worldwide occurrence and had long been prevalent in Korea, China, the USA, and others, has caused serious economic loss for field-grown tomatoes, and therefore is regarded as a major threat to tomato production. A wild tomato *S. pimpinellifolium* L3708, a resistant line, has been identified to be highly resistant to a wide range of *P. infestans* isolates overcoming *Ph* – 1, *Ph*2 and *Ph* – 3. Another cultivated tomato, *S. lycopersicum* Zaofen No. 2 was breeded by the Institute of Vegetables and Flowers, Chinese Academy of Agricultural Sciences, Beijing, China. Tomato Zaofen No. 2 is a susceptible accession to a variety of pathogens including *P. infestans*. In this study, we preformed comparative transcriptome profiling study on the differential expressed genes and lncRNAs (DEGs and DELs) between *S. pimpinellifolium* L3708 and *S. lycopersicum* Zaofen No. 2. More than 80 000 000 raw sequence reads were generated from two samples using the high-throughput sequencing. After expression levels (FPKM) for each gene and lncRNAs were calculated, a total of 1 037 genes and 688 lncRNAs showed significantly differential expression in leaves of Sp samples compared with Slz sample, including 463 up-regulated genes and 277 up-regulated lncRNAs. It was found that a total of 128 DEGs might be regulated 127 DELs by co-expressed analysis. GO analysis indicated that oxidoreductase activity, which might play an important role in regulating ROS to reduce damage to plant cells in plant-pathogen interactions, was highly represented, including 17 genes. In this GO term, 6 members of glutaredoxin (GRX) gene family was found, which were regulated 7 DELs. The results of qRT-PCR showed that the expression levels of these 6 GRXs and 7 DELs were consistent with them through RNA-Seq. These discoveries will provide us useful information to explain tomato resistance mechanisms against LB.

* Funding: This work was supported by grants from the National Natural Science Foundation of China (Nos. 31272167, 31471880 and 61472061)

** First author: CUI Jun, Ph. D candidate; E-mail: cuijun@ mail. dlut. edu. cn

*** Corresponding author: LUAN Yushi, Professor; E-mail: ysluan@ dlut. edu. cn

Roles of the PHT4 Family in Regulating Programmed Cell Death and Defense in Arabidopsis

LU Hua

(*Department of Biological Sciences*, *University of Maryland Baltimore County*, 1000 *Hilltop Circle*, *Baltimore*, *MD* 21250, *USA.*)

Abstract: Phosphorus (P) is an essential element for all living organisms and plays a crucial role in many physiological processes. Plants use phosphate transporters (PHTs) to acquire inorganic phosphate (Pi) directly from the soil and to reallocate Pi within the cell and between cells. In a genetic screen aimed to uncover new defense related genes in Arabidopsis, we identified a mutant impaired in the *PHT*4; 1 gene that showed altered immunity and cell death phenotypes. Further characterization of the *pht*4; 1 mutant and the corresponding gene indicate that *PHT*4; 1 is a negative regulator of plant defense, acting upstream of the key defense signaling mediated by salicylic acid. Expression of *PHT*4; 1 is likely regulated by the circadian clock gene *CCA*1. PHT4; 1 belongs to a six-protein family, five of which including *PHT4*; 1 were previously localized to the chloroplast while PHT4; 6 is the most distantly related member localized to the Golgi. Like *PHT*4; 1, *PHT*4; 6 also negatively regulates plant defense and cell death. Interestingly such function of *PHT*4; 6 is largely SA-independent but specifically requires defense genes *EDS*1 and *PAD*4 and jasmonic acid signaling. Other four chloroplast-localized PHT4 proteins (PHT4; 2 -4; 4) are not known to affect plant defense when individual corresponding genes were mutated, suggesting redundant function shared between these *PHT*4 genes and *PHT*4; 1. Together our results revealed the roles of members in the PHT4 family in cell death and disease control in Arabidopsis. They also indicate that PHT4; 6 has evolved to be functionally divergent from other PHT4 family members. Information obtained from our study could help to advance the understanding of mechanisms of defense signaling and eventually enable us to use such knowledge to improve disease-resistance traits in economically important crops.

Study on DNA Methylation Patterns of NB-LRR Family Encoding Genes in *Arabidopsis thaliana*

KONG Wei-wen*, LI Bin

(*School of Horticulture and Plant Protection*, *Yangzhou University*, *Yangzhou* 225009)

Abstract: Nucleotide-binding site leucine-rich repeat (NB-LRR) proteins in plants constitute a large family and play very important roles in disease resistance. The pivotal problems how NB-LRR genes in plants exist and how they are regulated at transcriptional level remain to be solved. Cytosine DNA methylation is a pervasive and conservative epigenetic mark in eukaryotes, and it is important in regulation of genes. In this study, NB-LRR encoding genes in model plant *Arabidopsis thaliana* (Columbia ecotype) are selected for study on the DNA methylation patterns in the wild type and mutants whose original genes are key factors regulating DNA methylation. Based on the results, this study explores further the issues between changes for gene expression at transcriptional levels and changes for DNA methylation patterns. The main results are as follows. 1. The vast majority of NB-LRR encoding genes in *A. thaliana* are modified by DNA methylation. DNA methylation can be found in the promoters as well as the gene bodies for the most NB-LRR encoding genes, and the main methylation type is CG methylation. Interestingly, there are three members including *At*1*g*58807, *At*1*g*59124 and *At*1*g*59218 without DNA methylation modification. 2. Loss of key regulators *AGO*4, *MET*1, *CMT*3, *DRM*1/2 and *DDM*1 function separately leads to decrease of the methylation counts of NB-LRR encoding genes. 3. The methylation levels of gene bodies of NB-LRR family in *A. thaliana* are higher than those of 500 - bp and 1000 - bp promoters. These are observed in the wild type and almost all the mutants, especially in the CG type methylation. 4. The average of methylation levels of NB-LRR encoding genes changes regularly according to the different regions of genes. 5. Individual genes encoded by NB-LRR family in *A. thaliana* are likely to be modulated by altered DNA methylation patterns. Analysis of transcriptome data from wild type and *met*1, *cmt*3, *drm*1/2 and *ddm*1 mutants reveals that differences at transcription levels between wild type and mutants are significant for 63 of NB-LRR encoding genes. Of these, 38 of NB-LRR encoding genes are up-regulated, and other 25 are down-regulated. There are two genes of NB-LRR family whose transcriptional levels are significantly higher in all the mutants than in the wild type *A. thaliana*. One of the family, however, whose expression at transcription level is significantly lower in all the mutants than in the wild type plants, is observed. The results in this study provide an important foundation for understanding the mechanisms of expression and regulation of NB-LRR encoding genes in *A. thaliana* in the view of epigenetics especially DNA methylation.

Key words: *Arabidopsis thaliana*; NB-LRR encoding genes; Cytosine methylation; Methylation pattern; Gene regulation

---

* First author: KONG Wei-wen; E-mail: wwkong@ yzu. edu. cn

红小豆种质资源抗锈性评价及组织学抗性机制研究*

韩　冬**，柯希望，殷丽华，左豫虎***

（黑龙江八一农垦大学农学院植物病理与应用微生物研究所，大庆　163319）

摘　要：由豇豆单胞锈菌（*Uromyces vignae*）引起的小豆锈病是红小豆生产中危害最为严重的病害之一，在我国红小豆种植区内普遍发生。该病发生严重时，导致叶片干枯、脱落、籽粒瘪瘦，给小豆生产带来了严重损失，培育和种植抗病品种是防治该病害最为经济、有效的手段之一，因此，明确小豆资源对锈病菌的抗性情况，明确病原菌侵染过程及抗病的组织学特征，可为深入探索寄主的抗性机制奠定基础，为小豆抗病品种培育提供材料和理论依据。

本研究采用人工接种的方法对50份小豆资源进行了抗病性鉴定，其中，抗病品种16份，占供试品种的32%；高抗品种6份，占供试品种的12%；感病品种32份，占供试品种68%。采用荧光显微技术观察了锈病菌夏孢子在不同抗性品种叶片上的侵染过程，结果表明，夏孢子在感病品种上于接种后6h即可萌发产生附着胞，24h在附着胞下方形成入侵栓并侵入寄主组织，产生气孔下囊和吸器母细胞，接种后48h形成胞内吸器，接种后5d可产生大量的胞间菌丝，接种8d后即在叶片表面产生夏孢子堆。在抗病品种上表现为萌发时间推迟，萌发率及侵染率降低，菌丝体附近的细胞产生胞壁沉积物以抵抗病菌的入侵，菌丝生长受抑制，夏孢子产生时间明显推迟。以上结果表明，我国小豆抗锈病资源较为丰富，且抗病品种对锈病菌的抗性主要表现为抑制病原菌的生长发育和扩展，本研究结果将为深入解析小豆抗锈病机理奠定基础，为小豆抗锈病相关基因挖掘、抗病品种培育及抗病品种合理利用提供重要依据。

关键词：红小豆锈病；种质资源；抗病性；组织学特征

*　基金项目：国家科技支撑计划项目（2014BAD07B05－06/H08）；国家自然科学基金（31501629）；中国博士后科学基金面上资助项目（2014M561378）

**　第一作者：韩冬，硕士研究生，主要研究方向为植物病理学；E-mail：13836789390@163.com

***　通讯作者：左豫虎，教授，主要从事植物病理学研究；E-mail：zuoyhu@163.com

水稻抗病蛋白 Pia CC 结构域的重组表达和纯化

刘　强*，郭力维，张轶琨，郭　娇，彭友良，刘俊峰**
（中国农业大学植物保护学院植物病理学系，北京　100193）

摘　要：水稻是世界三大粮食作物之一，常年受稻瘟病菌侵害，具有为害面积广、发病率高、损失重等特点。因此，深入解析水稻与稻瘟病菌互作机理对水稻抗病育种、稻瘟病菌防控具有重要的科学意义。前人研究表明，水稻抗病基因 *Pia* 由位置相连、方向相反的一对 NB-LRR 类型的 R 基因 *RGA4* 和 *RGA5* 组成，二者所编码的蛋白质通过 N 端的 Coiled-coil（CC）结构域相互作用，形成异源二聚体，而且 CC 结构域是这两个 R 蛋白行使功能所必需。比如，只有含有 CC 结构域的 RGA5 才能抑制 RGA4 所调节的细胞死亡。在植物体内，RGA4 和 RGA5 的 CC 结构域都能与自身相互作用，形成同源二聚体。

本研究利用原核表达系统，将编码 RGA4 和 RGA5 CC 结构域的基因分别构建到含有 GFP 标签的不同表达载体中，并尝试不同的诱导温度与表达菌株。通过分析菌体颜色以及对照诱导 PAGE 胶图，发现重组蛋白质未表达或者不可溶。为了获得可溶性蛋白，作者将编码 RGA4 和 RGA5 结构域的基因进行不同程度的截短，同样将其构建到不同的表达载体中，在多种不同表达菌株中进行表达分析。经过对比对照诱导和亲和层析初步纯化，编码 $RGA4^{1\sim137}$ 的基因在 pETGST－1b 表达载体与 BL21（DE3）表达菌株的表达组合、18°C 诱导条件下，可经初步纯化获得可溶性蛋白。下一步将进行表达条件的优化，提高其蛋白可溶性。以便于通过层析技术得到均一、稳定的适合晶体生长的 Pia CC 结构域蛋白质样品，为解析 Pia CC 结构域的结构和探究其作用机制奠定基础。

关键词：稻瘟病菌；抗病蛋白；CC 结构域

---

* 第一作者：刘强，硕士研究生，植物病理学专业；E-mail：17091007607@126.com
** 通讯作者：刘俊峰，教授；E-mail：jliu@cau.edu.cn

绿豆叶斑病菌的接种方法及种质资源抗病性评价*

张海涛**，柯希望，殷丽华，左豫虎***
（黑龙江八一农垦大学农学院植物病理与应用微生物研究所，大庆　163319）

摘　要：由变灰尾孢菌（*Cercospora canescens*）引起的绿豆叶斑病，又称红斑病，是我国及亚洲绿豆生产上危害最为严重的病害之一。国内外对该病的研究报道主要涉及田间防治药剂筛选和真菌的病原学研究，关于绿豆叶斑病病原菌的侵染机制方面的研究尚未见报道。因此，为明确绿豆资源对绿豆叶斑病菌的抗性情况，进行了变灰尾孢菌接种方法的优化和抗病性评价的研究。为深入探索病原菌的侵染机制奠定基础，为绿豆抗病品种培育提供材料和理论依据。

本研究采用人工接种的方法用菌丝块和孢悬液两种不同接种体进行接种，并筛选适宜的接种条件。结果表明，菌丝块为接种体接种叶片的发病率较高，分生孢子悬浮液为接种体的发病率不高。病原菌在20～30℃的温度范围内都可以使接种叶片发病，引起叶斑症状，但其最佳的接种温度为25℃。绿豆叶片在接种后保湿24h，36h，48h均能发病，且保湿时间越长，病斑面积越大。利用优化得到的接种方法对18份绿豆资源进行了抗病性鉴定，其中，抗病品种1份，占供试品种的5.6%；中抗品种11份，占供试品种的61.1%；中感品种4份，占供试品种的22.2%；高感品种2份，占供试品种的11.1%。以上结果表明变灰尾孢菌侵染的最佳温度为25℃，较长时间的叶面湿度有利于病菌的侵入。本研究结果将为绿豆叶斑病菌致病性的分化，田间侵染发病规律，田间防治以及侵染机制等方面的研究奠定基础。

关键词：绿豆尾孢叶斑病；接种方法；种质资源；抗病性

* 基金项目：科技部国家科技支撑计划（2014BAD07B05－06/H08）

** 第一作者：张海涛，硕士研究生，主要研究方向为植物病理学；E-mail：18245079648@163.com

*** 通讯作者：左豫虎，教授，博士研究生导师，主要从事植物病理学研究；E-mail：zuoyuhu@hlau.edu.cn

拟南芥 *AtBT4* 原核表达载体 pGEX4T－1－*BT4* 构建及其诱导表达的优化

郑　旭，黄聪聪，王冠宇，赵亚婷，邢继红，董金皋
（河北农业大学真菌毒素与植物分子病理学实验室，保定　071000）

摘　要：本实验室前期获得了拟南芥抗病新基因 *AtBT4*，该基因正调控拟南芥对灰葡萄孢的抗性。AtBT4 蛋白属于转录调节因子，并且具有 DNA 结合活性。通过酵母双杂交实验获得了7个可以与转录调节因子互作的蛋白。本研究以拟南芥野生型 cDNA 为模板克隆 *AtBT4* 基因，通过酶切、连接使其与含有 GST 标签蛋白的 pGEX4T－1 载体构成重组载体；经菌液 PCR、双酶切和测序鉴定正确后，将重组载体 pGEX4T－1－*BT4* 转化大肠杆菌 BL21。在不同培养温度、不同诱导时间条件下诱导表达 AtBT4 蛋白，SDS-PAGE 结果显示：pGEX4T－1－*BT4* 载体所表达目的蛋白大小约69kDa，预期大小一致，原核表达蛋白正确。在不同培养温度18℃、28℃、37℃条件下，37℃时表达量较高；在不同诱导培养时间2h、4h、6h条件下，4h时即可高效表达；由此确定了 pGEX4T－1－*BT4* 载体在大肠杆菌 BL21 中高效表达的条件。由于 pGEX4T－1－*BT4* 载体所表达的蛋白带有 GST 标签，我们可以进一步通过 pulldown 技术鉴定前期由酵母双杂交筛选得到的 AtBT4 互作蛋白；为阐明 AtBT4 调控拟南芥抗灰霉病的分子机制奠定基础。

关键词：拟南芥；AtBT4；互作蛋白；原核表达

大白菜 TuMV 抗性基因 *TuRBCS*01 两侧紧密连锁的分子标记应用研究*

李巧云**，张志刚，赵智中，刘栓桃，王立华，高会超，李溢真
（山东省农业科学院蔬菜花卉研究所，山东省设施蔬菜生物学重点实验室，
国家蔬菜改良中心山东分中心，山东省设施蔬菜生物学重点实验室，济南　250100）

摘　要： 分别以大白菜 TuMV 国家级抗源材料 8407 和高感 TuMV 的核心种质材料冠 291 为亲本构建分离群体，鉴定出的大白菜 TuMV 抗性基因 *TuRBCS*01 具有重要意义。已筛选的与该基因紧密连锁的分子标记可用于分子标记辅助选择。为了进一步验证其两侧最近两个分子标记 mBr4055 和 BrID10723 的检测准确率，对上述两个亲本的 F_2 代分离群体 110 个单株进行自交构建 $F_{2:3}$ 加系，并对每个加系的 TuMV 抗性进行了鉴定，以判断原 F_2 单株的 TuMV 抗性；利用上述两个标记引物扩增 F_2 代 110 个单株，根据扩增结果及 F_2 单株的 TuMV 抗性计算标记检测的准确率。结果表明，两标记均扩增为纯合抗病条带的有 21 株，其中，仅有 1 株经加系验证基因型为杂合抗病，其余均为纯合抗病，检测准确率为 95.2%。两标记均扩增为杂合抗病条带的有 19 株，其中，仍是仅有 1 株基因型为纯合抗病，其余均为杂合抗病，检测准确率为 94.7%。因此，上述两个标记可用于含有该基因的抗病植株的鉴定，并能区分纯合抗病和杂合抗病株。

* 基金项目：国家自然科学基金项目（No. 31301785）；“十二五”农村领域国家科技计划课题（No. 2012BAD02B01－6）；山东省自然科学基金项目（No. ZR2013CM035）；山东省农业良种工程项目“十字花科名产蔬菜新品种选育”

** 第一作者：李巧云，山东聊城人，副研究员，博士，研究方向：蔬菜生物技术；E-mail：liqiaoyun0606@126.com

Klebiella penumoniae SnebYK Induced System Resistance in Soybean Against Drought and Soybean Cyst Nematode*

LIU Dan, ZHAO Jing, WANG Yuan-yuan, ZHU Xiao-feng, LIU Xiao-yu, CHEN Li-jie, DUAN Yu-xi**

(*Nematology Institute of Northern China*, *College Plant Protection*, *Shenyang Agricultural University*, *Shenyang* 110866, *P. R. China*)

Abstract: Soybean (*Glycine max*) is an important economic crop around the world due to its nutritive and commercial value. On the one hand, soybean provides a large source of protein and oil. On the other hand, it can be ideal material to produce biodiesel. In the field conditions, soybean is subjected to various biotic and abiotic stresses, such as drought and soybean cyst nematode. These stresses will interfere with plant growth, result in a severe diminished output on soybean, and therefore its prevention is pressing. In addition, *Klebiella* sp. has been reported to can induce system resistance against biotic and abiotic stresses in some plants. The aim of this research was to verify *K. penumoniae* strain SnebYK can elicit induced system resistance of soybean against drought and soybean cyst nematode. Under drought stress, relative water content, growth parameter, and gas exchange of SnebYK-treated and untreated soybean were estimated. Split-root system was used to study SnebYK inducing system resistance against soybean cyst nematode. The results showed that soybean treated by SnebYK in drought showed higher relative water content (4.75%), root-top ratio (16.38%), assimilation rate (134.06%) and water use efficiency (92.70%) than untreated soybean. Meanwhile SnebYK caused 67.05% reduction in total soybean cyst nematode penetration. Hence, *K. penumoniae* SnebYK are potential to protect soybean from drought and soybean cyst nematode in an environmentally friendly way, and the biochemical and molecular mechanism of them is worth studying.

Key words: *Klebiella penumoniae*; Soybean; Induced system resistance; Drought; Soybean cyst nematode

---

* This work was supported by China Agriculture Research System CARS－04－PS13

** Corresponding author: DUAN Yu-xi; E-mail: duanyx6407@163.com

逆境胁迫下转 *N21* 基因烟草的抗性相关基因的表达*

丁亚燕**，郝　欣，陈丽丽，纪兆林***

（扬州大学园艺与植物保护学院，扬州　225009）

摘　要： 水稻白叶枯病菌 harpin 蛋白 $Hpa1_{Xoo}$N 端的卷曲螺旋结构域多肽 N21 能诱导烟草产生 HR，体外施用和在烟草体内表达 N21 都能使烟草获得对 TMV、胡萝卜软腐果胶杆菌、疫霉等病原菌的抗病特性。本文主要研究转 *N21* 基因烟草在干旱和 TMV 胁迫条件下，利用烟草 cDNA 微阵列测定并分析其转录组表达谱，了解抗性相关基因表达情况，有利于我们进一步解析 harpin 蛋白的结构和功能机制，也为进一步应用该类蛋白或多肽奠定基础。

对转录组谱数据进行两两比较，分析了抗性相关基因表达情况。

（1）未转基因亲本（CK）烟草和干旱胁迫（PEG）处理的 CK 烟草相比较，主要发现 20 个与抗性相关的基因差异表达（ratio > 3.0 或 ratio < 0.33），上调基因 8 个，下调基因 12 个。其中 AT5G46330 表现为极显著上调，是一种识别受体，可以调节植物对细菌病原的防御反应；AT1G32640 编码 MYC2，主要参与茉莉酸（JA）和脱落酸（ABA）的信号转导过程；AT5G23580、AT4G09570、AT2G17290、AT5G42380 都属于钙依赖型蛋白激酶家族蛋白。

（2）CK 烟草和接种 TMV 的 CK 烟草比较，有 18 个差异表达基因，其中，7 个上调，11 个下调。AT3G10190、AT5G53130、AT5G42380、AT1G76650 都是与钙结合相关的基因；AT3G17860 主要编码 JAZ6，是 JA 和 ZIM 的结构域蛋白，为茉莉酸反应基因转录抑制因子；AT2G38470 是植物对微生物感染的关键调控元件，调节 JA-SA 调控的抗性途径。

（3）转 *N21* 基因（N21）烟草和 PEG 处理的 N21 烟草，共计 25 个差异表达基因，13 个上调，12 个下调。AT5G56030 和 AT5G52640 都属于热激蛋白（HSP）基因；AT3G24110、AT5G53130、AT3G50770、AT5G42380、AT1G73630、AT4G26470、AT1G76650、AT1G18210、AT3G03000 都参与钙离子的结合调控过程；AT4G09570、AT1G66400 都属于钙依赖型蛋白激酶家族蛋白。

（4）N21 烟草和接种 TMV 的 N21 烟草，抗性相关差异表达基因 9 个，其中，3 个上调，6 个下调表达。这些基因都是与钙离子结合调控的相关基因，属钙依赖型蛋白激酶家族蛋白。

（5）PEG 处理的 CK 烟草和 PEG 处理的 N21 烟草，有 3 个抗性相关基因都表现为上调。AT3G50770、AT5G07390、AT1G66400 均与钙离子结合过程相关。

（6）接种 TMV 的 CK 烟草和接种 TMV 的 N21 烟草，5 个抗性基因差异表达，3 个上调，2 个下调。也都与钙离子的调控相关，通过钙离子调控，参与植物的诱导抗病过程。

关键词： 转 *N21* 基因烟草；卷曲螺旋；逆境胁迫；转录组表达谱；抗性相关基因

* 基金项目：国家自然科学基金资助项目（31101475）；扬州大学高层次人才科研启动基金资助项目（5018/137010407）

** 第一作者：丁亚燕，硕士研究生，主要从事分子植物病理学研究；E-mail：1720546087@ qq. com

*** 通讯作者：纪兆林，副教授，主要从事植物病害生物防治及分子植物病理学研究；E-mail：zhlji@ yzu. edu. cn

蜡质芽孢杆菌胞外多糖作为一类 MAMPs 激活植物系统免疫*

范志航**，蒋春号，郭坚华***
（南京农业大学植物保护学院，南京 210095）

摘　要：植物体响应根围定殖微生物诱导系统抗性依赖于寄主植物与定殖的根围菌的互作。然而定殖信号是什么以及植物体如何响应根围定殖微生物诱导系统抗性机制还不清楚。在本研究中，我们发现蜡质芽孢杆菌 AR156 的胞外多糖可以作为一种新的 MAMPs 在蜡质芽孢杆菌 AR156 诱导抗性过程中，植物体对根围菌的早期识别过程中起着重要的作用。本研究结果表明，AR156 分泌的胞外多糖能够诱导拟南芥产生对 *Pst* DC3000 的系统抗性；在此过程中，能够提升防卫相关基因 *PR*1、*PR*2、*PR*5 以及 MAPK 激酶 MPK6 的上调表达；另外，也能够激活植物体细胞防卫反应如活性氧暴发、胼胝质沉积以及防卫相关酶活性的提升。基因上位性分析结果显示，蜡质芽孢杆菌 AR156 胞外多糖仍能够在信号通路突变体 *jar*1、*etr*1 中诱导对 *Pst* DC3000 的系统抗性，与野生型 Col－0 相比，其在突变体 *jar*1、*etr*1 中诱导抗性能力略有降低；在 NahG 转基因植物以及 *npr*1 突变体中，AR156 胞外多糖的诱导抗性能力丧失。以上结果表明，蜡质芽孢杆菌 AR156 胞外多糖诱导的系统抗性依赖 MAMPs 信号识别和水杨酸信号通路转导，并且具有 NPR1 依赖性。综上所述，蜡质芽孢杆菌 AR156 胞外多糖在蜡质芽孢杆菌 AR156 诱导系统抗性过程中，作为一类 MAMPs，发挥着重要的作用。本研究首次报道蜡质芽孢杆菌 AR156 如何诱导植物产生系统抗性，也是首次解析植物根围定殖微生物如何被植物体识别，诱导植物体产生对病原菌的系统抗性。

关键词：诱导系统抗性；信号通路；微生物相关分子模式（MAMPs）；胞外多糖；蜡质芽孢杆菌 AR156；生物防治

---

* 基金项目：国家自然科学基金“蜡质芽孢杆菌基于改变番茄根系结构防治根结线虫病的机理研究”（31471812）；江苏省农业科技自主创新项目“高效设施蔬菜标准化、系列化生态友好型生产技术体系研究及示范”，项目编号：CX（15）1044

** 第一作者：范志航，福建福州人，在读博士；E-mail：1163144616@qq.com

*** 通讯作者：郭坚华，教授，主要从事植物病害生物防治；E-mail：jhguo@njau.edu.cn

转录因子 WRKY70 和 WRKY11 在调控蜡质芽孢杆菌 AR156 诱导系统抗性过程的作用机理研究*

蒋春号**，郭坚华***
（南京农业大学植物保护学院，南京 210095）

摘　要：研究发现，根围有益微生物通过同时激活植物体内水杨酸、茉莉酸/乙烯两条信号通路能够更有效的帮助植物体抵抗广谱性的抗性，但是为什么蜡质芽孢杆菌 AR156 能同时激活拟南芥体内这两条信号通路，而其他生防因子如荧光假单胞菌、部分芽孢杆菌只能激活其中的某一条信号途径。在本研究中，我们以拟南芥为模式植物探寻转录因子 WRKY70 和 WRKY11 在蜡质芽孢杆菌 AR156 诱导系统抗性过程中所起作用。结果表明，蜡质芽孢杆菌 AR156 处理拟南芥后，能够显著提升植物体内转录因子 WRKY70 的表达量，同时显著抑制 WRKY11 的表达。研究还发现转录因子 WRKY70 和 WRKY11 为蜡质芽孢杆菌 AR156 诱导细胞防卫反应以及防卫相关基因的表达所必须。过量表达转录因子 WRKY70 和 WRKY11 能够影响蜡质芽孢杆菌 AR156 诱导拟南芥产生对 *Pst* DC3000 的抗性。突变体验证结果显示：蜡质芽孢杆菌 AR156 诱导抗性的能力在转录因子 WRKY70 和 WRKY11 单突变体拟南芥植株中能够保留，但抗性能力略微下降，然而在 WRKY70 和 WRKY11 双突变体植株中则完全丧失。以上结果表明，转录因子 WRKY70 和 WRKY11 是通过两条不同途径调控蜡质芽孢杆菌 AR156 诱导系统抗性的，并且这两条途径之间具有协调作用。转录因子 WRKY70 和 WRKY11 靶标分析结果表明：WRKY70 通过水杨酸信号通路，WRKY11 通过茉莉酸信号通路来调控蜡质芽孢杆菌 AR156 诱导的系统抗性，而且这两条调控通路之间具有 NPR1 的依赖性。综合上述结果，发现转录因子 WRKY70 和 WRKY11 在调控蜡质芽孢杆菌 AR156 诱导系统抗性过程中起着重要的作用。本研究也是首次从转录因子角度出发，阐明根围有益微生物如何同时激活水杨酸、茉莉酸/乙烯两条信号通路，来诱导寄主植物对病原菌的抗性。

关键词：转录因子；诱导系统抗病；蜡质芽孢杆菌 AR156；生物防治；信号通路；拟南芥

---

* 基金项目：国家自然科学基金“蜡质芽孢杆菌基于改变番茄根系结构防治根结线虫病的机理研究”（31471812）；2015年江苏省研究生创新工程项目（KYLX15_ 0626）；江苏省农业科技自主创新项目“高效设施蔬菜标准化、系列化生态友好型生产技术体系研究及示范”，项目编号：CX（15）1044

** 第一作者：蒋春号，江苏盐城人，在读博士；E-mail：jchunhao@163.com

*** 通讯作者：郭坚华，教授，主要从事植物病害生物防治；E-mail：jhguo@njau.edu.cn

52 个甘蔗新品种（系）抗甘蔗褐锈病评价*

李文凤**，张荣跃，黄应昆***，单红丽，王晓燕，尹 炯，罗志明，仓晓燕
（云南省农业科学院甘蔗研究所，云南省甘蔗遗传改良重点实验室，开远 661699）

摘 要：由黑顶柄锈菌（*Puccina. melanocephala* H. Sydow & P. Sydow）引起的甘蔗褐锈病是世界性的甘蔗重要病害之一，目前在我国甘蔗主产区特别是湿热蔗区常年发生流行，造成巨大的经济损失，严重影响着蔗糖产业的可持续发展。利用抗病品种是控制该病害最经济有效的途径。为明确近年来国家甘蔗体系育成的新品种（系）对甘蔗褐锈病的抗性，确定其应用潜力，为合理布局和推广使用这些新品种（系）提供依据。本研究采用已报道的抗褐锈病基因 *Bru*1 稳定的分子标记对 52 个甘蔗新品种（系）进行抗褐锈病基因 *Bru*1 的分子检测，同时在云南甘蔗褐锈病高发蔗区德宏、保山 2 个国家甘蔗体系综合试验站连续 3 年对各测试品种进行田间自然抗病性调查分析。结果表明，52 个甘蔗新品种（系）中，29 个检测出含抗褐锈病基因 *Bru*1，23 个未检测到 *Bru*1；检测到 *Bru*1 的 29 个新品种（系）中，27 个连续 3 年田间调查均未发现有甘蔗褐锈病发生，而粤糖 42 号和桂糖 31 号 2 个品种则 3 年中有 1 年中度感染甘蔗褐锈病；23 个未检测出抗褐锈病基因 *Bru*1 的新品种（系）中，16 个连续 3 年田间调查结果甘蔗褐锈病中度至重度发生，而粤甘 34 号、福农 15 号、闽糖 01-77、桂糖 32 号、柳城 05-136、云蔗 06-80、甘蔗 02-707 等 7 个品种则未见发生褐锈病，预示这 7 个品种可能含有新抗褐锈病基因源。本研究结果显示，34 个新品种（系）具有抗褐锈病的应用潜力。

关键词：甘蔗；新品种（系）；褐锈病；抗病性

---

* 基金项目：现代农业产业技术体系建设专项资金资助（CARS-20-2-2）；云南省现代农业产业技术体系建设专项资金资助；云南省“人才培养”项目（2008PY087）

** 第一作者：李文凤，云南石屏人，研究员，主要从事甘蔗病害研究；E-mail：ynlwf@163.com

*** 通讯作者：黄应昆，研究员，从事甘蔗病害防控研究；E-mail：huangyk64@163.com

43 份甘蔗创新种质材料抗甘蔗花叶病鉴定评价*

李文凤**，王晓燕，黄应昆***，单红丽，张荣跃，罗志明，尹　炯
（云南省农业科学院甘蔗研究所，云南省甘蔗遗传改良重点实验室，开远　661699）

摘　要：甘蔗花叶病是一种重要的世界性甘蔗病害，近年来已成为中国蔗区发生最普遍、危害最严重的病害之一。前期研究表明引起中国蔗区甘蔗花叶病病原有甘蔗花叶病毒（*Sugarcane mosaic virus*，SCMV）、高粱花叶病毒（*Sorghum mosaic virus*，SrMV）和甘蔗线条花叶病毒（*Sugarcane streak mosaic virus*，SCSMV）3 种，其中 SrMV 和 SCSMV 是最主要致病病原。利用抗病品种是防治该病最经济有效措施，然而目前能同时兼抗几种病毒病原的品种和资源较为缺乏，发掘新抗病种质对选育抗病品种具有重要意义。本研究以甘蔗热带种路打士与滇蔗茅云南 95-19 进行远缘杂交获得的 41 份创新种质为材料，采用苗期人工接种和后期 RT-PCR 检测法，对包括亲本在内的 43 份创新种质材料进行了 SrMV、SCSMV 双重抗性鉴定评价。结果表明，43 份创新种质材料中，对 SCSMV 表现 1 级高抗到 3 级中抗的有 22 份，4 级感病到 5 级高感的有 21 份；对 SrMV 表现 1 级高抗到 3 级中抗的有 31 份，4 级感病到 5 级高感的有 12 份。综合分析结果显示，云 09-603、云 09-604、云 09-607、云 09-608、云 09-619、云 09-622、云 09-633、云 09-635、云 09-656、云滇 95-19 等 10 份材料对 SCSMV 和 SrMV 均表现 1-2 级抗病，占 23.26%，其中云 09-604、云 09-607、云 09-619、云 09-633、云 09-656、云滇 95-19 等 6 份材料对 2 种病毒均表现为 1 级高抗。研究结果明确了各创新种质材料对甘蔗花叶病 2 种主要致病病原的抗性，筛选出双抗 SCSMV 和 SrMV 的创新种质 10 份，为进一步利用抗病种质开展抗花叶病育种提供了科学依据和抗源材料。

关键词：甘蔗；创新种质；花叶病；抗性鉴定

---

* 基金项目：现代农业产业技术体系建设专项资金资助（CARS－20－2－2）；云南省现代农业产业技术体系建设专项资金资助

** 第一作者：李文凤，云南石屏人，研究员，主要从事甘蔗病害研究；E-mail：ynlwf@163.com

*** 通讯作者：黄应昆，研究员，从事甘蔗病害防控研究；E-mail：huangyk64@163.com

Molecular Cloning and Functional Characterization of the Tomato E3 Ubiquitin Ligase *SlBAH*1 Gene*

ZHOU S M**, WANG S M, LIN C, SONG Y Z, ZHENG X X, SONG F M, ZHU C X

(*State Key Laboratory of Crop Biology*, *Shandong Key Laboratory of Crop Biology*, *Shandong Agricultural University*, *Tai' an*, *Shandong* 271018, *P. R. China*)

Abstract: Emerging evidence suggests that E3 ligases play critical roles in diverse biological processes, including pathogen resistance in plants. In this study, an ubiquitin ligase gene (*SlBAH*1) was cloned from a tomato plant, and the functions of the gene were studied. The *SlBAH*1 gene contained 1002 nucleotides and encoded a protein with 333 amino acids. The SlBAH1 protein contained a SPX domain and a RING domain. SlBAH1 displayed E3 ubiquitin ligase activity in vitro. SlBAH1 was demonstrated to localize in the nucleus, cytoplasm and plasma membrane by a subcellular localization assay. The expression of *SlBAH*1 was induced by various hormones and *Botrytis cinerea* treatment. *SlBAH*1-silencing in plants obtained by Virus-induced gene silencing (VIGS) technology enhanced resistance to *Botrytis cinerea*, and the expression of pathogenesis-related (PR) genes, including *PR*1, *PR*2, *PR*4, *PR*5, and *PR*7, was significantly increased. These results indicated that the *SlBAH*1-dependent activation of defense-related genes played a key role in the enhanced fungal resistance observed in the *SlBAH*1-silenced plants and may be related to the SA-dependent and JA-dependent signaling pathways.

* Funding: This work was supported by the National Natural Science Foundation of China (No. 31272113) and the National Key Basic Research and Development Program (2009CB119005)

** Corresponding author: ZHU S M; E-mail: zhchx@ sdau. edu. cn

Isolation and Function Analysis of *NtLTP*4 from Tobacco*

ZHENG X X, CHEN C X, ZHOU S M, LIU H M,
SONG Y Z, WEN F J, ZHU C X**

(*State Key Laboratory of Crop Biology*, *Shandong Key Laboratory of Crop Biology*, *Shandong Agricultural University*, *Tai'an* 271018, *P. R. China*)

Abstract: Nonspecific lipid transfer proteins play an important rolein the plant biological and abiotic stress resistance. In the present study, *NtLTP*4 gene was isolated from common tobacco (*N. tabacum cv.* NC89). Sequence analysis revealed that the full-length cDNA consisted of 354 nucleotides, which encoded a protein of 116 amino acid residues. NtLTP4 protein was localized in the cell wall. Quantative RT-PCR analysis revealed that *NtLTP*4 transcript accumulation was up-regulated by high-salinity, drought stress, wounding, abscisic acid (ABA), Jasmine acid (JA), salicylic acid (SA), *R. Solanacearum*, and *Potato virus* treatment. To evaluate the functions of *NtLTP*4 in the regulation of stress responses, we produced transgenic tobacco lines over-expressing *NtLTP*4. After drought and salt stress treatments, the transgenic tobacco plants had a higher germination rate, lower malondialdehyde content, lowerlevels of reactive oxygen species (ROS) accumulation, and less growth inhibition than WT. The antioxidant enzyme activities were higher in the transgenic plants than in WT, which may be related to the up-regulated expression of some antioxidant genes via overexpression of *NtLTP*4. These results demonstrate the high toleranceof the transgenic plants to drought and salt stresses compared with the WT. In addition, The *NtLTP*4-overexpressing plants enhanced the resistance to the infection of *R. solanacearum* and PVY-O.

* Funding: This work was supported by the National Natural Science Foundation of China (No. 31272113)

** Corresponding author: ZHU C X; E-mail: zhchx@ sdau. edu. cn

A conserved *Puccinia striiformis* effector interacts with wheat NPR1 and reduces induction of *pathogenesis-relatedgenes* in response to pathogens

WANG Xiao-dong[1,2,3*], YANG Bao-ju[1*], LI Kun[1],
KANG Zhen-sheng[2], Dario Cantu[4], Jorge Dubcovsky[1,5**]

(1. *Department of Plant Science*, *University of California*, *Davis*, *CA* 95616, *USA*.; 2. *State Key Laboratory of Crop Stress Biology for Arid Areas and College of Plant Protection*, *Northwest Agriculture and Forestry University*, *Yangling*, *Shaanxi* 712100, *P. R. China*.; 3. *Department of Plant Pathology*, *Agriculture University of Hebei*, *Baoding*, *Hebei* 071000, *P. R. China*.; 4. *Department of Viticulture and Enology*, *University of California*, *Davis*, *CA* 95616, *USA*. 5. *Howard Hughes Medical Institute* (*HHMI*), *Chevy Chase*, *MD* 20815, *USA*.)

Abstract: In Arabidopsis, NPR1 is a key transcriptional co-regulator of systemic acquired resistance. Upon pathogen challenge, NPR1 translocates from the cytoplasm to the nucleus, where it interacts with TGA-bZIP transcription factors to activate the expression of several pathogenesis-related genes. In a screen of a yeast two-hybrid library from wheat leaves infected with *Pucciniastriiformis*f. sp. *tritici*, we identified a conserved rust effector that interacts with wheat NPR1 and named it *Puccinia* NPR1 interactor (PNPi). PNPi interacts with the NPR1/NIM1-like domain of NPR1 via its C-terminal domain. Using bimolecular fluorescence complementation assays, we detected the interaction between PNPi and wheat NPR1 in the nucleus of *Nicotiana benthamiana* protoplasts. A yeast three-hybrid assay showed that PNPi interaction with NPR1 competes with the interaction between wheat NPR1 and TGA2. 2. In barley transgenic lines over expressing *PNPi*, we observed reduced induction of multiple *PR* genes in the region adjacent to *Pseudomonas syringae*pv. *tomato* DC3000 infection. Based on these results, we hypothesize that PNPi has a role in manipulating wheat defense response via its interactions with NPR1.

Key words: Wheat; Stripe rust; NPR1; Pathogen effector; Yeast two-hybrid; Bimolecular fluorescence complementation; Transgenic barley.

* These authors contributed equally to this work.
** Corresponding author: Jorge Dubcovsky; E-mail: jdubcovsky@ucdavis.edu

Microarray Analysis on Differentially Expressed Genes Associated with Wheat *Lr*39/41 Resistance to *Puccinia triticina*

WANG Xiao-dong, BI Wei-shuai, GUO Yu-xin, LI Xing*, LIU Da-qun*

(*Department of Plant Pathology, Agriculture University of Hebei, Baoding, Hebei* 071000, *P. R. China*)

Abstract: Wheat leaf rust, caused by *Pucciniatriticina* (*Pt*), is one of the most severe fungal diseases threatening global wheat production. Utilization of leaf rust resistance (*Lr*) genes is the major solution for this issue. Wheat isogenic lines carrying *Lr*39/41 gene still showed a moderate or higher resistance to most of the leaf rust races detected in China. In our study, a typical hypersensitive response (HR) has been observed by microcopy in *Lr*39/41 isogenic lines inoculated with avirulent*Pt* race from 24 hours post inoculation, indicating a possible NBS-LRR and program cell death-related pathway for the mechanism of *Lr*39/41 gene. Two *Lr*39/41 resistance-associated suppression subtractive hybridization (SSH) libraries have been established. One with 3 456 clones constructed between *Pt*-inoculated and non-inoculated *Lr*39/41 isogenic lines. Another with 2 544 clones constructed between *Pt*-inoculated *Lr*39/41 isogenic lines and Thatcher susceptible lines. Microarray hybridization has been carried out on the established SSH libraries with RNAs extracted from *Pt*-inoculated and non-inoculated *Lr*39/41 isogenic lines and Thatcher susceptible lines, respectively. Differentially expressed genes (DEGs) were analyzed by significance analysis of microarrays (SAM) and a total of 386 clones with significant induction or suppression were further sequenced. Sequences were then annotated with GenBanknr database, URGI wheat genome database and gene ontology (GO) database. In total of 36 *Lr*39/41 resistance-related DEGs were found, many of which were previously reported in plant defense response. Four *Lr*39/41 resistance-related DEGs were also detected as Thatcher susceptible-related DEGs, indicating a potential involvement of basal resistance for these genes. Relative expression levels of 12 selected *Lr*39/41 resistance-associated DEGs during *Pt* infection were furthervalidated by qRT-PCR assay. A brief profile of DEGs associated with *Lr*39/41 resistance had been drafted.

Key words: Wheat leaf rust resistance; *Lr*39/41; Suppression subtractive hybridization; Microarray; QRT-PCR

* Corresponding authors: LI Xing; E-mail: lxkzh@163.com
Daqun Liu; E-mail: ldq@hebau.edu.cn

蛋白激发子Hrip1水稻互作蛋白的鉴定及功能研究*

李书鹏**，邱德文***

（中国农业科学院植物保护研究所，植物病虫害生物学国家重点实验室，北京 100081）

摘　要：蛋白激发子是一类能激发植物自身免疫、诱导植物获得性抗性的效应子。植物对激发子的识别是触发植物免疫信号传导的开关，候选受体的鉴定与功能研究将为深入了解植物与病原微生物的互作，揭示蛋白激发子诱导植物产生抗性的机制奠定坚实的基础。Hrip1是本实验室从极细链格孢菌（*Alternaria tenuissima*）中分离到的一个新蛋白激发子，具有激发烟草防御反应和提高对TMV系统抗性的功能。前期研究表明，Hrip1注射烟草叶片后能够引发烟草叶片的钙离子流入、介质碱化、激活水杨酸途径的蛋白激酶并且提高植物体内相关防卫基因的表达。本研究通过酵母双杂交，将Hrip1构建到含有酵母转录因子GAL4BD结构域的pDBleu载体上，融合表达作为诱饵，从水稻CDNA文库中，筛选到16个候选蛋白，经在酵母细胞中的重复验证，确定了一个可以和Hrip1互作的水稻蛋白CSN5。

生物信息学研究表明，CSN5是COP9信号复合体中的一个亚基，主要参与植物发育相关的生理过程，例如，光形态发生、植物生长素、茉莉酸的应答机制等。COP9参与类泛素蛋白NEDD8的代谢通路，而泛素蛋白的翻译后修饰已经被证实参与植物的抗性信号传导，免疫应答。CSN5蛋白是否参与Hrip1诱导水稻抗病作用，尚需要采用多种体内、体外蛋白互作技术和基因功能的研究方法进行深入研究。

关键词：蛋白激发子；Hrip1；蛋白互作；酵母双杂

---

* 基金项目：本研究项目受到国家自然科学基金（31371984）

** 第一作者：李书鹏，硕士研究生，主要从事蛋白激发子与植物互作的研究；E-mail：shupengli123@163. com

*** 通讯作者：邱德文，研究员，主要从事植物诱导苗裔及生物防治研究；E-mail：qiudewen@caas. cn

转蛋白激发子 PeaT1 提高水稻抗旱性[*]

史发超[**]，邱德文[***]
（中国农业科学院植物保护研究所，植物病虫害生物学国家重点实验室，北京　100081）

摘　要：蛋白激发子 PeaT1 来源于极细链格孢（*Alternaria tenuissima*），能提高烟草对 TMV 的抗性，并且叶片喷施 PeaT1 蛋白能够显著提高水稻的抗旱性。为了进一步研究 PeaT1 诱导水稻的抗旱机制，本研究用蛋白激发子 PeaT1 水稻过表达纯合株系（L1，L40，L43）四叶期进行抗旱测定，并利用蛋白质组学技术分析差异蛋白的功能。结果表明，与野生型日本晴（*Oryza sativa* L.）水稻相比，抗旱 12d 后复水 5d，转基因水稻的存活率高达 60%，比野生型水稻高 20%，达到显著水平，另外，转基因水稻叶片中 ABA 含量以及叶绿素含量在抗旱过程中均有显著提高。对转基因纯和株系进行 Label free quantitative 蛋白组学分析，获得 57 个显著差异的蛋白，包括上调蛋白 32 个，下调蛋白 25 个。Gene Ontology 分析发现，有 7 个差异蛋白参与了抗逆过程，其中 OsSKIPa 蛋白显著上调，参与 PeaT1 对干旱胁迫的响应，从而提高水稻的抗旱性；蛋白 *Os02g*0149800 参与 ABA 信号的转导，而 ABA 在抗旱中发挥着重要的作用；*Os07g*0141400 参与光系统 II 的过程，编码的蛋白提高水稻对条纹病的抗性。以上研究结果为深入探讨 PeaT1 诱导水稻抗病和抗旱作用机制奠定了基础。

关键词：激发子 PeaT1；抗旱性；蛋白组学；抗逆

* 基金项目：国家自然科学基金（31371984）
** 第一作者：史发超，博士研究生，主要从事蛋白激发子诱导抗性的研究；E-mail：sfchao1988@163.com
*** 通讯作者：邱德文，研究员，从事植物诱导免疫及生物防治研究；E-mail：qiudewen@caas.cn

稻瘟菌激发子 MoHrip1 和 MoHrip2 在水稻中的表达及功能研究*

王真真，韩 强，訾 倩，吕 顺，邱德文，曾洪梅**

（中国农业科学院植物保护研究所植物病虫害生物学国家重点实验室，北京 100081）

摘 要：MoHrip1 和 MoHrip2 是本实验室从稻瘟菌（*Magnaporthe oryzae*）分泌液中分离出来的两种新型蛋白激发子。前期研究表明，这两种蛋白激发子均能够引起烟草过敏反应，诱导烟草产生氧爆发和 NO 积累等早期反应，体外处理水稻可提高水稻对稻瘟病和白叶枯病的抗性。转基因技术可将外源基因导入植物体内，被认为是培育高产、抗逆品种的有效途径。本研究分别构建了 MoHrip1 和 MoHrip2 基因的融合表达载体，利用农杆菌介导法转化水稻，成功获得 T_1 代植株，并繁殖至 T_2 代。分别对 T_1、T_2 代植株进行分子检测，证明 MoHrip1 和 MoHrip2 基因在转基因水稻中能够正常表达。研究表明，与野生型水稻相比，MoHrip1 和 MoHrip2 转基因水稻能增强水稻的抗病性和抗旱性。荧光定量 PCR 表明，在 MoHrip1 和 MoHrip2 转基因水稻中，病程相关蛋白和水杨酸通路抗病相关基因被显著诱导表达，水稻防御反应被激活，说明两个激发子主要通过水杨酸（SA）信号通路增强水稻对稻瘟病的抗病；在干旱条件下，转基因水稻叶片中 ABA 含量显著高于野生型水稻的含量，而 GA 含量显著低于野生型的含量；在 MoHrip1 转基因水稻中，水稻叶片中抗旱相关基因 OsNCED2、OsNCED3、OsZEP 和 OsbZIP23 的表达水平分别是野生型的 17、6、28 和 2.9 倍；在 MoHrip2 转基因水稻中，它们分别是野生型的 35、3、52 和 1.6 倍。同时，MoHrip1 和 MoHrip2 还能够提高水稻农艺性状，如株高、千粒重等。抗体标签与荧光免疫结合的技术表明，MoHrip1 和 MoHrip2 主要分布在水稻表皮细胞的外周。

关键词：稻瘟菌；荧光定量；抗旱基因；抗体标签

* 基金项目：国家高技术研究与发展计划（“863”计划），项目（2012AA101504）

** 通讯作者：曾洪梅，研究员，主要从事植物病害生物防治研究；E-mail：zenghongmei@caas.cn

蛋白激发子 BcGS1 激活番茄免疫的功能及功能域鉴定*

杨晨宇**，张　易，杨秀芬***

（中国农业科学院植物保护研究所/植物病虫害生物学国家重点实验室，北京　100081）

摘　要：灰葡萄孢菌（*Botrytis cinerea*）能够引起多种作物的灰霉病，利用灰葡萄孢菌产生的激发子诱导植物抗性是减轻该病害发生发展的有效途径之一。病原菌分泌的细胞壁降解酶具有致病因子和激发植物免疫的双重功能。蛋白激发子 BcGS1 是从灰葡萄孢菌 BC-4-1-1 中分离的、具有诱导细胞坏死反应的分泌糖蛋白。功能研究表明：BcGS1 是一种细胞壁降解酶，葡聚糖 1，4-α-糖苷酶，该蛋白由 672 个氨基酸组成，含有 Glyco_-hydro_15（简称 GH15）和 CBM20_ glucoamylase（碳水化合物结合域）两个结构域，理论分子量为 70. 487kDa。BcGS1 能诱导番茄和烟草叶片产生类似 HR 的细胞坏死反应，并引起 H_2O_2 的积累；BcGS1 处理后能提高番茄和烟草系统叶片对假单胞杆菌和烟草花叶病毒的抗性。荧光定量 PCR 检测到防御相关基因 *PR*-1*a*，*TPK*1*b* 和 *Prosystemin* 的转录水平也大幅度提高。

为了进一步鉴定蛋白 BcGS1 的抗病机制，首先明确了蛋白激发子 BcGS1 的糖苷酶活性，用 3 –5 二硝基水杨酸定糖法，以可溶性淀粉、羧甲基纤维素、微晶纤维素和海带多糖为底物，测定了不同 pH（5-8）下 BcGS1 水解多糖底物产生还原性糖的含量，结果表明：BcGS1 只能水解淀粉，不能水解纤维素和海带多糖，推测灰葡萄孢菌侵染寄主初期分泌 BcGS1，并通过降解细胞壁产生的寡糖激发寄主的免疫防御反应。采用农杆菌介导的瞬时表达技术，分别在烟草细胞中表达了 BcGS1 全长，GH15 和 CBM20 结构域蛋白，荧光共聚焦显微镜观察到强烈荧光，说明三个蛋白均能在烟草细胞中融合表达；瞬时表达 BcGS1 的叶片有明显坏死斑，GH15 有轻微坏死斑，CBM20 没有明显坏死斑。为了进一步鉴定 BcGS1 的功能结构域，原核表达了 BcGS1 的两个结构域蛋白，结果表明两个结构域蛋白渗入叶片均不能引起明显的细胞坏死反应，说明 BcGS1 的激发子活性需要两个结构域的共同参与。

关键词：灰葡萄孢菌；蛋白 BcGS1；酶活性；细胞坏死；功能域鉴定

---

* 基金项目：北京市科技计划（D151100003915003）

** 第一作者：杨晨宇，在读研究生，主要从事蛋白激发子 BcGS1 诱导植物免疫机制的研究；E-mail：yangchenyu0825@163. com

*** 通讯作者：杨秀芬，研究员，主要从事生物农药的研究与利用；E-mail：yangxiufen@ cass. cn

蛋白激发子 MoHrip1 与水稻细胞的互作位点*

张 易**，杨秀芬，曾洪梅，郭立华，邱德文***

（中国农业科学院植物保护研究所，植物病虫害生物学国家重点实验室，北京 100081）

摘 要：MoHrip1 是从稻瘟菌（*Magnaporthe oryzae*）中分离出的一个蛋白激发子，前期研究证明原核表达的 MoHrip1 融合蛋白能诱导烟草早期防御反应，例如氧爆发，胼胝质积累，细胞质碱化等，并能诱导抗病相关基因的转录上调。说明稻瘟菌与植物互作过程中分泌的蛋白 MoHrip1 能诱导植物免疫反应，推测植物细胞膜上存在有感知 MoHrip1 的识别位点。作者构建了 *M. oryzae* 的 MoHrip1 过表达菌株，荧光显微镜镜检发现，MoHrip1 蛋白主要定位在孢子和菌丝细胞壁和分隔上。根据这一现象推测 MoHrip1 首先储存在细胞壁周围，在孢子萌发和侵染过程中，MoHrip1 被逐步释放到寄主细胞中发挥其生理功能。基于此推测，将 MoHrip1 过表达菌株孢子悬浮液接种于 PDB 液体培养基中震荡培养，分别在 1d，3d 和 7d 离心收集培养液并浓缩提取培养液中的总蛋白，Western blot 验证表明，随着培养时间的延长，发酵液中 MoHrip1 蛋白浓度逐渐增加，第 7d 分泌量最高，推测 MoHrip1 参与了稻瘟菌侵染寄主的过程。为了进一步明确 MoHrip1 参与寄主水稻细胞的互作。分别将 *GFP* 基因引入到到 MoHrip1 编码基因的 N 端和 C 端，毕赤酵母表达了 GFP-MoHrip1 和 MoHrip1-GFP 融合蛋白，将两种融合蛋白分别与水稻原生质体孵育，以 GFP 蛋白作为阴性对照，荧光显微镜观察发现，GFP-MoHrip1 和 MoHrip1-GFP 融合蛋白在水稻原生质体表面发出荧光，说明水稻细胞膜上存在与外源蛋白激发子 MoHrip1 互作的互作位点。

关键词：MoHrip1；激发子；蛋白表达；原生质体

* 基金项目：本研究项目受到国家自然科学基金（31371984）

** 第一作者：张易，博士研究生，主要从事蛋白激发子与植物互作的研究；E-mail：zhangyi825@outlook.com

*** 通讯作者：邱德文，研究员，主要从事植物诱导苗裔及生物防治研究；E-mail：qiudewen@caas.cn

iTRAQ Quantitative Proteomics Analysis of the Defense Response of Wheat Against *Puccinia striiformis* f. sp. *tritici*

YANG Yu-heng, YU Yang, BI Chao-wei

(*Collegeof Plant Protection*, *Southwest University*, *Chongqing* 400715)

Abstract: Wheat stripe rust, caused by *Puccinia striiformis* f. sp. *tritici* (*Pst*), is considered one of the most aggressive diseases to wheat production. In this study, we used an iTRAQ-based approach for the quantitative proteomic comparison of the incompatible *Pst* race CYR23 in infected and non-infected leaves of the wheat cultivar Suwon11. A total of 3 475 unique proteins were identified from three key stages of interaction (12, 24, and 48 h post-inoculation) and control groups. Quantitative analysis showed that 530 proteins were differentially accumulated by *Pst* infection (fold changes >1.5, $P<0.05$). Among these proteins, 10.54% was classified as involved in the immune system process and stimulus response. Intriguingly, bioinformatics analysis revealed that a set of reactive oxygen species metabolism-related proteins, peptidyl - prolylcis - trans isomerases (PPIases), RNA-binding proteins (RBPs), and chaperonins was involved in the response to *Pst* infection. Our results were the first to show that PPIases, RBPs, and chaperonins participated in the regulation of the immune response in wheat and even in plants. This study aimed to provide novel routes to reveal wheat gene functionality and better understand the early events in wheat - Pst incompatible interactions.

Key words: Quantitative proteomics; wheat; *Puccinia striiformis* f. sp. *tritici*; Incompatible reaction

Plant Innate and Phytohormone Immunities are Suppressed by Virulence Factors of Geminivirus to form Mutualism Between Virus and Vector

ZHAO Ping-zhi, YAO Xiang-mei, SUN yan-wei, WANG Ning, YE Jian

(*State Key Laboratory of Plant Genomics*, *Institute of Microbiology*, *Chinese Academy of Sciences*, *Beijing* 100101, *China*)

Abstract: Plants have evolved an intricate signaling system to senseinvasions by pathogens or insects. The plant innate immunity and phytohormone [such as jasmonic acid (JA), ethylene (ET) and salicylic acid (SA)] play pivotal roles in signaling plant defense against invasions. Although exogenous-phytohormonecould activated the mitogen-activated protein kinase (MAPK) cascades of plant immunity system, the understanding for which molecular components regulated crosstalk between plant immunity pathway and JA signaling pathway is still unknown. Here, we identified a novel plant resistance factor-WRKY20 as a vital cross point of these two plant basal immunity to pathogen and pest. Interestingly, this transcription factor WRKY20 has been hijacked by a virulence factor of geminivirus-βC1. we will update our new evidences to explain the mechanics of mutualism between virus and vector and how apply the knowledge to design better strategy to control geminivial diseases.

Disease Resistance Through Infection Site Production of a Toxic α-SNAP That Impairs Vesicle trafficking

Andrew Bent

(*Professor, Department of Plant Pathology, University of Wisconsin - Madison, Madison, WI, USA*)

Abstract: Current paradigms for mechanisms of plant disease resistance emphasize expression of R gene products or MAMP receptors that mediate pathogen detection, causing cell wall reinforcement, production of antimicrobial compounds, and in some cases, hypersensitive response (HR) cell death. Paradigms are less well developed for mechanisms of plant defense against biotrophic pathogens that overcome preformed plant defenses, escape detection by R gene products, and secrete effectors that block other typical plant immune responses. We have identified at least part of the mechanism for soybean *Rhg*1, a complex locus that confers resistance to soybean cyst nematode (SCN). Soybean is one of the world's four largest food crops. SCN is by far the most yield-reducing pathogen of soybean in the U. S. SCN often does not cause obvious above-ground disease symptoms, so many growers do not try enough to fight this disease. SCN is also one of the most damaging pathogens of soybean in China. SCN invade plant roots and establish a feeding site called a "syncitium" near the root vascular cylinder. This feeding site takes over and massively reprograms multiple surrounding root cells. For successful growth and reproduction, SCN must keep plant syncitium cells active and cooperative for at least three weeks. *Rhg*1 is widely used by plant breeders and soybean farmers to control SCN, and *Rhg*1 has been known for many years to function by causing early failure of the developing syncitium. We first reported in 2012 that one of the *Rhg*1 genes causing SCN resistance encodes an unusual alpha-SNAP protein. α-SNAP functions with NSF (an ATPase) to mediate vesicle trafficking by forming the core SNARE recycling machinery. α-SNAPs are conserved across eukaryotes. α-SNAP stimulates SNARE complex disassembly by NSF, which enables future rounds of vesicle fusion. We now have evidence that the soybean *Rhg*1α-SNAP interacts less well with NSF, impairs exocytosis and other vesicular trafficking, and is cytotoxic at high doses. This toxicity is dose-dependent, and can be prevented by shifting the cellular ratio toward wild-type α-SNAPs. *Rhg*1α-SNAP levels increase in the SCN-induced syncitium. Hence this very important disease resistance functions by an interesting mechanism: the host carries a toxic variant of a core housekeeping protein, the abundance of the protein is low enough in most tissues to not harm plant yields, but the biotrophic interface established by the pathogen is disrupted by elevated abundance of the protein, which in the case of *Rhg*1 disrupts cellular vesicle trafficking.

Increasing CK Content Enhances Rice Resistance to Sheath Blight Caused by Necrotrophic Pathogen *Rhizoctonia solani*

XUE Xiang[1], WANG Yu[1], LI Lei[1], ZHANG Gui-yun[2], ZHANG Ya-fang[1], CHEN Zong-xiang[1], SUN Ming-fa[1], PAN Xue-biao[1], ZUO Shi-min[1*]

(1. *Jiangsu Key Laboratory of Crop Genetics and Physiology/Co-Innovation Center for Modern Production Technology of Grain Crops, Key Laboratory of Plant Functional Genomics of the Ministry of Education, Yangzhou University, Yangzhou* 225009, *China*; 2. *Institute of Agricultural Sciences in Coastal Region of Jiangsu Province, Yancheng* 224002, *China*)

Abstract: Sheath blight (SB), caused by necrotrophic fungus *Rhizoctonia solani* Kühn (*R. Solani*), is one of the most destructive diseases in rice worldwide. Increasing studies have showed that plant hormone cytokinins (CKs) play an important role in regulation of plant disease resistance. However, its role in enhancing rice resistance to SB remains unknown. Here, we found that exogenous application of synthetic CK kinetin (KT) reduced the severity of *R. Solani*-induced symptoms. Transgenic rice lines with increased endogenous CK levels by engineering CK-biosynthesis gene *IPT* delayed senescence or a stay-green phenotype and significantly enhanced SB resistance in both field tests and detached leaf inoculation assays. In contrast, transgenic rice lines overexpressing *CKX*4, encoding a CK-degradation enzyme, contain significantly lower CKs content compared to wild type and show significantly enhanced susceptibility to SB. We also found a 'stay-green' rice mutant, which maintains high CK content due to loss of function of CK degradation, displays higher resistance to SB at the later developmental stage. We conclude from our microscopic examination that the mechanism leading to enhanced resistance by increasing CK content is due to stronger ability of the transgenic lines or the mutant plant to inhibit cell death caused by *R. solani*, which ultimately results in poor pathogen infection. Importantly, we found that the transgenic lines and the 'stay-green' mutant with increased CK levels are almost the same as WT in their morphological and yield traits. Taken together, our results demonstrate that modulating CK levels is a feasible approach to the development of new rice varieties with excellent SB disease resistance, which is of great importance in rice breeding toward SB resistance.

Key words: Rice; Cytokinins; Sheath blight; Stay-green; Disease resistance

* Corresponding author: ZUO Shi-min; E-mail: smzuo@yzu.edu.cn

Hfq in *Pectobacterum carotovorum* subsp. *cartovorum* Influences the Production of AHL and Carbapenem, Positive Regulates the Expression of T3SS and T6SS, and is Crucial to Virulence and Response of Host Plants

WANG Chun-ting, PU Tian-xin, YAO Pei-yan, LOU Wang-ying, FAN Jia-qin*

(*Department of Plant Pathology*, *Nanjing Agricultural University*, *Nanjing* 210095, *China*)

Abstract: Non-coding small RNAs (sRNAs) have important roles in modulating a wide range of cellular processes and physiological responses in bacteria. In this study, 22 predicted sRNAs in *Pectobacterium carotovorum* subsp. *carotovorum* strain PccS1 were identified and three (ArcZ, CsrB, SraG) of them regulated the pathogenicity. Hfq, the small RNA chaperone, is necessary for the function of its binding sRNAs. In this work, it was found that the mutant of *hfq* deleted fromPccS1 significantly reduced the virulence as well as the activities of the key virulence determinants of exo-enzymes, including pectatelyase, protease, and cellulose. Meanwhile, the production of *N*- (3-oxohexanoyl) -L-homoserine lactone (AHL) and 1-carbapen-2-em-3-acid carboxylic acid (carbapenem), biofilm formation, motility and stability of the cells were all decreased in the *hfq* deletion mutant compared with the wild-type PccS1. The results of quantitative reverse transcription polymerase chain reaction (qRT-PCR) indicated that the relative expression of related pathogenic genes in Δ*hfq* were remarkable attenuated. Furthermore, the examination of callose depositionin *Nicotiana benthamiana* revealed that the leaves inoculated with Δ*hfq* had a much higher level of callose deposition than those inoculated with the wilt-type PccS1. So we deduced that*hfq*was involved in the suppression of cell wall reinforcement by callose deposition. The data of qRT-PCR revealed that the transcript expression level of most the genes encoding the components of the clusters both the type III secretion system (T3SS) and type VI secretion system (T6SS) were seriously decreasedin Δ*hfq*. Meanwhile, the results of Western blot proved that the ability of transport outside of type VI secretion effector (T6SE) Hcp was almost gone in Δ*hfq*. So it is suggestion that Hfqin PccS1influences the production of AHL and carbapenem, positive regulates the expression of T3SS and T6SS, and is crucial to virulence and response of host plants.

Key words: Hfq; *Pectobacterium carotovorum*; Virulence; Regulation

* Corresponding author: FAN Jia-qin; E-mail: fanjq@ njau. edu. cn

Roles of the PHT4 Family in Regulating Programmed Cell Death and Defense in Arabidopsis

LU Hua

(*Department of Biological Sciences*, *University of Maryland Baltimore County*, 1 000 *Hilltop Circle*, *Baltimore*, *MD* 21250, *USA*.)

Abstract: Phosphorus (P) is an essential element for all living organisms and plays a crucial role in many physiological processes. Plants use phosphate transporters (PHTs) to acquire inorganic phosphate (Pi) directly from the soil and toreallocate Pi within the cell and between cells. In a genetic screen aimed to uncover new defense related genes in Arabidopsis, we identified a mutant impaired in the *PHT*4; 1 gene that showed altered immunity and cell death phenotypes. Further characterization of the *PHT*4; 1 mutantand the corresponding gene indicate that *PHT*4; 1is a negative regulator of plant defense, acting upstream of the key defense signaling mediated by salicylic acid. Expression of *PHT*4; 1is likely regulated by the circadian clock gene *CCA*1. PHT4; 1belongs to a six-protein family, five of which including PHT4; 1 werepreviously localized to the chloroplast while PHT4; 6 is the most distantly related member localized to the Golgi. Like *PHT*4; 1, *PHT*4; 6also negatively regulates plant defense and cell death. Interestingly such function of *PHT*4; 6 is largely SA-independent but specifically requires defense genes *EDS*1 and *PAD*4 and jasmonic acid signaling. Other four chloroplast-localized PHT4 proteins (PHT4; 2-4; 4) are not known to affect plant defense when individual corresponding genes were mutated, suggesting redundant function shared between these *PHT*4 genes and *PHT*4; 1. Together our results revealed the roles of members in the PHT4 family in cell death and disease control in Arabidopsis. They also indicate that PHT4; 6 has evolved to be functionally divergent from other PHT4 family members. Information obtained from our study could help to advance the understanding of mechanisms of defense signaling and eventually enable us to use such knowledge to improve disease-resistance traits in economically important crops.

*OsGLO*1 Mediates Disease Resistance Against Rice Blast Through the Jasmonic Acid-signaling Pathway

YU Dong-li[1,2*], SONG Xiao-ou[1,2*], WANG Jian-sheng[1,2], SHENG Cong[1,2], BAO Ya-lin[1,2], LIN Si-yuan[1,2], WANG Xiu-juan[1,2], LU Wei[1,2], Ayesha Baig[3], NIU Dong-dong[1,2], ZHAO Hong-wei[1,2**]

(1. *College of Plant Protection*, *Nanjing Agricultural University*, *Nanjing* 210095, *China*; 2. *Key Laboratory of Integrated Management of Crop Diseases and Pests* (*Nanjing Agricultural University*), *Ministry of Education*; 3. *Environmental Sciences*, *COMSATS Institute of Information Technology* (*CIIT*), *Abbottabad*, *Pakistan*)

Abstract: Glycolate-oxidase (GLO) mediated H_2O_2 formation is one of the two processes producing H_2O_2 during photosynthesis. It has been reported that GLO mediated H_2O_2 production is associated with abiotic stresses such as drought. Here we showed that *OsGLO*1 was specifically suppressed by the infection of a compatible *Magnorporthae* oryzae (*M. oryzae*) strain Guy11 at both RNA and protein level, while in rice infected with an incompatible strain (2539), expression of *OsGLO*1 remain unchanged. *In vivo* analysis revealed that expression of *OsGLO*1 was positively correlated with H_2O_2 accumulation and callose deposition, and resistance related genes such as *OsPBZ*1, *OsPAD*4, and *OsAOS*2. Our results indicate that *OsGLO*1 is a defense response component involved in disease resistance against rice blast, in which the Jasmonic acid (JA) signaling pathway may be employed.

Key words: Rice blast; Disease resistance; Glycolate-oxidase; Proteomic; Jasmonic acid; Hydrogen peroxide

---

* These authors contributed equally to this work

** Corresponding author: WANG Xiu-juan; E-mail: hzhao@ njau. edu. cn

Overexpression of *OsOSM*1 Gene Enhanced Rice Resistance to Sheath Blight Caused by *Rhizoctonia solani*

XUE Xiang, CAO Zi-xiang, ZHANG Xu-ting, WANG-Yu, ZHANG Ya-fang, CHEN Zong-xiang, PAN Xue-biao, ZUO Shi-min*

(*Jiangsu Key Laboratory of Crop Genetics and Physiology/Co-Innovation Center for Modern Production Technology of Grain Crops, Key Laboratory of Plant Functional Genomics of the Ministry of Education, Yangzhou University, Yangzhou* 225009, *China*)

Abstract: Sheath blight (SB), caused by *Rhizoctonia solani*, is one of the most destructive rice diseases worldwide. It has been difficult to generate SB resistant variety through conventional breeding because of typically quantitative nature of rice resistance to SB. Previously, we found that the gene Os12g0569500 (named *OsOSM*1), belongs to TLP-PA subfamily, was strongly induced expression by R. solani in SB resistant variety YSBR1. In the study, we identified five genes including *OsOSM*1 that showed high identity on amino acid sequence. Interestingly, only *OsOSM*1 was found highly expressed at rice booting stage and in leaf sheath, which was correlated with SB serious development in rice. Transcription of *OsOSM*1 was strongly induced by methyl jasmonate (MeJA) and reduced by JA biosynthesis inhibitor salicylhydroxamic acid (SHAM), while was not apparently affected by ethephon (ET), salicylic acid (SA) and kinetin (KT). OsOSM1 was localized to plasma membrane. Overexpression of *OsOSM*1 could significantly up-regulate the expression of pathogenesis-related genes and strongly enhanced rice resistance to SB. Intriguingly, we did not found apparently alternation of *OsOSM*1 overexpression lines compared with the wild type on morphologies and grain yield. Taken together, our results demonstrate that *OsOSM*1 plays an important role in rice SB resistance and can be used in rice breeding toward SB resistance.

Key words: Rice; *OsOSM*1; Gene expression; Sheath blight resistance; Agronomic traits

* Corresponding author: ZUO Shi-min; E-mail: smzuo@ yzu. edu. cn

The Interactions Between Some Commonly Antagonistic Microbes in Tobacco Fields of Luoyang*

DING Yue-qi**, SONG Xi-le, KANG Ye-bin***

(*College of Forestry*, *Henan University of Science and Technology*, *Luoyang* 471003, *China*)

Abstract: In order to screen out the antagonistic microorganisms with benign symbiosis and enhance the control effects of tobacco soil-borne diseases by using multiple biocontrol strains, seventeen species of antagonistic microorganisms (mainly included 6 species of *Trichoderma*, 4 species of bacteria and 7 species of actinomycetes), which were screened and proved that have strong inhibition effects on common tobacco root pathogenic fungus from tobacco growing soil of Luoyang, were used as the research object of this experiment on the basis of previous studies. In this test, we measured the interactions between these antagonistic microorganisms by means of a method called improved confrontation culture. The mainly results obtained are as follows:

(1) The confront culture results of *Trichoderma* strains showed that: most of *Trichoderma* strains ghave inhibition effects with each other. Among them, T1, T2 and T3 are selected for their well compatibilities, which suggested that these three species of strains have symbiotic abilities. The inhibition effects are usually expressed as the competition of growth space, such as the impaction of aerial mycelium growth and sporulation ability. At the same time, there was a certain relationship between the inhibition effect and the mycelial growth rate.

(2) The experimental results of the compatibilities between four tested antagonistic bacterial strains multi-performance to be compatible. Among them, the mutual inhibition rate of compatible combination B1-B2 and B3-B4 below 20%, which showed up complete compatibilities and could be used as mixed objects; however, as incompatible combination B3 has strong inhibitory effects on B1, whose inhibition rate could reach 100% and can not be mixed.

(3) The confrontation results between seven speices of antagonistic actinomycetes showed that there were some inhibitory effects on different strains. The interactions between antagonistic actinomycetes were often expressed through influencing hyphae growth as well as changing the color of the aerial hyphae. Through comparative analysis, we may safely draw the conclusion that the combination of LA5, SA57, LB27 and YB4 almost have no obvious inhibition effects.

(4) When analyze the compatibilities between these screened biocontrol strains of above tests, the results showed that T1, T2, T3, SA57, LA5 and YB4 were compatible with each other; T1, T2, T3, B1, B2 and B4 could also be used as mixed objects. The inhibitory effects of antagonistic actinomycetes to antagonistic bacteria were very strong, whose inhibitory rate could achieve 100%.

This study shows that, there were inhibition effects between antagonistic microbes, and these in-

* 基金项目：河南省烟草公司科技项目（2012M10）；洛阳市烟草公司项目（2014M06）

** 第一作者：丁玥琪，在读硕士研究生，主要从事植物免疫学研究

*** 通讯作者：康业斌，教授，博士，主要从事植物免疫学研究；E-mail：kangyb999@163.com

compatibilities can directly effect the control efficiency of mixed biological agents. The screened strains with high compatibilities in this experiment can provide theoretical basis for the further using of controlling the soil-borne diseases of tobacco.

Key words: Antagonistic microbes; Interactions; Confrontation culture

丁香假单胞大豆致病变种S1菌株的$HrpZ_{PsgS1}$蛋白诱导抗病性和促进植物生长的功能域研究*

伍辉军，张宏月，顾　沁，高学文**
（南京农业大学植物保护学院植物病理学系，农作物生物灾害综合治理教育部重点实验室，南京　210095）

摘　要：植物病原细菌丁香假单胞菌编码产生的HrpZ蛋白能够在非寄主植物和抗病植物上激发过敏（hypersensitive response，HR）反应，具有诱导植物抗病性及促进植物生长等有益生物学效应。本研究以丁香假单胞大豆致病变种S1菌株的$HrpZ_{PsgS1}$为对象，研究该HrpZ蛋白中决定激发HR、诱导抗病和促生等生物学效应的功能结构域。生物信息学分析表明，HrpZPsgS1含有1个β折叠和9个α螺旋，这9个α螺旋对蛋白质的空间结构的形成至关重要。本研究利用常规PCR和反向PCR等构建了11个$HrpZ_{PsgS1}$突变体，这些突变体缺失了不同的α螺旋区域或者连接区域。诱导烟草HR活性检测表明，缺失了$HrpZ_{PsgS1}$C-端的重组蛋白$HrpZ_{1\text{-}102}$、$HrpZ_{\triangle 195\sim 238}$、$HrpZ_{\triangle 241\sim 248}$、$HrpZ_{\triangle 254\sim 298}$、$HrpZ_{\triangle 290\sim 313}$诱导HR能力降低；除$HrpZ_{200\sim 346}$，缺失了N-端的重组蛋白诱导HR能力与全长没有差别；而$HrpZ_{74\sim 204}$和$HrpZ_{1\sim 194}$则丧失了诱导HR的能力，但仍能够激发烟草的微敏反应。这表明，$HrpZ_{PsgS1}$的C-端是负责调控HR的关键功能域，缺失C-端的重组蛋白HR能力减弱，而N-端的特定部分对HR具有调控作用。诱导植物抗病性和促进植物生长的活性检测结果表明，突变体$HrpZ_{\Delta 89\sim 124}$和$HrpZ_{\Delta 254\sim 298}$诱导烟草抗TMV、诱导水稻抗白叶枯病及促进水稻生长的活性都显著增强。同时Real-time PCR检测结果表明，$HrpZ_{\Delta 89\sim 124}$和$HrpZ_{\Delta 254\sim 298}$激活了比全长更高水平的HR、抗病、促生相关基因的转录表达。本研究表明，重组蛋白$HrpZ_{\Delta 89\sim 124}$和$HrpZ_{\Delta 254\sim 298}$具有开发成生物蛋白农药的潜力。

关键词：$HrpZ_{PsgS1}$；HR；抗病；促生

---

* 基金项目：国家863项目（2012AA101504）

** 通讯作者：高学文，博士，教授，从事生物防治与细菌分子生物学研究；E-mail：gaoxw@ njau. edu. cn.

假禾谷镰刀菌侵染诱导小麦抗感品种茉莉酸途径相关基因的差异表达分析*

李永辉，王利民，陈琳琳，李洪连，丁胜利**
（河南农业大学植物保护学院，郑州 450002）

摘 要：小麦茎基腐病是一种世界性病害，能够造成严重的损失。黄淮麦区引起小麦茎基腐病的优势病原菌是假禾谷镰刀菌（*Fusarium pseudograminearum*）。为了研究小麦受假禾谷镰刀菌侵染后的互作分子机理，我们选用假禾谷镰刀菌强致病力菌株WZ2-8A，小麦高感品种国麦301和中抗品种周麦24，采用转录组测序的方法，对小麦受到假禾谷镰刀菌侵染后基因的差异表达进行分析。基于茉莉酸途径在植物的抗病过程中起到的重要作用，针对茉莉酸途径中相关基因，我们对两个小麦品种在接菌后5d和15d的茉莉酸途径差异表达的基因进行生物信息学分析，初步探讨了小麦受假禾谷镰刀菌侵染后茉莉酸途径基因的表达动态。

小麦在假禾谷镰刀菌胁迫下涉及茉莉酸途径的26个基因，其中LOX基因有7个，有3个在两个品种中表达量均无明显变化；有2个在两个品种中表达量变化趋势一致。一个和编码脂肪氧合酶2.2（*LOX*2.2）的高度同源基因在国麦301中的第5d表达量上调50倍，第15d表达量下降3.9倍；在周麦24中第5d和对照比较表达量下调6.9倍，第15d和第5d比较表达量下调5.3倍。一个和编码脂肪氧合酶8（*LOX*8）的高度同源的基因在国麦301中第5d的表达量上调9.9倍，维持该水平到第15d；在周麦24中表达量持续上调，第5d上调6.1倍，第15d和第5d相比上调18.9倍。涉及13个12-OPDR基因，其中有9个基因在周麦24中第5d和第15d表达量持续上调，在这9个基因中有7个基因表达量变化均在5倍以上。这9个基因在国麦301中有6个基因第5d表达量上调，到第15d下降。2个基因在第5d和对照比较表达量基本不变，第15d下调表达。1个基因在第5d下调表达，第15d上调表达。涉及JAZ的基因有6个，这6个基因在周麦24中均持续上调表达，其中4个基因差异表达倍数均在5倍以上。

综合上述研究结果表明，假禾谷镰刀菌侵染主栽小麦品种国麦301和周麦24后，影响寄主小麦茉莉酸合成和代谢，诱导表达的这些基因可能参与小麦对假禾谷镰刀菌的抗病过程，具体作用还要进一步的实验进行研究。

关键词：假禾谷镰刀菌；差异表达；茉莉酸

---

* 国家公益性行业（农业）科研专项（201503112）；河南省科技计划项目（152300410073），河南农业大学人才引进项目启动基金（30600861）资助

** 通讯作者：丁胜利，河南农业大学特聘教授；E-mail：shengli－ding@163.com

大豆疫霉效应分子 PsCRN63 调控植物先天免疫及胞内二聚化的分子机制

李　琦[1,2]，张美祥[1]，沈丹宇[1]，刘廷利[1]，陈彦羽[1]，周俭民[2]，窦道龙[1]
（1. 南京农业大学植物保护学院，南京　210095；
2. 中国科学院遗传与发育生物学研究所，北京　100101）

摘　要：植物病原卵菌通过产生大量效应分子来侵染寄主，然而这些效应分子的作用方式在很大程度上是未知的。我们研究发现，大豆疫霉的效应分子 PsCRN63 能够抑制病原物相关分子模式（pathogen-associated molecular patterns，PAMP）触发的免疫（PAMP-triggered immunity，PTI）的标记基因，flg22 诱导的 *FRK*1 基因的表达。但是，PsCRN63 不能抑制 PTI 上游信号通路的相关事件，包括 flg22 诱导的 MAPK 的激活以及 BIK1 的磷酸化，这表明它作用于 MAPK 级联反应的下游。*PsCRN*63 转基因拟南芥植株对病原细菌丁香假单胞菌番茄致病变种（*Pseudomonas syringae patho* var *tomato*，*Pst*）DC3000 和病原卵菌辣椒疫霉（*Phytophthora capsici*）的敏感性增强。另外，与野生型植株相比，*PsCRN*63 转基因植株中 flg22 诱导的活性氧暴发和胼胝质沉积均受到抑制。同时，PTI 途径相关基因的表达在 *PsCRN*63 转基因植株中也受到下调。有趣的是，我们发现 PsCRN63 蛋白的 N 端和 C 端能够通过反向连接的方式在植物细胞内发生互作，从而形成一个同源二聚体。另外，形成二聚体所需的 N 端和 C 端结构域在 CRN 效应分子中非常保守，这暗示着疫霉 CRN 效应分子的同源/异源聚合物的形成为其发挥生物学功能所必需。事实证明，二聚体的形成确为 PsCRN63 行使 PTI 抑制及细胞死亡诱导功能所必需。

关键词：疫霉；效应分子；PsCRN63；植物免疫；二聚化

100个小麦品种资源抗条锈性鉴定及重要抗条锈病基因的SSR检测*

孙建鲁[1,2,3**]，王吐虹[2]，冯　晶[2,1***]，蔺瑞明[2]，王凤涛[2]，姚　强[1,3,4]，郭青云[1,3,4***]，徐世昌[2]

（1. 青海大学农牧学院，西宁　810016；2. 中国农业科学院植物保护所植物/病虫害生物学国家重点实验室，北京　100193；[3] 青海省农林科学院青海省农业有害生物综合治理重点实验室，西宁　810016；4. 农业部西宁作物有害生物科学观测实验站，西宁　810016）

摘　要：了解小麦品种资源对中国条锈菌生理小种抗病性水平及其所含重要抗条锈病基因，可为合理种植和利用小麦品种资源提供理论依据。选用中国小麦条锈菌毒性稳定小种CYR17、流行小种CYR32和条锈菌混合小种分别对100个小麦品种资源进行苗期和成株期抗条锈性鉴定，并采用SSR分子标记技术检测其所含重要抗条锈病基因 *Yr*5、*Yr*10、*Yr*15、*Yr*18、*Yr*24和 *Yr*26。结果表明，在苗期对CYR17和CYR32表现抗病的分别为84个和59个品种。在成株期对CYR17表现抗病的有67个品种，感病的有3个品种，表现慢锈性的有30个品种；对CYR32表现抗病的有68个品种，感病的有9个品种，表现慢锈性的有23个品种；对混合小种表现抗病的有72个品种，感病的有5个品种，表现慢锈性的有23个品种。在鉴定中保持稳定抗病性的有StarkeⅡ、兰天15号等11个品种。SSR检测发现，在供试品种中重要抗病基因 *Yr*5、*Yr*10、*Yr*15、*Yr*18、*Yr*24、*Yr*26分布频率不均衡，分别占供试品种的2.00%、37.00%、3.00%、19.00%、25.00%、24.00%。其中，*Yr*5、*Yr*15和 *Yr*18检出率相对较低，*Yr*10、*Yr*24和 *Yr*26检出率较高。供试品种对CYR17、CYR32和混合小种表现出的抗病性主要由 *Yr*10、*Yr*18、*Yr*24、*Yr*26及未知基因控制。StarkeⅡ、兰天15号等11个品种在试验中对供试小麦条锈菌生理小种均表现免疫，在抗病品种选育中有重要应用价值。

关键词：小麦条锈病；抗病性鉴定；*Yr*基因；SSR分子标记

---

* 基金项目：国家自然科学基金（31261140370、31272033）；国家重点基础研究发展计划（973计划）（2013CB127700）和“十二五”国家科技支撑计划（2012BAD19B04）

** 第一作者：孙建鲁，硕士研究生，青海大学，研究方向为小麦条锈病抗病遗传机制

*** 通讯作者：冯晶，副研究员，硕士生导师，主要从事小麦抗病性研究；E-mail：jingfeng@ippcaas.cn

郭青云，研究员，硕士生导师，从事植物保护研究；E-mail：guoqingyunqh@163.com

PR 蛋白参与小麦抗叶锈病防御反应的研究*

王海燕

摘　要：病程相关蛋白（pathogenesis related proteins，PRP/PRs）是防卫反应基因编码的重要产物，其分布广泛，在不同植物中均发现此类基因的存在，在植物抗病性和系统获得抗性中发挥着重要作用。PR 蛋白的相关报道较多，也常在植物 R 基因或抗病相关基因的研究中被作为报告基因。但 PR 作用的基本机制没有明确的研究结果，尤其在小麦抗叶锈病过程中的作用尚不明确。为此，本研究对 PRP 家族中的 PR1、PR2 和 PR5 基因开展研究。首先，利用同源克隆法结合 RACE 技术在小麦抗叶锈病近等基因系中获得 3 类 PR 基因，并分析基因结构特征；然后，在核酸水平上验证目的基因与小麦抗叶锈病的关系：利用 qPCR 技术从转录水平获得 PR 基因在小麦-叶锈菌亲和组合和非亲和组合中的表达模式，并分析其在小麦不同组织器官中的表达特异性；蛋白水平上验证目的基因与小麦抗叶锈病的关系：利用 Western 杂交技术从翻译水平获得候选基因在小麦-叶锈菌亲和组合和非亲和组合中的表达模式；最后，利用转基因和 VIGS 技术验证目的基因与小麦叶锈病抗性的相关性。研究结果为全面、详细地分析目的基因在小麦抗叶锈病中的作用提供实验依据，同时，这些基因的获得丰富了小麦的基因库，为小麦抗叶锈病育种提供了丰富的基因资源，具有重要的理论意义和应用价值。

* 国家自然科学基金青年基金项目“小麦 *TaLr19TLP1* 基因抗叶锈病功能鉴定及互作蛋白的筛选”（31501623）；河北省自然科学基金“叶锈菌与 TcLr19 小麦互作体系中病程相关蛋白基因的克隆与功能分析”（C2012204005）

植物抵御链格孢菌的植保素 Scopoletin 的激素调控机理

吴劲松*

（中国科学院昆明植物研究所资源植物和生物技术重点实验室，昆明 650201）

摘　要：赤星病是烟草属最重要的真菌性病害，它是由营腐生性生活的病原真菌链格孢菌（*Alternaria alternata* tobacco pathotype 或赤星病菌）致病产生。该病原菌主要通过叶片的气孔侵入。当植物感受到该真菌的入侵后，可以通过激活茉莉酸信号在真菌侵入部位大量的诱导抗菌的香豆素类化合物 Scopoletin 来抵御病原菌的入侵。链格孢菌的入侵，可以直接诱导植物积累更多的茉莉酸。而当茉莉酸合成缺陷的植物即使在感受到真菌入侵的情况下也不再能积累 Scopoletin，在外源施加茉莉酸甲酯的情况下可以部分恢复产生 Scopoletin 的能力。不能感受茉莉酸的植物和不能产生茉莉酸的植物一样，均不能产生 Scopoletin，而且该植物亦非常的感病。然而当植物被外源茉莉酸处理时，植物并不能直接合成 Scopoletin。这个结果意味着植物还需要别的信号通路，和茉莉酸途径一起来诱导 Scopoletin 的生物合成。结合转录组和乙烯信号通路缺失的转基因植物，我们发现植物还可以激活乙烯信号通路，与茉莉酸一起协同诱导 Scopoletin 的生物合成。

关键词：赤星病；茉莉酸；链格孢；激素调控

* 通讯作者：吴劲松；E-mail：jinsongwu@ mail. kib. ac. cn

GmCYP82A3, a Soybean Cytochrome P450 Family Geneinvolved in the JA Signaling Pathway, Enhancesplant Resistance to Biotic and Abiotic Stresses

YAN Qiang[1], CUI Xiao-xia[2], LIN Shuai[1], GAN Shu-ping[2], XING Han[2], DOU Dao-long[1*]

(1. *Department of Plant Pathology*, *Nanjing Agricultural University*, *Nanjing* 210095, *China*; 2. *National Center for Soybean Improvement*, *National Key Laboratory of Crop Genetics and Germplasm Enhancement*, *Nanjing Agricultural University*, *Nanjing* 210095, *China*)

The cytochrome P450 monooxygenases represent a large and important enzyme superfamily in plants. The enzymes catalyze numerous monooxygenation/hydroxylation reactions in biochemical pathways. They are also involved in metabolicpathways and participate in the homeostasis of plant hormones. The *CYP*82 family of genes specifically resides in dicots and is usually induced by distinct environmental stresses. However, their functions are largely unknown, especially in soybean (*Glycine max* L.) and *Phytophthora sojae*interaction. We report *GmCYP*82*A*3, a gene from the soybean *CYP*82 family. Its expression was induced by *P. sojae* infection, salinity and droughtstresses, and treatment withmethyl jasmonate (MeJA) or ethephon (ETH). Theexpression levels wereconsistently high in resistant cultivars. *Nicotianabenthamiana*transgenic plants expressing *GmCYP*82*A*3 exhibited strong resistance to *Botrytis cinerea*and *P. parasitica*, and enhanced tolerance to salinity and drought stresses. Furthermore, transgenic plants were less sensitive to jasmonic acid (JA), and the enhanced resistance was accompanied by increased expression of the JA/ETsignaling pathway-related genes and enhanced suberin deposition.

* Corresponding author: DOU Dao-long; E-mail: ddou@ njau. edu. cn

用酵母双杂交筛选和验证与 WYMV P1 互作的蛋白*

刘丽娟，戚文平，孙炳剑**
（河南农业大学植物保护学院，郑州 450002）

摘　要：小麦黄花叶病是冬小麦上重要的病毒病害之一，严重影响小麦的产量和品质。小麦黄花叶病毒（Wheat yellow mosaic virus，WYMV）是马铃薯 Y 病毒科病毒（Potyviridae）大麦黄花叶病毒属（Bymovirus）成员之一，由土壤进行传播。在我国江苏、河南、安徽、山东、陕西、湖北、四川等省均有分布，并且呈逐年蔓延趋势。WYMV 有两条正义的 RNA 链组成，RNA2 全长 3 656个核苷酸，经蛋白酶切割后产生 P1、P2 两个成熟蛋白。目前，关于 WYMV RNA2 基因功能的研究较少对其编码的蛋白功能也只是推测，已知与马铃薯 Y 病毒属成员 HC-Pro 的中心区域和 N 端区域有许多共有的保守区域和位点，但这些保守序列在蛋白功能方面的作用仍不完全清楚。目前有关 P1 蛋白在病毒侵染、复制、传播、以及寄主互作中的作用还未报道。为了筛选与 WYMV P1（Wheat yellow mosaic virus P1 protein）互作的小麦蛋白并验证其互作。本研究的主要结果如下：

根据 GenBank 公布的 WYMV-P1 的基因序列设计一对特异性引物，用感染 WYMV 的小麦为材料提取小麦总 RNA 后反转录成 cDNA，以 cDNA 为模板获得 P1 序列，构建诱饵载体 PGBKT7-P1。利用酵母双杂交从小麦 cDNA 文库中筛选出 8 个小麦候选蛋白，将筛选到的候选蛋白与诱饵载体 PGBKT7-P1 共转到酵母 Y2HGold 中初步验证这些候选蛋白与 WYMV-P1 互作。用 Westen 检测诱饵载体 PGBKT7-P1 能否在酵母 Y2HGold 中正常表达。从筛选出的小麦蛋白中选取 2 个即小麦叶绿体 Rubisco（Ribulose bisphosphate carboxylase oxygenase）大亚基命名为 RbcL，Rubisco 小亚基命名为 RbcS，设计引物，扩出全长，构建重组载体与 P1 重组载体共转酵母 Y2HGold 中验证 RbcL，RbcS 和 P1 的互作关系。研究结果表明：酵母双杂交初步验证结果表明利用酵母双杂交筛选到的 8 个小麦候选蛋白均与 WYMV-P1 互作；Westen 检测结果显示 PGBKT7-P1 诱饵载体能正常在酵母 Y2HGold 中表达；初步从 cDNA 文库中筛选出来的 RbcS 扩出全长后存在互作与 P1 存在互作。

关键词：小麦黄花叶病毒；酵母双杂交；蛋白互作

---

* 基金项目：NSFC－河南人才培养联合基金项目（U1304322）；国家公益性行业（农业）科研专项（201303021）
** 通讯作者：孙炳剑，副教授；E-mail：sbj8624@ sina. com

长春花 *CrCBSX3* 基因的克隆及功能初探

李　艳[1,2]，陈　旺[1]，刘胜毅[1]，吴云锋[2*]
（1. 中国农业科学院油料作物研究所，武汉　430062；2. 西北农林科技大学植物保护学院，旱区作物逆境生物学国家重点实验室，杨凌　712100）

摘　要：长春花［*Catharanthus roseus*（L.）G. Don］是植原体繁殖过程中广泛应用的繁殖寄主，在植原体侵染后通常会表现出丛枝、花绿变及花变叶等典型症状。胱硫醚 β 合成酶结构域蛋白（cystathionine-synthase（CBS）domain - containing proteins，CDCPs）含有进化上保守的 *CBS* 结构域，而且在古生物、细菌和真核生物中普遍存在。*CBS* 结构域位点的突变会导致人类的很多遗传性疾病，同时推测 *CBS* 结构域可能作为细胞能量代谢的传感器，维持细胞内的氧化还原。实验室之前分析的长春花在感染小麦蓝矮植原体后的差异表达基因中，*CBS* 基因在小麦蓝矮植原体侵染后的表达量上升。为了研究 *CBS* 基因在小麦蓝矮植原体与长春花互作中的功能，本实验通过 RACE 技术，克隆到了长春花的 *CBSX*3 基因，全长 1 051bp，ORF 为 618bp，编码 205 个氨基酸，由两个 CBS 结构域串联而成。荧光定量结果显示 *CrCBSX*3 基因在长春花嫁接小麦蓝矮植原体 10 天后受到强烈诱导表达；在花器官中的表达量最高，根中的表达量最低。在本氏烟叶片中瞬时表达该基因，结果显示 *CrCBSX*3 在本氏烟叶片中能够抑制 BAX 诱导的细胞坏死。以上实验结果表明 *CrCBSX*3 很可能在长春花与小麦蓝矮植原体的早期互作中发挥了作用。

关键词：长春花；*CrCBSX*3；小麦蓝矮植原体；互作

* 通讯作者：吴云锋；E-mail：wuyf@ nwsuaf. edu. cn

长链非编码 RNA 响应水稻稻瘟病和纹枯病的表达分析

牛冬冬，圣 聪，张 鑫，乔露露，金海翎，赵弘巍*
（南京农业大学植物保护学院植物病理系，江苏 210095）

摘 要：小分子 RNA 是一类 20～24 核苷酸长度的非编码 RNA，它可以与含 Argonaute（AGO）蛋白的核酶复合物结合形成 RNA 诱导沉默复合体，切割靶标 mRNA 或抑制其翻译。小分子 RNA 可以参与植物的免疫防卫机制。而长链非编码 RNA（Long non-coding RNA）是一类 24～40 核苷酸长度的 RNA 分子，参与植物抗病反应以及 DNA 的甲基化。

水稻稻瘟病和纹枯病是我国稻区的主要病害，为害严重，长链非编码 RNA 在水稻抗病中的作用还不清楚。本研究分别在稻瘟病菌 Guy11 和纹枯病菌 *R. solani* AG1 IA 侵染水稻日本晴后 0 天和 1 天取样，提取总 RNA。通过截取 24～40nt 的 RNA 片段，在 3′端和 5′端连接接头等手段，构建了小分子 RNA 文库。通过高通量测序发掘在病原菌侵染后变化显著的长链小分子 RNA。生物信息学分析发现，在 24～40nt 之间有大量的长链非编码 RNA，主要分布在基因组的重复序列区。通过 Northern blot 检测发现在稻瘟病侵染后，siR76113，siR67823 和 siR72573 表达水平下降；在纹枯病菌侵染后，siR51031 表达水平降低，而 siR118183 和 siR194568 表达水平升高。通过 qRT-PCR 检测发现，上述 siRNAs 的靶标基因的表达水平与对应 siRNAs 的表达呈现相反趋势，表明水稻长链 siRNAs 可能通过调控靶标基因的表达介导植物的抗病免疫反应。这些 siRNAs 的功能有待进一步研究。

关键词：长链非编码 RNA；稻瘟病；纹枯病

---

* 通讯作者：赵弘巍；E-mail：hzhao@ njau. edu. cn

Expression Analysis and Functional Characterization of a Pathogen Induced Thaumatin-like Protein Gene in Wheat Conferring Enhanced Resistance to *Puccinia triticina*

ZHANG Yan-jun*, ZHANG Jia-rui, WEI Xue-jun, WANG Hai-yan*, LIU Da-qun*

(*Authors' addresses: Center of Plant Disease and Plant Pests of Hebei Province, College of Plant Protection, Agricultural University of Hebei, Baoding* 071001, *China*)

Pathogenesis-related (PR) protein-5, also known as thaumatin-like protein (TLP), is encoded by a complex group of genes that are involved in host defense against biotic and abiotic stresses as well as regulation of physiological processes in numerous plant species. Our earlier studies reported the isolation of a full-length *TaLr*19*TLP*1 gene (516bp) from wheat infected by leaf rust (*Puccinia triticina*). Quantitative real-time polymerase chain reaction (qRT-PCR) analyses revealed that *TaLr*19*TLP*1 transcript was significantly induced and upregulated during incompatible interaction while a relatively low level of the transcript was detected during compatible interaction. Here, we demonstrate that the accumulation of *TaLr*19*TLP*1 transcript is significantly different in tested wheat organs. *TaLr*19*TLP*1 was induced by salicylic acid (SA), methyl jasmonic (MeJA), ethephon (ETH) and abscisic acid (ABA). The transcripts of *TaLr*19*TLP*1 accumulated at higher levels following pre-treatment with SA, MeJA, and ABA prior to infection with *P. triticina*. A slight induction was observed in ETH pre-treated seedlings compared with treatment without inoculation. In addition, *TaLr*19*TLP* was found to be predominately localized to extracellular spaces of onion epidermal cell. Knocking down the expression of *TaLr*19*TLP* through virus-induced gene silencing (VIGS) reduced wheat resistance against leaf rust pathogen. These results suggest that *TaLr*19*TLP*1 mediated disease resistance in wheat exposed to leaf rust pathogen.

* Corresponding authors: WANG Hai-yan; E-mail: ndwanghaiyan@163.com
LIU Da-qun; E-mail: ldq@hebau.edu.cn

本氏烟草 *AGO1* 基因的 amiRNA 表达载体构建及干扰效果分析

张其猛[1,2]*，哈　达[1,2]**
（1. 内蒙古大学生命科学学院，呼和浩特　010000；2. 内蒙古自治区牧草与特色作物生物技术重点实验室，呼和浩特　010000）

摘　要：AGO（Argonaute）是一类高度保守的蛋白家族，通过与 small RNAs 形成沉默复合体参与植物转录水平和转录后水平的基因沉默，对基因沉默和植物抗病性起重要调控作用。AGO1 是 AGO 蛋白家族的基础成员。本研究通过构建 amiRNA（artificial microRNAs）载体，旨在对 *AGO1* 基因进行特异性沉默，并分析 amiRNA 对 *AGO1* 基因的沉默效果。利用 WMD（Web microRNA designer）平台和 http：//benthgenome. qut. edu. au/网站设计针对 *AGO1* 的 amiRNA 序列，利用 PCR 技术扩增拟南芥 miR159b 前体序列，构建特异性沉默 *AGO1* 的 amiRNA 重组载体。本研究以模式植物本氏烟草（*Nicotiana benthamiana*）为研究材料，利用含有 amiRNA 重组载体的农杆菌介导法侵染本氏烟草叶片，分别在 2d，5d，10d，15d 收集侵染叶片。利用 RT-qPCR 分析 *AGO1* 基因的表达量。结果表明，2d，5d 收集的侵染叶片 *AGO1* 基因的表达量基本没有变化，10d 时 3 个 amiRNAs 对 *AGO1* 基因的沉默效率分别达到了 75. 2%，61. 5%，9. 0%，15d 时沉默效率不如侵染 10d 显著。由此我们得出结论 10d 时人工构建的 miRNA 在本氏烟草中成功表达，实现了对靶基因的切割，达到了对 *AGO1* 基因沉默的效果。为后期探索 *AGO1* 对植物抗病效果的研究奠定了基础。

关键词：*AGO1*；amiRNA；本氏烟草；RNAi

* 第一作者：张其猛，河北沧州人，硕士研究生，主要从事植物与病原相互作用研究；E-mail：lidez@ imu. edu. cn
** 通讯作者：哈达，博士，教授，主要从事植物基因工程与植物分子生物学研究；E-mail：nmhadawu77@ imu. edu. cn

miRx 调控水稻稻瘟病抗性和农艺性状*

李金璐**，赵胜利，肖之源，王　贺，曹小龙，杨　柟，
樊　晶，黄　富，李　燕***，王文明***
（四川农业大学水稻研究所，成都　611130）

摘　要： microRNAs（miRNAs）是一类相对保守，长度为 20～25 个碱基的非编码 RNA，广泛存在于原核和真核生物中，可以在转录水平上切割靶 mRNA 或者在翻译水平阻止靶基因的翻译从而调控基因的表达，是真核生物生长发育和抗性的重要调控因子。ETMs（endogenous microRNA target mimic）是 miRNAs 的内源模拟靶标，通过诱捕对应的 miRNAs 抑制 miRNAs 对其靶基因的调控作用。到目前为止，关于 miRNAs 的报道已经很多，但是关于水稻 miRNAs 对稻瘟病抗性的研究较少。在实验室的前期工作中，通过高通量测序在正常情况下和受稻瘟菌侵染下的感病水稻株系 LTH 和抗病株系 IRBLkm-Ts，筛选得到 30 余个在稻瘟菌侵染条件下表达量有显著变化的水稻 miRNAs。其中，miRx 在稻瘟病抗性中被预测为正调控因子。在 miRx 过量表达株系上，活体/离体接稻瘟菌后鉴定病斑，喷菌检测抗性基因转录水平，DAB 染色鉴定 H_2O_2 累积，和 Trypan Blue 染色观察菌丝等一系列抗性检测实验证明，过量表达 miRx 增强了水稻对稻瘟病的抗性。

另外，miRx 过表达株系表现出明显的农艺性状改变，其分蘖角度变大，籽粒宽度增加，千粒重增加。为了进一步探究 miRx 对水稻稻瘟病抗性与农艺性状的调控机理，我们构建了 miRx 的内源模拟靶标（ETM）的过表达水稻材料。通过实时荧光定量 PCR 检测，证明在不同的株系中，miRx 的靶基因表达显著高于对照材料，说明 ETM 有效消弱了 miRx 对靶基因转录的抑制。过表达 ETM 的植株在农艺性状上也发生了改变，表现为生长周期显著延长，结实率显著降低。同时，我们构建了 miRx 靶基因的过表达转基因植株和缺失突变体植株。下一步的实验将通过对 miRx 靶基因过表达和沉默的转基因植株进行抗病性鉴定和发育表型分析，从而勾画 miRx 调控水稻稻瘟病抗性和农艺性状的初步图谱。

关键词： miRNA；ETM；稻瘟病；内源模拟靶标；miRx

---

*　基金项目：国家自然科学基金（31430072 和 31471761）
**　第一作者：李金璐，硕士研究生，主要研究水稻 microRNA 调控稻瘟病抗性的分子机制；E-mail：617271733@ qq. com
***　通讯作者：李燕，E-mail：jiazaihy@ 163. com
王文明，E-mail：j316wenmingwang@ 163. com

Identification of Broad Spectrum Rice Blast Resistance Genes with IRRI Rice Monogenic Lines

WANG Ji-chun [1,2*], KIM Dong-run [3], WU Xian [1,2], REN Jin-ping [1,2], LIU Xiao-mei [1,2], JIANG Zhao-yuan [1,2], Y. Jia [4], GAO Zeng-gui [3]

(1. *Institute of Plant Protection*, *Jilin Academy of Agricultural Sciences*, *Changchun* 130033, *China*; 2. *Key laboratory of Integrated Pest Management on Crops in Northeast. Ministry of Agriculture. P. R. China*; 3. *College of Plant Protection*, *Shenyang Agricultural University*, *Shenyang* 110161, *China*; 4. *U. S. Department of Agricultural Research Service*, *Dale Bumpers National Rice Research Center*, 2890 *Hwy* 130*E*, *Stuttgart*, *AR* 72160, *USA*)

Abstract: Rice blast disease, caused by the fungus *Magnaporthe oryzae*, is the most destructive rice disease worldwide. This disease is managed with a combination of the use of resistant cultivars, application of fungicides and improved cultural practices. Among them, the use of resistant cultivars is the most economical and environmentally sound approach to control blast disease. In this report, 218 isolates from commercial rice fields of Northeast China, South Korea and the South USA were collected, and inoculated with 24 monogenic lines carrying 24 major resistance genes, *Pia*, *Pib*, *Pii*, *Pik*, *Pik-s*, *Pik-h*, *Pik-m*, *Pik-p*, *Pish*, *Pit*, *Pita*, *Pita*-2, *Piz*, *Piz-t*, *Piz*-5, *Pi*1, *Pi*3, *Pi*5 (*t*), *Pi*7 (*t*), *Pi*9, *Pi*11 (*t*), *Pi*12 (*t*), *Pi*19, *Pi*20 and the susceptible recurrent parent Lijiangxintuanhegu (LTH), under greenhouse conditions. The percentages of resistant reaction of *Piz-t*, *Pi*9, and *Pi*12 (*t*) containing monogenic lines were about 80% against the isolates of Northeast China and South Korea, of *Pita*-2 and *Piz*-5 were about 80% against isolates of South Korea, and of *Pi*9 and *Pita*-2 were 92.9% and 78.6% against isolates of the Southern USA, respectively. These findings suggest that *Piz-t*, *Pi*9 and *Pi*12 (t) are effective in areas where blast were collected in northeast china, and *Piz-t*, *Pi*9, *Pi*12 (t) *Pita*-2 and *Piz*-5 are effective in areas of where blast were collected in South Korea, and *Pi*9 and *Pita*-2 are effective in the Southern USA where blast were collected, respectively.

Key words: Rice blast; resistance gene; identification; IRRI rice monogenic lines

* Corresponding author: WANG Ji-chun, E-mail: jlsnkyzbs@126.com

大豆抗病毒基因 *GmNH23* 的筛选鉴定及结构域分析

崔秀琦，哈　达*
（内蒙古大学生命科学学院，内蒙古自治区牧草与特色作物
生物技术重点实验室，呼和浩特　010000）

摘　要：大豆花叶病毒（Soybean Mosaic Virus，SMV）是一种在全球传播范围较广的病毒。目前大豆的一些 SMV 抗性相关的遗传连锁标记（QTL）在染色体上已经定位，但是具体的抗性基因（resistance gene，*R*）仍未被发现。在烟草中已经发现了抗 TMV 的 N 蛋白，N 蛋白属于 NBS-LRR 家族蛋白。我们在大豆基因组上发现了 35 个 *N* 基因的同源基因，将其命名为 *GmNH*（*Glycine max N-Homolog*）基因。当抗病基因过表达时，会与病原相互作用产生植物细胞程序性死亡的超敏反应（Hypersensitive Response，HR）。实验利用超敏反应对 23 个 *GmNH* 基因的 SMV 抗性进行了筛选鉴定。同时因为 *GmNH* 基因是 *N* 基因的同源基因，所以检测了 *GmNH* 基因对 TMV 的抗性。结果表明，在大豆 *N-homolog* 基因中只有 *GmNH23* 对 SMV 和 TMV 均具有抗性，而其余的 *N-homolog* 基因对这两种病毒都没有抗性。另外，我们把 *GmNH23* 的 5 个结构域，即 TIR、NB-ARC、DUF、TN、ND 进行了分析。其中，通过超敏反应初步鉴定 TN 结构域对 TMV 具有抗性。为进一步鉴定 SMV-大豆互作的分子机制打下了基础。

关键词：TMV；SMV；超敏反应；*GmNH23*

* 通讯作者：哈达

第七部分
病害防治

1株柑橘溃疡病生防内生细菌的鉴定、发酵条件优化及生防潜力研究*

刘　冰**，吴思梦，宋水林，乐美玲，李　恩
（江西农业大学农学院，南昌　330045）

摘　要：溃疡病是柑橘上一种重要的细菌性病害，对柑橘生产造成了很大的威胁。在前期试验中通过离体和活体筛选获得1株能够有效防治柑橘溃疡病的赣南脐橙内生细菌GN223，本研究对其进行了鉴定、发酵条件优化和生防潜力测定。通过形态特征、培养性状、生理生化特性和16S rDNA测定，将菌株GN223鉴定为*Enterobacter cowanii*；发酵条件优化结果表明，当接种量为1%、装瓶量50mL/250mL时，菌株GN223在pH值为8的YPD培养液中发酵60h，其发酵滤液对柑橘溃疡病菌的抑制效果最佳；盆栽试验表明，菌株GN223对水稻白叶枯病和细菌性条斑病均有一定程度的防效，且菌悬液的防效要高于发酵滤液，此外该菌株对水稻生长还表现出较为明显的促进作用。由以上结果可以看出，菌株GN223对植物细菌性病害有较好的防治效果，是一株具有开发潜力的生防菌株。研究结果可为植物细菌性病害的生物防治以及开发具有自主知识产权的生物农药奠定基础。

关键词：柑橘溃疡病；鉴定；发酵条件；细菌性病害

* 基金项目：国家自然科学基金（31460139）；江西省科技支撑计划项目（20121BBF60024）
** 第一作者：刘冰，博士，副教授，主要从事植物病理学方面的教研工作；E-mail：lbzjm0418@126.com

GC-MS Analysis and Antibacterial Activity of Volatile Oils from the Leaves and Fruits of *Taxodium distichum*[*]

ZHANG Wei-hao[**], TANG Xiang-you, QIN Kai, WANG Jun, SHAN Ti-jiang[***]

(*Guangdong Key Laboratory for Innovative Development and Utilization of Forest Plant Germplasm, College of Forestry and Landscape Architecture, South China Agricultural University, Guangzhou* 510642, *China*)

Abstract: *Taxodium distichum* is an important economic and ecological tree species in South China. This study was toanalyze and identify the chemical composition and antibacterial activity of volatile oil extracted from the leaves and fruits of *Taxodium distichum*. The volatile oil was extracted by hydro-distillation and filter paper diffusion method was used to detect the inhibitory activity against seven different bacteria. The yield of volatile oil from leaves and fruits were 0.211% (w/fw) and 0.657% (w/fw), respectively. The chemical compositions of the volatile oil wereanalyzed by gas chromatography-mass spectrometry (GC-MS). The relative percentages of different chemical components was measured by peak area normalization. The results show that the volatile oil from the leaves and fruits were different in the chemical composition and relative content, but *D*-limonene was the highest component in both of them. Thirty-eight components were identified from the volatile oil of the leaves and eighteen components in the fruits. *D*-limonene (21.16%), 1 - methyl - 3 - 1 - methyle-Cyclohexen (14.03%) and Caryophyllene oxide (11.26%) were the major compounds in the volatile oil of leaves, while the major compounds in the fruits were *D*-Limonene (31.08%), Ferruginol (11.82%), 5 - ethylidene - 1 - methyl-Cycloheptene (9.77%). The volatile oil from leaves and fruits showed strongest inhibitory activity against *Staphylococcus haemolyticus* and the inhibition zone diameter were (11.00 ± 1.00) mm and (20.00 ± 1.00) mm, respectively. The followed was *Ralstoniasolanacearum*, which the inhibition zone diameter were (10.67 ±0.58) mm and (17.67 ±0.58) mm. It suggested that the antibacterial activity of the fruits volatile oil was significantly stronger than the leaves. The volatile oils were rich in *Taxodium distichum* leaves and fruits and they showed a certain antibacterial activity against *Staphylococcus haemolyticus* and *Ralstonia solanacearum*. These results would provide an important theoretical basis for the comprehensive development and utilization of *Taxodium distichum*.

Key words: *Taxodium distichum*; Volatile oil; GC-MS; Antibacterial activity

---

* 基金项目：广东省林业科技创新项目（2015KJCH043）

** 第一作者：张伟豪，硕士，研究方向：植物和微生物的次生代谢；E-mail：vhaozhang@ foxmail. com

*** 通讯作者：单体江，博士，讲师，研究方向：植物和微生物的次生代谢；E-mail；tjshan@ scau. edu. cn

皂苷对三七种子的自毒活性及其与结构的关系*

钏有聪**，罗丽芬，袁　也，朱书生，杨　敏***
（云南农业大学农业生物多样性与病虫害控制教育部重点实验室，昆明　650201）

连作障碍是制约名贵中药材三七生产的主要限制因子。三七根系分泌的自毒皂苷是导致三七连作障碍的主要原因之一。三七在生长过程中可产生 70 余种 20（S）－原人参三醇型皂苷和 20（S）－原人参二醇型皂苷，总含量可高达 12% 以上。

为了探明三七主要皂苷的自毒活性及其与皂苷结构的关系，试验测定了 8 种 20（S）－原人参三醇型皂苷（20S－原人参三醇、Rg_1、Rg_2、Rh_1、F_1、R_1、R_f、Re）和 6 种 20（S）－原人参二醇型皂苷（Compound K、Rb_1、Rd、F_2、Rb_3和 Rg_3）在根际浓度范围内（0.01～5.0 mg/L）对三七种子萌发和生长的影响，并深入分析了毒性与皂苷结构的关系。结果表明，皂苷 20（S）－原人参三醇、F_1和 R_f无明显的自毒活性，而皂苷 Rg_1、Rg_2、Rh_1、R_1和 Re 对三七种子萌发、株高、根长、整株鲜重和根重均具有显著的自毒活性。皂苷 Re 和 Rg_1在供试浓度范围内其抑制作用随着浓度的升高而增加。皂苷 R_1、Rg_2和 Rh_1在浓度供试浓度范围内，抑制活性未呈现明显的随浓度增加而升高的趋势。20（S）－原人参二醇型皂苷 Compound K、Rb_1、Rd、F_2、Rb_3和 Rg_3对三七种子萌发和幼苗生长的抑制活性测定结果显示，皂苷 Compound K、F_2、Rb_3和 Rg_3无明显的自毒活性，而皂苷 Rb_1和 Rd 在供试浓度下对三七种子萌发、株高、根长、整株鲜重和根重均具有显著的抑制活性。

自毒活性与皂苷结构的关系分析结果表明，20（S）－原人参三醇型皂苷的自毒活性可能与 C_6位上连接的糖苷种类和数量有关。20（S）－原人参二醇型皂苷的自毒活性与 C_3和 C_{20}位上糖苷数量和种类密切相关。该研究结果为三七连作土壤中自毒物质的消解技术研究提供了理论支撑。

---

* 基金项目：云南农业大学青年基金；文山三七现代农业工程技术产业示范项目

** 第一作者：钏有聪；E-mail：chuanyoucong@163.com
*** 通讯作者：杨敏；E-mail：yangminscnc@126.com

抗咪鲜胺的田间水稻恶苗菌适合度及其抗性机制研究*

周俞辛**，于俊杰，俞咪娜，尹小乐，杜　艳，齐中强，宋天巧，刘永锋***
（江苏省农业科学院植物保护研究所，南京　210014）

水稻恶苗病是水稻生产上的一种重要病害，咪鲜胺是用于防治该病害的一类咪唑类广谱内吸性杀菌剂，目前江苏省已经检测到田间对咪鲜胺产生抗性的水稻恶苗病菌菌株。本研究以不同抗性倍数的抗咪鲜胺田间水稻恶苗菌株和两株敏感菌株为供试材料，分别测定了其适合度参数包括抗性稳定性、温度敏感性、菌丝生长速率、产孢和孢子萌发能力，结果表明，部分抗性菌株适合度与敏感菌株相当甚至高于敏感菌株。采用菌丝生长速率法测定其对不同作用机制的杀菌剂包括 2－氰基丙烯酸酯类杀菌剂氰烯菌酯和 DMI 类杀菌剂丙环唑、苯醚甲环唑及氟环唑的敏感性，对结果进行 spearman 分析后，表明咪鲜胺对氰烯菌酯、丙环唑、苯醚甲环唑和氟环唑均不具有交互抗药性。扩增各菌株的 *CYP51B* 全长，通过序列比对分析，结果表明，各抗性菌株 CYP51B 的 511 位氨基酸均为丝氨酸（S），而敏感菌株该位点均为苯丙氨酸（F）。本研究结果表明水稻恶苗菌对咪鲜胺一旦产生抗性，将具有很高的抗性风险，需要轮换不同作用机制的杀菌剂以延缓田间抗性的发生和发展。水稻恶苗病菌对咪鲜胺产生抗性的分子机制可能与 CYP51B 上的点突变相关。

---

* 基金项目：国家自然科学基金（31501675）；江苏省农业科技自主创新基金（CX（15）1054）
** 第一作者：周俞辛；E-mail：rita1720@126.com
*** 通讯作者：刘永锋；E-mail：liuyf@jaas.ac.cn

3种种衣剂防治芸豆根腐病试验研究*

曲建楠**，申永强**，左豫虎，刘　震，韩　冬，崔　佳，刘　铜***
（黑龙江八一农垦大学农学院植物病理与应用微生物所，大庆　163319）

摘　要： 芸豆（*Phaseolus vulgaris*）是普通菜豆和多花菜豆的总称，为豆科（Leguminosae）菜豆属（*Phaseolus*）的小宗杂粮作物，其具有丰富的营养和极高的药用价值。近年来，随着芸豆种植面积的扩大和重、迎茬面积的增加，芸豆根腐病呈逐年加重趋势，并成为生产中危害严重的病害之一。然而关于防治芸豆根腐病的药剂筛选试验鲜有报道，尚缺乏有效的防治方法。为了筛选对芸豆生长安全且对根腐病有较好防治效果的药剂，本研究选取38%多·福·克、25g/L咯菌腈和25%噻虫·咯·霜灵三种种衣剂进行筛选试验。结果表明，25%噻虫·咯·霜灵悬浮种衣剂、25g/L咯菌腈悬浮种衣剂和38%多·福·克种衣剂在试验设定的包衣药种比处理种子，对芸豆种子发芽和秧苗生长安全。田间试验结果表明，在芸豆出苗后10d，药种比为2 g/kg的25g/L咯菌腈悬浮种衣剂处理的防治效果最好，达到92.92%。其次是25%噻虫·咯·霜灵悬浮种衣剂，其各处理的防治效果均在80%以上。芸豆出苗后30d，药种比为2g/kg的25%噻虫·咯·霜灵悬浮种衣剂的防治效果最好，为59.83%，增产幅度为29.43%。但25g/L咯菌腈悬浮种衣剂处理防治效果持效期较短，芸豆出苗30d后防效均降到50%以下。结果表明25%噻虫·咯·霜灵悬浮种衣剂对芸豆根腐病防治效果最好且安全，这将为生产中芸豆根腐病防控提供技术支持。

关键词： 芸豆；根腐病；种衣剂；噻虫·咯·霜灵悬浮种衣剂

---

* 基金项目：国家科技支撑项目杂“豆病虫草安全高效防控技术”（2014BAD07B05－06）

** 第一作者：曲建楠，硕士研究生，从事植物病理学研究；E-mail：18345638704@163.com
申永强为同等贡献作者

*** 通讯作者：刘铜，副教授，从事植物病理学研究；E-mail：liutongamy@sina.com

6种铜制剂对苹果腐烂病菌的抑制作用持效期及影响因素的研究*

郭永斌**，任立瑞，唐兴敏，王亚南，胡同乐，王树桐***，曹克强***
（河北农业大学植物保护学院，保定 071001）

摘 要：为明确苹果生产中常用铜制剂的特性，筛选出其中表现突出的药剂，制定相应施用技术，进一步精细指导药剂的田间使用，本研究选取了生产上常用的80% 波尔多液可湿性粉剂、77% 硫酸铜钙可湿性粉剂、37.5% 氢氧化铜悬浮剂、12% 松脂酸铜乳油、29% 石硫合剂水剂、倍量式波尔多液200倍液（Cu：CaO：H_2O =1：2：200）、80% 代森锰锌可湿性粉剂和99% 分析纯硫酸铜晶体6种铜制剂，其中代森锰锌和硫酸铜作为对照药剂。采用改良的孢子萌发法，测定了不同铜制剂对苹果腐烂病菌分生孢子萌发抑制作用的持效期，并探究光照和降雨对各药剂持效期的影响。同时，研究了各药剂在果树叶片上的展着性。结果表明：6种供试铜制剂中，倍量式波尔多液对苹果腐烂病菌分生孢子萌发抑制效果最佳，且施药后20d抑制率仍高于85%，光照和降雨对其抑菌作用无显著性影响，药剂在叶正面和叶背展着性适中。其次是氢氧化铜和石硫合剂，持效期同样长达20d以上，抗光照和降雨能力仅次于倍量式波尔多液。硫酸铜钙和松脂酸铜虽具有较长的持效期，但降雨和光照对其防效有显著的影响。两种对照药剂中，硫酸铜表现出与倍量式波尔多液同样高的防效和抗光照及抗降雨能力，但其在试验过程中造成了幼果药害。而代森锰锌则持效期短，光照和降雨对其抑菌作用影响显著。因此，本研究推荐倍量式波尔多液用于腐烂病的田间防控。

关键词：苹果腐烂病；铜制剂；持效期；影响因素

---

* 基金项目：国家苹果产业技术体系（CARS－28）
** 第一作者：郭永斌，硕士研究生，主要从事植物病害流行与综合防治研究；E-mail：435451255@qq.com
*** 通讯作者：王树桐，博士，教授，主要从事植物病害流行与综合防治研究；E-mail：bdstwang@163.com
曹克强，博士，教授，主要从事植物病害流行与综合防治研究

Biocontrol Efficacy of Biocontrol Agents on Bacterial Leaf Streak Caused by *Xanthomonas oryzae* pv. *oryzicola* in Field Trials*

ZHANG Xiao-fang[1 **], FU Lina[1 **], GU An-yu[2]; LI Xiao-lin[2], JI Guang-hai[1 ***]

(1. *Key Laboratory of Agriculture Biodiversity for Plant Disease Management under the Ministry of Education*, *Yunnan Agricultural University*, *Kunming* 650201, *China*; 2. *Institute of Food Crop Research*, *Yunnan academy of agricultural sciences*, *Kunming* 650201, *China*)

Abstract: [Objective] The present experiment was conducted to screen out the most effective biological agents which could establish rapidly in the rhizosphere and leaves of rice and inhibit the growth of plant pathogenic bacteria, in order to provide reference for Biological control of bacterial leaf streak. [Methodology] The prevention and control test of rice bacterial leaf streak was conduted in Ruili city of Dehong prefecture and Mengwang township of Jinghong city in Yunnan province with 10 sifted-out strains of bio-control bacteria. Spraying different concentrations of bacteria suspension on the leaves of the rice plant, respectively after 3 weeks, to measure the control efficiency of the 10 strains of bio-control bacteria on bacterial leaf streak in two test points. [Results] The field trials reveal that the biocontrol efficacy of L6 (the bacterial suspension of *Lysobacter antibioticus* L6) is the highest and most prominent reaches to 72.00%; the biocontrol efficacies of Phenazino – 1 – carboxylic acid in the two test points are approximate, both near to 71.00%; the biocontrol efficacies of chemical pesticide Ye Banning are a little different, the biocontrol efficacy is 85.9% in Mengwang township, while it is 66.7% in Ruili city. Biocontrol agents compared with the pesticides, the biocontrol efficacies of chemical pesticide Ye Banning and Ye Qingshuang, the same effective components are "Bismerthiazol", are both near to 64.3%. [Conclusion] In consequence, this research showed that strain L6 had better control effect on bacterial leaf streak, equally, optimal concentration of 3×10^8 cfu/mL. It is of great potential to be a biocontrol agents in suppressing bacterial leaf streak disease.

Key words: Bacterial leaf streak; Biocontrol agents; Biocontrol efficacy

* 基金项目：国家自然基金会（31460458，31360002）；农业部公益性行业专项（201303015）

** 第一作者：张小芳，硕士研究生；E-mail：384952241@qq.com
付丽娜，硕士研究生；E-mail：1024422659@qq.com

*** 通讯作者：姬广海，博士，教授，从事分子植物病理等；E-mail：jghai001@aliyun.com；550356818@qq.com

制磷脂菌素在枯草芽孢杆菌 9407 菌株防治苹果轮纹病中的作用*

范海燕**，张占伟，李　燕，汝津江，王　琦***
（中国农业大学植物病理学系，农业部植物病理学重点实验室，北京　100193）

摘　要：枯草芽孢杆菌 9407（B9407）是本实验室从苹果上分离获得的有益细菌，在室内和田间条件下对苹果轮纹病菌具有明显的防治效果，而且对苹果等植物的多种病原真菌具有抑菌作用，应用前景广阔。为深入研究该菌株防病及调控机制，本实验室成功构建了 B9407 Tn*YLB* – 1 转座子随机插入突变体库。采用平板颉颃法筛选了 14 300株突变子，获得 1 株抑菌活性完全丧失的突变子，通过反向 PCR 确定转座子插入位点为制磷脂菌素合成关键基因 *ppsB* 上。为了验证制磷脂菌素在 B9407 菌株防治苹果轮纹病菌中的功能，本研究对制磷脂菌素合成基因操纵子中的 *ppsB* 基因进行缺失突变获得 Δ*pps*。抑菌活性测定结果显示，与 B9407 菌株相比，Δ*pps* 的脂肽提取物对苹果轮纹病菌分生孢子完全丧失抑菌活性。薄层层析结合生物自显影结果表明，B9407 菌株的脂肽提取物能在 Rf = 0. 1 ~ 0. 2（以三氯甲烷：甲醇：水 = 65：25：4 作为展开剂）位置处产生一个抑制苹果轮纹病菌的抑菌圈，而 Δ*pps* 的提取物则不能产生抑菌圈。利用高效液相色谱检测发现 Δ*pps* 完全丧失了制磷脂菌素的合成能力。苹果果实上的生防效果测定结果表明，接种病原菌 10d 后，Δ*pps* 处理的苹果发病率显著高于 B9407 菌株的处理。以上结果表明制磷脂菌素在 B9407 菌株防治苹果轮纹病菌中起重要作用。在 B9407 菌株中，制磷脂菌素的调控途径尚不清楚，还需要通过分子生物学等手段明确其调控途径。本研究结果为深入研究 B9407 菌株的防病机制及其调控途径奠定了基础。

关键词：枯草芽孢杆菌；抑菌活性；突变体；制磷脂菌素

---

* 基金项目：国家苹果产业技术体系（CARS – 28）
** 第一作者：范海燕，博士，中国农业大学植物病理学系；E-mail：fanhaiyan6860@ 163. com
*** 通讯作者：王琦，教授，主要从事植物病害生物防治与微生态学研究；E-mail：wangqi@ cau. edu. cn

辣根素对植物病原菌的抑菌活性及其作用机制初探*

王彦柠**，黄小威，罗来鑫，李健强***

（中国农业大学植物病理学系/种子病害检验与防控北京市重点实验室，北京 100193）

摘 要：辣根素是十字花科植物体内的一种组成型天然化合物，主要成份为异硫氰酸烯丙酯（Ally Isothiocyanate，AITC），具有刺激性辛辣气味及熏蒸活性辣根素作为一种环境友好的植物源活性抑菌物质近年来在病害防治研究领域备受关注，国际上已经实现商业化，并已成为推荐替代甲基溴的重要代表性产品之一，在美国已经许可使用于有机农产品生产中病害的防控与管理。为了探究辣根素的生物活性，本研究以中国农业大学研究开发的92% AITC原药为材料，以平板熏蒸法结合菌丝生长法、孢子萌发法及细菌菌落生长法等生物测定技术，检测了其对常见的26种植物病原真菌、卵菌及细菌的抑制作用，并在此基础上研究了AITC对导致玉米根腐病的轮枝样镰刀菌（*Fusarium verticillioides*）、导致棉花立枯病的立枯丝核菌（*Rhizoctonia solani*）及导致棉花黄萎病的大丽轮枝菌（*Verticillium dahliae*）不同生长阶段的抑菌活性及对其膜通透性、呼吸强度及能量合成等生理生化指标的影响。主要研究结果如下。

（1）室内毒力测定结果表明，AITC对供试的茄链格孢（*Alternaria solani*）、灰葡萄孢（*Botrytis cinerea*）、禾谷镰刀菌（*Fusarium graminearum*）、立枯丝核菌（*Rhizoctonia solani*）等21种植物病原真菌及瓜果腐霉（*Pythium aphanidermatum*）、辣椒疫霉（*Phytophthora capcisi*）等卵菌的菌丝生长具有显著的抑制作用，其EC_{50}分布在0.94～24.64μg/mL；对其中12种病原真菌的孢子萌发亦具有显著抑制作用，EC_{50}分布在0.26～0.69μg/mL；对番茄溃疡病菌（*Clavibacter michiganensis* subsp. *michiganensis*）、十字花科黑腐病菌（*Xanthomonas campestris* pv. *campestris*）等5种植物病原细菌的生长具有显著抑制作用，EC_{50}分布在0.81～8.61μg/mL。

（2）在离体条件下，AITC对*F. verticillioides*、*R. solani*及*V. dahliae*菌株3T、TZ4、TZ7不同生长阶段的抑菌作用呈现相同趋势：表现为显著抑制5株供试病菌的孢子或菌核的萌发及菌丝的生长，而对孢子或菌核的形成无显著影响；与对菌丝生长的抑制作用相比，AITC对五株供试靶标菌孢子或菌核的萌发表现出更显著的抑制作用。

（3）扫描电子显微镜观察表明，10μg/mL AITC熏蒸处理*F. verticillioides* 5d、*R. solani* 3d及*V. dahliae* 10d后，各供试靶标菌菌丝表面均出现严重的皱褶、缢缩及凹陷或膨大，新生菌丝可见分支畸形等变化。

（4）AITC影响*F. verticillioides*及*R. solani*的生理生化过程，表现为供试靶标真菌的膜通透性增大，呼吸耗氧量降低，细胞内高能磷酸化合物含量减少。

关键词：辣根素；熏蒸；大丽轮枝菌；轮枝样镰刀菌；立枯丝核菌

---

* 基金项目：高等学校博士学科点专项科研基金（编号：20120008130006）

** 第一作者：王彦柠，硕士研究生，主要从事植物源杀菌剂作用机制研究；E-mail：wangyn@cau.edu.cn

*** 通讯作者：李健强，教授，主要从事种子病理学及杀菌剂药理学研究；E-mail：lijq231@cau.edu.cn

响应曲面法优化解淀粉芽孢杆菌 T429 高产脂肽抗生素培养基及发酵条件*

乔俊卿**，张荣胜，刘邮洲，刘永锋***

（江苏省农业科学院植物保护研究所，南京 210014）

摘　要：由真菌引起的稻纹枯病，稻瘟病，稻曲病对水稻的高产、稳产造成严重的威胁。生物防治已然成为水稻病害综合防治的重要措施之一，尤其是芽孢杆菌属，极具有应用前景。江苏省农业科学院植物保护研究所稻病与生防研究室前期筛选获得一株对水稻纹枯病、稻瘟病和稻曲病均具有较好防效的解淀粉芽孢杆菌 T429，为了进一步提高 T429 产生脂肽类抗生素的能力，本研究通过响应曲面法对生防菌 T429 的高产抗生素发酵培养基成分和发酵培养条件进行了优化。

本研究以水稻纹枯病菌为指示菌，利用酸沉淀法，基于 LB、Landy 及常用的 10 种发酵培养基为出发培养基，进行 T429 发酵液脂肽类抗生素粗提液抑菌（水稻纹枯病）效果和发酵菌含量的分析，Plackett-Burman 结果显示，影响生防菌 T429 抑菌效果和菌含量的主要因子为黄豆饼粉、酵母粉和蛋白胨；随后的中心组合实验设计（CCD）结果表明，当黄豆饼粉 6.08g/L、蛋白胨 4.73g/L、酵母粉 2.03g/L 时，T429 产生的脂肽类抗生素抑制水稻纹枯病的效果最好，抑菌带宽为 10.12mm，据此，确定 T429 高产脂肽类抗生素的发酵培养基配方：黄豆饼粉 6.08g/L，蛋白胨 4.73g/L，酵母粉 2.03g/L，小麦粉 5g/L，玉米粉 5g/L，阳离子 205mg/L。基于此培养基成分，本研究继续优化了发酵培养条件（装液量、接种量、温度、转速、发酵时间）的最优组合，通过 Plackett-Burman 实验首先确定了装液量、接种量和发酵时间是影响 T429 产生脂肽类抗生素的重要因素，随后的中心组合实验设计（CCD）结果显示，当装液量是 105mL/500mL 发酵三角瓶、接种量为 0.87% 和发酵时间为 42.79h 时，T429 所产生的抗生素对纹枯病菌的抑菌带最宽，可达 11.27mm。最终我们确定 T429 的发酵培养条件：装液量为 105mL、接种量为 0.87%、发酵时间为 42.79h、发酵温度为 28℃，转速为 180r/min。本研究通过响应曲面法优化了生防解淀粉芽孢杆菌 T429 高产脂肽类抗生素的发酵培养基成分及培养条件，这为发酵生防菌 T429 的成本最优化提供了理论指导。

关键词：解淀粉芽孢杆菌；脂肽类抗生素；响应曲面法；发酵培养基；发酵培养条件

---

* 基金项目：江苏省现代农业研究示范类项目（BE2015254）；江苏省农业科技自主创新资金项目［CX（13）3024］；江苏省科技支撑计划项目（BE2014386）

** 第一作者：乔俊卿，山西大同人，博士，副研究员，主要从事植物病害及生物防治研究；E-mail：junqingqiao@hotmail.com

*** 通讯作者：刘永锋，研究员，江苏省农业科学院植物保护研究所；E-mail：liuyf@jaas.ac.cn

The *Shewanella algae* Strain YM8 Produces Volatiles with Strong Inhibition Activity Against *Aspergillus* Pathogens and Aflatoxins

GONG An-dong[1,2], LI He-ping[3], ZHANG Jing-bo[1], WU Ai-bo[4], LIAO Yu-cai[1]

(1. *College of Plant Science and Technology, Huazhong Agricultural University, Wuhan, China*; 2. *College of Life science, Xinyang Normal University, Xinyang, China*; 3. *College of Life Science and Technology, Huazhong Agricultural University, Wuhan, China*; 4. *Key Laboratory of Food Safety Research Institute for Nutritional Sciences, Shanghai Institutes for Biological Sciences, Chinese Academy of Sciences, Shanghai, China*)

Abstract: Aflatoxigenic *Aspergillus* fungi and associated aflatoxins are ubiquitous in the production and storage of food/feed commodities. Controlling these pathogen is a challenge. In this study, the *Shewanella algae* strain YM8 was found to produce volatiles that have strong antifungal activity against *Aspergillus* pathogens. Gas chromatography-mass spectrometry profiling revealed 15 volatile organic compounds (VOCs) emitted from YM8, of which dimethyl trisulfide was the most abundant. We obtained authentic reference standards for six of the VOCs; these all significantly reduced mycelial growth and conidial germination in *Aspergillus*; dimethyl trisulfide and 2, 4 – bis (1, 1 – dimethylethyl) -phenol showed the strongest inhibitory activity. YM8 completely inhibited *Aspergillus* growth and aflatoxin biosynthesis in maize and peanut samples stored at different water activity levels, and scanning electron microscopy revealed severely damaged conidia and a complete lack of mycelium development and conidiogenesis. YM8 also completely inhibited the growth of eight other species of phytopathogenic fungi: *A. parasiticus*, *A. niger*, *Alternaria alternate*, *Botrytis cinerea*, *Fusarium graminearum*, *Fusarium oxysporum*, *Monilinia fructicola*, and *Sclerotinia sclerotiorum*. This study demonstrates the susceptibility of *Aspergillus* and other fungi to VOCs from marine bacteria and indicates a new strategy for effectively controlling these pathogens and the associated mycotoxin production in the field and during storage.

水稻纹枯病菌生防菌的筛选*

罗文芳**，魏松红***，刘志恒，李　帅，陈筱萌，李思博，张　优，王海宁
（沈阳农业大学植物保护学院，沈阳　110866）

摘　要：水稻纹枯病（*Rhizoctonia solani*）是一种世界性水稻病害，也是我国水稻生产上三大病害之一。由于水稻种质资源中缺乏高抗纹枯病的品种，所以水稻纹枯病的防治仍在很大程度上依赖化学药剂防治。化学农药污染环境、不易降解、靶标选择性不高，长期、大范围的应用易引起抗药性，因此，生物防治已成为防治水稻纹枯病的新途径。本研究从辽宁（沈阳、铁岭、盘锦、丹东）、吉林（公主岭）、黑龙江（哈尔滨、七星农场、八五八农场、绥滨农场、二九一农场）等10个地区采集水稻田土样33份，共分离出真菌30株，细菌114株，放线菌67株。试验采用平板对峙法、菌丝生长速率法和菌核浸泡法筛选对水稻纹枯病有明显抑制作用的生防菌。采用三点平板对峙培养法，筛选出24株抑菌圈达30.00～42.50mm的高效颉颃菌（真菌13株、细菌7株、放线菌4株），其中HF1－2的抑菌圈达42.50mm。显微镜观察被抑制的菌丝变形，细胞内含物有明显的溶解作用，导致菌丝细胞空洞、原生质分布不均匀；应用菌丝生长速率法筛选得到无菌发酵上清液对水稻纹枯病菌具有较强颉颃活性的真菌1株和细菌6株，菌丝生长校正抑制率为71.76%～92.24%，其中JLN3－7的校正抑制率达92.24%；应用菌核浸泡法筛选到无菌发酵上清液对菌核萌发具有明显抑制作用的细菌2株，真菌1株，抑制率分别为70.46%、75.03%和80.19%。

关键词：水稻纹枯病菌；生防菌；筛选

---

* 基金项目：国家水稻产业技术体系项目（CARS－01）；辽宁水稻产业体系专项资金项目（辽农科〈2013〉271号）

** 第一作者：罗文芳，在读硕士，植物病理学专业；E-mail：576263465@qq.com
*** 通讯作者：魏松红，博士，教授，研究方向：植物真菌病害及水稻病害研究；E-mail：songhongw125@163.com

嘧菌酯对石榴干腐病菌的生物学活性

杨　雪[1]，张爱芳[1]，郭遵守[2]，李　澜[3]，陈　雨[1]，徐义流[4]

（1. 安徽省农业科学院植物保护与农产品质量安全研究所，农业部有害生物合肥科学观测实验站，农业部农产品质量安全风险评估实验室，合肥　230031；2. 砀山县农业委员会，砀山　341321；3. 怀远县荆涂山石榴科技有限公司，怀远　233499；4. 安徽省农业科学院，园艺作物种质创制及生理生态安徽省重点实验室，合肥　230031）

摘　要：为探究石榴干腐病的病原菌种类及嘧菌酯对石榴干腐病的生物学活性，采用单孢分离法对石榴病果中的病原菌进行分离纯化，然后进行分子鉴定和致病性测定，并通过室内生测法测定了嘧菌酯对其菌丝生长和孢子萌发的影响，且连续2年开展了大田防治试验。经形态学、分子生物学鉴定及致病性测定结果表明，石榴干腐病的病原菌为石榴壳座月孢（*Pilidiella granati*）。嘧菌酯在水杨肟酸（SHAM）的协同作用下，对石榴干腐病菌的菌丝生长和孢子萌发具有强烈的抑制作用，对其菌丝生长和孢子萌发的平均 EC_{50} 分别为 0.202μg/mL 和 0.007μg/mL（含 100μg/mL SHAM）；在大田防治试验中，嘧菌酯对石榴干腐病具有良好的防效，其中 1 000倍 25%嘧菌酯 SC 稀释液在 2013 和 2014 年的防效分别可达 90.85%和 81.91%，显著高于其他处理。表明嘧菌酯可作为防治石榴干腐病的候选药剂。

关键词：石榴干腐病；石榴壳座月孢；嘧菌酯；抑菌活性；田间防效

枯草芽孢杆菌 DZSY21 抗玉米纹枯病的研究

苏　博[1]，顾双月[2]，丁　婷[2]
（1. 安徽省作物生物学重点实验室，合肥　230036；
2. 安徽农业大学植物保护学院，合肥　230036）

摘　要：本试验以前期获得的一株对玉米纹枯病菌具有较高抑菌活性的杜仲内生细菌菌株 DZSY21（*Bacillus subtilis* DZSY21）为试验菌株，分析其对玉米纹枯病抗性。结果表明在玉米叶鞘引入生防细菌 DZSY21，明显提高玉米植株对纹枯病的抗性，其病情指数在整个生长时期均处于较低水平。

关键词：内生细菌；玉米纹枯病；抗性

玉米纹枯病（maize sheath blight）是由立枯丝核菌（*Rhizoctonia solani*）引起的土传性病害，已成为为害玉米的严重病害。目前，主要是利用化学药剂防治该病，然而化学农药在使用过程中容易造成环境污染、药害、农产品农药残留超标、农田生态平衡和生物多样性被破坏等等问题[2]。近年来，围绕植物内生细菌及其代谢产物控制植物病害已成为研究的热点和重点。本试验以前期获得的一株对玉米纹枯病菌具有较高抑菌活性的杜仲内生细菌菌株 DZSY21（*Bacillus subtilis* DZSY21）为试验菌株，分析其对玉米纹枯病抗性，以期利用颉颃菌株开发出绿色的微生物农药，为以后枯草芽孢杆菌 DZSY21 的开发应用提供科学依据。

1　材料与方法

1.1　菌种来源

菌种为本实验前期分离筛选的一株对玉米纹枯病菌具有较好抑菌活性的杜仲内生细菌，菌株编号为 DZSY21（*Bacillus subtilis* DZSY21），NCBI 登录号为 KP777560。

1.2　供试玉米品种

玉米自交系品种昌 7－2。

1.3　颉颃菌株 DZSY21 的菌液制备

颉颃菌株 DZSY21 接种于牛肉膏蛋白胨液体培养基中（1L 三角锥形瓶装牛肉膏蛋白胨液体培养基 350mL），搅拌速率为 160r/min，于 35～37℃ 培养 8～10h，调节发酵液中枯草芽孢杆菌 DZSY21 的菌体浓度至 1.0×10^{6}CFU/mL，获得 DZSY21 发酵液。

1.4　颉颃菌株 DZSY21 引入玉米植株对玉米纹枯病的防效测定

1.4.1　玉米纹枯病菌的制备

将玉米茎秆剪为长约 5cm、宽约 2cm 的节段，用 1% 的蔗糖溶液将其浸泡 30min 后，倒掉多余的溶液，把玉米茎秆分装在三角瓶中，121℃ 灭菌 30 min 后接种纹枯病菌，28℃，培养 4d。

1.4.2　颉颃菌株 DZSY21 对玉米植株叶鞘部位的不同处理

将玉米种子（昌 7－2）依次用 1% 的次氯酸钠表面消毒 10min，用无菌水冲洗 3～5 遍，然后置于铺有无菌湿滤纸的培养皿内，于 25℃ 的恒温培养箱内催芽，待种子萌发后备用。实验田设于安徽农业大学教学实习基地农萃园（土壤为黄棕壤土）中，垄栽种植，垄面宽 70cm，垄距 30cm，株距 30 cm，行距 60cm，每垄栽植 30 粒处理过的玉米种子，共栽植 5 垄。待 5 垄玉米均长至大喇叭口期，分别于每垄选取长势相当的玉米植株进行如下 5 种处理：①空白对照处理（CK）：在健康玉米植株贴近地面的第一个叶鞘上接种灭菌的玉米茎秆（灭菌的玉米茎秆不接种玉米纹枯病菌），保湿培养 24h；②植物病原菌玉米纹枯病菌处理（B）：在健康玉米相同部位接种玉米纹枯病菌（取 1.5.1 培养好的带有玉米纹枯病菌的玉米茎秆放入玉米分蘖的叶鞘与茎之间。下同），保湿培养 24h；③颉颃菌株 DZSY21 发酵液处理组（S）：将枯草芽孢杆菌 DZSY21 发酵液（含菌量为 1.0×10^6CFU/mL）直接喷洒于健康玉米植株相同部位，每个叶鞘表面喷洒 5mL（下同），保湿培养 24h；④在玉米叶鞘上先进行颉颃菌株 DZSY21 发酵液处理，再接种玉米纹枯病菌的混合处理组（SB）：将枯草芽孢杆菌 DZSY21 发酵液直接喷洒于叶鞘表面，保湿培养 24h 后，然后将玉米纹枯病菌分别接种于玉米植株的相同叶鞘部位，保湿培养 24h 后，对玉米植株进行正常水肥管理；⑤在玉米叶鞘上先进行井冈霉素处理，再接种玉米纹枯病菌的混合处理组（Y）：将井冈霉素溶液（50μg/mL）5mL 喷在健康玉米植株贴近地面的第一个叶鞘表面，保湿培养 24h 后，然后将玉米纹枯病菌分别接种于相同叶鞘部位，保湿培养 24h 后，对玉米植株进行正常水肥管理。

每种处理 15 棵重复，分别于接种玉米纹枯病菌后的第 4d、第 6d、第 8d、第 10d、第 12d 调查纹枯病的发病情况，根据玉米纹枯病害分级标准[3]统计发病的严重度，计算病情指数和发病率。

2　结果与分析

颉颃菌株 DZSY21 引入玉米叶鞘部后，分别于接种玉米纹枯病菌后的第 4d、6d、8d、10d、12d 进行玉米纹枯病病害调查，计算病情指数。结果如图所示，随取样时间的延长，“玉米纹枯病菌（B）”处理组玉米患病逐渐加重，第 12d 病情指数达到 75.41%；而引入 DZSY21 菌株的“颉颃菌株 DZSY21 + 玉米纹枯病菌（SB）”处理组玉米患病较轻，截至接种玉米纹枯病菌后的第 12d，其病情指数为 33.26%，与“井冈霉素 + 玉米纹枯病菌（Y）”处理组玉米病情指数（31.65%）无显著差异。上述结果说明颉颃菌株 DZSY21 在玉米叶鞘的引入能够降低玉米纹枯病的发病率，减少纹枯病对玉米的危害，明显提高玉米植株对纹枯病的抗性。此外，经颉颃菌株 DZSY21 发酵液（S）单独处理的玉米植株，5 个时期调查的病情指数均为 0，说明颉颃菌株 DZSY21 在玉米叶鞘中的介入对玉米的生长没有影响，从一定程度上也反映了 DZSY21 菌株对玉米没有致病作用。

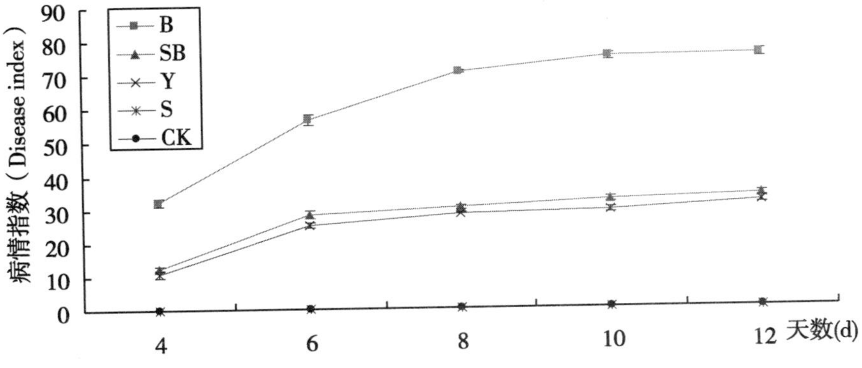

图　DZSY21 引入玉米叶鞘对不同生长期玉米纹枯病病情指数的影响

Figure　The effect of DZSY21 on disease index of maize sheath blight

3 讨论

目前，所报道的关于生物防治玉米纹枯病所用菌株多为玉米根际或内生微生物。如邱小燕[4]等人利用在玉米根际土壤分离得到的根际细菌“515－126”开展田间玉米纹枯病的防治效果测定，结果表明该菌株在田间对玉米纹枯具有一定防治效果，相对防效达到49.37%；毛腾霄[5]利用从玉米茎基部叶鞘分离得到的芽孢杆菌菌株BS－8D接入玉米植株茎基部，能延缓病菌对植株的侵染。本研究中，利用前期从杜仲体内获得的一株对玉米纹枯病菌具有较强颉颃作用的内生枯草芽孢杆菌DZSY21，研究其对玉米纹枯病的防治效果，结果表明枯草芽孢杆菌DZSY21的接种在较大程度上能降低玉米纹枯病的发病率及其严重程度。生防芽孢杆菌要充分发挥其在防治植物病害中的作用，在施入环境后，能否与定殖在寄主植物上的病原菌竞争，对其作用的发挥至关重要，已成为筛选、评价生防菌的一个重要指标[6-7]。后续研究表明菌株DZSY21具有较强的生物膜形成能力，能够在玉米植株体表及体内有效定殖（数据未发表），这更进一步表明，枯草芽孢杆菌DZSY21具有被开发成用于防治玉米纹枯病害的生防制剂的潜力。

参考文献

［1］罗茂，彭华，高健，等．玉米纹枯病抗性相关miRNA的鉴定与功能分析［J］．中国生物化学与分子生物学报，2012，28（12）：1 122－1 132.

［2］姚金保，马鸿翔，张平平，等．小麦优良亲本宁麦9号的研究与利用［J］．核农学报，2012，26（1）：17－21.

［3］陈文生．玉米纹枯病抗性资源筛选及抗性QTL元分析［D］．雅安：四川农业大学，2013，6.

［4］邱小燕，张敏，胡晓，等．枯草芽孢杆菌的定殖能力及对玉米纹枯病的防治效果［J］．植物病理学报，2010，28（4）：492－496.

［5］毛腾霄．枯草芽孢杆菌BS－8D防治玉米纹枯病（*Rhizoctonia solani*）的研究［D］．成都：四川农业大学，2006，6.

［6］Nicolás Pastor，Evelin Carlier，Javier Andrés，et al. Characterization of rhizosphere bacteria for control of phytopathogenic fungi of tomato［J］. Journal of Environmental Management，2012（95）：332－337.

［7］Jia-Hong Ren，Hao Li，Yan-Fang Wang，et al. Biocontrol potential of an endophytic *Bacillus pumilus* JK-SX001 against poplar canker［J］. Biological Control，2013（67）：421－430.

番茄灰霉病生防芽孢杆菌筛选、评价及鉴定*

鹿秀云**，商俊燕，李社增，郭庆港，张晓云，马　平***
（河北省农林科学院植物保护研究所，河北省农业有害生物综合防治工程技术研究中心，
农业部华北北部作物有害生物综合治理重点实验室，保定　071001）

摘　要：番茄灰霉病是由半知菌亚门灰葡萄孢菌（*Botrytis cinerea*）引起的一种重要病害。该病菌主要为害番茄叶片、果实，对保护地番茄生产构成极大威胁，已经成为番茄设施栽培的限制性障碍。在我国防治番茄灰霉病主要依靠化学药剂，然而，化学农药对易产生抗药性的番茄灰霉病却难以持续奏效，而且长期大量不合理的使用化学农药，还引发了一系列严重的社会和生态问题。芽孢杆菌（*Bacillius* sp.）是一种受到广泛关注的生防细菌，以其分布广、易分离培养、能产生抗逆性较强的芽胞、贮藏期长和使用方便等特点，成为一种理想的生防微生物。本研究开展了番茄灰霉病生防芽孢杆菌筛选、评价及鉴定研究。首先应用对峙培养法从1 942株土壤细菌中筛选到灰葡萄孢菌颉颃菌656株，其中抑菌率>80%或抑菌带半径>7.0mm的颉颃菌有52株。采用离体叶片法通过两批次重复试验从52株优秀颉颃菌中筛选到6株防治番茄灰霉病的效果大于70%的菌株，分别是菌株HMB27698、HMB27648、HMB27636、HMB27633、HMB27746和HMB27688，防治效果分别为78.36%、73.58%、72.77%、71.60%、71.16%和70.42%。分别以番茄和黄瓜为寄主，通过盆栽试验评价6株生防菌防治灰霉病的效果。结果表明，在番茄上，供试的6株生防菌在两批次试验中平均防效从64.00%到77.59%，与化学药剂对照250g/L嘧菌酯悬浮剂（先正达苏州作物保护有限公司生产，800倍水稀释液喷雾处理）的防效（74.47%）相当；在黄瓜上，供试的6株生防菌在两批次试验中平均防效从61.91%到93.48%，有4株生防菌的防效优于化学药剂对照250g/L嘧菌酯悬浮剂的防效（78.70%），其中菌株HMB27688的防效（93.48%）最高。开展田间小区试验评价了4株生防菌防治番茄灰霉病的效果。两次调查结果发现，菌株HMB27633和HMB27688防治效果分别为93.62%、85.19%和85.11%、81.48%，防效均优于化学药剂250g/L嘧菌酯悬浮剂的防效（72.34%、74.07%）。通过形态特征观察、16S rDNA和*gyr*B基因序列同源性比对，将HMB27633和HMB27688鉴定为解淀粉芽孢杆菌（*Bacillus amyloliquefaciens*）。

关键词：番茄；灰霉病；芽孢杆菌；筛选；鉴定

---

* 基金项目：公益性行业（农业）科研专项（201303025）；国家高技术研究发展计划（863计划）（2011AA10A205）

** 第一作者：鹿秀云，硕士，副研究员，主要从事植物病害生物防治研究；E-mail：luxiuyun03@163.com
*** 通讯作者：马平，博士，研究员，主要从事植物病害生物防治研究；E-mail：pingma88@126.com

放线菌 JY－22 对烟草赤星病菌的抑菌控病作用*

邓永杰[1**]，张　旺[1]，黄国联[2]，李　斌[3]，魏周玲[1]，陈德鑫[4***]，孙现超[1***]

（1. 西南大学植物保护学院，北碚　400716；2. 湖南省烟草公司郴州市公司，郴州　423000；3. 中国烟草总公司四川省公司，成都　610000；4. 中国农业科学院烟草研究所，青岛　266101）

摘　要：自 2007 年本实验室对缙云山土壤颉颃放线菌资源进行调查研究，获得一株颉颃效果显著的放线菌 JY－22，鉴定为链霉菌属吸水链霉菌（*Streptomyces hygroscopicus*）。优化了该菌发酵条件，此外 JY－22 菌株对温度与紫外具有很好的稳定性，80℃水浴 3h，紫外照射 30min 均不会对发酵滤液抑菌活性造成影响。对真菌病原菌抑菌谱广，抑菌率达 41.9%～75.7%。为明确放线菌 JY－22 的活性代谢产物抑菌活性机理和控病效果。本试验将 10 块培养 4d 的烟草赤星病菌菌饼加入 100mL PD 培养基中，摇床培养 36h 后加入无菌发酵滤液，使其浓度分别为 10%、5%、2.5%、1%、0%，24h 后采用电导率法测对细胞膜渗透性影响，96h 后采用菌丝湿重法和紫外吸收法测对菌丝生长、可溶性蛋白、麦角甾醇及丙二醛含量的影响；通过离体叶片控病试验分析了菌株 JY－22 无菌发酵滤液对烟草赤星病控病效果。结果表明，JY－22 无菌发酵滤液对烟草赤星病菌菌丝生长量及孢子萌发均有显著的抑制作用分别达到 73.3% 和 73.2%，同时培养皿对峙颉颃培养观察造成菌丝畸形，扭曲，膨大，细胞质聚集外漏。无菌发酵滤液处理后，电导率增大电解质外漏，丙二醛含量显著升高，麦角甾醇与可溶性蛋白含量显著降低。JY－22 无菌发酵滤液对烟草赤星病害有防治作用，治疗控病效果和保护控病效果分别为 73.54% 和 84.52%。结论表明放线菌 JY－22 对烟草赤星病具有显著颉颃作用，通过作用于烟草赤星病菌细胞膜抑制麦角甾醇合成，引起细胞膜脂质过氧化致使透性改变，导致菌丝畸形；JY－22 无菌发酵滤液具有很好的防控作用。

关键词：放线菌 JY－22；抑菌作用；烟草赤星病菌；机理

---

* 基金项目：国家自然科学基金（30900937）；中央高校基本科研业务费专项资金（XDJK2016A009，2362015xk04）；中国烟草总公司四川省公司科技项目（201202007）；中国烟草总公司湖南省公司科技项目（14－16ZDAa02）

** 第一作者：邓永杰，硕士研究生，从事植物病理学研究

*** 通讯作者：陈德鑫，博士，副研究员；E-mail：13963973187@126.com

孙现超，博士，研究员；E-mail：sunxianchao@163.com

放线菌 LG－9 发酵液对棉花黄萎病菌抑菌作用及稳定性分析*

穆凯热姆·阿卜来提[1**]，陈　明[1]，刘　政[2]，王晓东[1***]
（1. 石河子大学绿洲农作物病害防控重点实验室，石河子　832000；
2. 兵团农垦科学院，石河子　832000）

摘　要：棉花黄萎病是由大丽轮枝菌（*Verticillium dahliae* Kleb.）引起的一种土传性维管束病害，具有寄主范围广、土壤传播快、抗逆性强和变异性大等特点，严重影响新疆棉花的产量和品质，给新疆棉花生产造成极大的危害，已成为制约新疆棉花产业可持续发展的重要的限制因素。目前，一些防治方法防效不佳，利用生物防治棉花黄萎病已经成为棉花黄萎病综合防治中最具潜力的防治措施之一。2014—2015 年作者从新疆不同棉田土壤中大量分离放线菌，经平板对峙和活体筛选获得 1 株对棉花黄萎病原菌抗生作用显著的放线菌菌株 LG－9。通过菌株 LG－9 发酵液对棉花黄萎病菌抗生作用测定和抑菌活性稳定性分析，试验结果表明菌株 LG－9 发酵液对棉花黄萎病菌菌丝、分生孢子萌发和菌核萌发具有显著的抑制作用，对菌丝抑制率为 70%，且易造成菌丝畸形，顶端膨大等现象；对分生孢子和微菌核萌发抑制率分别为 46.4% 和 100%。LG－9 发酵液对紫外线和酸碱稳定，在 pH 值小于 8 的条件下均能保持较强的抑菌活性；发酵液经 20℃ 处理 30min，抑菌活性最大，抑菌圈直径为 1.37cm，但高于 100℃，发酵液抑菌活性丧失。表明放线菌菌株 LG－9 对棉花黄萎病具有很好的生防潜质和应用开发的前景。

关键词：棉花黄萎病；放线菌；发酵粗提液；抑菌作用

---

* 基金项目：国家自然科学基金（项目编号 31360452）
** 第一作者：穆凯热姆·阿不来提，在读研究生，从事植物病害生物防治研究；E-mail：1808123234@qq.com
*** 通讯作者：王晓东，博士，副教授，从事植物病害生物防治研究；E-mail：wxdong11@163.com

木美土里复合微生物菌剂对苹果再植病害的生防效果*

赵 璐**，刘 欣，王树桐***，曹克强***
（河北农业大学植物保护学院，保定 071001）

摘 要：苹果再植病害是苹果产区的一种重要的土传病害，严重威胁到我国苹果产业的健康发展。本研究通过室内对峙试验和盆栽试验，研究了木美土里生物菌肥对苹果再植病害的防治效果。室内试验结果表明木美土里复合微生物菌剂对尖孢镰刀菌（*Fusarium oxysporum*）HS2 的抑制率达到 68.59%；从木美土里复合微生物菌剂中分离得到 7 株可培养真菌，3 株可培养细菌。对分离菌株的抑菌试验结果表明，所有分离菌株均可不同程度的抑制尖孢镰刀菌（*F. oxysporum*），其中效果最好的真菌为哈茨木霉（*T. harzianum*），抑制率为 81.48%，颉颃效果高于复合微生物菌剂的细菌菌株有 3 株，其中，2 株鉴定为解淀粉芽孢杆菌（*B. amyloliquefaciens*），1 株鉴定为地衣芽孢杆菌（*B. licheniformis*），3 个细菌菌株对病原镰刀菌的抑制率依次为 82.22%、80.74% 和 71.11%。盆栽试验结果表明，木美土里复合微生物菌剂对海棠苗再植病害的防治效果达到 87.78%，显著优于恶霉灵对照处理 5.55%；海棠苗生长明显得到促进，株高是对照处理的 1.51 倍，茎粗是对照的 2 倍，鲜重是对照的 5.74 倍，干重是对照的 5.21 倍，叶面积是对照的 2.83 倍。以上研究结果表明，木美土里复合微生物菌剂对苹果再植病害病原真菌有较强的抑制作用，对盆栽海棠幼苗的再植病害有较好的生防效果。

关键词：复合微生物菌剂；苹果再植病害；生物防治

---

* 基金项目：国家苹果产业技术体系（CARS－28）；河北省自然科学基金（c2016204140）
** 第一作者：赵璐，在读硕士研究生，研究方向：植物病害流行与综合防治；E-mail：1391492647@ qq. com
*** 通讯作者：王树桐，博士，教授，从事植物病害流行与综合防治研究；E-mail：bdstwang@ 163. com
曹克强，博士，教授，从事植物病害流行与综合防治研究；E-mail：ckq@ hebau. edu. cn

荔枝霜疫霉和稻瘟病菌对 SYP－9069 的敏感性检测*

林　东[1**]，王秋实[1]，薛昭霖[1]，李慧超[2]，刘长令[2]，刘西莉[1***]
（1. 中国农业大学植物病理学系　北京，100193；2. 沈阳化工研究院
新农药创制与开发国家重点实验室，沈阳　110021）

摘　要：SYP－9069 是沈阳化工研究院自主创制研发的具新颖结构的化合物，已有试验结果表明，该化合物对部分植物病原卵菌和真菌具有较好的抑菌作用，同时对蚜虫、红蜘蛛等也有较好的活性。本研究分别测定了 46 株荔枝霜疫霉（*Peronophythora litchi*）和 108 株稻瘟病菌（*Magnaporthe oryzae*）对 SYP－9069 的敏感性，建立了相应的敏感基线。研究结果如下：

本研究采用菌丝生长速率法测定了分离自福建、广东、广西和海南 4 个荔枝主产区的 46 株荔枝霜疫霉以及分离自安徽、福建、湖北、黑龙江、吉林、海南、云南、江苏和四川 9 个水稻产区的 108 株稻瘟病菌对 SYP－9069 的敏感性。结果表明，SYP－9069 对供试荔枝霜疫霉的 EC_{50} 介于 0.017 2μg/mL 至 0.916 9μg/mL，平均 EC_{50} 为（0.228 7±0.709 3）μg/mL；SYP－9069 对供试稻瘟病菌的 EC_{50} 介于 0.073 2μg/mL 至 1.379 3μg/mL，平均 EC_{50} 为（0.333 0±0.300 0）μg/mL。表明 SYP－9069 对供试的荔枝霜疫霉和稻瘟病菌均具有良好的抑制活性。

上述研究结果为进一步开展 SYP－9069 的抗性风险评估和作用机制研究奠定了基础，为 SYP－9069 登记用于防治田间荔枝霜疫霉及稻瘟病害提供理论依据。

关键词：SYP－9069；荔枝霜疫霉；稻瘟病菌；敏感基线

---

* 基金项目：公益性行业（农业）科研专项（编号：201303023）

** 第一作者：林东，在读博士研究生；E-mail：lindong2012@ cau. edu. cn

*** 通讯作者：刘西莉，教授，主要从事杀菌剂药理学及病原菌抗药性研究；E-mail：seedling@ cau. edu. cn

一株萎缩芽孢杆菌 Bacillus atrophaeus YL3 的鉴定及其脂肽类化合物分析*

刘邮洲[1**]，陈夕军[2]，梁雪杰[1]，钱亚明[3]，乔俊卿[1]，刘永锋[1]
（1. 江苏省农业科学院植物保护研究所，南京 210014；2. 扬州大学园艺与植保学院，扬州 220009；3. 江苏省农业科学院园艺研究所，南京 210014）

摘　要：萎缩芽孢杆菌（*Bacillus atrophaeus*）属芽孢杆菌，在含酪氨酸培养基上产生黑色素。萎缩芽孢杆菌能产生芽胞，耐受不良环境，无致病性，广泛应用于医疗卫生、工业和农业等领域。现已发现萎缩芽孢杆菌对尖孢镰刀菌（*Fusarium oxysporum*）、核盘菌（*Sclerotinia sclerotiorum*）、大丽轮枝菌（*Verticillium dahliae*）、胶孢炭疽菌（*Colletotrichum gloeosporioides*）等植物病原菌具有颉颃作用。

实验室从江苏省徐州市贾汪区耿集镇草莓的根围土壤中，分离出对草莓枯萎病菌（*F. oxysporum*）具有较强颉颃作用的芽孢杆菌 YL3。经过生理生化反应特征、16S rDNA 及 *gyrB* 基因序列比对，鉴定为萎缩芽孢杆菌。采用室内平板对峙生长法测定其抑菌活性，发现菌株 YL3 对黄单胞菌（*Xanthomonas oryzae* pv. *oryzicola*）的抑制作用最强，抑菌圈直径为 55.67mm，可形成明显的抑菌圈，其次对欧文氏菌（*Erwinia amylovora*）也具有较好的抑制作用，抑菌圈直径为 36.50mm。菌株 YL3 对草莓枯萎病菌和甘蓝黑斑病菌菌丝生长有较强的抑制作用，抑制率分别为 49.63% 和 51.30%，对草莓炭疽病菌和油菜菌核病菌菌丝生长的抑制作用相对较弱，抑制率分别为 43.15% 和 45.74%。萎缩芽孢杆菌 YL3 水剂 10 倍稀释液处理后，对草莓枯萎病防效与对照药剂多菌灵 800 倍液（即田间推荐使用浓度）效果相当，100 倍稀释液对“宁玉”品种和“红颊”品种的枯萎病防效分别为 67.62% 和 59.26%，而化学药剂 50% 多菌灵 1 600倍稀释液的防效较差，分别为 29.52% 和 34.89%。采用酸沉淀法提取菌株 YL3 脂肽类化合物并进行高效液相色谱串联质谱分析（LC-MS）。结果表明，菌株 YL3 发酵液中脂肽类粗提物对草莓枯萎病菌菌丝生长有抑制作用，随着剂量加大抑制效果显著。LC-MS 检测到菌株 YL3 脂肽类粗提物中含有表面活性素和泛革素的离子峰出现，由此推断菌株 YL3 室内抑菌作用可能是产生了对细菌有强烈颉颃作用的表面活性素和对病原真菌有强烈颉颃作用的泛革素。

关键词：萎缩芽孢杆菌；分离鉴定；脂肽类化合物；颉颃

---

* 基金项目：江苏省农业科技自主创新基金项目 CX（15）1037

** 第一作者：刘邮洲，副研究员，主要从事园艺作物病害生物防治研究；E-mail：shitouren88888@163.com

Biological Control, Growth Promotion, and Host Colonization of European Horticultural Plants by Endophytic *Streptomyces* spp. *

CHEN Xiaoyulong[1], Maria Bonaldi[1], Armin Erlacher[2], Andrea Kunova[1], Cristina Pizzatti[1], Marco Saracchi[1], Gabriele Berg[2], Paolo Cortesi[1]**

(1. *Department of Food, Environmental and Nutritional Sciences* (*DeFENS*), *Università degli Studi di Milano, Via Giovanni Celoria* 2, 20133 *Milano, Italy*; 2. *Institute of Environmental Biotechnology, Graz University of Technology, Petersgasse* 12, 8010 *Graz, Austria*)

Introduction: Yield losses caused by phytopathogens should be minimized to maintain the food quality and quantity for the demand of massively growing human population. At the same time, yield limitation due to soil fertility and nutrition deficiencies add extra constraints to plant production. Thus, searching for sustainable solutions to suppress phytopathogens, as well as to increase the yield is gaining high public interests. Plant root systems are colonized by vast amounts of microbes, some of which facilitate biological control and plant growth promotion activities. Streptomyces, abundant in soil, are a group of filamentous bacteria producing a variety of beneficial secondary metabolites, gifting them the potential to be developed as bio-pesticides and bio-fertilizers. We labeled two bioactive Streptomyces strains with EGFP marker to investigate their interactions with lettuce using confocal laser scanning microscopy (CLSM), and evaluated their biocontrol activities against Sclerotinia sclerotiorum on lettuce, as well as PGP activities on several economically important horticultural plants. Additionally, we performed scanning electron microscopy (SEM) observations to verify the endophytic colonization of lettuce by Streptomyces.

Results: The abundant colonization of young lettuce seedling by two Streptomyces strains demonstrated their capability to interact with the host from early stages of seed germination and root development up to two weeks. Plant-strain specific PGP activity was observed; e. g., S. cyaneus ZEA17I promoted the growth of lamb lettuce but not that of tomato. When were applied to S. sclerotiorum inoculated substrate in growth chamber, S. exfoliatus FT05 Wand S. cyaneus ZEA17I significantly reduced lettuce basal drop incidence by 44.8% and 27.6%, respectively, compared to the inoculated control ($P < 0.05$). Interestingly, under field conditions, S. exfoliatus FT05W reduced the disease incidence by 40%, and there was only 10% protection of lettuce after S. cyaneus ZEA17I application. Our results indicate the greatly promising potential of Streptomyces for exploitations in agriculture as biocontrol agents.

* Funding: The authors thank Prof. Mervyn Bibb (John Innes Centre, UK), for kindly providing the plasmid pIJ8641, as well as prof. Flavia Marinelli (University of Insubria, Italy), for donor strain *E. coli* ET12567 for *Streptomyces* transformation

** Corresponding author: Paolo Cortesi; E-mail: paolo. cortesi@ unimi. it

微黄青霉ZF1对玉米丝黑穗病的作用机理与防治效果研究*

苏前富**，贾 娇，张 伟，孟玲敏，李 红，晋齐鸣***
（吉林省农业科学院/农业部东北作物有害生物综合治理重点实验室，公主岭 136100）

摘 要：玉米丝黑穗病（head smut）是世界玉米产区普遍发生且为害严重的土传病害。目前，生产上防治该病害主要依靠化学防治。为减少化学农药的使用、降低环境污染，本研究通过微黄青霉菌（*Penicillium minioluteum*）ZF1与病菌混合接种抗病自交系黄早四的方法，调查生防菌株对丝黑穗病的防治效果，进一步分析了生防菌ZF1培养液对丝孢堆黑粉菌（*Sporisorium reilianum*）冬孢子的萌发和菌丝的影响。结果发现，单独接种丝孢堆黑粉菌后黄早四的平均发病率为62.3%，而施用微黄青霉ZF1颗粒剂与丝孢堆黑粉菌混合接种后丝黑穗病平均发病率为24.5%，比单独接种丝孢堆黑粉菌低37.8%，防治效果为60.67%；在10%～100%微黄青霉培养液中丝孢堆黑粉菌孢子萌发率极低，而PD培养基中冬孢子几乎全部萌发，表明微黄青霉ZF1对丝孢堆黑粉菌菌丝和孢子萌发均具有较强的抑制作用，对丝黑穗病具有一定的防治效果。文献报道，28%灭菌唑悬浮剂对玉米丝黑穗病的防治效果为62.44%，2%的戊唑醇种衣剂的防治效果为80.8%～90.4%，本试验选用的生防菌微黄青霉ZF1颗粒剂对玉米丝黑穗病的防治效果与28%灭菌唑悬浮剂相同，但与2%的戊唑醇种衣剂相比，差距还较大。刘洪亮等在土壤中筛选到抑制丝孢堆黑粉菌的生防菌SF-1，其发酵液对丝孢堆黑粉菌冬孢子的抑制率为76.56%，生防菌ZF1对丝孢堆黑粉菌冬孢子的抑制率高达100%，远高于生防菌SF-1的抑菌效果。研究结果表明生防菌ZF1在防治玉米丝黑穗病上具有良好的应用前景，下一步将通过改进剂型和提高孢子成活率等方式对菌株进行改良，通过改进施用方法和调节使用量等方面提高生防菌ZF1对玉米丝黑穗病的防治效果。

关键词：微黄青霉；丝黑穗病；生防菌

---

* 项目基金：国家玉米产业技术体系（CARS-02）

** 第一作者：苏前富，副研究员，博士，研究方向：玉米病虫害综合防治
*** 通讯作者：晋齐鸣，本科，研究员；E-mail：qiming1956@163.com

没食子酸抑制水稻细菌性条斑病菌的机制*

魏昌英**，陈媛媛，黎芳靖，汪锴豪，袁高庆，林 纬，黎起秦***
（广西大学农学院，南宁 530004）

摘 要：植物中的酚类化合物没食子酸（Gallic acid，GA）具有抗菌、抗氧化和抗肿瘤等多种生物功能，我们的前期研究发现，GA 对水稻细菌性条斑病菌（*Xanthomonas oryzae* pv. *oryzicola*）具有较强的抑制作用，目前尚未有 GA 对水稻细菌性条斑病菌抑菌机制方面的研究报道。本研究对 GA 影响水稻细菌性条斑病菌的细胞结构、功能和生理生化等方面进行测定。结果表明：在菌体浓度为 10^7 CFU/mL 时，GA 对水稻细菌性条斑病菌的最低抑菌浓度（MIC）为 200μg/mL。在电镜下观察 MIC 的 GA 对水稻细菌性条斑病菌的形态结构影响，发现经 GA 处理后的菌体变形，细胞壁破损；测定病菌培养液电导率、乳酸脱氢酶活性以及大分子物质含量的变化情况的结果发现，GA 处理后的菌体电解质外渗，大分子物质渗漏到培养液中，乳酸脱氢酶活性增加，说明 GA 不仅引起水稻细菌性条斑病菌细胞膜通透性的改变，还破坏了细胞膜的完整性；GA 可抑制病菌的呼吸作用，用 GA 处理菌体后，其培养液中有大量的丙酮酸累积，菌体中与呼吸作用相关的酶类（苹果酸脱氢酶、琥珀酸脱氢酶和 NADH 氧化酶）活性受到抑制，GA 与丙二酸、碘乙酸、磷酸钠抑制病菌的呼吸作用的叠加抑制率分别为 11.34%、21.88% 和 37.45%，由于 GA 与丙二酸抑制病菌的呼吸作用的叠加率最小，推测 GA 可能影响病菌的呼吸代谢三羧酸循环途径（TCA）；不同浓度的 GA 对菌体胞外多糖的产生均没有影响，GA 处理菌体后，菌体胞外酶（蛋白酶、纤维素酶和果胶酶）的活性受到影响，且 GA 作用的浓度越高，胞外水解酶的活性越显著性降低。

关键词：水稻细菌性条斑病菌；没食子酸；抑菌机制

* 基金项目：广西自然科学基金（2014GXNSFAA118073）
** 第一作者：魏昌英，硕士研究生，研究方向为植物细菌性病害及其防治；E-mail：734001458@qq.com
*** 通讯作者：黎起秦，教授，研究方向为植物病害及其防治；E-mail：qqli5806@gxu.edu.cn

狭叶十大功劳抑菌物质分离及其对水稻细菌性条斑病的防治作用*

黎芳靖**，陈媛媛，周荣金，黎起秦，林　纬，袁高庆***

（广西大学农学院，南宁　530004）

摘　要：狭叶十大功劳是一种药用植物和观赏性植物，在我国资源丰富。本文笔者前期研究发现，十大功劳提取物对黄单胞杆菌属植物病原细菌有较强的抑制作用，但其对植物病害的防治作用研究未见报道。本研究以水稻细菌性条斑病菌［*Xanthomonas oryzae* pv. *oryzicola*（Fang et al.）Swings et al.］为供试菌种，从狭叶十大功劳中分离和纯化具有较好抑菌活性的化合物，并对其抑菌活性、在水稻叶片中的传导性以及对水稻细菌性条斑病的盆栽防治效果进行测定。结果表明：从狭叶十大功劳中获得 2 个单体化合物，经波谱鉴定该两种化合物分别为小檗碱和药根碱。室内毒力测定发现，小檗碱对水稻细菌性条斑病菌的抑菌活性（EC_{50} = 2. 9008mg/L）强于药根碱（EC_{50} = 16. 055 3mg/L）。小檗碱抑菌谱较广，对常见 15 种植物病原菌物（*Exserohilum turcicum*、*Bipolaria maydis*、*Alternaria brassicae*、*Magnaporthe grisea*、*Curvularia lunata*、*Fusarium oxysporum* f. sp. *niverum*、*Corynespora cassiicola*、*Pestalotiopsis mangiferae*、*Botrytis cinerea*、*Phoma putaminum*、*Phyllosticta jasminicola*、*Phytophthora nicotianae*、*Colletotrichum musae*、*Penicillium digitatum*、*Sclerotium rolfsii* 以及 *Phomopsis citri*）和 5 种植物病原细菌（*Xanthomonas oryzae* pv. *oryzicola*、*Xanthomonas oryzae* pv. *oryzae*、*Xanthomonas axonopodis* pv. *citri*、*Xanthomonas campestris* pv. *pruni*、*Xanthomona campestris*）均具有抑菌作用，其中，对玉米小斑病菌（*B. maydis*）、香蕉弯孢霉叶斑病菌（*C. lunata*）、水稻白叶枯病菌（*X. oryzae* pv. *oryzae*）和水稻细菌性条斑病菌（*X. oryzae* pv. *oryzicola*）的抑制作用最好，但对生防细菌枯草芽孢杆菌 *Bacillus subtilis* B47 以及 B196 菌株的生长基本没有影响。小檗碱可渗入水稻叶片中，并可在植株中上下输导，在水稻叶片中持效期长达 21d。小檗碱在 0. 1 g/L 浓度下对水稻细菌性条斑病的防效（75. 68%）与 20% 叶枯唑可湿性粉剂有效成分 0. 44g/L 的防效（79. 42%）相当。

关键词：水稻黄单胞菌栖稻致病变种；水稻细菌性条斑病；狭叶十大功劳；小檗碱；药根碱

---

* 基金项目：国家自然科学基金项目（31560523）

** 第一作者：黎芳靖，硕士研究生，研究方向为植物细菌性病害及其防治；E-mail：1789430758@ qq. com

*** 通讯作者：袁高庆，副教授，研究方向为植物病害及其防治；E-mail：ygqtdc@ sina. com

白僵菌对油菜菌核病菌的颉颃作用研究

齐永霞，陈方新
（安徽农业大学植物保护学院，合肥　230036）

摘　要：在实验室条件下，通过对峙培养法、杯碟法等方法研究了球孢白僵菌不同菌株对油菜菌核病菌的抑制作用。平板对峙培养结果表明，白僵菌不同菌株对油菜菌核病菌的菌丝生长均具有一定的抑制作用，在对峙培养处，油菜菌核病菌不产生菌核。杯碟法研究结果表明，球孢白僵菌分生孢子悬浮液对油菜菌核病菌的菌丝生长及菌核产生量具有较好的抑制作用。测定了不同培养液培养的球孢白僵菌代谢液对油菜菌核病菌菌丝生长和菌核产生量的抑制作用，结果表明，查氏培养液培养得到的球孢白僵菌代谢液对油菜菌核病菌菌丝生长的抑制作用最好，菌核产生量最少，SDAY 培养液培养得到的球孢白僵菌代谢液对油菜菌核病菌菌丝生长的抑制作用最差。

关键词：白僵菌；油菜菌核病菌；抑制作用；对峙培养；代谢液

油菜菌核病俗称麻秆、烂秆，是油菜生产中的重要病害之一[1]。在我国，油菜菌核病居油菜三大病害之首，严重制约了长江流域及东南沿海地区油菜的产量和品质[2]。近年来，国内外学者对油菜菌核病的防治工作进行了广泛深入的研究，由于目前尚未发现免疫和高抗油菜品种，因此对油菜菌核病的防治仍然采用以化学农药防治为主的措施，但长期大量使用化学农药不仅会引起环境污染，还会危害人类的健康，正因如此生物防治日益受到人们的重视[3]。

球孢白僵菌（*Beauveria bassiana*）是广泛应用于生物防治的虫生真菌之一，长期以来，球孢白僵菌作为一种微生物杀虫剂备受国内外生物学者和生产单位的关注，研究证实球孢白僵菌对蚜虫、松墨天牛等害虫确实起到了很好的防治效果[4-11]。目前，尽管人们对球孢白僵菌的生防用途做了很多研究，但研究重点及主要方向均集中在害虫生物防治方面，有关球孢白僵菌防治植物病害的研究甚少，特别是球孢白僵菌对油菜菌核病菌的生物防治国内尚未见报道。作者通过平板对峙培养等方法研究了球孢白僵菌不同菌株对油菜菌核病菌的抑制作用，研究结果对生产和应用该菌株进行防治植物病害具有重要的指导意义。

1　材料与方法

1.1　供试菌株

油菜菌核病菌（*Sclerotinia sclerotiorum*）、球孢白僵菌（*Beauveria bassiana*）（编号分别为 Bb234、Bb509、Bb510、Bb2092、Bb2024、Bb1174）、布氏白僵菌（*Beauveria brongniar*）（编号为 Bbr81）、多形白僵菌（*Beauveria amorpha*）（编号为 Ba12）。

1.2　供试培养基

SDAY 培养基、PDA 培养基、查氏培养液（Czapek 培养液）、PDB 培养液、SDAY 培养液。

1.3　平板对峙培养抑菌效果

采用平板对峙培养法测定白僵菌不同菌株对油菜菌核病菌的抑制作用。用十字交叉法测量供试菌株菌丝生长直径，按下列公式计算菌丝生长抑制率。

$$\text{菌丝生长抑制率（\%）} = \frac{\text{对照菌落直径} - \text{处理菌落直径}}{\text{对照菌落直径}} \times 100\%$$

1.4　白僵菌无菌代谢液对油菜菌核病菌的抑制作用

将供试白僵菌菌株分别接种到 SDAY 培养液、PDB 培养液、查氏培养液中，黑暗条件下振荡（25℃、150r/min）培养 20d 后，用灭菌滤纸过滤去除菌丝，5 000r/min 离心 10min，去沉淀后，取上清液，用旋转蒸发仪浓缩 10 倍后取出，再用孔经 0.22μm 的细菌过滤器将上清液进行过滤，滤液经 PDA 平板培养，48h 无菌落出现视为无菌代谢液。取白僵菌无菌代谢液 2mL（浓缩 10 倍后），分别加入 18mL 的 PDA 培养基摇匀后制成平板，以不加代谢液的平板作对照，然后接菌碟，25℃培养，每隔 24h 测一次菌落直径，直至对照长满整个培养皿。第 10d 观察记载不同处理的油菜菌核病菌菌核产生数量。

2　结果与分析

2.1　白僵菌与油菜菌核病菌平板对峙培养结果

由表 1 可以看出，白僵菌不同菌株和油菜菌核病菌对峙培养 3d 后，油菜菌核病菌的菌丝生长均受到一定程度的抑制作用，其中，Bb2092 菌株对油菜菌核病菌的抑制作用最好。培养至第 3d 时，菌丝生长抑制率达 37.83%

表 1　平板对峙培养白僵菌对油菜菌核病菌菌丝生长的抑制作用

白僵菌菌株	菌丝生长抑制率（%）（3 d）
Bb234	36.55
Bb509	34.89
Bb510	35.12
Bb2092	37.83
Bb2024	36.82
Bb1174	33.98
Bbr81	30.56
Ba12	28.05

2.2　白僵菌无菌代谢液对油菜菌核病菌菌丝生长及菌核产生量的影响

由表 2 可以看出，白僵菌无菌代谢液对油菜菌核病菌菌丝生长及菌核产生具有一定的的抑制作用，其中，Czapek 培养液培养得到的代谢液对油菜菌核病菌菌丝生长及菌核产生量的抑制效果最好，其次是 PDB 培养液，SDAY 培养液培养得到的代谢液的抑制效果最差。在供试的白僵菌菌株中，Bb2024 菌株的抑菌效果最好。

表 2　白僵菌代谢液对油菜菌核病菌菌丝生长及菌核产生量的影响

白僵菌菌株	菌丝生长抑制率（%）（3d）			菌核产生量（个）		
	PDB 培养液	SDAY 培养液	Czapek 培养液	PDB 培养液	SDAY 培养液	Czapek 培养液
Bb234	42.34	33.15	49.02	20	27.67	14.33
Bb509	41.08	31.38	50.08	18.67	24.33	16.67
Bb510	43.79	34.05	51.28	20.67	23.67	16.33
Bb2092	45.26	34.26	51.58	18.67	22.67	12.67
Bb2024	46.22	35.26	52.64	16.67	25.33	10.67

3　小结与讨论

化学农药防治植物病害是植物病害综合防治措施之一，具有见效快、防病效果好等优点，近几十年来由于化学农药的施用不当，而造成环境污染，影响了生态环境。所以生物源农药的研究被提上了日程。笔者通过平板对峙培养法、杯碟法等方法研究了白僵菌不同菌株对油菜菌核病菌的抑制作用。平板对峙培养法实验结果表明，Bb2092 菌株对油菜菌核病菌的抑制作用最好，抑制率达到 37.83%。白僵菌代谢液对油菜菌核病菌菌丝生长的抑制实验结果表明，不同培养液培养得到的白僵菌无菌代谢液对油菜菌核病菌均具有一定的的抑制作用。其中，Czapek 培养液培养得到的代谢液对油菜菌核病菌菌丝生长及菌核产生量的抑制效果最好，其次是 PDB 培养液，SDAY 培养液培养得到的代谢液的抑制效果最差。在供试的白僵菌菌株中，Bb2024 菌株的抑菌效果最好。

参考文献

[1] 李国庆．作物菌核病病源—核盘菌的多样性研究［M］．武汉：华中农业大学，1996：188－189.

[2] 肖兰英．油菜菌核病综合防治关键技术［J］．经济作物，2010，2：129－130.

[3] 余素红，曾明森，吴光远．球孢白僵菌的研究与展望［J］．茶叶科学技术，2009（3）：8－11.

[4] 张立钦，刘军．松墨天牛优良白僵菌菌株筛选［J］．南京林业大学学报，2000，24（2）：33－37.

[5] 马良进，杨毅，张立钦．松墨天牛寄生白僵菌的优良菌株筛选［J］．东北林业大学学报，2006，34（5）：4－6.

[6] Ownley B H，Dee M M，Gwinn K D. Effect of conidial seed treatment rate of entomopathogenic *Beauveria bassiana* 11－98 on endophytic colonization of tomato seedlings and control of Rhizoctonia disease［J］. Phytopathology，2008，98：S118.

[7] Clark M M，Gwinn K D，Ownley B H. Biological control of *Pythium myriotylum*［J］. Phytopathology，2006，96：S25.

[8] 徐福元，张培，赵菊林，等．利用小蠹虫释传白僵菌技术防治松材线虫病的研究［J］．林业科学研究，2000，13：63－68.

[9] 段彦丽，张永安，王玉珠．温度和增效剂对苏云金杆菌杀虫活性的影响［J］．中国森林病虫，2007，26（1）：1－4.

[10] 胡强，贝纳新，高萍，等．高毒力杀蚜白僵菌的分离及筛选［J］．现代农药，2008，7（2）：44－46.

[11] 孙继美，丁珊，肖华，等．球孢白僵菌防治松墨天牛的研究［J］．森林病虫通讯，1997（3）：16－18.

种衣剂副作用防控技术研究与应用

于思勤，刘　一
（河南省植物保护植物检疫站，郑州　450002）

摘　要：种衣剂具有预防苗期病虫害，促进作物生长的作用，已在农业生产中广泛应用。本文概述了种衣剂的组成及种衣剂副作用的表现，系统分析了种衣剂副作用产生的原因，提出了预防控制种衣剂副作用的技术措施，为提升种衣剂质量，促进种子产业健康发展提高提供了依据。

关键词：种衣剂；副作用；防控技术

1926年美国的Thornton和Ganulee首先提出种子包衣问题，20世纪30年代英国Germains种子公司在禾谷类作物上首次成功地研制出种衣剂，1976年美国R. C. McGinnis进行了小麦包衣种子田间试验，获得了抗潮、抗冷、抗病、出苗快、长势好的效果。到20世纪80年代，世界上发达国家种子包衣技术已基本成熟。中国于20世纪90年代初开始研发种衣剂，经过近30年的研究和推广应用，已开发出农药型、药肥型、生物型和特异型种衣剂500多种。广泛应用于农业生产，为控制农作物苗期种传和土传病害、地下害虫和苗期病虫害发挥了重要作用。随着种衣剂大面积推广应用，因种衣剂质量差异、使用技术不规范及极端环境条件造成的药害事件频频发生，在一定程度上影响了种衣剂的推广应用，因此，如何提高种衣剂的使用效果，有效避免副作用发生，已成为生产上亟待解决的问题。

1　我国种子处理剂的发展及应用

使用农药处理种子，是从源头控制种传病害、土传病害和苗期害虫的有效措施。在原始农业中，为了保护种子萌发和健康生长，人们使用草木灰、升汞、砷、胶泥等处理种子，预防作物苗期病虫害发生。自从化学农药问世以来，逐渐使用药剂处理种子，防治苗期病虫害，随着农药研发及使用技术的完善，种子处理剂受到普遍欢迎，使用范围和应用面积迅速扩大，尤其是种衣剂的发明和应用，为农作物苗期病虫害防治提供了重要手段。

1.1　20世纪80年代以前

没有专门用于拌种的药剂，主要使用75%萎锈灵、75%五氯硝基苯、50%福美双、50%多菌灵、70%敌克松等，按照一定的剂量拌种，防治种子和土壤传播的病害；使用有机磷拌种或六六六粉剂处理土壤，防治地下害虫。缺点是：拌不匀、毒性大、防效不高，拌后不能储存，需要稍晾后播种。

1.2　20世纪90年代以前

防治药物、防治方法、防治效果基本上没有太大变化。主要使用75%萎锈灵、75%五氯硝基苯、50%福美双、50%苯来特、50%多菌灵、50%甲基托布津、15%三唑酮等，按照一定的剂量拌种，防治种子和土壤传播的病害；使用有机磷拌种或林丹粉剂处理土壤，防治地下害虫。

1.3　20世纪90年代以后

随着植物保护理念的发展，我国开始引进国外的种衣剂，并且着手研制国产种衣剂。试验结果证明，种衣剂是一个新生事物，能够从源头控制农作物苗期病虫害，防治效果和增产作用显著优于以前的拌种药剂。

国内企业在模仿的同时，根据国内生产的农药种类和剂型，将杀菌剂、杀虫剂、微肥复配在一起，陆续开发出了小麦种衣剂、玉米种衣剂等。大多数配方是杀虫剂甲基异柳磷、克百威、辛硫磷与杀菌剂福美双、多菌灵、三唑酮的复配剂，优点是价格便宜，使用范围广泛；缺点是防治效果一般，持效期短，使用不当易出现药害。

国外农化公司利用先进的农药研发技术和推广优势，引进推广了2.5%适乐时（咯菌腈）悬浮种衣剂、3%敌委丹（苯醚甲环唑）悬浮种衣剂、70%锐胜（噻虫嗪）干种衣剂、58.5%高巧（福美双+戊菌隆+吡虫啉）湿拌种剂、40%卫福合剂（萎锈灵 +福美双）、2%立克秀（戊唑醇）干拌剂等。

1.4 当前推广应用的种衣剂

进口杀菌型种衣剂有适乐时、敌委丹、适麦丹、满适金、金阿普隆、顶苗新、立克秀、卫福、武将、亮盾、亮穗、扑力猛、全蚀净等。进口杀虫型种衣剂有锐胜、帅苗、劲苗、高巧等。

我国种衣剂登记的品种虽然很多，但有效成分多局限在克百威、吡虫啉、福美双和多菌灵、三唑酮、戊唑醇等品种。

1.5 种衣剂的成分组成

种衣剂（seed coated with a pestcide）是由相关药剂、营养元素、植物生长调节剂、成膜剂、湿润剂、分散剂、警戒色和助剂加工而成，具有一定的强度和通透性，可直接包覆于种子表面的活性悬浮剂。我国目前研发种衣剂多为复合型，少量为单元型，有效成分主要是农药（包括杀菌剂、杀虫剂）、微量元素和植物生长调节剂。

杀菌剂主要是多菌灵、福美双、三唑酮、戊唑醇、五氯硝基苯、三唑醇、咪鲜胺、萎锈灵、苯醚甲环唑、立枯磷、咯菌腈、甲霜灵、克菌丹等，以多菌灵、福美双、三唑酮、戊唑醇、三唑醇最为普遍。这些杀菌剂主要用于防治种传和土传病害、苗期病害等，促进作物幼苗正常生长。

杀虫剂主要是克百威、丁硫克百威、吡虫啉、辛硫磷、甲拌磷、甲基异柳磷、氯氰菊酯、阿维菌素、噻虫嗪等，以克百威、丁硫克百威、吡虫啉、辛硫磷、甲基异柳磷最为普遍，主要用于防治苗期地下害虫、蚜虫、蓟马、飞虱、线虫等。

植物生长调节剂主要用于促进幼苗发根和生长。像加赤霉酸促进生长，加萘乙酸促进发根等。如用于潮湿寒冷土地播种时，种衣剂中加入萘乙烯（Styrene）可防止冷害。如种衣剂中加入半透性纤维素类可防止种子吸水过快而造成吸胀损伤。如靠近种子的内层加入活性炭、滑石粉和肥土粉，可防止农药和除草剂的伤害。如种衣剂中加入氧化钙，种子吸水后放出氧气，促进幼苗发根和生长等等。

微肥包括锌、镁、钼、硼等，主要用于促进种子发芽和幼苗植株发育。

2 种衣剂的副作用表现

2.1 部分农药有抑制出苗的作用

唑类杀菌剂有抑制种子发芽的作用，含有唑类杀菌剂的种衣剂使用量大或包衣不均匀，能造成种子出苗迟缓、苗弱，长势差；使用“喷雾”用吡虫啉制剂拌种，造成小麦不出苗或出苗迟缓，严重时影响出苗率，造成缺苗断垄。

2.2 不良环境条件造成种衣剂副作用

包衣种子播种后，遇到干旱、低温、土壤湿度过大、播种过深等不良环境条件，使种子出苗历期延长，种衣剂副作用加重，造成种子不发芽或出苗率明显下降，严重影响幼苗生长。

2.3 种衣剂质量差异大

2003年国家实行种子补贴后，高中低档种衣剂在市场上呈现泛滥之势，许多厂家仓促上马，技术不过关，产品质量不稳定或者以假充真，不仅达不到防治效果，而且副作用严重，给农业生产带来很大损失，影响了种衣剂的推广应用。

2.4 种衣剂的药害症状

经过包衣后种子出苗期较未包衣种子一般晚出苗1~3d，个别作物在出苗后叶片有皱缩现象，大约持续1周，如大豆种子。主要原因是种衣剂内含有对作物敏感成分，使用不当就会使种子发芽受到影响，长时间不能拱土，即使拱土幼苗也不能很快恢复生长，而形成小老苗，对作物后期产量有严重影响。

3 种衣剂副作用产生的原因分析

3.1 种衣剂选择不当

目前的种衣剂不能包治所有种传、土传病害，如玉米常见土传病害有7种，现在登记使用的种衣剂主要针对丝黑穗病和黑粉病防效好，而对霜霉病和斑点病等效果较差。小麦种衣剂对纹枯病、黑穗病、根腐病有效，对全蚀病、禾谷孢囊线虫等防效较差。因些在生产上表现为盲目使用，不能根据不同地区、不同病虫发生情况选择合适的种衣剂品种，造成防效较差，甚至出现药害。

3.2 使用剂量控制不严

种子处理药剂都有一定毒性，在生产、销售、使用过程中由于使用量不准，用量少导致使防效不高，用量大导致种衣剂副作用发生。

3.3 环境因素变化大

包衣种子播种需要良好苗床，包衣种子贮藏性能的关键因素受含水量和药剂比例大小影响较大，含水量高可能导致包衣种子发芽率下降。另外，土壤湿度过大或过小，低洼盐碱地或黏重土壤，覆土过深或过浅，气温过低等，都可能造成药害，影响包衣种子的正常生长。

3.4 国产种衣剂产品技术落后，产品结构不合理

我国种衣剂产品与发达国家相比，在产品理化性状、悬浮性、稳定性等方面有较大差距，影响有效成分的发挥和商品性。同时国内种衣剂企业产品技术基本处于同一水平，企业规模小而分散，种衣剂生产、销售较为混乱，严重影响了种子市场的健康发展。

3.5 使用技术不规范

种子公司出于成本考虑，往往选择价格低、质量差的种衣剂进行种子包衣，个别种子加工企业甚至将旧种、陈种、坏种包衣后销售，造成种子发芽率低，保苗效果差，使农民对包衣种子产生误解。农民购买种衣剂自行包衣时，往往使用量偏大，整地质量差、播种过深及播种后遇到不良环境条件等均造成种子出苗率下降，药害事件时有发生。

4 种衣剂副作用的预防与控制

4.1 有效成分向低毒方向发展

种衣剂中的杀虫剂如克百威、甲基异硫磷甲拌磷等均属于高毒农药，对人、畜安全隐患极大，人、畜中毒事故时有发生，迫切需要用丁硫克百威、吡虫啉、噻虫嗪、辛硫磷等高效低毒杀虫剂来取代；严格控制唑类杀菌剂的用量。

4.2 研究开发新型种衣剂

根据不同作物、不同生态区域，研发有针对性的种衣剂，如在西北地区推广抗旱型种衣剂，在东北地区推广抗寒型种衣剂；在农药、微量元素、激素、成膜剂、分散剂和助剂的选择上要确保对种胚不造成伤害，不用在水中溶解度大的溶剂，农药的颗粒细度要大于种皮孔，避免堵塞种皮孔。针对不同配方的种衣剂制定产品质量技术标准，通过强制执行和产品质量抽查，规范种衣剂生产和销售。使同一类型的种衣剂具有可比性，避免因质量标准不统一而造成的安全使用

隐患。

4.3　将生物防治剂引入种衣剂

化学农药的使用带来环境污染、对有益生物产生不利影响、导致有害生物对农药产生抗药性等问题，为生物防治的应用提供了广阔的前景。已经用于生物防治的微生物有真菌、细菌、放线菌等，目前我国已经登记的生物型种衣剂有枯草芽孢杆菌、苏云金杆菌、阿维菌素、淡紫拟青霉悬浮种衣剂，在生产上展现出广阔应用前景，随着研究和开发的深入，将会探索出越来越多的生物防治方法。

4.4　开发种子丸粒化种衣剂

种子丸粒化技术因其具有防治病虫害、方便在其中加入微肥、激素和保水剂等、提高种子质量、实现精量播种、节省费用等优点而成为当前农业生产中的热点。油菜、烟草、甜菜、蔬菜、牧草等小粒或不规则种子的成功丸粒化及在生产中的广泛应用，为它的深入研究和开发提供了极好的范例，开辟了新的思路，以前被认为的技术难点将逐步突破。包括玉米种子、带绒棉种、水稻种子在内的丸粒化技术均需要进行田间试验和推广应用，促进丸粒化技术的完善并推动它的迅速发展。

4.5　研制特异型种衣剂

异型种衣剂是根据不同作物和目的而专门设计的种衣剂类型。如 Sladdin 等人用过氧化钙包衣小麦种子，使播种在冷湿土壤中的小麦出苗率从 30% 提高到 90%；江苏为水稻旱育秧而设计的高吸水种衣剂，中科院气象所研制的高吸水树脂抗旱种衣剂；浙江大学研制的直播稻专用种衣剂、油菜专用种衣剂、玉米抗寒型种衣剂，中国农科院植保所研制的防治花生地下害虫的缓控释种衣剂等，为在不同条件下大面积推广应用，避免种衣剂副作用发生、提高防控效果提供了技术和物质保障。

4.6　规范种衣剂使用技术措施

根据不同地区、不同作物及主要防控对象，合理选择利用种衣剂，在试验示范的基础上，推广对农作物安全，防治病虫效果好的种衣剂。统一包衣时，要严格控制种子的含水量、种衣剂用量和包衣质量，尽量缩短包衣种子存放时间；农民自行拌种时，要根据说明书要求使用，不要盲目增加用量，拌种后要尽快播种，播种不宜过深，以免出现副作用。种衣剂仅限种子包衣使用，严禁对水喷雾；催芽播种时，先催芽，后拌药剂。

4.7　研究种衣剂副作用应急防控技术措施

根据不同作物、不同类型种衣剂副作用的田间表现症状，建立种衣剂副作用识别和快速诊断标准，组织开展种衣剂安全使用技术研究和推广应用，根据副作用严重程度和发生的频率，采取不同的应对措施，例如：在西北地区推广的种衣剂要添加聚谷氨酸抗旱剂，在东北地区推广的种衣剂要添加防寒剂，在南方地区推广的种衣剂要考虑湿度过大引起的药害问题；在除草剂残留量大或除草剂药害严重的地区，推广使用含有奈安的种衣剂，预防除草剂药害。盐碱较重的地块，请勿使用种衣剂；黏性土壤保肥、保水条件好，应根据气候和病虫害情况选择使用种衣剂；沙性土壤地通透性、保肥，保水性差，遇低温最容易造成烂种、死苗，可选用不含唑类杀菌剂的种衣剂。

为了解决生产中的疑难问题，提高种衣剂的研究和应用水平，农业部 2013 年设立了“种衣剂副作用安全防控技术研究与示范”公益性行业专项，组织有关科研、教学、推广和生产企业的专家开展协作攻关，随着研究成果的推广应用，将为种衣剂副作用安全防控提供重要的技术支撑。

参考文献（略）

地衣芽孢杆菌 W10 对桃枝枯病的生物防治研究*

高汝佳[1**]，黄弘樑[1]，戴慧俊[2]，赵文静[1]，纪兆林[1***]，董京萍[1]，童蕴慧[1]，徐敬友[1]
（1. 扬州大学园艺与植物保护学院，扬州 225009；
2. 无锡太湖阳山水蜜桃科技有限公司，无锡 214155）

摘 要：桃枝枯病是由桃拟茎点霉（*Phomopsis amygdali*）引起的一种桃树重要病害。目前，控制桃枝枯病的主要措施是化学杀菌剂，但是，化学药剂大量使用给环境、生态平衡和人们的健康带来了一定的影响，也使病菌产生抗药性，为减少化学农药的使用，生物防治已受到广泛的关注，成为一种重要、安全的防治方法。地衣芽孢杆菌 W10 是本实验室筛选出的一株生防细菌，对灰霉病菌、纹枯病菌、轮纹病菌、炭疽病菌、菌核病菌、褐腐病菌等多种植物病原真菌有较强的抑制作用，而且具有定殖和诱导抗病性能力，田间防效与化学药剂腐霉利相当，该菌株生防机制主要是产生一种 46kDa 的抗菌蛋白。本文研究了生防细菌地衣芽孢杆菌（*Bacillus licheniformis*）W10 及其抗菌蛋白对桃枝枯病菌的抑制及其与化学农药复配对病害的防治作用。

本研究表明，W10 菌液及其抗菌蛋白对桃枝枯病菌有很强的颉颃能力，尤其是能明显抑制病菌菌丝生长、产孢、孢子萌发和芽管伸长。浓度为 2×10^8 CFU/mL 菌液和 3.064 mg/mL 抗菌蛋白对枝枯病菌的生长抑制率都达到了 100%。5×10^8 CFU/mL 菌液和 7.660mg/mL 抗菌蛋白对病菌产孢抑制率分别为 94.9% 和 100%。1×10^9 CFU/mL 菌液和 3.064mg/mL 抗菌蛋白能完全抑制病菌孢子萌发和芽管伸长，其中，5×10^8 CFU/mL 菌液和 1.625mg/mL 抗菌蛋白的抑制率达到了 80% 以上。此外，W10 菌液及抗菌蛋白处理后，病菌菌丝出现畸形甚至断裂、细胞原生质渗漏、细胞壁破损等现象。田间试验表明，地衣芽孢杆菌 W10 菌液对桃枝枯病也有一定的防治效果。因此，地衣芽孢杆菌 W10 对桃枝枯病具有较好的生防潜力。

为了减少农药使用和提高防治效果，也进行了 W10 菌液与化学农药咪鲜胺、多菌灵和烯唑醇的复配研究，结果表明菌液与咪鲜胺配比 1∶1、1∶2 时对桃枝枯病菌有明显的增效作用，2∶1 时为颉颃作用；菌液与烯唑醇的 1∶1、1∶2、2∶1 配比对桃枝枯病菌都具有明显的增效作用，但菌液与多菌灵复配表现为颉颃或相加作用。田间试验表明，在田间枝枯病发生严重流行情况下，1×10^9 CFU/mL W10 菌液与 45% 咪鲜胺水乳剂 2 000倍液 1∶1 复配时对枝枯病病枝率防效达到了 50% 以上；而 1×10^9 CFU/mL W10 菌液与 50% 烯唑醇悬浮剂 5 000倍液 1∶1 复配时的防效也达到了 50% 以上。上述结果为地衣芽孢杆菌 W10 及其抗菌蛋白防治桃枝枯病奠定了基础。

关键词：桃枝枯病；桃拟茎点霉；地衣芽孢杆菌；抑制作用；复配；田间防效

---

* 基金项目：国家现代农业产业技术体系建设专项（CARS－31－2－02）；江苏省农业科技自主创新资金项目［CX（14）2015，CX（15）1020］；江苏省无锡市农业科技支撑项目（CLE01N1410）

** 第一作者：高汝佳，硕士研究生，主要从事植物病害生物防治研究；E-mail：287161545@qq.com
*** 通讯作者：纪兆林，副教授，主要从事植物病害防控及分子植病研究；E-mail：zhlji@yzu.edu.cn

美花红千层挥发油 GC-MS 分析及其抑菌活性研究*

祝一鸣**，段志豪，，王小晴，王　军，单体江***
（华南农业大学，林学与风景园林学院/广东省森林植物种质创新
与利用重点实验室，广州　510642）

摘　要：植物挥发油又称植物精油，被誉为液体黄金，具有多种生物活性。为了研究了美花红千层叶片和果实挥发油的化学成分及其抑菌活性，本研究采用水蒸汽蒸馏法分别提取美花红千层叶片和果实中的挥发油，其得率分别为0.16% 和0.48%（以鲜重为基础）；通过气相色谱与质谱联用（GC-MS）分析挥发油的化学成分，并采用抑菌圈法测定了挥发油的抗菌活性。从美花红千层叶片挥发油中鉴定出桉叶油醇（50.67%）、（1*R*）－（＋）－α－蒎烯（19.27%）、2－莰烯（7.85%）、*D*－柠檬烯（3.32%）和*p*－伞花烃（2.26%）等30种化学组分，占总相对含量的92.98%；从果实挥发油中鉴定出（1*R*）－（＋）－α－蒎烯（40.91%）、桉叶油醇（38.57%）、柠檬烯（3.96%）和间异丙基甲苯（2.21%）等27种化学分，占总相对含量的91.71%。其中叶片与果实挥发油中共有组分只有7种，且同一种组分在相对含量上差异较大。美花红千层叶片挥发油对根癌农杆菌的抑制活性最强，其次为溶血葡萄球菌，抑菌圈直径分别为（34.0±0.5）mm和（33.8±2.8）mm，而对枯草芽孢杆菌的抑制活性最弱，抑菌圈直径为（19.3±1.2）mm，但仍强于阳性对照硫酸链霉素［抑菌圈直径（18.0±0.2）mm］。果实挥发油对桉树青枯病菌的抑菌圈直径为（23.0±1.0）mm，明显强于叶片挥发油［（19.3±1.2）mm］和硫酸链霉素［（18.0±0.2）mm］，而对其他供试细菌的抑制活性均弱于叶片挥发油。

关键词：美花红千层；挥发油；GC-MS分析；抑菌活性

* 基金项目：广东省林业科技创新项目（2015KJCH043）
** 第一作者：祝一鸣，硕士，研究方向：植物和微生物的次生代谢；E-mail：zhu_ yiming1992@ foxmail. com
*** 通讯作者：单体江，博士，讲师，研究方向：植物和微生物的次生代谢；E-mail：tjshan@ scau. edu. cn

葡萄灰霉病产挥发性抑菌物质酵母菌的筛选与鉴定*

张　迪**，穆凯热姆·阿卜来提，王晓东***
（石河子大学绿洲农作物病害防控重点实验室，石河子　832000）

摘　要：灰霉病是葡萄生产和贮藏过程中的重要病害，严重影响葡萄的产量和品质。目前防治该病以化学防治为主，但长期使用化学杀菌剂易造成灰霉病菌的抗药性及食品安全问题。因此迫切需要寻找一种安全有效的防病保鲜技术。笔者从新疆不同植物的叶、花和果实表面进行分离获得酵母菌150株，经平板对峙法和活体接种法，成功筛选出1株产挥发性抑菌物质且对葡萄灰霉病菌具有显著颉颃作用的酵母菌株PT－8，活体防效达70.0%。菌株PT－8在PDA平板上菌落呈卵圆形，乳白色，边缘较整齐，繁殖方式为出芽生殖，能形成假菌丝。对多种糖能够发酵，不能发酵半乳糖、乳糖和淀粉。对D－阿拉伯糖、乳酸、甲醇、乙醇和亚硝酸钠则不能同化。能产生芳香的酯类物质，不产生淀粉。可以水解尿素。耐较高渗透压。37℃未见生长。菌株PT－8通过PCR扩增后，获得ITS－5.8S rDNA片段大小为346bp。根据酵母菌菌株PT－8的形态特征、生理生化特性及ITS－5.8S rDNA序列分析，将酵母菌菌株PT－8鉴定为*Metschnikowia* aff *pulcherrima*。

关键词：葡萄灰霉病；酵母菌；挥发性抑菌物质；鉴定

---

* 基金项目：公益性行业（农业）科研专项（201303025）

** 第一作者：张迪，在读硕士研究生，研究方向为植物病害生物防治；E-mail：1538971356@qq.com

*** 通讯作者：王晓东，博士，副教授，从事植物病害生物防治研究；E-mail：wxdong11@163.com

蛋白质组学用于化合物对病原菌作用机制研究*

梅馨月**，杨　敏，丁旭坡，朱书生***
（云南农业大学农业生物多样性与病害控制教育部重点实验室，昆明　650201）

摘　要：近年来医学领域蛋白质组学被广泛运用于靶标的分析，特别是药剂的作用模式的研究及药剂毒理学的评价。添加外源化合物后蛋白质组学能鉴定出显著差异的蛋白，因此能为杀菌剂对病原物的作用模式提供许多有用的信息。然而，用蛋白质组的方法探索杀菌剂的作用机制方面的研究还较少。

为了探索蛋白质组学在杀菌剂或化合物对病原菌作用机制的研究是否可行，本研究选用了一个作用靶标位点已知的杀菌剂即苯酰菌胺来进行研究。苯酰菌胺是一种高效作用于卵菌的苯甲酰胺类杀菌剂。它通过结合 *β-tubulin* 来抑制病原菌的微管蛋白聚合和有丝分裂从而引起有丝分裂阻滞，但是它对 *β-tubulin* 的靶标位点还尚未知道。本研究选择蛋白质组学中的双向电泳的技术来鉴定苯酰菌胺处理后的三七疫霉菌中蛋白质的上下调情况，明确其作用机制并探索蛋白质组学技术在化合物作用机制研究中的可行性。

药剂处理后，通过双向电泳共检测到 21 个差异显著的蛋白点，其中上调蛋白质有 14 个，下调的蛋白质有 7 个。与细胞骨架/细胞运动相关蛋白明显下调，如 β－微管蛋白、肌动蛋白等；与代谢解毒相关、糖代谢相关、线粒体相关、氧化应激反应的蛋白显著上调，如儿茶酚甲氧基转移酶、葡糖激酶、锰超氧化物歧化酶等。

综上，蛋白质组的数据分析表明，苯酰菌胺处理降低了三七疫霉中与细胞骨架相关蛋白的表达并导致了细胞死亡。然而，与糖类、线粒体和代谢解毒相关的酶表达的上调可能与到三七疫霉对苯酰菌胺的抗性相关。这些结果也表明蛋白质组学不仅对研究杀菌剂作用有重要的帮助，也为我们研究外源化合物对病原菌的作用机制提供了新的手段。

关键词：双向电泳；苯酰菌胺；细胞骨架；作用机制

---

* 基金项目：云南省博士研究生学术新人奖

** 第一作者：梅馨月，云南文山人，博士；E-mail：meixinyuemm123@126.com

*** 通讯作者：朱书生，教授；E-mail：shushengzhu79@126.com

解淀粉芽孢杆菌 Lx－11 诱导水稻防卫反应基因表达和抗氧化酶系研究*

张荣胜**，戴秀华，陈志谊，刘邮洲，刘永锋***
（江苏省农业科学院植物保护研究所，南京 210014）

摘 要：由稻黄单胞菌稻生致病变种（*Xanthomonas oryzae* pv. *oryzicola*，*Xooc*）引起水稻细菌性条斑病（Bacterial leaf streak）是我国四大水稻病害之一，一般年份可导致水稻产量损失 15% ~25%，严重时可达 40% ~60%，对水稻的高产稳产造成严重的威胁。利用生防微生物防治植物病害是极具有应用前景的防治途径，是世界各国植病专家的研究热点。本研究室前期筛选获得一株对水稻细菌性条斑病具有良好防效的解淀粉芽孢杆菌 Lx－11，现已进入农药登记程序，为了进一步探究解淀粉芽孢杆菌 Lx－11 防治水稻细菌性条斑病生防机制，我们通过对生防菌 Lx－11 诱导植株防卫反应基因表达和抗氧化酶系进行初步研究。

研究表明：喷施生防菌 Lx－11 后水稻植株中防卫反应基因 *POX*、*PAL*、*LOX*、*PR10a*、*PR1a* 和 *PR4* 表达水平与空白对照相比都有不同程度的提高，其中 *POX*、*PR4* 和 *PR10a* 基因表达水平最大值分别是对照处理的 9.06 倍、5.35 倍和 9.25 倍，这些基因可能参与寄主植株抵御条斑病菌入侵过程中起着重要作用。单独接种病原菌 *Xooc* b5－16 后，水稻植株中仅 *POX* 和 *PR4* 基因表达水平有所提高，但表达强度也明显低于生防菌处理。此外通过对植株体内抗氧化作用的酶 POD，SOD 和 CAT 研究发现，单喷生防菌 Lx－11 处理和单独接种病原菌 *Xooc* b5－16 处理后，两处理都分别在 24h，12h 和 48h 时达到活性高峰，但接种病原菌 *Xooc* b5－16 处理酶活低于生防菌处理，表明 POD，SOD 和 CAT 同样参与寄主植株抵御条斑病菌入侵过程。通过生防菌 Lx－11 与水稻植株之间互作研究，为有益微生物诱导寄主植株如何抵御病原细菌入侵提供了科学依据。

关键词：解淀粉芽孢杆菌；细菌性条斑病；防卫反应基因；抗氧化酶系

---

* 基金项目：公益性行业（农业）科研专项经费项目（201303015）；国家自然科学基金（31501691）；江苏省农业科技自主创新资金项目［CX（13）3024］

** 第一作者：张荣胜，博士，助理研究员，主要研究方向为植物细菌病害生物防治及其机理研究；E-mail：r_ szhang@163. com

*** 通讯作者：刘永锋，研究员，江苏省农业科学院植物保护研究所；E-mail：liuyf@ jaas. ac. cn

河北省小麦茎基腐病发生及防治*

周　颖**，杨文香，张毓妹，张　娜***，刘大群***
（河北农业大学植物保护学院植物病理系/河北省农作物病虫害生物防治工程技术研究中心/国家北方山区农业工程技术研究中心，保定　071001）

小麦茎基腐病（Wheat crown rot）又称作旱地脚腐病，是世界性的重要土传病害，近年来全国各大麦区普遍发生。该病害不仅对小麦的产量和质量造成严重影响，同时也危害人畜的健康和安全。本文对河北省沧州献县、邯郸临漳、石家庄赵县、保定中铁苗木 4 个地方，于小麦苗期、拔节期、灌期进行取样调查，发现各地小麦不同时期小麦茎基腐病均有一定程度发生，病情指数为 12. 58 ~28. 54。

分离获得 178 个菌株分离物，通过菌落培养性状、形态学观察和分子生物学鉴定，共鉴定出 8 种镰刀菌病原，分别为禾谷镰刀菌（*Fusarium graminearum*）、假禾谷镰刀菌（*F. pseudograminearum*）、尖孢镰刀菌（*F. oxysporum*）、厚垣镰刀菌（*F. chlamydospores*）、层出镰刀菌（*F. proliferatum*）、锐顶镰刀菌（*F. acuminatum*）、腐皮镰刀菌（*F. solani*）、燕麦镰刀菌（*F. avenaceum*）。通过拌土法进行分离物柯氏法则验证，发现其中禾谷镰刀菌、假禾谷镰刀菌、尖孢镰刀菌、层出镰刀菌、厚垣镰刀菌为河北省小麦茎基腐病主要致病菌，且各地优势致病菌类群及致病力有所不同，禾谷镰刀菌、假禾谷镰刀菌为致病性最强。其中，沧州只分离到假禾谷镰刀菌，临漳镰刀菌致病类群最为丰富。

以禾谷镰刀菌、假禾谷镰刀菌、尖孢镰刀菌、厚垣镰刀菌和层出镰刀菌为指示菌进行室内毒力测定，结果表明立克秀、敌委丹、酷拉斯、适乐时对 5 种菌菌丝生长抑制效果均较好。以这 4 种药剂对禾谷镰刀菌、假禾谷镰刀菌的温室盆栽防效和小区防效结果表明，酷拉斯药剂拌种防效均优于适乐时药剂拌种，防效可达 60% 以上。沧州献县及邯郸临漳大田防效结果表明，采用酷拉斯药剂拌种防效优于适乐时药剂拌种，与小区药效结果一致。

对 23 个小麦微核心种质和 50 个国外小麦材料进行田间抗性鉴定，结果表明，所检测国内外品种中，23 个小麦微核心种质对禾谷镰刀菌、假禾谷镰刀菌抗性水平为中感至中抗，无高抗材料，5 个国外小麦种质对 2 个菌都具有高抗作用，包括 JWCHI F2000、chapco、Weebill、Avocet、tonichis81。

用赵县分类的禾谷镰刀菌菌悬液对小麦穗部进行注射接种，可引起小麦白穗，与赤霉病症状相同，对麦穗进行组织分离获得与接种物性状一致的病原，说明赵县分离物禾谷镰刀菌可以引起小麦赤霉病。

* 基金项目：河北省小麦产业技术体系创新团队建设项目

** 第一作者：周颖，江苏宿迁，硕士研究生
*** 通讯作者：张娜，副教授，研究方向：植物病害生物防治与分子植物病理学；E-mail：zn0318@ 126. com
刘大群，教授，研究方向：植物病害生物防治与分子植物病理学；E-mail：ldq@ hebau. edu. cn

防治小麦根腐病药剂筛选*

王茹茹**，张毓妹，范学锋，杨文香，张立荣***，刘大群***
（河北农业大学植物保护学院植物病理系/河北省农作物病虫害生物防治工程技术研究中心/国家北方山区农业工程技术研究中心，保定 071001）

小麦根腐病（Wheat root rot）是小麦生产中严重的真菌性病害之一，随着研究的不断深入，近年来发现该病害是由多种病原混合引起的，且在种植小麦地区不断传播蔓延，控制困难，造成严重的经济损失。因此，筛选有效防治药剂对于控制该病害尤为重要。研究通过多点多地调查、室内化学药剂筛选、田间防效比较，以期获得有效控制该病害的药剂。从 24 种化学药剂中筛选出 9 种对小麦根腐病防治效果较好的制剂，分别是苯甲嘧菌酯、咯菌腈、噻霉酮、1% 申嗪霉素、72% 霜脲·猛锌、甲基硫菌灵、氟硅唑、27% 酷拉斯（噻虫嗪、苯醚甲环唑、咯菌腈）及氟环唑。室内毒力测定表明 EC_{50} 为 0.005 ~ 37.151mg/L，抑菌效果最好的是氟硅唑。EC_{50} 分别是 0.015mg/L 和 0.011mg/L。在大田试验中所使用 9 种药剂适乐时（25% 咯菌腈）、27% 酷拉斯、1% 申嗪霉素、井冈霉素、欧博（氟环唑）、戊唑醇·菌核净、45% 戊唑醇、阿米妙收（苯醚甲环唑·丙环唑）、巴斯夫欧博（氟环唑）及阿米妙收（苯甲嘧菌酯）中。防治小麦根腐病效果较好的药剂是 27% 的酷拉斯和阿米妙收（苯甲嘧菌酯），其次是欧博（氟环唑）和 1% 的申嗪霉素，再次是苯醚甲环唑·丙环唑。研究为生产应用及病害的有效控制提供依据。

* 基金项目：河北省现代农业产业体系——小麦产业创新团队建设项目

** 第一作者：王茹茹，河北张家口，硕士研究生
*** 通讯作者：张立荣，副教授，研究方向：植物病害生物防治与分子植物病理学；E-mail：zlr139@126.com
刘大群，教授，研究方向：植物病害生物防治与分子植物病理学；E-mail：ldq@hebaiu.edu.cn

解淀粉芽孢杆菌 JT84 发酵工艺的优化*

王法国[1,2**]，张荣胜[1]，于俊杰[1]，俞咪娜[1]，齐中强[1]，
杜　艳[1]，尹小乐[1]，宋天巧[1]，陆　凡[1]，刘永锋[1***]
（1. 江苏省农业科学院植物保护研究所，南京　210014；
2. 南京农业大学植物保护学院，南京　210095）

摘　要：干悬浮剂（DF，dry flowable）又被称为水分散粒剂（WG，water dispersible granules），是一种加水后能迅速崩解并分散成悬浮液的粒状制剂。近年来，干悬浮剂凭借其崩解性、分散性、悬浮性好，有效成分含量高，流动性好，贮运化学物理性状稳定，包装费用低，逐渐替代水剂、可湿性粉剂和悬浮剂而成为大力发展且深受欢迎的产品剂型之一。江苏省农业科学院植物保护研究所稻病与生防研究室分离得到一株对水稻纹枯病、稻曲病、稻瘟病都具有良好抑菌效果的解淀粉芽孢杆菌 JT84。为了提高解淀粉芽孢杆菌 JT84 干悬浮剂中的芽孢和抑菌活性物质含量，我们对发酵工艺进行了优化。

通过观察解淀粉芽孢杆菌芽孢转化特点和常用培养基中的芽孢含量，以 YPG（葡萄糖 5g、胰蛋白胨 5g、酵母膏 5g）为基础，进行培养基优化。通过单因素试验依次确定碳源、氮源、大量元素和微量元素后，利用 Plackett-Burman 试验设计及 Box-Behnken 设计的响应曲面法（RSM，response surface methodology）对解淀粉芽孢杆菌 JT84 的发酵培养基和培养条件进行了优化。确定了培养基 3 个主要因子的质量浓度分别为葡萄糖 3.59g/L、大豆蛋白胨 21.89g/L 和 K_2HPO_4 0.49g/L。培养条件优化试验得出：培养温度、装液量和发酵时间是影响发酵芽孢数的主要因子，由所得响应曲面方程预测出这 3 个主要因子分别为 30.67℃、68.17mL/250mL 和 65.12h。在初始 pH 值为 6.7，转速 180r/min，接种量 3% 的条件下，发酵芽孢数为 1.67×10^9，对水稻纹枯病菌的抑菌带宽较基础培养基提高 30% 左右。经摇瓶发酵试验和抑菌活性验证该理论预测值与实际值无显著差异。通过对解淀粉芽孢杆菌摇瓶发酵工艺进行优化，为将来解淀粉芽孢杆菌商品化生产提供基础。

关键词：解淀粉芽胞杆菌；Plackett-Burman 试验设计；响应曲面法；培养基优化

---

* 基金项目：江苏省农业科技自主创新资金项目［CX（13）3024］；江苏省科技支撑项目（BE2014386）
** 第一作者：王法国，硕士研究生，主要从事水稻病害的生物防治研究工作；E-mail：2015802155@njau.edu.cn
*** 通讯作者：刘永锋，博士，研究员，主要从事水稻病害及其生物防治研究工作；E-mail：liuyf@jaas.ac.cn

Identification of Physiological Races of *Bipolaris maydis* and Their Sensitivities to Three Fungicides in Fujian Province*

SHI Niu-niu[1,2**], DU Yi-xin[1,2], RUAN Hong-chun[1,2], GAN Lin[1,2], YANG Xiu-juan[1,2], CHENG Fu-ru[1,2***]

(1. *Institute of Plant Protection, Fujian Academy of Agricultural Sciences, Fuzhou* 350013, *China*; 2. *Fujian Key Laboratory for Monitoring and Integrated Management of Crop Peste, Fuzhou* 350013, *China*)

Abstract: Corn southern leaf blight was one of the major diseases of maize. In the present study, physiological races of 214 isolates of *Bipolaris maydis* were identified under the condition of spray inoculating and the sensitivity of their to tebuconazole, pyraclostrobin and meptyldinocap were determined by mycelium growth rate method. The result showed that four physiological races O, C, S and T all existed in Fujian with the Occurrence frequency of 63. 55%, 15. 42%, 11. 21%, and 9. 81%, respectively, and the race O was dominant. The occurrence frequency of strong pathogenic strains, medium pathogenic strains and wake pathogenic strains in race O were 56. 62%, 33. 82% and 9. 56%, respectively. EC_{50} values of 214 isolates of *Bipolaris maydis* to tebuconazole, pyraclostrobin and meptyldinocap ranged from 0. 0249μg/mL to 21. 5823μg/mL, 0. 0321μg/mL to 0. 7249μg/mL and 0. 1463μg/mL to 3. 4127μg/mL, respectively, and the results of sensitivity frequency analysis revealed that sensitivity of some isolates to tebuconazole had decreased, but almost all isolates still kept higher sensitivity to pyraclostrobin and meptyldinocap. Thus, the mean EC_{50} values of 0. 2662μg/mL and 1. 3406μg/mL could be used as sensitivity baseline to isoprothiolane and iprobenfos for fields resistance monitoring, respectively. The sensitivity correlation analysis showed that there were no correlation between the sensitivity of isolates to three fungicides. The present study will provide the basis for the maize variety distribution and the control of corn southern leaf blight.

Key words: *Bipolaris maydis*; Physiological races; Fungicides; Sensitivity

* 基金项目：福建省种业工程项目（FJZZZY－1526）；福建省科技厅项目（2009N2005）；福建省省属公益类科研院所专项（2014R1024－5）

** 第一作者：石妞妞，助理研究员，主要从事植物真菌病害及其防治研究；E-mail：niuniushi@ 126. com

*** 通讯作者：陈福如，研究员，主要从事植物真菌病害及其防治研究；E-mail：chenfuruzb@163. com

氟噻唑吡乙酮在黄瓜植株内的吸收传导活性研究*

迟源冬**，苗建强，董　雪，刘西莉***
（中国农业大学植物病理学系，北京　100193）

摘　要： 氟噻唑吡乙酮（oxathiapiprolin）是美国杜邦公司2015年上市的具有全新化学结构及作用靶标的新型杀菌剂。本研究结合生物测定法和高效液相色谱法（HPLC）探究了氟噻唑吡乙酮的向顶性传导活性，为该药剂的合理使用提供理论依据，同时建立了我国黄瓜霜霉病菌对氟噻唑吡乙酮的敏感基线，为田间抗药性监测提供理论支持。

生物测定法结果显示，在离体黄瓜叶片的叶柄处施用200μg/mL氟噻唑吡乙酮药液后，均匀接种孢子囊悬浮液并保湿培养8d后观察，发现在叶柄周围半径约为3cm的范围内没有病斑出现，与对照相比，供试药剂对病斑面积的抑制率为74.82%；在叶柄施处施用400μg/mL氟噻唑吡乙酮药液后，仅在叶尖观察到少量病斑，与对照相比，药剂对病斑面积的抑制率为93.56%。采用叶柄处施用400μg/mL烯酰吗啉作为阳性对照，发现黄瓜叶片上无病斑出现，而施用400μg/mL百菌清药液作为阴性对照的黄瓜叶片，几乎全叶发病。

经HPLC的分析检测，在黄瓜幼苗根部施药后，氟噻唑吡乙酮可以被根部吸收并传至地上部的茎和叶，叶片上检测到的药剂浓度约为1μg/g，说明该药剂具有一定的向顶性内吸传导活性，但吸收传导的药剂量较少，其中，根部吸收、茎部积累和叶部积累药剂的最大值分别是对照药剂烯酰吗啉的1/5、1/20和1/165。由上述研究结果可以看出，氟噻唑吡乙酮具有一定的向顶性内吸传导活性。

采用叶盘漂浮法测定了77株采自全国11个省市的黄瓜霜霉病菌对氟噻唑吡乙酮的敏感性。试验结果表明，EC_{50}分布在$2.40\times10^{-5}\sim9.63\times10^{-4}$μg/mL，平均$EC_{50}$为$2.23\times10^{-4}$μg/mL，77株黄瓜霜霉病菌对氟噻唑吡乙酮的敏感性呈连续分布，表明从田间采集的菌株未出现敏感性下降的亚群体。因此，可将该平均EC_{50}值作为黄瓜霜霉病菌对氟噻唑吡乙酮的敏感性基线，为田间抗性监测提供参考。

关键词： 黄瓜霜霉病菌；氟噻唑吡乙酮；敏感基线；内吸传导

---

* 基金项目：国家自然科学基金项目（31471791）

** 第一作者：迟源冬，在读硕士研究生；E-mail：chiyuandong@ cau. edu. cn

*** 通讯作者：刘西莉，教授，主要从事杀菌剂药理学及病原菌抗药性研究；E-mail：seedling@ cau. edu. cn

解淀粉芽孢杆菌 HAB－2 抑菌化合物分离鉴定及关键基因调控机制研究*

靳鹏飞**，王皓楠，刘文波，郑服丛，缪卫国***
（海南省热带生物资源可持续利用重点实验室/海南大学环境与植物保护学院，海口　570228）

摘　要：HAB－2 菌株是本实验室从新疆棉花根际土壤中分离得到一株生防细菌，初步试验表明其对多种植物病原菌具有较好的抑菌活性，本研究通过形态学、现代分离技术和分子生物学等技术方法，对 HAB－2 菌株的分类地位、抑菌活性物质的分离与结构分析、脂肽类物质关键基因调控机制，以及其防治植物病原菌和促进植物生长的作用进行了研究。结果表明：通过形态学观察和分子生物学手段鉴定菌株为解淀粉芽孢杆菌，并且发现该菌株对真菌细菌具有较好的抑菌效果，抑菌谱广，为进一步研究活性成分，通过各种方法提取菌株发酵液，发现正丁醇提取物活性好稳定性强，进一步对正丁醇提取物进行分离，得到一种新的脂肽类抗生素，命名为 C_{14} bacillomycin DC，其活性强度是咪酰胺（99.7%）的 131 倍，斑马鱼毒性测试，该化合物毒性仅为微毒。分子方面对该菌株代谢途径进行研究，发现该菌株缺失脂肽类物质合成的关键酶基因 *sfp*，然而菌株内有一条 *lpa* 基因，对其功能进行研究，结果显示 *lpaHAB*－2 基因在脂肽类物质合成途径中起着与 *sfp* 基因同等重要的作用，并且发现合成新化合物的基因具有碱基突变，这可能与该菌株可以产生新型脂肽类抗生素提供一个依据。

关键词：解淀粉芽孢杆菌 HAB－2 菌株；脂肽类物质；C_{14} bacillomycin DC；*lpaHAB*－2；抑菌活性

---

* 基金项目：海南省产学研专项一体化资金（CXY20140038）；国家自然基金（31360029，31160359）；海南省自然科学基金（20153131）

** 第一作者：靳鹏飞，在读博士，研究方向为热带植物病理学；E-mail：283321431@qq.com
*** 通讯作者：缪卫国，博士，教授，研究方向为分子植物病理等；E-mail：weiguomiao1105@126.com

黄芪根腐病多功能颉颃芽孢杆菌的筛选与鉴定

郝　锐[1]，秦雪梅[2]，高　芬[3]
（1. 山西大学生物技术研究所，太原　030006；2. 山西大学中医药现代研究中心，
太原　030006；3. 山西大学应用化学研究所，太原　030006）

摘　要：根际促生菌（PGPR）可以促进矿质营养吸收和利用，产生促进植物生长的次生代谢物以及抑菌物质，从而促进植物生长或抑制病原微生物。芽孢杆菌作为一类重要的PGPR菌，具有很强的环境适应性和友好性，在土传病害的防治中显示出了巨大潜力。黄芪作为山西道地中药材素有“十药八芪”之称，近年来根腐病普遍发生，严重影响了其产量和质量，亟待开发符合GAP种植规范要求的有效防控技术。目前，利用PGPR防治黄芪根腐病仅有个别报道。本课题前期以山西黄芪根腐病优势病原菌 *Fusarium solani* 和 *Fusarium acuminatum* 为靶标，筛选获得了具有抑菌作用的颉颃菌30株，但未对其进行促生功能的测定与评价。基于此，我们以分布最广泛的代表性病原菌 *F. solani* 为靶标，采用同步－平板对峙和发酵液—牛津杯法对上述颉颃菌进行抑菌活性稳定性评价，并对抑菌效果显著且稳定性良好的9株颉颃菌进行室内促生性能测定，结果显示：菌株G10无论是活体平板对峙试验，还是牛津杯法测定发酵液活性，均表现出了明显且稳定的抑菌效果，发酵液抑菌圈透明、清晰，直径达18.38mm（孢子悬液浓度：20～30个/视野，10×15倍数下；400mL孢子悬液/500mL PDA培养基；20mL/皿；发酵液加样：200mL）；同时，菌株G10还具有溶磷能力、固氮能力、淀粉酶活性、产IAA、产HCN等5项促生长特性。参照《伯杰氏细菌鉴定手册（第8版）》和《常见细菌系统鉴定手册》对菌株G10的菌体形态和菌落特征进行观察，并测定其生理生化特性，结果表明：菌株G10与枯草芽孢杆菌（*Bacillus subtilis*）特征相符；16SrDNA序列同源性分析显示，该菌与 *B. subtilis* 同源性最高，达98%，且在系统发育树上聚为一枝，亲缘关系最近，可确定该菌为 *B. subtilis*；为了进一步验证鉴定结果，又采用BBL Crystal™细菌鉴定仪进行鉴定，结果与形态及分子鉴定相符。因此，最终将菌株G10的分类地位归属为枯草芽孢杆菌 *B. subtilis*。综上所述，多功能枯草芽孢杆菌G10不仅对黄芪根腐病优势病原菌具有较强的颉颃作用，且具多项促生能力，是开发黄芪根腐病专用微生物农药的有效物质基础。

关键词：黄芪根腐病；多功能颉颃菌；筛选；鉴定

24%烯肟·戊唑醇油悬剂对水稻纹枯病、稻曲病的防治效果*

徐　赛[1**]，李　娟[1]，陈　宇[2]，韩　翔[2]，石绪根[1***]
（1. 江西农业大学农学院，南昌　330045；2. 中化作物保护品有限公司，上海　200121）

摘　要：本文分别于 2014 年和 2015 年进行了中化作物保护品有限公司提供的 24%烯肟·戊唑醇油悬剂防治水稻纹枯病、稻曲病田间药效试验，结果显示，试验设置的 24%烯肟·戊唑醇油悬剂 3 个剂量中，中、高两个剂量对纹枯病具有较好的防治效果，中间剂量（40mL/667m²）的防效（86.43%，82.42%）与对照药剂爱可（20%烯肟·戊唑醇悬浮剂）53mL/667m² 的防效（83.77%，82.26%）相当，不存在显著性差异（P<0.05，下同）；高剂量（50mL/667m²）的防效（89.70%，87.82%）与对照药剂拿敌稳（75%肟菌·戊唑醇水分散粒剂）10g/667m² 的防效（92.09%，89.81%）相当，不存在显著性差异；而低剂量（30mL/667m²）的防效（84.00%，80.42%）虽然不及中、高剂量防效优异，但也显著高于对照药剂好力克（430g/L 戊唑醇悬浮剂）15mL/667m² 的防效（78.85%，74.42%）。3 个试验剂量对稻曲病均有较好的防治效果，高、中、低 3 个剂量的防效与爱可 53g/667m² 的防效（89.73%，90.44%）相当，差异不显著；高试验剂量的防效（91.48%，92.03%）略低于拿敌稳 10g/667m² 的防效（92.49%，93.70%），但差异不显著，中、低试验剂量的防效则显著低于拿敌稳 10g/667m² 的防效；但与好力克的防效（80.33%，82.68%）相比，低试验剂量的防效（87.59%，88.64%）明显较好，且差异显著。试验期间未观察到药害发生，并且未对水稻生长造成其他不良影响。以上试验结果证明，24%烯肟·戊唑醇油悬剂在合适的施用剂量下对水稻纹枯病、稻曲病均有较好的防治效果，且对水稻安全，值得在生产上推广应用。

关键词：24%烯肟·戊唑醇油悬剂；水稻纹枯病；稻曲病；防治效果

---

* 基金项目：江西省教育厅科技计划项目（GJJ150381）
** 第一作者：徐赛，硕士研究生；E-mail：2540599349@QQ.com
*** 通讯作者：石绪根，讲师；E-mail：sxgjxnd@126.com

Genotypes and Characters of Phenamacril-resistance Mutants in *Fusarium asiaticum*

LI Bin, ZHENG Zhi-tian, LIU Xiu-mei, CAI Yi-qiang, MAO Xue-wei, ZHOU Ming-guo

(*College of Plant Protection, Nanjing Agricultural University, Key Laboratory of Pesticide, Nanjing, Jiangsu Province* 210095, *China*)

Abstract: *Fusarium asiaticum* is a critical pathogen of wheat Fusarium head blight (FHB) in the southern China. Selective phenamacril fungicide has been extensively used for controlling FHB in recent years, which reduced both the FHB severity and mycotoxin effectively. Our previous report indicated that resistance of *F. asiaticum* to phenamacril evolutes with myosin5 mutation. Present paper revealed the variety of phenamacril-resistance genotypes and their frequency associated with resistance degree. Of 82 resistant mutants randomly selected from domestic resistant mutants, 25.6%, 7.3% and 67.1% showed low resistance (LR), moderate resistance (MR) and high resistance (HR), respectively, to phenamacril determined by EC_{50} values. A135T, V151M, P204S, I434M, A577T, R580G/H or I581F responded LR. S418R, I424R and A577G appeared HR. K216R/E, S217P/L, or E420K/G/D showed HR. Interestingly, all of the mutations concentrated in myosin5 motor domain and mutations conferring to high resistance occurred at codon 217 and 420, which we called that 'core region'. Homology modeling revealed that mutations far away from the 'core region' lead to lower resistance degree. Growth rate of hypha and asexual reproduction did not change in phenamacril resistants except some of high resistance appeared abnormal colony.

玉米杂交种对茎腐病的抗性评价*

李　红**，晋齐鸣***
（吉林省农业科学院，公主岭　136100）

摘　要：玉米茎腐病又称玉米茎基腐病，是世界玉米产区普遍发生的一种土传病害，严重影响到玉米的产量。近年来，由于耕作制度、种植结构及气候环境的影响，茎腐病在我国有逐年加重的趋势。一般年份发病率为10%～20%，严重年份发病率高达50%以上，给农业生产带来极大损失。选育和推广抗病品种是防治玉米茎腐病的最为有效的措施。2015年我们对东北春玉米区生产上的主栽品种进行了玉米抗茎腐病鉴定评价，监测品种的抗性变化，指导农业生产。

收集玉米杂交种179份。对照品种为掖478和齐319。试验设在公主岭市吉林省农科院植保所农作物抗病性鉴定圃内。采用田间人工接种鉴定技术方法。播种时，将禾谷镰刀菌在高粱粒上扩繁的培养物30g撒在种子旁边。在玉米乳熟后期进行病株率调查。抗性鉴定评价标准：病株率0～5.0%，高抗（HR）；病株率5.1%～10.0%，抗病（R）；病株率10.1%～30.0%，中抗（MR）；30.1%～40.0%，感病（S）；病株率40.1%～100%，高感（HS）。

试验结果：茎腐病对照材料齐319发病率为0（HR）、掖478发病率为63.7%（HS）。人工接种成功，鉴定结果可靠。179份材料中，对茎腐病表现高抗（HR）75份，占41.9%；抗病（R）42份，占23.5%；中抗（MR）51份，占28.5%；感病（S）7份，占3.9%；高感（HS）4份，占2.2%。可以看出生产上的多数品种为抗病品种，对茎腐病表现抗病（高抗HR、抗病R、中抗MR）的有168份，占总数的93.9%；表现感病（感病S、高感HS）的有11份，占总数的6.1%。感病品种仍然对生产存在威胁，应注意对品种的选择应用，避免造成严重损失。

在抗病育种工作中，应选择优良抗病自交系作亲本，以获得抗病的后代。由于耕作模式变化、气候变化等因素，病原菌可能出现新的生理小种，导致原来抗病的品种丧失抗性。因此，应加强抗源的筛选与利用、生理小种监测与抗病性鉴定，为品种合理布局提供参考。

---

* 基金项目：国家玉米产业技术体系项目

** 第一作者：李红，副研究员，从事玉米病害研究；E-mail：lihongcjaas@163.com

*** 通讯作者：晋齐鸣，研究员，从事玉米病害研究；国家玉米产业技术体系病虫害防控研究室岗位专家

香蕉枯萎病生防菌的鉴定、发酵及生物菌肥创制*

李松伟[1,2**]，黄俊生[1***]，邱德文[2***]

(1. 中国热带农业科学院环境与植物保护研究所 农业部热带农林有害生物入侵检测与控制重点实验室，海口　570228；2. 中国农业科学院植物保护研究所，植物病虫害生物学国家重点实验室，北京　100081)

摘　要：香蕉枯萎病是由土壤真菌尖镰孢古巴专化型（*Fusarium oxysporum* f. sp. *cubense*）引起的维管束系统性病害。在我国广大香蕉种植区均有发病，其发病传染迅速且极难防治，对我国的香蕉产业造成巨大的危害。通过从海南、广州、云南、四川等地采集土壤以及香蕉植株组织样品中，进行颉颃生防菌的分离纯化，通过平板对峙和发酵产物颉颃香蕉枯萎病菌 1、4 号生理小种实验，筛选出具有较强颉颃活性生防菌 726 株，生防菌之间共存实验，与有益菌共存实验，筛选意向菌株 25 株，分别对筛选菌株进行发酵液盆栽防治香蕉枯萎病防治实验，筛选出目的菌株 6 株：2 株芽孢杆菌、1 株木霉菌、2 株放线菌、1 株淡紫拟青霉。分别对生防菌株进行培养，观察其形态特征，测定其生理生化特性，16S rDNA 序列同源性分析和 ITS 序列同源分析，进行菌株鉴定。

为提高生防菌株发酵液中活菌或孢子含量，通过培养基筛选，发酵条件优化，获得菌株在三角摇瓶（2L）和发酵罐（100L、500L）的最佳发酵优化组合；为降低生防菌株发酵成本，分别对有益生防菌群进行三角瓶固体发酵和固体发酵罐（100L、1 000L）发酵优化，分别确定最佳发酵优化组合。进一步测试生防菌在不同介质中的存活习性和变化规律，分别对生防菌株进行抗生素标记，确定生防菌株在有机肥中的存活规律，初步建立生物菌肥的创制技术体系。

生物菌肥盆栽和小区实验，分别进行土壤消毒，生防复合菌肥，进行盆栽和小区实验，在香蕉枯萎病的防治中，具有良好的效果，巴西蕉种植盆栽和小区防治效果为 56. 86% 和 36. 41%；抗（耐）病品种“农科一号”盆栽和小区防治效果为 91. 45% 和 67. 79%。确定香蕉抗（耐）枯萎病品种种植，土壤消毒处理以及生防菌肥、菌剂使用，是防治香蕉枯萎病的有效途径。

关键词：香蕉枯萎病；颉颃生防菌；发酵优化；生物菌肥；防治实验

---

* 基金项目：国家自然科学基金项目（No. 31201467）

** 第一作者：李松伟，在读研究生，主要从事微生物发酵及生物菌肥创制研究；E-mail：lear9999@163. com

*** 通讯作者：黄俊生，主要从事微生物农药研究与利用；E-mail：H888111@126. com

邱德文，主要从事生物农药的研究与利用；E-mail：qiudewen@cass. cn

Screening of Resistance to Fungicides in *Botrytis cinerea* Isolates from Tomato in Hubei province, China

M. S. Hamada[1,2 *], Muhammed Adnan[1], LUO Chao-Xi[1 **]

(1. *College of Plant Science and Technology and the Key Lab of Crop Disease Monitoring & Safety Control in Hubei Province, Huazhong Agricultural University, Wuhan, People's Republic of China*; 2. *Pesticides department, Faculty of Agriculture, Mansoura University, Mansoura* 35516, *Egypt.*)

Abstract: One hundred ninety two *Botrytis cinerea* isolates were obtained during 2012 and 2013 growing seasons from greenhouse tomatoes in eight different locations in Hubei Province and assayed for the sensitivity to several fungicides (azoxystrobin, kresoxim-methyl, pyrimethanil, fenhexamid and iprodione). The mean EC_{50} values for sensitive isolates were 0. 011, 0. 009, 0. 152, 0. 232 and 0. 108μg/mL, respectively. Among the eight locations, six had a high frequency (more than 50%) of resistance to azoxystrobin which ranged from 50. 0% to 91. 67%. It gives very strong warning about the continuous usage of this fungicide in the management strategies in Hubei Province. It was interesting to observe that Azo^R-pyr^R multiple-resistant phenotype was detected in five locations among the eight locations tested (Huanggang, Xiaogan, Jingzhou, Qianjiang and Jingmen) with high frequency in Jingmen (66. 67%) and Qianjiang (30%), respectively. The lowest frequency was detected to fenhexamid that is 19. 79%. The point mutation G143A was the predominant mutation detected in *Cytb* gene in *B. cinerea* isolates resistant to QoI fungicide azoxystrobin. Genetic analysis of fenhexamid-resistant isolates reveled that several point mutations occurred in Erg27 gene and P57A and A378T were the most frequent point mutations detected. Moreover, new substitution in the amino acid position 375 was observed in which Glutamic acid changed to Glycine (E375G) instead of lysine (E375K) as it was reported previously. The information obtained in this study is useful in monitoring and managing fungicide resistance in *B. cinerea* populations in Hubei Province.

Key words: *Botrytis cinerea*; Fungicide resistance; Greenhouse; Tomato

* First author: M. S. Hamada, male, Postdoctoral fellow; E-mail: m_ sobhy@ mans. edu

** Corresponding author: LUO Chao-Xi, Professor, Research interests include interaction between rice and rice false smut fungus, fungicide resistance; E-mail: cxluo@ mail. hazu. edu. cn

Baseline Sensitivity and Cross-resistance of *Cochliobolus heterostrophus* to three DMI Fungicides Propiconazole, Diniconazole and Prochloraz, and Their efficacy in Controlling Southern Corn Leaf Blight in Fujian Province, China*

DAI Yu-li[1**], GAN Lin[1], RUAN Hong-chun[1], DU Yi-xin[1], SHI Niu-niu[1], WEI Zhi-xia[1], LIAO Lei[2], CHEN Fu-ru[1], YANG Xiu-juan[1***]

(1. *Fujian Key Laboratory for Monitoring and Integrated Management of Crop Pests, Institute of Plant Protection, Fujian Academy of Agricultural Sciences, Fuzhou* 350013; 2. *College of Plant Protection, Fujian Agriculture and Forestry University, Fuzhou, Fujian* 350002)

Abstract: Southern corn leaf blight (SCLB), caused by *Cochliobolus heterostrophus*, is one of the most important diseases affecting the cultivation and yield of corn in China. To date, as most hybrid varieties with high yielding and good quality are susceptible or even highly susceptible to SCLB in most corn-growing ecological regions, especially in Fujian Province, chemical control with fungicides would be an important measure for the control of this disease. The demethylation inhibitor (DMI) fungicides, such as propiconazole, diniconazole and prochloraz, are extensively used in China for the control of corn diseases, such as corn head smut and corn smut. In this study, a total of 276 isolates of *C. heterostrophus* from seven locations in Fujian Province of China were tested for sensitivity to the demethylation inhibitor (DMI) fungicides during the stage of mycelial growth. The EC_{50} ranges of values for propiconazole, diniconazole and prochloraz inhibiting mycelial growth of the 276 isolates of *C. heterostrophus* were 0.045 5 ~ 4.864 9, 0.046 0 ~ 4.014 4 and 0.020 7 ~ 2.441 5μg mL, with the mean EC_{50} values of 0.546 7 ± 0.034 5, 0.600 1 ± 0.035 7 and 0.508 8 ± 0.024 2μg mL, respectively. These values suggested that the most tested *C. heterostrophus* isolates were very sensitive to these three DMI fungicides. Results of cross-resistance experiments showed that significant positive correlations between the sensitivity to propiconazole and diniconazole ($r = 0.955\ 7$, $P < 0.000\ 1$), propiconazole and prochloraz ($r = 0.9529$, $P < 0.000\ 1$) and between diniconazole and prochloraz ($r = 0.901\ 8$, $P < 0.000\ 1$) provide a direct evidence for cross-resistance among these three closely related fungicides. However, there was no positive cross-resistance between the three DMI fungicides and the other six tested fungicides of different modes of action, including carbendazol, chlorothalonil, mancozeb, iprodione, fluazinam and pyraclostrobin. Results of greenhouse pot experiments showed that application of propiconazole exhibited greater control efficacy than diniconazole and prochloraz for the control of SCLB. One sprays of 25% propiconazole

* 基金项目：福建省属公益类科研院所专项（2014R1024－5），福建省农业科学院博士启动基金（2015BS－4），福建省农业科学院青年科技英才百人计划项目（YC2016－4）

** 第一作者：代玉立，博士，助理研究员，研究方向：真菌学及植物真菌病害；E-mail：dai841225@126.com

*** 通讯作者：杨秀娟，硕士，研究员，研究方向：植物病害防治；E-mail：yxjzb@126.com

EC at 250μg/mL afforded the best control of SCLB at both two independent corn life stages, the period of seedling and tasseling respectively, with the control efficacy ranging from 80. 31% to 84. 85%. These results suggested that propiconazole will be a good alternative fungicide for control of SCLB. This work provides information on the development of DMI resistance in populations of *C. heterostrophus* in Fujian Province of China and methodologies for future resistance monitoring for this pathogen.

Key words: The demethylation inhibitor fungicides; *Cochliobolus heterostrophus*; Baseline sensitivity; Cross-resistance; Control efficacy.

Biological Control of Sclerotinia Stem Rot of Oilseed Rape Using *Bacillus subtilis* Strain RSS-1*

DAI Yu-li, WU Ya, ZHENG Ting, ZHANG Qian-ru,
PAN Yue-min, ZHANG Hua-jian, GAO Zhi-mou**
(*Department of Plant Pathology, Anhui Agricultural University, Hefei* 230036)

Abstract: Sclerotinia stem rot, caused by *Sclerotinia sclerotiorum* (Lib.) de Bary, is one of the most important diseases of oilseed rape (*Brassica napus* L.) in the world. The *bacterial* strain RSS-1 which was isolated from the rhizospheric soil of oilseed rape, proved to be a useful biocontrol strain for application. Morphology, physiological and biochemical tests and 16S rDNA analysis demonstrated that it was *Bacillus subtilis*. RSS-1 showed strong antifungal activity in dual-culture assay. RSS-1 significantly inhibited mycelial growth of *S. sclerotiorum* by 93.4% ($P<0.01$). Both RSS-1 cell suspension and cell-free filtrate suppressed sclerotial production by 97% ~ 100%. In greenhouse experiments, all three tested concentrations, 1×10^6, 1×10^7 and 1×10^8 CFU/mL of cell suspension or cell-free culture filtrate significantly reduced sclerotinia stem rot severity of oilseed rape ($P<0.05$). The most effective concentration of RSS-1 was 1×10^8 CFU/mL, which reduced the severity by 91.1%. Both pretreatment with RSS-1 and treatment with RSS-1 post inoculation at 24 h significantly reduced ($P<0.05$) the sclerotinia stem rot severity of rapeseed, compared with the control. The *B. subtilis* RSS-1 was more effective in reducing disease severity when applied at 24 h before inoculation with *S. sclerotiorum* than at 24 h post inoculation with *S. sclerotiorum*. In greenhouse conditions, the population density of *B. subtilis* strain RSS-1 on rapeseed leaves significantly decreased ($P<0.05$) by 5.55 to 2.10 log units over 6 days. However, the population density of RSS-1 in rhizospheric soil of rapeseed increased by 4.05 to 4.75 log units over 30 days. *B. subtilis* strain RSS-1 significantly inhibit the polygalacturonases, cellulose activities and oxalic acid accumulation ($P<0.01$) during the *S. sclerotiorum* infection. These experimental results provided evidence that RSS-1 could be an excellent alternative biological resource for biocontrol of rapeseed Sclerotinia stem rot caused by *S. sclerotiorum*.

* This work was supported by the Common wealth Specialized Research Fund of China Agriculture (Grant No. 201103016)

** Corresponding author: GAO Zhi-mou; E-mail: gaozhimou@126.com

Screening of Pepper Germplasm Resources Resistance to Root Rot*

LI Xue-ping[1,2]**, LIU Dan[2], LI Huan-yu[2], QI Yong-hong[1,2],
GUO Wei[2], LI Xiao[2], LI Min-quan[1,2]***
(1. *Gansu Academy of Agricultural Science*, *Lanzhou* 730070, *China*;
2. *Pratacultural College*, *Gansu Agricultural University*, *Lanzhou* 730070, *China*)

Abstract: Resistance of 21 varieties of pepper large-sized planted in Gansu province were measured in order to breed pepper varieties that were resistant to pepper root rot, results indicate that no immune variety was found in the 21 materials. The number of varieties found to be resistant and susceptible to *F. oxysporum* was 14 and 7, respectively. Among the 14 resistant ones, differences were presented in their resistance and only 1 high-resistant variety was found, changmei8. In terms of *F. solan*, 13 varieties were found to be resistant and other 8 varieties were all susceptible or high-susceptible. Although the resistance of the 13 varieties varied, no high-resistant variety was found. There were 11 varieties presented resistant to *F. semitectum*, 90% of which were moderate-resistant, and the rest 10 varieties were susceptible to *F. semitectum*. hangjiao8, longjiao8, jinfu, lacui12 and jimei118 showed resistant to most of pathogenic bacteria, while hangjiao5, jinjiao1, dayingxiong et al showed susceptible or extreme susceptible to them. The resistance of other tested varieties to diverse pathogenic bacteria was not completely the same.

Key words: Pepper; Root rot; Resistant germplasm resources

* 基金项目：国家公益性行业（农业）计划项目（201503112）
** 第一作者：李雪萍，甘肃庆阳人，博士研究生，研究方向为植物病理学；E-mail：lixueping0322@126.com
*** 通讯作者：李敏权，甘肃庆阳人，研究员，博导，研究方向为植物病理学；E-mail：lmq@gsau.edu.cn

湖北省草莓保护地灰霉病菌的抗药性研究*

范　飞[1**]，李　娜[1]，李国庆[1,2]，罗朝喜[1,2***]
(1. 华中农业大学植物科技学院，武汉　430070；2. 湖北省作物病害监测和安全控制重点实验室，武汉　430070)

摘　要：灰霉病作为草莓的重要病害之一，给草莓生产造成了十分严重的经济损失。防治灰霉病生产上以化学防治为主。由于杀菌剂的长期大量使用，抗药性问题日益突出，已成为病害防治中新的挑战。在2012年和2013年的初夏，从湖北省10个地级市的草莓保护地中采集分离了240个灰霉单孢菌株，并检测了其对多菌灵、乙霉威、啶酰菌胺、咯菌腈和嘧菌环胺的敏感性。结果表明，多菌灵、乙霉威和嘧菌环胺的抗性分布广泛，相反啶酰菌胺抗性只局限在武汉和随州。尚未发现咯菌腈的抗性。此次报道是国内首次发现田间灰霉菌株对啶酰菌胺抗性的报道。就抗性表型而言，一共发现了6种，分别为抗多菌灵，抗多菌灵和乙霉威，抗多菌灵和嘧菌环胺，抗多菌灵、乙霉威和嘧菌环胺，抗多菌灵和啶酰菌胺及抗多菌灵、啶酰菌胺和嘧菌环胺。其中，抗多菌灵及抗多菌灵和嘧菌环胺的表型十分普遍。两个啶酰菌胺抗性菌株、3个多菌灵抗性菌株和6个多菌灵、乙霉威和嘧菌环胺的三抗菌株在不含药的培养基上继代培养10代，表明这些抗性是稳定的。此外，还测定了以上菌株的环境适合度参数，包括菌丝生长速率、对NaCl的渗透压敏感性、致病力和离体、活体产孢量。结果表明，多菌灵抗性菌株、多菌灵、乙霉威和嘧菌环胺的三抗菌株和敏感菌株在各参数之间无显著差异。然而，啶酰菌胺抗性菌株在离体草莓叶片上的致病力显著小于敏感菌株，预示啶酰菌胺抗性菌株在寄主体内扩展处于劣势。然而，其他适合度参数与敏感菌株没有显著差异，表明啶酰菌胺抗性菌株仍具有一定的竞争力。为了调查抗性菌株的抗性机理，克隆并测序了13个多菌灵抗性菌株和5个多菌灵和乙霉威抗性菌株的β微管蛋白基因片段。结果表明，13个多菌灵抗性菌株均含有E198V或E198A突变，而5个抗多菌灵和乙霉威的抗性菌株均含有E198K突变。此外，克隆并测序了2个啶酰菌胺抗性菌株的琥珀酸脱氢酶B亚基（SdhB）的基因片段。结果表明，2个啶酰菌胺抗性菌株均含有H172R突变。

关键词：灰霉病；草莓保护地；杀菌剂；抗药性

* 基金项目：本研究受公益性行业（农业）科研专项经费（201303025）
** 第一作者：范飞，博士研究生
*** 通讯作者：罗朝喜，教授，主要从事稻曲病与水稻互作及病原真菌抗药性分子机理研究；E-mail: cxluo@mail.hazu.edu.cn

猕猴桃果实熟腐病生防细菌的筛选及鉴定*

欧阳慧**，王园秀，蒋军喜***
（江西农业大学农学院，南昌 330045）

摘 要：由葡萄座腔菌（*Botryosphaeria dothidea*）引起的猕猴桃果实熟腐病是猕猴桃果实近成熟期至贮藏期的一种重要病害，近年来在江西省奉新县大量发生，损失惨重。为了探索对该病有效的生物防治途径，我们从奉新县猕猴桃种植区采集30份土壤样品，按照土壤芽孢杆菌的分离方法共分离获得267株细菌，通过平板对峙培养，发现34株细菌对猕猴桃果实熟腐病菌具有颉颃作用，其中13株细菌颉颃效果明显。镜检显示，抑菌圈边缘的菌丝表现异常，出现菌丝细弱、稀疏及局部断裂或膨大现象。采用形态学观察、生理生化试验和16S rDNA序列分析方法，对7株抑制效果最好的细菌进行种类鉴定，表明它们均为甲基营养型芽孢杆菌（*Bacillus methylotrophicus*）。未来将对颉颃菌的培养条件和颉颃活性物质做进一步的研究。

关键词：猕猴桃果实熟腐病；颉颃细菌；筛选；鉴定

---

* 国家自然科学基金（31460452）；江西省科技计划项目（20141BBF60019）

** 第一作者：欧阳慧，硕士研究生，研究方向为分子植物病理学；E-mail：jxndoyh2014@126.com

*** 通讯作者：蒋军喜，教授，博士，主要从事植物病害综合治理研究；E-mail：jxau2011@126.com

菌药合剂协同防治烟草黑胫病研究*

牟文君**，奚家勤，薛超群，胡利伟，宋纪真***
（中国烟草总公司郑州烟草研究院，郑州　450001）

摘　要：由疫霉菌烟草变种（*Phytophthora parasitica* var *nicotianae*）引起的烟草黑胫病（Tobacco black shank）是烟草主要土传病害。病原菌在烟株苗期至成株期均可侵染，在茎基形成水渍状黑斑，皮层组织变黑凹陷，髓部黑褐色坏死并干缩呈碟片状，导致全株萎蔫死亡。化学防治是烟草黑胫病防控的重要措施，烯酰吗啉已登记用于烟草黑胫病的防治，在生产中使用较为广泛。本实验室前期研究表明，烟草黑胫病菌对烯酰吗啉的抗性突变体具有一定的生存适合度，一旦抗性菌株发展为优势群体，将会降低烯酰吗啉的防治效果。氟吡菌胺是由拜耳公司研发的苯甲酰胺类杀菌剂，对病原卵菌高效，通过作用于细胞膜的血影蛋白而达到杀菌效果，其独特的作用机制使其与多种杀菌剂均无交互抗药性。生防菌株具有良好的环境相容性，利用生防菌株防治烟草黑胫病具有安全性好、成本低等优点。本研究前期获得了一株解淀粉芽孢杆菌，该烟草内生菌对烟草根结线虫和黑胫病均具有良好的防治效果。生物、化学协同防治病害可减轻对环境的压力，延缓对化学药剂产生抗性，使药效更稳定和持久。因此，本研究首先通过室内毒力测定筛选获得了烯酰吗啉与氟吡菌胺复配的最佳配比，两者按比例复配具有相加效果，增效系数为0.98。在复配杀菌剂的 EC_{50}、EC_{90}、$3\times EC_{90}$ 浓度下，杀菌剂对解淀粉芽孢杆菌的生长抑制率均小于10%，说明复配杀菌剂与生防菌株具有良好的相容性。解淀粉芽孢杆菌与复配杀菌剂协同使用时对烟草黑胫病菌的菌丝生长抑制率显著提高，单剂解淀粉芽孢杆菌浓度为 10^{10} cfu/mL 时的抑制率为57.41%，单独使用复配杀菌剂浓度为0.2μg/mL 时的抑制率为50.85%，但当两者协同使用时对菌丝生长的抑制率为80.34%，根据协同增效评价方法，烯酰吗啉、氟吡菌胺与解淀粉芽孢杆菌混用具有增效作用，为防治烟草黑胫病提供了一种新型备选药剂。

关键词：烟草黑胫病；防治；杀菌剂

---

* 基金项目：郑州烟草研究院院长科技发展基金项目（122015CA0190），河南省烟草公司重点项目（D2103204）

** 第一作者：牟文君，工程师，主要研究方向为烟草病害化学防治；E-mail：muwenjun@126.com
*** 通讯作者：宋纪真，研究员；E-mail：songjz@ztri.com.cn

抗病品种及其健康保护在防控香蕉枯萎病上的应用*

甘　林[1,2**]，杜宜新[1,2]，郑加协[2]，杨秀娟[1,2]，阮宏椿[1,2]，石妞妞[1,2]，陈福如[1,2***]
（1. 福建省农业科学院植物保护研究所，福州　350013；
2. 福建省作物有害生物监测与治理重点实验室，福州　350003）

摘　要：为了有效防治土传性香蕉枯萎病，本研究在田间进行了抗病品种和微生物菌剂（肥）的筛选和应用。结果表明，种植不同香蕉品种对枯萎病的防治效果存在显著差异，其中种植粉杂 1 号和贡蕉的防效最好，分别为 78.62% 和 72.41%。而施用微生物菌肥和生物有机肥对香蕉枯萎病均表现出较好的防治效果，其防效分别为 87.51% 和 77.51%，两者之间达显著性差异。在此基础上提出以种植抗病品种和应用防病微生物菌剂（肥）为主、对环境安全的化学杀菌剂苗期土壤处理为辅的病害综合防控技术，并进行示范。3 个示范片连续 3 年对病害的总体防控效果达 90% 以上，从而有效地控制了香蕉枯萎病的为害，减少了香蕉产量的损失。本研究结果为化学杀菌剂的减量使用及香蕉枯萎病的绿色防控提供了理论依据。

关键词：香蕉枯萎病；抗病品种；微生物菌肥；综合防治效果

---

* 基金项目：国家公益性行业（农业）科研专项（“由尖孢镰刀菌引起的作物土传病害综合防控技术研究”200903049－08）；福建省属公益类科研院所专项（“生防菌 XJ－6 在防治香蕉枯萎病中的定殖动态研究”2016R1023－2）；国家农业科技成果转化资金项目（“香蕉枯萎病疫情监测控制技术示范”2009GB2C400174）

** 第一作者：甘林，硕士，助理研究员；研究方向：植物病害防治；E-mail：millergan@ yeah. net

*** 通讯作者：陈福如；E-mail：chenfuruzb@ 163. com

13 种杀菌剂对玉米大斑病菌和弯孢霉叶斑病菌的毒力测定*

甘　林[1,2**]，代玉立[1,2]，杨秀娟[1,2]，杜宜新[1,2]，阮宏椿[1,2]，石妞妞[1,2]，陈福如[1,2***]
（1. 福建省农业科学院植物保护研究所，福州　350013；
2. 福建省作物有害生物监测与治理重点实验室，福州　350003）

摘　要： 玉米叶斑病是玉米生产上一类主要的病害。为了筛选出高效、低毒的杀菌剂，本文采用生长速率法和孢子萌发法比较了 13 种杀菌剂对玉米大斑病菌和弯孢霉叶斑病菌的毒力作用。结果表明，不同杀菌剂对 2 种病菌菌丝生长和孢子萌发的抑制活性之间存在显著差异。其中，氟啶胺、苯醚甲环唑、烯唑醇、丙环唑、腈苯唑、咪鲜胺锰盐等对病菌菌丝生长具有较强的抑制效果，其 EC_{50} 值均低于 1μg/mL，而代森锰锌的抑制作用较差，EC_{50} 值分别为 5.3745μg/mL 和 6.4650μg/mL；6 种不同作用机理的杀菌剂中，氟啶胺和异菌脲对大斑病菌孢子萌发抑制效果较好，EC_{50} 值分别为 0.3147μg/mL 和 1.5156μg/mL，而丙环唑和吡唑醚菌酯则对弯孢霉叶斑病菌孢子萌发抑制作用较佳，EC_{50} 值分别为 5.0490μg/mL 和 18.8439μg/mL。本研究结果为有效控制玉米叶斑病提供了试验依据。

关键词： 玉米；大斑病菌；弯孢霉叶斑病菌；杀菌剂；毒力测定

---

* 基金项目：福建省属公益类科研院所专项（2014R1024－5）；福建省农业科学院博士启动基金（2015BS－4）
** 第一作者：甘林，硕士，助理研究员，主要从事植物病害防治；E-mail：millergan@ yeah. net
*** 通信作者：陈福如，研究员；E-mail：chenfuruzb@ 163. com

Mechanism of Action of the Benzimidazole Fungicide on *Fusarium graminearum*: Interfering with Polymerization of Monomeric Tubulin but not Polymerized Microtubule

ZHOU Yu-jun[1], XU Jian-qiang[2], ZHU Yuan-ye[1], DUAN Ya-bing[1], ZHOU Ming-guo[1]

(1. *College of Plant Protection, Nanjing Agricultural University, Key Laboratory of Monitoring and Management of Crop Diseases and Pest Insects, Ministry of Agriculture, Nanjing, China*; 2. *College of Forestry, Henan University of Science and Technology, Tianjing Rd No* 70 471003, *Luoyang, People's Republic of China*)

Abstract: Tubulins are the proposed target of clinically relevant anticancer drug, anthelmintic, and fungicide. β_2-tubulin of plant pathogen *Fusarium graminearum* was considered as the target of benzimidazoles compounds by homology modeling in our previous work. In this study, α_1-, α_1-, and β_2-tubulin of *F. graminearum* were produced in *E. coli*. Three benzimidazole compounds (carbendazim, benomyl, and thiabendazole) interacted with the recombinant β_2-tubulin and reduced the maximum fluorescence intensity of 2μM β_2-tubulin 47, 50, and 25%, respectively, at saturation of compound-tubulin complexes. Furthermore, carbendazim significantly inhibit the polymerization of α_1-/β_2-tubulins and α_2-/β_2-tubulins 90.9% ±0.4% and 93.5% ±0.05%, respectively *in vitro*. The similar result appeared with benomyl on the polymerization of α_1-/β_2- tubulins and α_2-/β_2-tubulins were 89.9% ± 0.1% and 92.6% ±1.2% inhibition ratio, respectively. Besides, thiabendazole inhibited 81.6% ±1% polymerization of α_1-/β_2-tubulins, whereasit had less effect on α_2-/β_2-tubulins polymerization with 20.1% ± 1.9% inhibition ratio. However, the three compounds can not destabilize the polymerized microtubule. Toilluminates the issue, mapping the carbendazim binding sites and β/α subunit interface on β/α-tubulin complexes by homology modeling shown the two domains were closed to each other. Understanding the nature of the interaction between benzimidazole compounds and *F. graminearum* tubulin is fundamental for the development of tubulin specific anti-*F. graminearum* compounds.

Key words: *F. graminearum*; Tubulin; Benzimidazole; Polymerization; Homology modeling

葡萄酸腐病发生条件和抗病性相关因素研究*

王彩霞[1**]，李　红[1]，董二容[1]，李兴红[2]，任争光[1]，魏艳敏[1***]
（1. 北京农学院植物科学技术学院农业应用新技术北京市重点实验室，
北京　102206；2. 北京市农林科学院植保环保所，北京　100097）

摘　要：葡萄酸腐病近年来已成为葡萄生产后期的重要病害之一，造成大量果实腐烂，影响葡萄和葡萄酒的产量和质量。本实验采用室内人工接种的方法进行葡萄酸腐病发病条件研究，实验明确了葡萄酸腐病室内接种的适宜方式为菌悬液滴加，发病温度范围为 25～30℃，湿度范围为≥70%，混合接种重要致病菌株 sf-19、sfyg3-2、sfb-18 和 sf-24 发病较为严重，不同接种浓度影响侵染速度，葡萄自身的愈合能力对病原菌侵入或扩展也有较大的影响，果蝇幼虫对葡萄酸腐病的病情发展具有促进作用，果蝇成虫对葡萄酸腐病的扩展未见明显影响，田间实验测定的发病温湿度值与室内测定结果基本一致。另外，从田间葡萄酸腐病发生情况来看，不同葡萄品种对酸腐病的抗病性存在差异。为了明确品种抗病性相关因素，实验测定了人工接种葡萄果实的糖含量和单宁含量，并结合田间调查数据，结果发现含糖量高的品种‘世纪无核’、‘黄意大利’、‘美人指’和‘亚都蜜’，发病较重；而糖含量低的品种‘秋黑’、‘巨玫瑰’和‘魏可’，不发病或发病较轻，糖含量与抗扩展能力有一定的相关性。单宁的测定显示其含量与抗病性呈明显正相关。

关键词：葡萄酸腐病；致病性；发病条件；单宁；含糖量

---

* 项目资助：现代农业产业技术体系建设专项资金资助（CARS－30－bc－2）

** 第一作者：王彩霞，在读硕士，主要研究方向植物病害防治；E-mail：1970232122@ qq. com
*** 通讯作者：魏艳敏，教授，主要研究方向植物病害防治；E-mail：yanminwei@ 139. com.

壳寡糖对灰葡萄孢的抑制作用及对草莓果实灰霉病的控制效果*

王晓莹**，赵思霁，张国珍***
（中国农业大学植物保护学院植物病理学系，北京　100193）

摘　要：草莓灰霉病是草莓采后最常发生且最有破坏性的病害之一。壳寡糖是一种对人畜无毒、价廉、可生物降解、具有很强抑菌作用、不存在毒性残留问题的生物保鲜剂。为了解壳寡糖对草莓果实灰霉病的控制效果，本试验用大连中科格莱科生物科技有限公司生产的壳寡糖（分子量 ≤1 000），首先测定了壳寡糖对草莓灰霉病的病原菌灰葡萄孢菌丝生长和孢子萌发的抑制作用，在此基础上，用壳寡糖溶液处理新鲜草莓果实，研究其对草莓采后灰霉病的抑制作用。试验结果表明：在含有质量分数为 0.004%、0.02%、0.1%、0.5%、1.0% 壳寡糖的 PDA 和 WA 培养基上，灰葡萄孢的菌丝生长和分生孢子萌发受到显著地抑制，且随着质量分数的增加，抑制效果增强。当壳寡糖浓度为 1% 时，对灰葡萄孢菌丝生长的抑制率为 89.5%，对孢子萌发的抑制率为 98.7%，用 1% 壳寡糖溶液处理草莓果实后，接种浓度为 1×10^5 个/mL 的孢子悬浮液，4d 后观察病斑扩展情况。与清水处理相比，果实病斑扩展抑制率为 35.8%，并且壳寡糖对草莓果实的感官品质没有影响。可见，壳寡糖处理对草莓采后灰霉病害具有较明显的控制效果，对草莓果实采后保鲜和灰霉病的防控中具有一定的应用前景。

关键词：草莓果实灰霉病；壳寡糖；孢子萌发；菌落生长

---

* 基金项目：公益性行业（农业）科研专项（201303025）
** 第一作者：王晓莹，在读硕士研究生，植物病理学专业；E-mail：1757851373@qq.com
*** 通讯作者：张国珍，教授；E-mail：zhanggzh@cau.edu.cn

小麦纹枯病菌对噻呋酰胺抗性机制研究*

孙海燕**，李　伟，邓渊钰，陈怀谷***
（江苏省农业科学院植物保护研究所，南京　210014）

摘　要：小麦纹枯病又称小麦尖眼斑病（wheat sharp eyespot），是在世界温带麦区广泛分布的土传真菌病害，主要由禾谷丝核菌［*Rhizoctonia cerealis* van der Hoeven（AG-DI 融合群）］侵染所致。从20世纪70年代中后期开始，随着小麦品种的更替及高产栽培措施（如早播、密植、高肥等）的推广，该病在我国冬麦区普遍发生，现已成为长江流域及黄淮平原麦区的重要病害。目前，生产上推广的小麦品种对纹枯病的抗性普遍较差，药剂防治成了控制该病害的重要手段。

琥珀酸脱氢酶抑制剂类杀菌剂（SDHIs）是被杀菌剂抗性行动委员会新划分出来的一类作用机制和抗性机理相似的化合物，作用机制主要为抑制病原菌琥珀酸脱氢酶活性，从而干扰病菌的呼吸作用。由于该类杀菌剂作用位点单一，其抗性风险备受关注。已有研究结果表明，病原菌对该类药剂的抗性机制主要是琥珀酸脱氢酶 B、C、D 亚基（*sdhB*，*sdhC*，*sdhD*）发生点突变。噻呋酰胺是一种 SDHI，对丝核菌有特效。申请者前期研究结果表明噻呋酰胺对小麦纹枯病菌有很好的离体抑制活性，对病害的田间防治效果也优于常用杀菌剂，说明其在小麦纹枯病防治上具有良好的应用前景。*R. cerealis* 无有性阶段，不产孢子，胞内双核，其对噻呋酰胺抗性机制未有报道。

本研究通过紫外诱变的方法获得20株噻呋酰胺抗性突变体，扩增和比较这20株抗性突变体以及其亲本菌株的琥珀酸脱氢酶 B、C、D 亚基，结果表明：有1株抗性突变体 *RCSdhB* 基因的246位氨基酸发生点突变，有6株抗性突变体 *RCSdhC* 基因的139位氨基酸发生点突变，还有11株抗性突变体 *RCSdhD* 基因的116位氨基酸发生点突变。这些突变体的氨基酸突变都是由组氨酸（His，H）突变为酪氨酸（Tyrosine，Y），其中4株为纯合突变体，14株为杂合突变体。还有2株抗性突变体在 *RCSdhB*、*RCSdhC* 和 *RCSdhD* 基因上均无突变。测定了这20株抗性突变体对噻呋酰胺、啶酰菌胺、联苯吡菌胺、氟唑菌苯胺和氟酰胺这5种 SDHIs 杀菌剂的敏感性，结果表明：*RCSdhB* 上的 H246Y 突变菌株对噻呋酰胺、啶酰菌胺、氟唑菌苯胺都表现中抗，抗性倍数分别为11.0倍，39.0倍和15.3倍，对联苯吡菌胺表现低抗，抗性倍数为4.8；*RCSdhC* 上的 H139Y 和 *RCSdhD* 上的 H116Y 突变菌株对噻呋酰胺表现中抗，抗性倍数范围为11.5 ~ 31.2，对啶酰菌胺、联苯吡菌胺和氟唑菌苯胺都表现为低抗，抗性倍数范围分别为1.3 ~ 6.4，2.9 ~ 9.5和3.0 ~ 7.4。这20株抗性突变体对氟酰胺都表现为敏感。与亲本菌株相比，不同抗性突变体表现出不同的适合度。不同突变体之间的适合度差异与突变基因类型之间无相关性。以上结果表明：小麦纹枯病菌对噻呋酰胺的抗性与 *RCSdhB*、*RCSdhC* 和 *RCSdhD* 基因点突变相关。

关键词：小麦纹枯病菌；噻呋酰胺；琥珀酸脱氢酶；抗性机制

---

* 项目来源：国家自然科学基金（31301702）；国家小麦产业体系（CARS－3－1－17）

** 第一作者：孙海燕，副研究员，主要从事小麦病害防控研究；E-mail：sunhaiyan8205@126.com
*** 通讯作者：陈怀谷，研究员，主要从事小麦病害防控研究；E-mail：huaigu@hotmail.com

油菜无菌苗移栽对根肿病防效的研究

陈　旺，曾令益，刘　凡，任　莉，徐　理，陈坤荣，方小平*

（农业部油料作物生物学与遗传育种重点实验室，中国农业科学院油料作物研究所，武汉　430062）

摘　要：为了寻找一种低成本，高效、环保的根肿病防治措施应用于油菜农业生产。本研究通过田间试验研究油菜无菌苗移栽对根肿病的防效，并对移栽的最适苗龄和移栽时间进行探索。试验根据湖北地区每年油菜正常播种期，提前 20 天开始育苗，采用分批育苗（7 天间隔、7 批次）同时移栽和同时育苗分批移栽（7 天间隔、5 批次）的方式进行。研究结果表明油菜无菌苗移栽对根肿病的防效显著。不同移栽日期的油菜根肿病的发病率和病情指数存在显著的差异，移栽日期越晚，根肿病发病越轻。移栽日期控制在 11 月 4 日时，此时日平均气温不高于 20℃，根肿病的发病率可以控制在 10% 以下。同时研究还发现不同移栽时期，移栽油菜苗龄越大，根肿病发病率和病情指数越高。本研究为生产上预防油菜根肿病提供了一种经济有效的措施。

关键词：油菜；根肿病；无菌苗移栽

* 通讯作者：方小平；E-mail：xpfang2008@163. com

植物内生细菌 YY1 菌株的酶活性分析*

赵　莹，李　盼，赵玉兰，郝志敏**，董金皋**
（河北农业大学真菌毒素与植物分子病理学实验室，保定　071000）

摘　要：玉米大斑病是重要的玉米叶部病害之一，引起该病的病原菌为玉米大斑病菌（*Setosphaeria turcica*），严重发生时会极大地影响玉米的产量和品质。目前防治大斑病主要是利用抗病品种，并辅以适当的药剂防治。但由于抗病品种的单一大面积应用往往导致抗性迅速丧失，化学药剂使用又常常带来严重的负面影响。因此，寻求新的病害防治途径势在必行。生物防治是指利用有益生物及其代谢产物和基因产品等控制有害生物的方法。利用植物内生菌对植物病害进行生物防治具有巨大的发展潜力，YY1 是在健康玉米植株内分离得到的内生细菌，YY1 菌株为枯草芽孢杆菌（*Bacillus subtilis*）。经对其生物学特性、抗菌性能、抗菌物质及抗菌机理进行分析与研究，发现 YY1 菌株在玉米体内有一定的传导特性，同时 YY1 菌株产生的蛋白类物质及脂肽类物质均有较强抑菌活性。生物酶是由活细胞产生的具有催化作用的有机物，大部分为蛋白质，为了进一步研究 YY1 菌株产生的蛋白类物质，本研究首先对 YY1 产生的生物酶类进行分离鉴定，发现 YY1 能够产生淀粉酶及纤维素酶活力，纤维素作为真菌细胞壁的组分，纤维素酶可以降解植物细胞壁，已达到抑菌效果。淀粉作为一种多糖，真菌细胞壁中含量复杂，含有多糖成分，YY1 具有淀粉酶力，能够将大斑病菌细胞壁中淀粉水解，产生的单糖供自身利用。本试验测定 YY1 纤维素酶活力 73.8 U/mL，淀粉酶活力 62.9 U/mL，由于时间有限，未对 YY1 的发酵条件进行优化，下一步将对 YY1 产酶进行发酵条件优化。通过优化 YY1 产酶情况，可以应用于玉米大斑病菌的田间生物防治。

* 基金项目：国家农业部现代化农业产业技术体系项目（CARS-02）
** 通讯作者：郝志敏；E-mail：hzm_ 0322@ 163. com
董金皋；E-mail：dongjingao@ 126. com

一株萎缩芽胞杆菌 *Bacillus atrophaeus* YL3 的鉴定及其脂肽类化合物分析*

刘邮洲[1**]，陈夕军[2]，梁雪杰[1]，钱亚明[3]，乔俊卿[1]，刘永锋[1]

（1. 江苏省农业科学院植物保护研究所，南京　210014；2. 扬州大学园艺与植保学院，扬州　220009；3. 江苏省农业科学院园艺研究所，南京　210014）

摘　要：萎缩芽胞杆菌（*Bacillus atrophaeus*）属芽胞杆菌，在含酪氨酸培养基上产生黑色素。萎缩芽胞杆菌能产生芽胞，耐受不良环境，无致病性，广泛应用于医疗卫生、工业和农业等领域。现已发现萎缩芽胞杆菌对尖孢镰刀菌（*Fusarium oxysporum*）、核盘菌（*Sclerotinia sclerotiorum*）、大丽轮枝菌（*Verticillium dahliae*）、胶孢炭疽菌（*Colletotrichum gloeosporioides*）等植物病原菌具有颉颃作用。

实验室从江苏省徐州市贾汪区耿集镇草莓的根围土壤中，分离出对草莓枯萎病菌（*F. oxysporum*）具有较强颉颃作用的芽胞杆菌 YL3。经过生理生化反应特征、16S rDNA 及 *gyrB* 基因序列比对，鉴定为萎缩芽胞杆菌。采用室内平板对峙生长法测定其抑菌活性，发现菌株 YL3 对黄单胞菌（*Xanthomonas oryzae* pv. *oryzicola*）的抑制作用最强，抑菌圈直径为 55.67mm，可形成明显的抑菌圈，其次对欧文氏菌（*Erwinia amylovora*）也具有较好的抑制作用，抑菌圈直径为 36.50mm。菌株 YL3 对草莓枯萎病菌和甘蓝黑斑病菌菌丝生长有较强的抑制作用，抑制率分别为 49.63% 和 51.30%，对草莓炭疽病菌和油菜菌核病菌菌丝生长的抑制作用相对较弱，抑制率分别为 43.15% 和 45.74%。萎缩芽胞杆菌 YL3 水剂 10 倍稀释液处理后，对草莓枯萎病防效与对照药剂多菌灵 800 倍液（即田间推荐使用浓度）效果相当，100 倍稀释液对“宁玉”品种和“红颊”品种的枯萎病防效分别为 67.62% 和 59.26%，而化学药剂 50% 多菌灵 1 600倍稀释液的防效较差，分别为 29.52% 和 34.89%。采用酸沉淀法提取菌株 YL3 脂肽类化合物并进行高效液相色谱串联质谱分析（LC-MS）。结果表明，菌株 YL3 发酵液中脂肽类粗提物对草莓枯萎病菌菌丝生长有抑制作用，随着剂量加大抑制效果显著。LC-MS 检测到菌株 YL3 脂肽类粗提物中含有表面活性素和泛革素的离子峰出现，由此推断菌株 YL3 室内抑菌作用可能是产生了对细菌有强烈颉颃作用的表面活性素和对病原真菌有强烈颉颃作用的泛革素。

关键词：萎缩芽孢杆菌；鉴定分离；脂肽类化合物

* 基金项目：江苏省农业科技自主创新基金项目 CX（15）1037

** 第一作者：刘邮洲，副研究员，主要从事园艺作物病害生物防治研究；E-mail：shitouren88888@163.com

生防菌 TB2 对甘蔗叶片抗病相关酶活的诱导作用*

梁艳琼[1**]，唐 文[2]，吴伟怀[1,3]，习金根[1]，郑肖兰[1]，李 锐[1]，
郑金龙[1]，贺春萍[1***]，易克贤[1***]
（1. 中国热带农业科学院环境与植物保护研究所/农业部热带农林有害生物入侵检测与控制重点开放实验室/海南省热带农业有害生物检测监控重点实验室，海口 571101；
2. 海南大学环境与植物保护研究学院，海口 570228；3. 农业部橡胶树生物学与遗传资源利用重点实验室/省部共建国家重点实验室培育基地—海南省热带作物栽培生理学重点实验室/农业部儋州热带作物科学观测实验站，儋州 571737）

摘 要：以多酚氧化酶（PPO）、过氧化物酶（POD）、超氧化物歧化酶（SOD）、苯丙氨酸解氨酶（PAL）和过氧化氢酶（CAT）等5种防御酶作为植物抗病性反应指标，研究生防菌 TB2 和甘蔗赤腐病病菌对甘蔗植株抗病相关酶的影响。结果表明，经先接病原菌后喷施生防菌 TB2、先喷施生防菌后接病原菌、病原菌处理、生防菌处理等处理后，与植物防御抗病相关的 PPO、POD、SOD、PAL、CAT 防御酶活性均比对照组高，表明 TB2 和赤腐病菌均能诱导甘蔗叶片中的防御酶活性增强。先喷施生防菌后接病原菌的防御酶活性比同期先接病原菌后喷施生防菌的高，说明 TB2 诱导能力强于赤腐病菌。两者共同处理的甘蔗叶片中 5 种防御酶活性高于单独处理，可见 TB2 和赤腐病菌共同诱导具有协同增效作用。

关键词：甘蔗；生防菌；防御酶系；诱导抗性

---

* 资助项目：中央级公益性科研院所基本科研业务费专项（NO. 2015hzs1J014、NO. 2012hzs1J012、NO. 2014hzs1J012）；中国热带农业科学院橡胶研究所省部重点实验室．科学观测实验站开放课题（RRI－KLOF201506）

** 第一作者：梁艳琼，苗族，助理研究员；研究方向：植物病理；E-mail：yanqiongliang@126. com
*** 通讯作者：贺春萍，硕士，研究员；研究方向，植物病理；E-mail：hechunppp@163. com
易克贤，博士，研究员；研究方向：分子抗性育种；E-mail：yikexian@126. com

16 种杀菌剂对新疆红枣黑斑病的室内毒力测定*

刘础荣[1**]，阮小珀[1]，董　玥[1]，罗来鑫[1]，朱天生[2]，李志军[2]，李健强[1***]
（1. 中国农业大学植物病理学系/种子病害检验与防控北京市重点实验室，北京　100193；
2. 塔里木大学，阿拉尔　843300）

摘　要：新疆是我国红枣的主产区，红枣黑斑病是该地区红枣生产中常见的病害之一；该病害主要为害红枣果实，导致果实黑斑，严重影响红枣的商品品质，也可侵染叶片和花导致叶片黑斑和落花现象。自 2010 年新疆阿克苏地区红枣黑斑病大暴发以来，新疆各地区该病害均有报道，细极链格孢（*Alternaria tenuissima*）是引起新疆地区红枣黑斑病的主要病原菌之一。

目前生产中防控主要以化学防治为主，筛选防控红枣黑斑病的安全高效药剂以及研究施药技术具有重要意义。本研究以抑制呼吸作用、信号转导、细胞膜甾醇合成和氨基酸及蛋白质合成的 4 种不同作用机制的 16 种杀菌剂原药为供试药剂，采用平皿菌丝生长抑制法进行了室内毒力测定。结果表明，吡唑醚菌酯、啶酰菌胺、嘧菌环胺、嘧霉胺、咯菌腈和戊唑醇等 15 种杀菌剂对红枣黑斑病菌表现出较高抑菌活性，其中吡唑醚菌酯和嘧菌环胺对供试的细极链格孢（*A. tenuissima*）室内抑菌效果最好，其 EC_{50} 值分别为 0. 033 7μg/mL 和 0. 100 6μg/mL。

关键词：红枣黑斑病；细极链格孢（*Alternaria tenuissima*）；杀菌剂；室内毒力

---

* 基金项目：国家科技支撑计划专题（2014BAC14B04－3）

** 第一作者：刘础荣，硕士研究生；E-mail：x8281288@163. com

*** 通讯作者：李健强，教授，主要研究方向为种子病理与杀菌剂药理学；E-mail：lijq231@cau. edu. cn

辣根素对植物病原菌的抑菌活性及其作用机制初探*

王彦柠**，黄小威，罗来鑫，李健强***

（中国农业大学植物病理学系/种子病害检验与防控北京市重点实验室，北京　100193）

摘　要：辣根素是十字花科植物体内的一种组成型天然化合物，主要成分为异硫氰酸烯丙酯（Ally Isothiocyanate，AITC），具有刺激性辛辣气味及熏蒸活性辣根素作为一种环境友好的植物源活性抑菌物质近年来在病害防治研究领域备受关注，国际上已经实现商业化，并已成为推荐替代甲基溴的重要代表性产品之一，在美国已经许可使用于有机农产品生产中病害的防控与管理。为了探究辣根素的生物活性，本研究以中国农业大学研究开发的92% AITC原药为材料，以平板熏蒸法结合菌丝生长法、孢子萌发法及细菌菌落生长法等生物测定技术，检测了其对常见的26种植物病原真菌、卵菌及细菌的抑制作用，并在此基础上研究了AITC对导致玉米根腐病的轮枝样镰刀菌（*Fusarium verticillioides*）、导致棉花立枯病的立枯丝核菌（*Rhizoctonia solani*）及导致棉花黄萎病的大丽轮枝菌（*Verticillium dahliae*）不同生长阶段的抑菌活性及对其膜通透性、呼吸强度及能量合成等生理生化指标的影响。主要研究结果如下。

（1）室内毒力测定结果表明，AITC对供试的茄链格孢（*Alternaria solani*）、灰葡萄孢（*Botrytis cinerea*）、禾谷镰刀菌（*Fusarium graminearum*）、立枯丝核菌（*Rhizoctonia solani*）等21种植物病原真菌及瓜果腐霉（*Pythium aphanidermatum*）、辣椒疫霉（*Phytophthora capcisi*）等卵菌的菌丝生长具有显著的抑制作用，其EC_{50}分布在0.94～24.64μg/mL；对其中12种病原真菌的孢子萌发亦具有显著抑制作用，EC_{50}分布在0.26～0.69μg/mL；对番茄溃疡病菌（*Clavibacter michiganensis* subsp. *michiganensis*）、十字花科黑腐病菌（*Xanthomonas campestris* pv. *campestris*）等5种植物病原细菌的生长具有显著抑制作用，EC_{50}分布在0.81～8.61μg/mL。

（2）在离体条件下，AITC对*F. verticillioides*、*R. solani*及*V. dahliae*菌株3T、TZ4、TZ7不同生长阶段的抑菌作用呈现相同趋势：表现为显著抑制5株供试病菌的孢子或菌核的萌发及菌丝的生长，而对孢子或菌核的形成无显著影响；与对菌丝生长的抑制作用相比，AITC对5株供试靶标菌孢子或菌核的萌发表现出更显著的抑制作用。

（3）扫描电子显微镜观察表明，10 μg/mL AITC熏蒸处理*F. verticillioides* 5d、*R. solani* 3d及*V. dahliae* 10d后，各供试靶标菌菌丝表面均出现严重的皱褶、缢缩及凹陷或膨大，新生菌丝可见分支畸形等变化。

（4）AITC影响*F. verticillioides*及*R. solani*的生理生化过程，表现为供试靶标真菌的膜通透性增大，呼吸耗氧量降低，细胞内高能磷酸化合物含量减少。

关键词：辣根素；熏蒸；大丽轮枝菌；轮枝样镰刀菌；立枯丝核菌

* 基金项目："高等学校博士学科点专项科研基金"（编号：20120008130006）

** 第一作者：王彦柠，硕士研究生，主要从事植物源杀菌剂作用机制研究；E-mail：wangyn@ cau. edu. cn

*** 通讯作者：李健强，教授，主要从事种子病理学及杀菌剂药理学研究；E-mail：lijq231@ cau. edu. cn

植物有益微生物广谱抗病性的研究*

王大成**，郭坚华***
（南京农业大学植物保护学院，南京 210095）

摘 要：自然界中，病原菌和食草型昆虫通过侵染植物，使植物产生病害进而影响植物的生长。而植物根围生长着众多的非致病性细菌，部分非致病性细菌能够促进植物生长并保护植物免受病原菌和食草型昆虫的侵染。植物促生细菌（plant growth-promoting rhizobacteria，PGPR）能通过降解土壤污染元素，产生促生素，或抑制植物病害、虫害的发生来促进植物生长。根围促生菌也能通过诱导系统抗病产生广谱抗病性，抵抗植物病害和食草型昆虫的侵染。

实验室前期在中国山东省莱芜市生姜、江苏扬州黄瓜以及广东东莞香蕉的不同生境筛选了大量菌株，通过体外颉颃活性以及与生防相关的几种酶活性筛选，总计得到 968 株有效菌株，进一步利用 ARDRA（核糖体限制性多态性）和 BOX-PCR 技术，通过指纹图谱分析将菌株进行分类，968 株菌株分为 17 属 42 种。16S rRNA 序列分析显示，芽孢杆菌属（*Bacillus* sp.）、肠杆菌属（*Enterobacter* sp.）和泛菌属（*Pantoea* sp.）占据前三的位置，总共比例达到了 85.12%。并且这 3 种不同的属在防治真菌、细菌以及病毒病害上都有大量报道。由此可表明，植物促生细菌的防病效果具有很强光谱性，而这一特性主要通过生防菌株诱导植物产生抗病性，调节植物防卫机制来阻止多种病害的发生。通过在不同地区、不同作物上进行大量的生防菌株的筛选，可对今后生防菌株的筛选及菌株生防效果的评价提供一个可参考的评估系统。

关键词：植物促生细菌；生物防治；诱导抗病性

* 基金项目：江苏省农业科技自主创新项目“高效设施蔬菜标准化、系列化生态友好型生产技术体系研究及示范”，项目编号：CX（15）1044；江苏省农业科技支撑项目，项目名称：“防治茄果类蔬菜土传病害的微生物农药创制及示范推广”，项目编号：BE2015364

** 第一作者：王大成，河南信阳人，在读博士；E-mail：942201408@qq.com

*** 通讯作者：郭坚华，教授，主要从事植物病害生物防治；E-mail：jhguo@njau.edu.cn

Seed Treatment with Plant Beneficial Fungi *Trichoderma longibrachiatum* T6 Enhances Tolerance of Wheat Seedling to Salt Stress

ZHANG Shu-wu[1,2,3,4], XU Bing-liang[1,2,3,4 *], LIU Jia[1,2,3,4], HOU Bao-hong[1,2,3,4]

(1. *College of Grassland Science, Gansu Agricultural University*; 2. *Key Laboratory of Grassland Ecosystems, the Ministry of Education of China*; 3. *Pratacultural Engineering Laboratory of Gansu Province*; 4. *Sino-U. S. Centers for Grazingland Ecosystems Sustainability, Lanzhou* 730070, *China*)

Abstract: Salinity is a major environmental factor and is responsible for the loss of crop production worth billions of dollars every year. A number of strategies have been applied to alleviate the negative effects of salt stress on plant growth, but they were always existed in some disadvantages. Soil-borne *Trichoderma* strains are versatile beneficial fungi which can stimulate growth and enhance whole-plant resistance to plant pathogens, but no information is available regarding to the strain of *T. longibrachiatum* T6 enhances tolerance of wheat seedling to NaCl stress. Here we determined the effect of NaCl stress on *T. longibrachiatum* T6 growth and the role of plant-growth-promoting fungi *T. longibrachiatum* T6 in wheat seedlings growth and developments under salt stress, and investigated the role of *T. longibrachiatum* T6 in the resistance to salt stress at physiological levels. The strain of *T. longibrachiatum* T6 tolerated lower NaCl stress well but differential inhibitory effects were observed with the higher salt treatment. Wheat seedlings were inoculated with the strain of *T. longibrachiatum* T6 and were compared with non-inoculated control. Shoot height, root length, and shoot and root weights were measured on 15 days old of wheat seedlings grown under 150 mM or in control. A number of colonies were re-isolated from the roots of wheat seedling under salt stress. The relative water content in the leaves and roots, chlorophyll content and root activity were significantly increased, and the accumulation of proline content in leaves was markedly accelerated with the plant growth parameters, but the content of leaf malondialdehyde (MDA) under saline condition was significantly decreased. The antioxidant enzymes-superoxide dismutase (SOD), peroxidase (POD), and catalase (CAT) in wheat seedlings were increased by 29, 39 and 19%, respectively, with the application of the strain of *T. longibrachiatum* T6 under salt stress.

* Corresponding author: Bingliang Xu; E-mail address: xubl@gsau.edu.cn

第八部分　其　他

Isolation and Characterization of Antagonistic Endophyte in *Areca catechu* L. *

SONG Wei-wei **, ZHOU Hui, YU Feng-yu, NIU Xiao-qing, TANG Qing-hua, QIN Wei-quan ***

(*Coconut Research Institute*, *Chinese Academy of Tropical Agricultural Science*, *Wenchang* 571339)

Abstract: *Areca catechu* L. is the main economic crop in Hainan, which is one of the most important southern herbal medicine resources. For founding new type of agro-chemicals controlling plant disease, a total of 91 endophytes were isolated from the healthy leaves and flowers of areca plants with traditional cultural methods. Of which, 85 strains were isolated from leaves, and 6 strain was isolated from flowers. Then we screened the high efficient endophytes against areca anthracnose by the plate confrontation tests. The results showed that there were 4 strains with obivious antifungal bioactivity against areca anthracnose. Furthermore, the 4 strains of endophyte were identified according to the characteristics of morphology, physiology and biochemistry tests and 16S rDNA sequence analysis. About 1 500bp fragments of 16S rDNA were amplified by PCR method using the extracted genomic DNA as the template. According to sequences analysis and comparison with the data of GenBank by BLAST, three were classed to *Bacillus* genera and one was classed to *Streptomyces* genera, the identities were from 97% to 99%.

Key words: Endophyte; *Areca catechu* L. ; Isolation; Antagonism; Identification

* Fund Project: the Key Project of Hainan Province (ZDYF2016058); Natural Science Foundation of Hainan Province (314144)

** First author: SONG Wei-wei, Doctor; E-mail: songweiwei426@ sohu. com

*** Corresponding author: QIN Wei-quan, Professor; E-mail: QWQ268@ 163. com

药用植物内生菌的分离及颉颃菌株的筛选*

于　森[1]，刘淑艳[1,2**]

（1. 吉林农业大学农学院，长春　130118；2. 吉林农业大学
食药用菌教育部工程研究中心，长春　130118）

摘　要：药用植物具有丰富的内生菌资源，其内生菌由于与药用植物的协同进化，可能具有与宿主植物相似的活性成分，本文对采自吉林农业大学药园的藿香（*Agastache rugosa* ）、千叶蓍（*Achillea milleflium*）、黄芩（*Scutellaria baicalensis*）、萱草（*Hemerocallis fulva*）、轮叶党参（*Codonopsis lanceolatae*）、牛蒡（*Arctium lappa*）、费菜（*Sedum aizoon*）利用研磨法和组织块法分别进行了内生细菌和内生真菌的分离，并采用活体筛选法和平板对峙法，筛选了对黄瓜白粉病病原菌和甜瓜枯萎病病原菌具有抑菌活性的内生菌株。结果表明：在 7 种药用植物叶片中共分离获得 122 株内生细菌和 334 株内生真菌，活体筛选结果有 33 株细菌对黄瓜白粉病病原菌有抑菌效果；平板对峙法筛选结果有 5 株真菌对甜瓜枯萎病病原菌有抑菌效果。

关键词：药用植物；内生菌；分离；筛选

* 基金项目：中青年科技创新领军人才及团队项目

** 通讯作者：刘淑艳

不同施肥水平对三七生长和根腐病的影响

魏　薇，黄惠川，尹兆波，杨　敏，朱书生
（云南农业大学植物保护学院，云南生物资源保护与利用国家重点实验室，昆明　650201）

摘　要：三七［*Panax notoginseng*（Burk.）F. H. Chen］属五加科人参属多年生草本植物，因其具有散瘀止血，消肿定痛、止血、降脂和保护肝脏等功效，是我国特有的名贵中药材。矿质营养是植物生长发育的基础，合理施肥则是作物取得优质高产的关键因子之一。目前，对三七合理施肥的研究较少，因此，在三七种植过程中普遍存在着盲目和过量施肥的现象。我们研究了不同施肥水平对三七种苗生长和根腐病的影响，旨在探索不同施肥水平对三七产量、质量和抗病性的影响，为三七规范化生产（GAP）和提高药材产量和质量提供科学的理论支持。

本文研究了氮磷钾肥料不同的施用水平对三七种苗存苗率、产量和根腐病的影响。结果表明氮肥浓度在40kg/亩时三七种苗的存苗率显著低于10kg/亩和20kg/亩，三七种苗地上、地下及整株干重均显著低于20kg/亩；磷肥浓度在48kg/亩时三七种苗的存苗率显著低于12kg/亩和24kg/亩,三七种苗地上、地下及整株干重均显著低于24kg/亩；钾肥浓度在72kg/亩时三七种苗的存苗率显著低于18kg/亩和36kg/亩，三七种苗地上、地下及整株干重均显著低于24kg/亩。过量施肥会使三七种苗的存苗率和产量显著降低。根腐病病害调查的结果表明氮肥浓度在40kg/亩、磷肥浓度在48kg/亩、钾肥浓度在72kg/亩时，三七种苗根腐病的发病率和病情指数分别显著高于20kg/亩、24kg/亩和36kg/亩的施肥浓度。进一步对三七根际微生物进行了分离鉴定和统计，结果表明随着氮磷钾施肥水平的增加，引起三七根腐病的镰刀菌的分离频率显著增加，这表明过量施肥可能会影响三七根际土壤微生物区系失衡，使导致三七根腐病的病原微生物增加，从而加重了根腐病的发生。

综上所述，施用不同浓度的氮磷钾肥料不但对三七的生长、产量具有较大的影响，不同处理间三七生长期内根腐病的发病率和病情指数存在较大的差异。不合理的施肥会使三七土壤养分失衡，三七的生长受阻，自身的抗逆能力下降，从而导致根腐病的发生，且发病率和病情指数上升。因此，合理施肥在三七的生产种植过程中至关重要，本研究对指导种植三七合理施肥具有非常重要的意义。

关键词：旆肥水平；三七；存苗率；根腐病

Diagnostics and Detection of Different Groups Phytoplasmas in China Using an Oligonucleotide Microarray on the Platform of ArrayTube

WANG Sheng-jie*, LIN Cai-li, YAN Dong-hui**, YU Shao-shuai, LI Yong, WANG Lai-fa, PIAO Chun-gen, GUO Min-wei, HUAI Wen-xia, TIAN Guo-zhong**

(*Key Laboratory of Forest Protection of State Forestry Administration, Research Institute of Forest Ecology, Environment and Protection, Chinese Academy of Forestry, Beijing* 100091, *China*)

Abstract: The 16S rDNA-specific oligonucleotide probes designed to detect phytoplasma have been reported. In order to find optimal specific probe and to develop the detection technique using oligonucleotide microarray on the platform of ArrayTube to detect and identify phytoplasmas associated plant disease in China. PCR amplification and microarray hybridization were used to detected 15 symptomatic plants probably infected with phytoplasma and asymptomatic plants as healthy controls collected from different regions in China. Phytoplasma 16S rDNA were detected in 13 of 15 symptomatic plants but not in all their healthy controls. Thirteen phytoplasmas were classified into 16Sr I, 16Sr II, 16Sr V and 16Sr XIX groups. Among 17 tested probes, the universal probe designated Pp – 502 could be used to detect all phytoplasmas associated with plant disease. The specific probe designated Pp I – 465 for 16Sr I group could be used to detected four phytoplasma strains of 16Sr I group associated with paulownia witches' -broom, chinaberry witches' -broom, mulberry dwarf and lettuce yellows. The probe Pp II – 629 for 16Sr II could be used to detect three phytoplasma strains 16Sr II group associated with peanut witches' – broom, sweet potato witches'-broom and cleome witches'-broom. Three phytoplasma strains of 16Sr V associated with jujube witches' -broom, cherry lethal yellows and Bischofia polycarpa witches broom and chestnut yellows crinkle phytoplasma of 16Sr XIX could also be detected by specific probes, but they have obvious cross hybridization with other group probes. Compared with PCR amplification, the sensitivity of microarray to detect phytoplasma in plant was increased 1000-fold. Phytoplasmas of 16SrI and 16SrV group respectively were detected in periwinkle with symptoms of phyllody and witches' -broom collected from Fujian province and Robinia hispida with symptom of witches' -broom collected from Henan province. While no phytoplasma was detected in peony with symptom of yellowing collected from Beijing province and willow with symptom of witches' broom collected from Inner Mongolia Autonomous Region. The oligonucleotide microarray on the platform of ArrayTube could be used as a method to investigate phytoplasmas in China, and could provide reliable foundation for phytoplasma identification and classification.

Key words: Phytoplasma; Detection and identification; 16S rDNA gene; ArrayTube

* 第一作者：王圣洁，博士研究生，主要方向：分子植物病理；E-mail：wsjguoyang@ 126. com

** 通讯作者：严东辉，研究员，硕士生导师，从事树木微生物及寄主—病原分子互作研究；E-mail：yandh@ caf. ac. cn

田国忠，研究员，博士生导师，从事分子植物病理研究；E-mail：tian3691@ 163. com

Selection and Validation of Reference Genes for Gene Expression Analysis in *Vigna angularis* Using Quantitative Real-Time RT-PCR*

SHEN Yong-qiang**, KE Xi-wang, YIN Li-hua, HAN Dong, ZUO Yu-hu***

(*Heilongjiang Bayi Agricultural University*, *Daqing* 163319, *China*)

Abstract: Adzuki bean (*Vigna angularis*) is one of the most important legume crops in Asia due to its nutritious protein and starch contents, as well as its useful potential for cropping structure adjustment in China. In spite of its economic importance, gene expression analysis system for gene function verification of adzuki bean is still absent. Therefore, reference genes for gene expression analysis based on the quantitative real time PCR (qRT-PCR) were screened in current study. A total of nine general housekeeping genes, including *ACT*, *Fbox*, *ZMPP*, *GAPDH*, *EF*, *PP2A*, *UBC*, *UBN* and *PTB* were evaluated for their expression stability by qRT-PCR in four adzuki bean cultivars, three different tissues, four abiotic stress and a biotic stress. The best group of candidates as reference genes were as follows: *PP2A* and *PTB* for different cultivars; *EF*, *PP2A* and *UBN* for different tissues; *ACT*, *UBC* and *UBN* for biotic stress; *ACT* and *ZMPP* for waterlogging stress; *Fbox*, *UBC* and *UBN* for salinity-alkalinity stress; and *Fbox*, *ACT* and *PTB* for drought stress. Our results will provide a more accurate and reliable normalization of qRT-PCR data in adzuki bean.

Key words: Adzuki bean; Quantitative real time PCR; Reference genes

* Funding: National Science and Technology Support Plan (2014BAD07B05 - 06/H08); National Natural Science Foundation of China (31501629); China Postdoctoral Science Foundation (2014M561378)

** First author: SHEN Yong-qiang; E-mail: ShenYQ_ ZY@ 163. com

*** Corresponding author: ZUO Yu-hu, Professor, Phytopathology; E-mail: zuoyuhu@ byau. edu. cn

Colonization of Rice Plant by *Bacillus subtilis* Strains

SHA Yue-xia, WANG Qi*, LI Yan

(*Department of Plant Pathology, China Agricultural University;*
Key Laboratory of Plant Pathology, Ministry of Agriculture,
China Agricultural University, Beijing 100193, *China*)

Abstract: Rice blast, caused by *Magnaporthe oryzae*, is one of the most devastating diseases of cultivated rice (*Oryza sativa*) and a constant threat to global food security. Strains *B. subtilis* SYX04 and *B. subtilis* SYX20 are two endophytic bacterium obtained from rice leaves, the evaluating the antifungal and plant-growth-promoting properties have been deployed to control *M. oryzae* infection in rice. The present study sought to clarify its specific colonization patterns in rice plant using GFP reporter. The marked GFP-labled *B. subtilis* strains were inoculated onto the rice seedlings under axenic conditions. *B. subtilis* SYX04 and SYX20 used 5 days to finish colonization in rice root completely and the cells were able to multiply and spread, mainly to the intercellular spaces of different tissues. Observation by confocal laser scanning microscope (CLSM) showed that the bacteria were located into exodermis cell, cortex and concentrate mainly in the vascular bundle in rice roots, and that the bacteria migrated slowly from roots to stems and leaves. In addition, microscopic observation also revealed clearly that epidermis, ground tissue and vascular bundle were the three colonization sites for *B. subtilis* SYX04 and *B. subtilis* SYX20 strains in rice stems. Three days after spraying suspended inoculation, the GFP-labeled bacterial cells located into rice leaves through epidermal hair, epidermis cell, stoma, parenchymal cell, sclerenchyma, bulliform cell and vascular bundle. No bacterial cells were found into mesophyll cell of rice leaves. The results showed that *B. subtilis* SYX20 and SYX04 can colonize in rice plant rapidly and efficiently. As far as we know, this is the first detailed report of the colonization pattern for *Bacillus subtilis* in rice leave.

* Corresponding author: WANG Qi; E-mail: wangqi@ cau. edu. cn

植物的力敏感探测器表皮毛：力学刺激表皮毛诱发的防御反应*

高文强**，曹志艳，董金皋，周丽宏***
（河北农业大学生命科学学院，保定　071001）

摘　要：植物叶面上的表皮毛形态结构各异，如针状、钩状、扁平状、伞状和锚状等。结构各异的表皮毛常常影响着其功能。肾豆（Kidney bean）和 *M. pumila* 叶面上钩状和锚状的表皮毛可以钩住昆虫腿，从而困住在其表面爬行的昆虫，如飞蛾（moths）、蜜蜂（andrenid bee）、叶蝉（leafhopper）甲虫（bruchid beetle）等。南瓜属叶面上多细胞形态的表皮毛影响着叶柄的向性运动；拟南芥叶面上的表皮毛底部形态与奶嘴结构相似，外界力学碰触时易使其在底部发生屈曲形变和应力集中。拟南芥表皮毛常作为研究细胞周期调控、细胞极性、细胞增大等方面的模型。然而，关于表皮毛形态结构防御功能的研究鲜见报道。前期研究表明当力学触碰表皮毛分支顶部时，会引发底部支持细胞内的 Ca^{2+} 发生震荡变化和细胞间质 pH 值升高。植物受到生物和非生物胁迫时，Ca^{2+} 浓度变化、pH 值升高是植物启动防御反应的早期信号事件和基础。这些早期信号事件是否可以引发下游的防御反应？本研究中探索了力学刺激拟南芥表皮毛，所诱发的下游防御反应，利用软毛毛笔轻抚拟南芥叶片，轻抚后，拟南芥叶片内花青素含量升高（图 1），诱发表皮毛和与其相连的细胞内产生过氧化氢，如图 2 所示。

探索表皮毛形态结构的功能可以更深层次地理解和发掘植物与环境的内在联系并提供新的研究思路。

关键词：拟南芥；表皮毛；力学刺激；防卫化合物

* 基金项目：河北农业大学科研启动项目（ZD201610）
** 第一作者：高文强，本科生，主要研究方向为表为毛形态结构功能；E-mail：gwenqiang@ outlook. com
*** 通讯作者：周丽宏，讲师，主要从事植物物理研究；E-mail：lihongzhou12@ 126. com

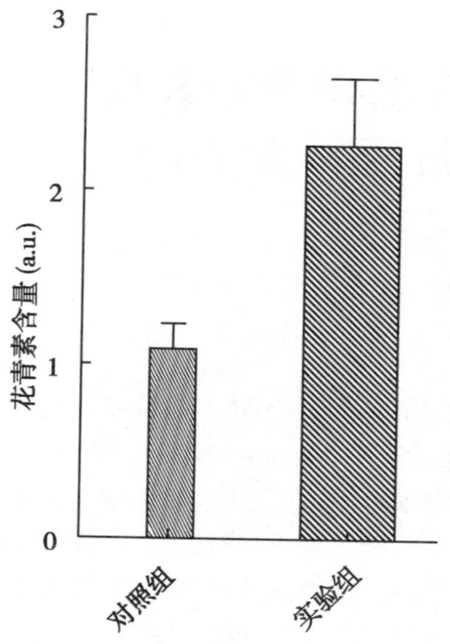

图 1　轻抚表皮毛引起叶片内花青素含量升高

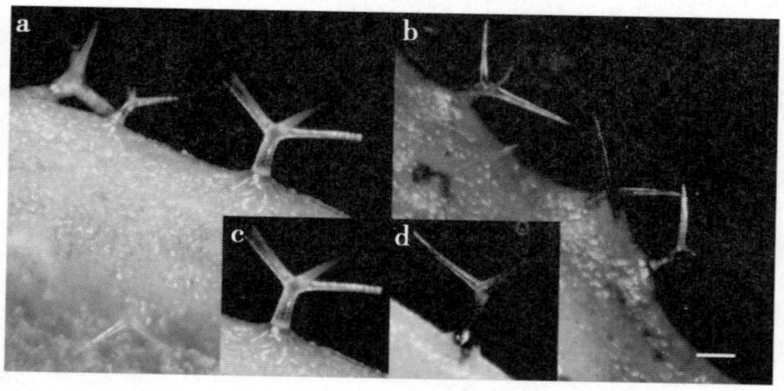

图 2　轻抚表皮毛诱发表皮毛细胞内和与其相连细胞内产生过氧化氢

利用 diaminobenzidine（DAB）染色检测细胞内产生的过氧化氢：a. 未被轻抚的表皮毛；b. 轻抚后的表皮毛；c. 和 d. 为局部放大的表皮毛

一株对柑橘木虱具强致病力的虫生真菌的分离鉴定*

宋晓兵**，彭埃天***，程保平，凌金锋，陈　霞
（广东省农业科学院植物保护研究所/广东省植物保护新技术重点实验室，广州　510640）

摘　要：柑橘木虱（*Diaphorina citri* Kuwayama）隶属半翅目（Hemiptera）木虱科（Psyllidae），主要为害九里香和柑橘等芸香科植物，是柑橘黄龙病的重要传播虫媒，柑橘木虱的频繁发生是柑橘黄龙病流行的重要因素。当前生物防治以其绿色、环保、持效、无污染、不易产生抗药性等优点被认为是最具应用前景的防治手段。从柑橘木虱罹病虫体分离到一株的虫生真菌，命名为 GZMS－28，菌落在 PDA 培养基上为乳白色絮状，菌落蓬松、呈放射状生长，圆形或不规则形状。菌株产孢细胞轮生或单生，分生孢子梗向顶部轴式产孢，形成弯曲的具有凸起的产孢轴。分生孢子梗呈穗状，分生孢子透明，球形或卵形，壁薄，大小为 2.5μm × 3.5μm。提取菌株 GZMS－28 的总 DNA，以通用引物 ITS1 和 ITS4 扩增菌株的 rDNA ITS 序列，其 rDNA ITS 序列与 GenBank 中白僵菌属多个球孢白僵菌菌株的对应序列相似性达到 99%。综合菌株 GZMS－28 的培养特性、形态特征及 ITS 序列比对结果，将菌株 GZMS－28 初步鉴定为 *B. bassiana*。菌株GZMS－28 对柑橘木虱具有较强的致病力，以 1.0×10^{8} 个孢子/mL 悬浮液处理柑橘木虱成虫后 3 d，校正死亡率为 62.0%；处理后 7 d，校正死亡率达 95.7%，表明菌株 GZMS－28 具有开发成生物制剂的潜能。

关键词：柑橘木虱；球孢白僵菌；致病性；生物防治

* 基金项目：广东省农业领域引导项目（2013B020309004）；广东省科技计划项目（广东省农作物病虫害绿色防控技术研究开发中心建设）；广东省公益研究与能力建设项目（2014B020203003）

** 第一作者：宋晓兵，助理研究员，研究方向：果树病害防控；E-mail：xbsong@126.com
*** 通讯作者：彭埃天，研究员；E-mail：pengait@163.com

烟草悬浮细胞先天免疫反应检测体系的建立及分析*

陈萌萌**，房雅丽，范　军***

（中国农业大学植物病理学系，北京　100193）

摘　要：分离和鉴定病原菌中诱导植物免疫反应的激发子对研究植物先天免疫反应机理具有重要意义。为了建立适于大规模筛选研究的植物细胞先天免疫反应检测体系，本文利用BY－2烟草悬浮细胞比较了以氯酚红、Luminol及5－氨基水杨酸作为反应底物的3种免疫反应检测体系。结果表明，氯酚红检测方法需要在较低的细胞浓度下进行，该反应迅速而直观，但不适于蛋白类激发子的大规模筛选；Luminol化学发光检测方法的干扰因子较多，不适于细胞粗提物的分析；相比之下，5－氨基水杨酸检测方法在BY－2细胞浓度较高时反应最明显，该体系具有反应迅速、稳定、显色反应可随时间延长而增强以及检测成本低等适于高通量筛选的诸多优势。该方法为病原菌激发子的功能基因组研究奠定了基础。

关键词：烟草悬浮细胞；BY－2；植物先天免疫；PAMPs；激发子

---

* 基金项目：教育部博士点基金（20130008110005）

** 第一作者：陈萌萌，博士研究生，主要研究方向植物病理学

*** 通讯作者：范军，教授，主要研究方向植物病理学；E-mail：jfan@ cau. edu. cn

薄层层析—生物自显影法快速筛选活性植物资源*

张伟豪**，翁道玥，宋慧云，段志豪，王　军，单体江***
（华南农业大学，林学与风景园林学院/广东省森林植物种质创新
与利用重点实验室，广州　510642）

摘　要：我国华南地区植物资源丰富，其中大戟科、夹竹桃科、樟科和芸香科等植物具有很好的生物活性。为快速筛选出具有较好生物活性的植物资源，本文通过薄层层析—MTT—生物自显影法测定了华南地区常见的36种植物对7种供试细菌的抑制活性，同时采用薄层层析—DPPH—生物自显影法进测定了其抗氧化活性。抗菌活性的结果表明36种供试植物中，有31种植物的乙酸乙酯层粗提物表现出抗菌活性，其中，有8种植物的抗菌活性较强，对7种供试细菌均表现出抑菌活性，抑菌圈直径在5～10mm，分别为红背桂、红背山麻杆、蝴蝶果、桂木、红果仔、锡叶藤、山油柑和豺皮樟。抗氧化活性的结果表明，有35种植物次生代谢产物表现出抗氧化活性，其中，有15种植物的抗氧化活性明显高于其他植物。红背山麻杆，红果仔、锡叶藤、山油柑和豺皮樟等5种植物抗菌与抗氧化活性都较强，有潜在的应用和开发前景，可作为首选的候选植物资源。此外，黄花夹竹桃、银柴、粪箕笃、海金沙和小蜡等未表现出抗菌活性，无抗氧化活性的植物仅有羊角拗，这6种植物是否有其他活性，还需要进一步的研究。薄层层析—生物自显影法不仅可以快速筛选具有较好活性的植物资源而且对植物活性化合物的数量和极性有初步的了解，为后续活性植物资源的开发和利用奠定基础。

关键词：植物资源；薄层层析—生物自显影法；抗菌活性；抗氧化活性

---

*　基金项目：广东省林业科技创新项目（2015KJCH043）
**　第一作者：张伟豪，硕士，研究方向：植物和微生物的次生代谢；E-mail：vhaozhang@foxmail.com
***　通讯作者：单体江，博士，讲师，研究方向：植物和微生物的次生代谢；E-mail：tjshan@scau.edu.cn

根肿菌侵染拟南芥根部的代谢变化研究*

何璋超[1,2]**，陈　桃[2]，毕　凯[1,2]，高知枭[1,2]，赵　莹[1,2]，姜道宏[1,2]***
（1. 湖北省作物病害监测和安全控制重点实验室（华中农业大学），武汉　430070；
2. 农业微生物学国家重点实验室（华中农业大学），武汉　430070）

摘　要：芸薹根肿菌（*Plasmodiophora brassicae*）属于原生动物界、根肿菌门，是一种活体专性寄生的土传病原菌，可以为害 100 多种十字花科植物，引起寄主根部肿大。本研究利用基于 UPLC-MS 的代谢组学分析技术探索拟南芥受根肿菌侵染时根部代谢物的动态变化，通过 PCA、PLS、O-PLS 和 SVM 等多变量模式识别方法，结合相关数据库信息，以期筛选到重要的生物标记物和代谢通路。

本研究通过油菜根部肿块进行组织培养获得大量含有根肿菌的愈伤组织，采用 1/2 MS 培养基培养拟南芥 Col-0，接种根肿菌后，在根毛吸附的侵染早期、根毛侵染期、皮层侵染期等不同时期采取拟南芥根部样品，提取代谢物后进行质谱分析。目前已得到接种 50d 后的侵染后期拟南芥根部代谢组数据，运用 R version 3. 2. 3xcms 程序对质谱原始数据进行预处理，共得到 39 428个峰，先用 T-statistics 单变量分析方法，设置变化倍数 fold >4，p value <0.05，筛选出 2 736个峰。运用 R CAMERA 对 xcms 处理后的数据进行进一步注释，识别加合物峰、同位素峰、碎片峰等，剔除掉同一代谢物的其他形式的峰后，用 SIMCA-P + 11. 0 软件进行 PLS-DA 多变量统计分析，结果显示接种根肿菌后的处理组同对照组具有很好的分离趋势，说明接种根肿菌后拟南芥代谢发生了显著变化。综合运用 CAMERA 处理结果、Metaboanalyst 代谢组学分析平台、Mummichog 软件进行代谢物的鉴定和代谢通路富集分析，分析结果显示色氨酸代谢、赖氨酸代谢、不饱和脂肪酸 β 氧化、多聚不饱和脂肪酸生物合成等通路得到显著富集。

关键词：芸薹根肿菌；拟南芥；代谢组

---

* 基金资助：现代农业产业技术体系建设专项资金（CARS-13）
** 第一作者：何璋超，湖北孝感人，在读硕士研究生，研究方向为分子植物病理学；E-mail：he_ zhangchao@ 163. com
*** 通讯作者：姜道宏，教授，从事植物病理学相关研究；E-mail：daohongjiang@ mail. hzau. edu. cn

海南槟榔黄化病疫情监测网络信息平台的研究与应用*

罗大全[1**]，车海彦[1]，曹学仁[1]，胡　杰[2]
（1. 中国热带农业科学院环境与植物保护研究所，农业部热带作物有害生物综合治理重点实验室，海南省热带农业有害生物监测与控制重点实验室，海口　571101；
2. 中国热带农业科学院科技信息研究所，海口　571101）

摘　要：槟榔（*Areca catechu* L.）为棕榈科多年生常绿乔木，是海南重要的热带经济作物，也是国家重要的南药资源之一，位于四大南药之首，槟榔种植业已成为海南省热带作物产业中仅次于天然橡胶的第二大支柱产业。据海南省农业厅统计，截至2013年全省槟榔种植面积已达136.33万亩，收获面积90.24万亩，鲜果产量近90万t（干果约22万t），占全国产量的95%，是海南省东部、中部和南部山区等少数民族地区200多万农民增加家庭收入、脱贫致富的重要途径。

黄化病是一种严重危害槟榔生产种植的毁灭性传染病害，目前文献资料报道全球仅在印度、斯里兰卡和中国海南的槟榔种植区发生为害。据省林业厅森防站2012年初步调查统计，槟榔黄化病在琼海、万宁、陵水、三亚、保亭、定安、乐东、五指山、琼中、屯昌等10个市县都有发生，发病面积达5.1万亩。目前发病地区及面积仍在不断扩大，严重威胁着海南省槟榔产业的健康发展和下游相关产业的生存发展。由于该病害早期症状极易与生理性黄叶相混淆，加上目前针对植原体病害尚无有效的防治药剂，及时清除田间发病植株是防控槟榔黄化病最经济和有效的措施之一，开展病害监测对槟榔黄化病的科学防控有十分重要的作用。

本研究利用关系型数据库管理系Microsoft SQL Server 2005，将症状识别和槟榔黄化病病原检测技术相结合得到病害普查数据，形成槟榔黄化病疫情数据库；通过GPS定位数据采集点信息，以WebGIS地理信息平台为基础，基于电子地图展示网格能力数据，提供各类分析资源数据的分布图和渲染图，为共享平台提供资源能力视图；利用JavaScript语言动态更新客户端的信息，通过Web浏览器显示信息，动态表现疫情覆盖范围，疫情扩散情况；通过数据交换引擎、动态表单引擎、流程管理引擎等标准通用的功能模块，实现可订制、可配置的构建技术，建立海南省槟榔黄化病疫情监测平台。下一步将在全省槟榔主要种植区科学设立槟榔黄化病长期定位监测点，对黄化病发生动态开展实时监测，并通过此平台及时发布疫情信息。该平台的构建可有效地指导海南省槟榔黄化病的综合防控工作，从而保障海南槟榔产业的健康可持续发展。

关键词：槟榔黄化病；疫情监测；网络信息平台

* 基金项目：海南省重大科技计划项目（ZDZX2013019）、农业科技成果转化资金项目（2014GB2E200114）

** 第一作者：罗大全，研究员，研究方向为热带作物植原体病害；E-mail：luodaquan@163.com

板蓝根抗菌肽 Li-AMP1 的分离鉴定*

吴　佳**，董五辈***

（华中农业大学植物科技学院，武汉　430070）

摘　要：板蓝根（Isatis indigotica Fort）是一种传统的中药材，含有多种抗菌抗病毒物质，具有抗肿瘤、抗白血病等作用。抗菌肽（antibacterial peptides）是由生物自身合成的具有广谱抗性的小分子多肽，多数文献中趋向于称之为抗微生物肽（antimicrobial peptides）或肽抗生素（peptide antibiotics）。20 世纪 80 年代，瑞典科学家 Boman 等发现由北美天蚕产生的具有抑菌作用的多肽类物质-天蚕素（Cecropins）是第一个真正意义上的抗菌肽。目前，已有 1 200多种抗菌肽被陆续发现，其中，70 多种抗菌多肽的结构已被揭示，抗菌肽的范围和概念都得到了很大的扩展。抗菌肽对细菌有显著的抑制和杀伤作用，尤其是对那些抗药性病原菌的杀灭作用引起了广泛的重视。本课题组以中草药菘蓝为实验材料，经过对其蛋白分离纯化抗性测试发现一个候选抗菌基因。经过测序及其生物信息学分析表明该基因序列全长为 51bp（含终止密码子），编码 16 个氨基酸。该序列在 NCBI（National Center for Biotechnology Information）核酸数据库和蛋白数据库中都检索比对不到任何结果，是一个新的有抗菌活性的短肽，属于抗菌肽类，并将其命名为 Li-AMP1。我们利用枯草芽胞杆菌表达系统对 Li-AMP1 进行了克隆表达，表达产物对多种病原细菌如青枯病原菌、番茄溃疡病原菌、水稻细菌性条斑病菌等都有抑菌作用。表达产物抑菌试验还表明，Li-AMP1 加上组氨酸标签构建的融合蛋白则失去了抑菌作用。Li-AMP1 的作用靶标及其作用机制研究正在进行中。

关键词：板蓝根；抗菌基因；抗菌肽；枯草芽胞杆菌表达系统

---

* 基金项目：农业部转基因生物新品种培育科技重大专项，抗除草剂、抗纹枯病等转基因育种新材料获得（课题编号：2016ZX08003-001）

** 第一作者：吴佳，2014 级硕士研究生，分子植物病理学

*** 通讯作者：董五辈；E-mail：dwb@ mail. hzau. edu. cn

不同除草剂对黄芪田间杂草封闭处理的防效研究*

王丽婷[1]，赵莉霞[1]，史　娟[1**]，王　俊[2]，黄小灵[3]，辛学发[4]，米银花[4]
（1. 宁夏大学农学院，银川　750021；2. 宁夏大学民族预科学院，银川　750002；
3. 隆德县国隆中药材科技有限公司，4. 隆德县科技局，隆德　756500）

摘　要：为了筛选出对黄芪出苗和生长安全的播后苗前土壤封闭处理除草剂，本文探讨不同化学除草剂对隆德县黄芪田间主要杂草的封闭处理效果。采用田间小区法，选用施田补、嗪草酮和乙氧氟草醚3种药剂在中药材作物黄芪田间进行播后苗前土壤封闭处理。结果表明：嗪草酮对供试黄芪出苗和生长的影响较小，安全性最高；施田补对黄芪出苗没有影响但是对幼苗有一小部分药害，20d后药害逐渐解除；且施田补起到一定程度的增产作用。施田补和嗪草酮对黄芪田间杂草的总鲜重防效达到82%以上，乙氧氟草醚对部分杂草防效达80%以上，均与对照有显著的差异。

关键词：除草剂；土壤封闭处理；除草效果

隆德县位于六盘山西麓、宁南边陲。依托六盘山，自古以来隆德县就以中药材而闻名全国。近年来，隆德县把中药材产业作为调整农村产业结构、增加农民收入、发展县域经济的主要途径，列入三大支柱产业[1]。但在大田种植中药材时，发现杂草对于中药材种植的影响非常严重，因此，寻找方法控制中药材田间杂草以提高中药材种植的收益是现如今摆在当地亟需解决的难题[2-3]。目前，化学除草在农业种植中的运用越来越广泛，市场上存在着大量不同类型的除草剂，其作用机理、使用方式、防除对象和应用作物范围各不相同[4-5]。为了能够合理、最大限度的利用优良化学除草剂，充分发挥其除草作用，我们选用了3种田间常用化学除草剂，在黄芪田间进行了杂草封闭处理试验。

一直以来，很多科学工作者都在为中药材行业蓬勃发展贡献各自的力量，但是，对黄芪化学除草的研究不多，孙立晨等先后试验了异恶草松、咪哇乙烟酸对黄芪安全无害[6]。贾永、金晓华等在黄芪田对氟乐灵、拿扑净、拉索、利谷隆和灭草松等进行了安全性测定[7]。近年来，甘肃、陕西和内蒙古等地在黄芪田除草剂筛选及使用剂量上已有一定的研究报道[8]，但限于药材种植地域不同，田间杂草种类和草相的差异，以及除草剂较强的地域性等特点，这些研究结果并不能在隆德县有效的应用。因此，我们希望借此实验筛选出最优的药剂类型及剂量能够帮助当地农民进行中药材的种植并促进当地中药材产业的发展。

* 基金项目：自治区重点攻关项目（宁夏六盘山优势中药材规范化种植关键技术研究与示范）；2013宁夏科技支撑计划项目六盘山中药材规范化种植病虫害草害综合防控及连作障碍解除关键技术研究与示范

** 通讯作者：史娟，山东鄄城人，教授，博士，主要从事草地植物保护基础理论和应用技术方面的研究；E-mail：shi_ j@nxu. edu. cn

1 材料与方法

1.1 除草剂

金喹·嗪草酮：有效成分金喹 4% - 嗪草酮 26%，剂型乳油（Metribuzin），大连松辽化工公司；乙氧氟草醚（Fluoroglycofen）：有效成分乙氧氟草醚，含量 240g/L，剂型乳油，美国陶氏益农公司；

施田补（Pendimethalin）：有效成分二甲戊灵，含量 330g/L，剂型乳油，江苏龙灯化学有限公司。

1.2 作物品种

黄芪（*Astragalusmembra-naceus*）。

1.3 防治对象

隆德县中药材田间杂草有别于其他地区以禾本科杂草为主，该地区中药材田间杂草种类主要为藜（*Chenopodium album*）、苣荬菜（*Sonchus brachyotus* DC.）、打碗花（*Calystegia hederacea* Wall.）、野荞麦（*Fagopyrum esculentum* Moench）、小蓟（*Cirsium setosum*）、黄蒿（*Artemisia annua* Linn）等阔叶杂草以及部分禾本科杂草[9]。

1.4 田间试验设计与方法

黄芪地每种除草剂用 3 种浓度梯度进行处理，同时设空白对照，共计 3 个处理。随机区组排列，小区面积为 $20m^2$，3 次重复见表 1。试验设在隆德县神林乡庞庄。

黄芪试验地土壤为黄绵土，土壤有机质含量约 15.3%，pH 值 8.38 左右（数据来源于隆德县农技推广站）。机械旋耕后平整土地，于 2013 年 5 月 12 日播种，株行距为 10cm，条播方式播种。播种后覆土 2cm 左右，用铁锹将土面拍平。5 月 13 日进行土壤封闭处理。

施药方法为播后苗前土壤喷雾。施药器械为电动喷雾器。要求均匀一次性喷雾，不可重复踩踏施药面。

表 1 黄芪播后苗前除草剂处理浓度

供试除草剂	处理	公顷施药量	注意事项
施田补	T1	3 000mL	保证土壤平整，避免大土块或者植物的残渣，以免影响药效。避免浇灌及大雨天气
	T2	3 750mL	
	T3	4 500mL	
乙氧氟草醚	T1	450mL	为触杀型除草剂，喷药时要求均匀周到，施药剂量要准
	T2	600mL	
	T3	750mL	
嗪草酮	T1	825g	药效受土壤类型、有机质含量、湿度温度等影响较大。主要对一年生杂草和部分阔叶杂草有效
	T2	975g	
	T3	1125g	
CK	清水		

1.5 调查与计算方法

1.5.1 黄芪安全性调查

药后 15d、20d 和 30d 后观察黄芪出苗是否出现药害情况，并记录药害症状和程度。施药后 30d，每个小区随机选定 3 点，调查和统计每点 30cm ×3 行黄芪出苗数，施药后 50d，每小区各随机挖取30株黄芪幼苗观察是否有药害症状。施药后50d，每小区各随机挖取30株黄芪幼苗，测量株高和根长，称量鲜重，以确定试验药剂对黄芪生长及药害情况的影响。

1.5.2　黄芪生长状况调查

于成药收获期，从各试验小区随机挖取30株黄芪成药，测量根长、根头粗、株鲜重，并称量各试验小区的成药产量，以确定试验药剂对黄芪成药根系形态及产量的影响。

1.5.3　杂草调查

喷药后50d进行田间杂草种类、株数和鲜重调查。每小区用$1m^2$的竹框随机调查1点，分别计算各种杂草株防效和总鲜重防效；杂草调查方法采用绝对值调查法，每小区随机选取3点，每点$0.25m^2$（0.5m×0.5m）。采用SAS v8.01统计分析软件进行数据统计分析。计算公式：

株防效（%）＝（对照区杂草株数－施药区杂草株数）/对照区杂草株数×100

鲜重防效（%）＝（对照区杂草鲜重－施药区杂草鲜重）/对照区杂草鲜重×100

2　结果与分析

2.1　不同除草剂对黄芪安全性测定

表2　不同药剂处理对黄芪出苗和生长的影响以及对杂草的防效

供试除草剂	公顷施药量	出苗率（%）	根长（cm）	株高（cm）	鲜重（g/株）	杂草鲜重（kg）	防效（%）
施田补	3 000mL	82.5	9.11 abc	5.27 a	0.33 a	0.102	73.10 ab
	3 750mL	81.8	7.95 c	5.10 ab	0.31 a	0.084	77.87 ab
	4 500mL	81.2	8.51 bc	4.75 ab	0.30 a	0.055	85.45 ab
乙氧氟草醚	450mL	82.1	8.45 bc	4.66 ab	0.32 a	0.150	60.32 b
	600mL	69.3	8.72 abc	4.37 b	0.27 a	0.065	82.80 ab
	750mL	68.5	9.41 ab	4.72 ab	0.27 a	0.067	82.41 ab
嗪草酮	825g	79.2	9.77 a	5.34A	0.30 a	0.090	76.19 ab
	975g	69.2	9.73 a	5.13AB	0.32 a	0.080	78.84 an
	1 125g	83.2	9.43 ab	4.98AB	0.29 a	0.020	94.71 a
CK	清水	82.5	9.26 ab	5.04AB	0.35 a	0.378	

由表2可知，对照出苗率为82.5%，供试除草剂不同使用剂量处理土壤黄芪出苗率为69.2%～82.5%不等，其中，施田补3种浓度处理黄芪出苗率与对照接近，即不影响黄芪出苗；乙氧氟草醚低浓度处理组出苗率接近对照组，无影响，随施用浓度的升高出苗率低于对照组，即一定程度上抑制出苗；嗪草酮低浓度和高浓度处理组出苗率接近对照组，而中浓度处理组则表现出一定程度的抑制作用。观察发现3种浓度对黄芪子叶产生药害，表现为子叶边缘黄化，受害率为5%左右，真叶略有徒长现象，随着黄芪的生长，子叶逐渐恢复正常，真叶生长正常。随着黄芪进入三叶一心期，也就是施药后40d左右，供试各除草剂不同试验小区有5%左右的黄芪复叶出现轻微药害现象，复叶叶片略有收缩，以施田补最严重，表现为复叶和心叶皱缩，严重的心叶卷曲为团状，土壤2cm根茎出现肿大，有的出现缢缩，受害程度随药剂浓度的增加而增大。

嗪草酮各浓度处理组根长均大于对照组，即对黄芪苗期根长的生长具有促进作用，尤以低浓度处理组效果明显；施田补低浓度处理组根长量接近对照组，作用不明显，中浓度和高浓度处理组则出现一定的抑制作用；乙氧氟草醚处理组表现为抑制作用。嗪草酮和施田补低浓度和中浓度处理组在黄芪株高方面优于对照组，高浓度处理组效果低于对照组，且随施用浓度的增加，株高量呈下降趋势；乙氧氟草醚处理组株高明显低于对照组，即起抑制株高生长的作用。3种除草剂处理组的黄芪鲜重均低于对照组，但差异不明显，表明对鲜重影响不明显。

2.2 不同除草剂对黄芪生长的影响

表3 不同药剂处理对黄芪成药根系形态及产量的影响

供试除草剂	处理	公顷施药量	根长（cm）	根头粗（mm）	鲜重（g/株）	产量（kg/亩）
施田补	T1	3 000mL	25.7 ±3.75 ab	5.41 ±0.42 b	8.02 ±1 a	710 ±3.92 a
	T2	3 750mL	29.8 ±2.55 ab	6.43 ±0.87 ab	11.7 ±3.31 a	718 ±7.31 a
	T3	4 500mL	32.1 ±7.81 b	7.13 ±1.01 ab	11.3 ±1.96 a	731 ±10.97 a
乙氧氟草醚	T1	450mL	19 ±2.88 a	5.09 ±0.71 b	7.53 ±1.27 a	630 ±12.1 b
	T2	600mL	20.9 ±1 a	7.12 ±0.75 b	9.3 ±0.95 a	375 ±8.66 f
	T3	750mL	21.4 ±6.18 a	5.90 ±0.91 ab	8.65 ±1.71 a	394 ±15.3 ef
嗪草酮	T1	825g	23.0 ±2.71 a	6.98 ±0.92 b	9.25 ±0.84 a	460 ±11.16 c
	T2	975g	20.4 ±1.46 a	6.63 ±1 a	7.32 ±2.96 a	431 ±15.3 cd
	T3	1125g	22.8 ±3.6 a	7.44 ±0.13 ab	9.37 ±1.65 a	425 ±8.14 cd
CK	清水		22.9 ±0.8 a	6.30 ±0.56 a	8.7 ±0.28 a	640 ±5.77 b

由表3可知，各处理组亩产量都较对照组差异显著。其中施田补3个处理组的产量分别为710kg、718kg和731kg，都显著高于对照组的产量，即起增产作用；乙氧氟草醚低浓度处理组产量较对照组差异不显著，中浓度和高浓度处理组均低于对照组，即起减产作用；嗪草酮3个处理组的产量都显著低于对照组，即起减产作用。根长数据中，只有施田补3个处理组数据相对于对照组差异显著，且优于对照；乙氧氟草醚和嗪草酮处理后的根长相对于对照组差异不显著。根头粗数据中施田补3个处理组均较对照组差异显著，且随浓度升高跟头变粗，其中，低浓度处理组根头粗低于对照组；乙氧氟草醚3个处理组根头粗较对照差异显著，仅中浓度处理组显著粗于对照组；嗪草酮低浓度和高浓度处理组较对照差异显著，且均优于对照。黄芪株鲜重数据中，所有处理组相对于对照组差异不显著，即各除草剂对黄芪成药株鲜重无影响。

2.3 不同除草剂对各类主要杂草土壤封闭防除的效果

表4 不同药剂对黄芪田间主要杂草防治效果

供试除草剂	公顷施药量	荠菜		刺盖		苦菊		灰藜		打碗花		蒿		稗草		西伯利亚蓼		反枝苋	
		株数	防效（%）	株数	防效（%）	株数	防效（%）	株数	防效（%）	株数	防效（%）	株数	防效（%）	株数	防效（%）	株数	防效（%）	株数	防效（%）
嗪草酮	3 000mL	1	90.9	0	100	13	53.6	3	83.3	6	33.3	0	100	10	72.2	4	42.8	1	90.9
	3 750mL	3	75	5	66.7	9	64.3	3	83.3	3	66.6	3	88	8	77.8	3	57.1	1	90.9
	4 500mL	0	100	0	100	0	100	0	100	0	100	0	100	0	100	0	100	0	100
施田补	450mL	12	0	5	66.7	7	75	3	83.3	2	77.8	21	16	6	83.3	3	57.1	4	63.6
	600mL	7	41.7	6	60	4	85.7	0	100	0	100	2	92	2	94.4	2	71.4	0	100
	750mL	10	16.7	1	93.3	2	92.8	2	88.9	0	100	0	100	1	97.2	2	71.4	7	36.4
乙氧氟草醚	825g	7	41.7	2	86.7	20	28.6	9	50	0	100	32	0	5	86.1	3	57.1	0	100
	975g	2	83.3	4	73.3	2	92.8	5	72.2	0	100	1	96	8	77.8	1	85.7	1	90.9
	1 125g	4	66.7	5	66.7	4	85.7	7	61.1	0	100	2	92	4	88.9	2	71.4	0	100
CK	清水	12		15		28		18		9		25		36		7		11	

由表4可知，在黄芪田中嗪草酮对灰藜、蒿类和反枝苋的防效均在83.3%以上，均明显高

于对照。对于荠菜和稗草的防效也在72%以上，也明显高于对照。对苦菊、打碗花和西伯利亚蓼随着药剂的浓度增高防效增加，均最高浓度的时候防效最好，明显高于对照。施田补对苦菊、灰藜、打碗花和稗草的防效都超过75%，与对照相比差异显著。对刺盖需用高浓度效果显著，蒿类需用中、高浓度效果显著。乙氧氟草醚对打碗花、稗草和反枝苋的防效均超过了80%，与对照相比差异显著。对苦菊、蒿类和西伯利亚蓼只有在中高浓度防效显著。由表2可知，试验田杂草鲜重防效中施田补高浓度处理组、乙氧氟草醚中浓度处理组和高浓度处理组、嗪草酮高浓度处理组的防效都超过了82%，其中，嗪草酮高浓度处理组防效最好，高达94.71%。

3 结论与讨论

试验结果表明，在黄芪田中乙氧氟草醚对黄芪出苗率有一定的影响，且对黄芪苗期子叶产生一定的药害，但不影响黄芪的生长。施田补对黄芪复叶产生药害，低剂量药害影响小。嗪草酮对黄芪出苗和生长基本没有影响。

从试验田中主要杂草的防治效果上可以看出，在黄芪田中嗪草酮对灰藜、蒿类、反枝苋、荠菜和稗草的防效明显，且对苦菊、打碗花和西伯利亚蓼在最高浓度的时候防效最好。施田补对苦菊、灰藜、打碗花和稗草的防效优良，对刺盖需用高浓度效果显著，蒿类需用中、高浓度效果显著。乙氧氟草醚对打碗花、稗草和反枝苋的防效明显，对苦菊、蒿类和西伯利亚蓼只有在中高浓度防效显著。

从各试验田黄芪成药产量可得出，施田补处理起到一定程度的增产作用，乙氧氟草醚低浓度处理对产量无影响，乙氧氟草醚中浓度处理和高浓度处理及嗪草酮处理组则起到减产作用。施田补处理组和嗪草酮低浓度处理组促进成药根长的生长，施田补低浓度处理组及乙氧氟草醚低浓度和高浓度处理组抑制成药根头的生长，施田补低浓度处理组、乙氧氟草醚低浓度和高浓度处理组及嗪草酮中浓度处理组抑制成药鲜重的增长。

本试验结果初步筛选出了对黄芪出苗和生长比较安全，且对杂草具有一定防除效果的除草剂单剂，这为进一步深入研究化学除草剂除草活性和使用技术提供了前期研究结果。但仍需对筛选药剂的施用剂量、不同土壤有机质的施用剂量、安全施用方法以及除草剂高效减量的复配技术，以及对黄芪残留药害等开展系列研究，阐明这些问题，对进一步明确使用药量和施药方法，明确除草剂安全施用的必要条件和施用范围，建立对环境和中药材植物安全的低风险除草剂应用技术，以及建立黄芪田间杂草除草剂应用技术规程提供依据。

农田化学除草在我国农业生产中正发挥着越来越重要的作用。尽管已有黄芪田间杂草化学除草剂应用技术的研究报道[8]，但由于黄芪生长地域条件的差异、气候条件的不同，更重要的是田间杂草种类、群落组成特点以及草相的不同，导致中药材田间杂草的复杂程度不同。这也是限制中药材田间杂草化学除草技术应用、推广以及普及的因素之一。由于因化学除草技术具有较强的地域性和技术性，受当地土壤条件、气候状况以及土壤有机质等因素的影响较大，必须根据隆德县气候特点和土壤条件，进行除草剂田间活性筛选试验、对药材的安全性评价[10]，全面掌握使用技术后方可推广和使用。

隆德县中药材产业发展面临着田间杂草种群变化、群落演替加速，难治、恶性农田杂草和多年生杂草日趋加重的严峻现实，开展一次性化学除草配套技术与生态调控技术和人工防除有机结合的综合治理关键技术为产业之急需。本研究结果仅仅是一个初步的试验结果，后续相关研究有待进行。

参考文献

[1] 李巧弟，张密珍，杜巧霞. 隆德县中药材产业发展现状与问题 [J]. 宁夏农林科技，2007 (4)：64－65.

[2] 王亮，王丽丽，吴明根．当归适宜栽培化学除草剂筛选［J］．延边大学农学学报，2008，30（1）：46－51.
[3] 张忠民，谢柏龄，张远义．桔梗田间杂草化学防治试验［J］．山西农业科学，2001（10）：7－9.
[4] 张玉聚，孙建伟，王全德，等．中国除草剂应用技术大全［M］．北京：中国农业科学技术出版社，2009，12－13.
[5] 王宏富，韩忻彦．中国农田杂草可持续治理的现状与展望［J］．山西农业大学学报，2002，3：274－277.
[6] 孙立晨，何娟．黄立坤，等，董世臣．咪唑乙烟酸在豆科作物间作田的除草效果及其对豆科作物生长的影响［J］．中国植保导刊，2010，30（3）：39－42.
[7] 金晓华，丁建云，杨建国，等．应用25%灭草松水剂防除黄芪苗期双子叶杂草［J］．植物保护，2002，28（4）：53－54.
[8] 孙立晨，董世臣，黄立坤，等．几种除草剂在豆科作物田除草效果及安全性测定［J］．大豆科学，2009，28（5）：931－934.
[9] 史娟，王俊，辛学发，等．宁夏隆德县中药材田间杂草为害现状及防治对策［J］．农业科学研究，2014（2）：18－20.
[10] 王兆振，毕亚玲，丛聪，等．除草剂对作物的药害研究［J］．农药科学与管理，2013，34（5）：68－73.

宁夏中药材产区黄芪主要病虫草害种类及发生趋势

史　娟[1]，任　斌[1]，李文强[1]，黄小灵[3]，王　俊[2]，辛学发[4]，米银花[4]
（1. 宁夏大学农学院，银川　750021；2. 宁夏大学民族预科学院，银川　750002；
3. 隆德县国隆中药材科技有限公司；4. 隆德县科技局，隆德　756500）

摘　要：黄芪为豆科多年生草本植物，以根入药，是大宗药材之一。近年来，随着中药材野生资源日益枯竭以及国内外对中草药日益增长的需求增加，国内道地产区大力开展了黄芪的人工栽培及基地建设。宁夏隆德县地处黄土高原西部，土壤类型和气候条件适宜多种中药材的生长，是宁夏自治区人民政府确定的重点“优质中药材基地”。黄芪（*Astragalus mongholicus*）作为我国大宗中药材种，是目前隆德县中药材产业发展中主要的种植品种之一。随着人工栽培面积的增加和基地建设的扩大，黄芪病虫害草为害逐年上升成为制约黄芪产业发展和安全生产的重要问题。2012—2015 年期间，对宁夏六盘山地区种植的地产大宗药材黄芪病虫害草害发生进行了系统调查。结果表明，宁夏回族自治区黄芪生产中为害严重的主要病虫害有 7 种，其中，病害 3 种，害虫 4 种，杂草主要以苦菊、大小蓟、灰黎、苦荞麦等阔叶杂草为主。为害严重的病害主要是白粉病和霜霉病，对黄芪育苗和移栽生产影响较大，黄芪育苗生产中为害较严重的害虫有麦蚜、蒙古灰象甲、麦秆蝇和金龟甲，病害有白粉病、霜霉病和根腐病；黄芪药材移栽生产中危害严重的病虫害有豆蚜、白粉病和霜霉病、根部斑点病（暂定名）。田间恶性杂草主要是苦菊、大小蓟、苦荞麦和灰黎等阔叶杂草，禾本科杂草主要以冰草为主。其中根部斑点病（暂定名）（文献报道为根腐病）发病程度与重迎茬、连作种植密切相关，尤其是随着连作年限的增加发病呈递增趋势，表现为病斑数量的增加，严重时根组织皮层上的病斑数可达 200 个以上。是目前黄芪规范化生产中难以控制的主要病害之一。

关键词：中药材；黄芪；病虫草害

Response of Fungal Communities in Watermelon Basal Stems, Roots, and Rhizosphere to Different Fusarium-resistant varieties

XU Li-hui, ZENG Rong, GAO Shi-gang, DAI Fu-ming*

(*Institute of Eco-Environmental Protection, Shanghai Academy of Agricultural Sciences, Shanghai* 201403, *China*)

Abstract: Dynamics of microbial community structure including pathogens and non-pathogens can influence the development of watermelon soil-borne disease. The use of resistant watermelon variety is the main effective measure for controlling watermelon wilt, however, the impact of different resistant varieties on watermelon stem, root, and rhizosphere microbial communities diversity is limited. Characterization of fungal communities associated with healthy and diseased watermelon including three different Fusarium-resistant varieties were investigated through paired-end Illumina MiSeq sequencing in three spatial compartments: basal stems, roots, and rhizosphere. Thirty watermelon plants with high-resistant, moderate-resistant, and suceptible varieties were collected from the same field, half of which showing clear wilt symptoms. Species richness was highest in rhizosphere and lower in basal stems and roots, however, no notable differences in richness were observed between samples associated with diseased and healthy plants. Health status and variety both had significant effects on fungal community structures in basal stems and roots, whereas only variety had significant effects on communities in rhizosphere. A number of fungi were identified more abundant in the three compartments with high-resistant variety. Pathogens such as *Fusarium oxysporum* and *Monosporascus cannonballus* were abundant in diseased stems and roots with susceptible and moderate-resistant varieties. In conclusion, patterns of watermelon wilt disease were different among the three varieties, which were also reflected in fungal communities. Health status of plants was barely reflected in rhizosphere fungal communities, whereas health status was more important for shaping basal stem and root communities.

Key words: Fungal community; Watermelon wilt disease; Resistant variety

* Corresponding author: DAI Fu-ming; E-mail: fumingdai@163.com

不同耕作方式对我国主要玉米种植区土壤真菌种类及数量的影响*

赵丽琨[1**]，肖淑芹[1]，刘　畅[1]，刘雨佳[1]，薛春生[1***]，陈　捷[2***]
（1. 沈阳农业大学植物保护学院，沈阳　110866；
2. 上海交通大学农业与生物学院，上海　200240）

摘　要：我国玉米主产区大力推广深松和秸秆还田等保护性耕作技术，改善土壤结构，提升土壤蓄水抗旱的能力，提高玉米对水分的利用率和土壤肥力进而达到增产增收的效果。土壤微生物一方面可以促进作物生长发育，对土壤的形成发育、物质循环和肥力演变等均有重大影响，另一方面又影响着土传病害的发生。本研究对我国主要玉米种植区土壤真菌进行了分离和鉴定，明确不同耕作方式下玉米田土壤真菌种群结构和数量，为在我国玉米主要种植区实施深松和秸秆还田等保护性耕作措施提供依据。

从玉米田土壤中分离鉴定出镰孢菌属（*Fusarium*）、腐霉属（*Pythium*）、根霉属（*Rhizopus*）、弯孢菌属（*Curvularia*）、刺盘孢属（*Colletotrichum*）、枝孢属（*Cladosporium*）、葡萄孢属（*Botrytis*）、丝核菌属（*Rhizoctonia*）、链格孢属（*Alternaria*）、木霉属（*Trichoderma*）、肉座菌属（*Hypocrea*）、粘帚霉属（*Gliocladium*）、绿僵菌属（*Metarhizium*）、青霉属（*Penicillium*）、曲霉属（*Aspergillus*）、毛霉属（*Mucor*）、被孢霉属（*Mortierella*）、接霉属（*Zygorhynchus*）和毛壳菌属（*Chaetomium*）等真菌。

深松和秸秆还田处理的土壤真菌种群密度高于常规垄作，随着处理年限的增加，深松和秸秆还田处理的潜在致病菌种群密度降低，有益菌数量显著增加，中性菌数量波动较大。

关键词：耕作方式；玉米；土壤；真菌

感谢黑龙江哈尔滨、吉林洮南、辽宁沈阳、内蒙古自治区蒙西、河北邯郸、河南漯河、山东济宁、浙江东阳、陕西咸阳、四川绵阳、广西壮族自治区南宁、云南曲靖、山西忻州等玉米产业技术体系综合试验站的大力支持。

* 基金项目：国家现代农业（玉米）产业技术体系（CARS－02）
** 第一作者：赵丽琨，在读硕士研究生，从事玉米病害研究
*** 通讯作者：薛春生；E-mail：cshxue@ sina. com
陈捷；E-mail：jiechen59@ sjtu. edu. cn

转录因子和蛋白互作在同一个实验流程中的系统筛选*

汤　旋**，史军伟，董五辈***
（华中农业大学植物科技学院，武汉　430070）

摘　要：基于对酵母单杂交和双杂交的整合改进，我们建立了一项 One-for-All 酵母文库杂交技术。本技术可在一个实验流程中同时系统高效筛选全基因组中的转录因子和蛋白互作。在典型的转录因子中，DNA 结合结构域和激活结构域是共价相连的。酵母单杂交是基于 DNA 结合结构域筛选特定转录因子的，无法同时筛选各种不同的转录因子。本技术是基于保守的激活结构域筛选转录因子的。我们的研究表明，在酵母的质粒表达系统中，各种各样的转录因子都能激活报告基因。转录因子的激活结构域在进化上是保守的，在功能上是非特异性的。转录因子功能的特异性主要是其 DNA 结合结构域决定的。转录因子的主要功能是激活 RNA 聚合酶促进基因表达。RNA 聚合酶在所有的生命体里都是非常保守的且种类极少。RNA 聚合酶这种高度保守的特性也决定了转录因子的激活结构域的保守性和非特异性。我们证明了转录因子的这种特性并利用这种特性在基因组水平上系统筛选了所有可能的转录因子。对于蛋白互作的筛选，我们进行了文库对文库的杂交，系统筛选了可能互作的蛋白。本研究整合了转录因子与蛋白互作的筛选，使二者的筛选融入到同一个实验流程中。

关键词：转录因子；蛋白互作；系统筛选

---

* 基金项目：农业部转基因生物新品种培育科技重大专项，抗除草剂、抗纹枯病等转基因育种新材料获得（课题编号：2016ZX08003-001）

** 第一作者：汤旋，硕士，分子植物病理学
*** 通讯作者：董五辈，E-mail：dwb@ mail. hzau. edu. cn

拟南芥对烟草白粉菌侵入后抗性调控网络分析*

李　冉**，张凌荔，赵志学，王宇秋，王　贺，
樊　晶，李　燕，杨雪梅，王文明***
（四川农业大学水稻研究所，成都　611130）

摘　要：植物对白粉菌的抗性可简单分为侵入前抗性和侵入后抗性两类。从田间分离得到的烟草白粉菌 *Golovinomyces cichoracearum* SICAU1 对拟南芥具有弱致病力，能完全克服侵入前抗性，但其繁殖受植物侵入后抗性限制，而拟南芥基础抗性和 SA 信号通路双突变体 *pad*4-1/*sid*2-1 植株中侵入后抗性完全丧失。与野生型拟南芥 Col-0 相比，*pad*4-1/*sid*2-1 双突变体在高感白粉菌情况下能较长时间保持叶片新绿。接菌后 5 天，分生孢子梗的数量在 *pad*4-1/*sid*2-1 双突变体和野生型 Col-0 间有显著性差异；接菌后 8 天，双突变体叶片上即长出大量白粉菌，而野生型拟南芥开始出现可见坏死斑点。台盼蓝染色结果表明，野生型 Col-0 有白粉菌的部位出现大量细胞死亡，而双突变体 *pad*4-1/*sid*2-1 则有大量分生孢子，未观察到细胞死亡。定量 PCR 分析结果表明，接种白粉菌后野生型拟南芥 Col-0 主要是 SA 信号通路和基础抗性基因明显上调，衰老相关基因也比上调多；而双突变体 *pad*4-1/*sid*2-1 中 SA 信号通路基因被明显抑制，JA 和 ET 信号通路基因上调。差异转录组分析结果表明，接种白粉菌后野生型拟南芥 Col-0 主要是 SA 信号通路，而在双突变体 *pad*4-1/*sid*2-1 中营养运输相关基因被明显上调。根据基因表达模式的差异我们将基因进行了分类，其中在 Col-0 中明显上调而在 *pad*4-1/*sid*2-1 中基本不变的主要为 SA 信号通路相关基因；在 Col-0 中明显上调而在 *pad*4-1/*sid*2-1 中轻微上调的主要为不依赖 SA 信号通路相关基因；在 Col-0 中几乎不变而 *pad*4-1/*sid*2-1 中明显上调的则多数为营养运输相关基因，且这些基因的上调时间多数为接种后 6 天，可能对于后期维持植物的生长有关。这些结果说明拟南芥对白粉菌的侵入后抗性主要依赖于 SA 和营养运输相关基因，对认识白粉菌侵入后抗性的分子机理具有意义，并为进一步研究白粉菌病原与寄主互作机制打下了基础。

关键词：白粉病；侵入后抗性；SA 信号；JA 信号；ET 信号

---

* 基金项目：国家自然科学基金（31371931 和 31471761）

** 第一作者：李冉，硕士研究生，主要研究植物白粉病抗性的分子机制；E-mail：1634457605@qq.com

*** 通讯作者：王文明，研究员，主要从事植—病原菌相互作用机制研究；E-mail：j316wenmingwang@163.com

对虾下脚料碱性发酵物对香蕉枯萎病菌抑菌测定*

刘月廉**，吕庆芳，林巧玲，张德涛
（广东海洋大学农学院，湛江 524088）

摘 要：通过对对虾下脚料进行碱性调制，发酵后获得 pH 值为 8.21 的碱性产物。对该发酵产物分别设置原液、10 倍液、50 倍液、100 倍液及 200 倍液 5 个浓度梯度，以自然发酵物原液为对照，经平板培养，测定其对香蕉枯萎病菌 4 号小种的抑制效果。结果表明，碱性发酵物的 5 个浓度均对香蕉枯萎病菌有极显著或显著的抑制作用，其中原液和 10 倍液对香蕉枯萎病菌的抑制率分别为 78.51% 和，71.98%，达极显著水平（$P<0.01$）；50 倍液、100 倍液和 200 倍液的抑制率分别为 51.43%，39.58% 和 26.56%，达显著水平（$P<0.05$）。

关键词：香蕉枯萎病；对虾；下脚料；碱性发酵；抑制

---

* 基金项目：广东省科技厅科技计划项目（2014A020208119）

** 第一作者及通讯作者：刘月廉，广东雷州人，博士，教授，从事植物病理学研究；E-mail：mushwoman@126.com

Biological Control, Growth Promotion, and Host Colonization of European Horticultural Plants by Endophytic *Streptomyces* spp. *

Xiaoyulong Chen[1], Maria Bonaldi[1], Armin Erlacher[2], Andrea Kunova[1], Cristina Pizzatti[1], Marco Saracchi[1], Gabriele Berg[2], Paolo Cortesi[1] **

(1. *Department of Food, Environmental and Nutritional Sciences (DeFENS), Università degli Studi di Milano, Via Giovanni Celoria* 2, 20133 *Milano, Italy*;
2. *Institute of Environmental Biotechnology, Graz University of Technology, Petersgasse* 12, 8010 *Graz, Austria*)

Abstract: Yield losses caused by phytopathogens should be minimized to maintain the food quality and quantity for the demand of massively growing human population. At the same time, yield limitation due to soil fertility and nutrition deficiencies add extra constraints to plant production. Thus, searching for sustainable solutions to suppress phytopathogens, as well as to increase the yield is gaining high public interests. Plant root systems are colonized by vast amounts of microbes, some of which facilitate biological control and plant growth promotion activities. *Streptomyces*, abundant in soil, are a group of filamentous bacteria producing a variety of beneficial secondary metabolites, gifting them the potential to be developed as bio-pesticides and bio-fertilizers. We labeled two bioactive *Streptomyces* strains with EGFP marker to investigate their interactions with lettuce using confocal laser scanning microscopy (CLSM), and evaluated their biocontrol activities against *Sclerotinia sclerotiorum* on lettuce, as well as PGP activities on several economically important horticultural plants . Additionally, we performed scanning electron microscopy (SEM) observations to verify the endophytic colonization of lettuce by *Streptomyces*.

The abundant colonization of young lettuce seedling by two *Streptomyces* strains demonstrated their capability to interact with the host from early stages of seed germination and root development up to two weeks. Plant-strain specific PGP activity was observed; e. g. , *S. cyaneus* ZEA17I promoted the growth of lamb lettuce but not that of tomato. When were applied to *S. sclerotiorum* inoculated substrate in growth chamber, *S. exfoliatus* FT05W and *S. cyaneus* ZEA17I significantly reduced lettuce basal drop incidence by 44.8% and 27.6%, respectively, compared to the inoculated control ($P < 0.05$) . Interestingly, under field conditions, *S. exfoliatus* FT05W reduced the disease incidence by 40%, and there was only 10% protection of lettuce after *S. cyaneus* ZEA17I application. Our results indicate the greatly promising potential of *Streptomyces* for exploitations in agriculture as biocontrol agents.

---

* Acknowledgement: The authors thank Prof. Mervyn Bibb (John Innes Centre, UK), for kindly providing the plasmid pIJ8641, as well as prof. Flavia Marinelli (University of Insubria, Italy), for donor strain *E. coli* ET12567 for *Streptomyces* transformation.

** Corresponding author: E-mail paolo. cortesi@ unimi. it

Transgenic Rice Expressing *Chitinase* Specific dsRNAs Coferred Resistance to *Mythimna separata*

BAO Wen-hua[1 *], Hada Wuriyanghan[1 **]

(University of Inner Mongolia, Hohhot, China 010021)

Abstract: *Mythimna Separata* is major agricultural pests that cause significant yield losses of crop plants each year. Expression of double-stranded RNA (dsRNA) directed against suitable insect target genes in transgenic plants has been shown to give protection against pests through plant-mediated RNA interference (RNAi). Thus, as a potential effective strategy for insect pest management in agricultural practice, RNAi for *Mythimna separata* control has received close attention in recent years.

We propose here to develop *Mythimna separata*-resistant plants using RNAi technology. Our primary hypothesis is that if we can identify and deliver to *Mythimna separata* interfering RNAs that down-regulate expression of *Mythimna separata* genes such as *Chitinase* genes which are essential for their growth and development, they will result in a debilitating phenotype. To do this, we designed and carried out the following experiments. First, using Gateway cloning technology to construct Plant binary expression vector corresponding for interfering sequences against the target genes *chitinase*1 and *chitinase*2. Transgenic rice expressing the hairpin dsRNAs was verified by southern blot and RT-PCR. When the transgenic rice was fed to *mythimna separata*, target *Chitinase* genes transcript levels were substantially decreased, also resulting in retarded larval growth and increased mortality.

Key words: *Mythimna separata*; RNAi; *Chitinase* gene; transgenic rice

---

* First author: 包文化，研究生，工作于内蒙古大学生命科学学院

** Corresponding author: Hada Wuriyanghan; Email: nmhadawu77@ imu. edu. cn

De Novo Sequencing and Assembly of the Transcriptome for Oriental Armyworm *Mythimna separata* (Lepidoptera: Noctuidae)

LIU Ya-juan*, CHI Yu-chen, Hada WU riyanghan**

(*Inner Mongolia University*, *Hohhot*,, *China* 010021)

Abstract: Mythimna separata walker (Lepidoptera: Noctuidae) is a polyphagous pest of nearly100 families of more than 300 kinds of food and industrial crops. In recent years, outbreaks of M. separata in north China caused severe loss to the corn production. Although the representative species in other families of lepidopteran order have been sequenced at genomic or transcriptomic level, only several insect species have been sequenced at transcriptomic level in noctuidae family, although most of which are economically important crop pests. So far, only mitochondrial genome of *M. separate* has been sequenced, and both nucleotide and protein or EST sequence information is rarely available in database, strictly limiting molecular biology research in this insect species. In ours' study, we carried out a transcriptome sequencing for *M. separata*. The sequencing and subsequent bioinformatics analysis yielded 69, 238 unigenes, in which 45, 227 unigenes were annotated to corresponding functions by blasting with high homologous genes in database, giving annotation rate of 65. 32%. 26 selected unigenes encoding 19 different proteins were successfully amplified and expression profiles of these unigenes were analyzed at different life stages and in different tissues. The Unigenes we successfully amplified are proteins related to developmental and insecticide resistance. Our study provides most comprehensive transcriptome data for *M. separata* to date, and also provides reference sequence information for noctuidae family insects.

Key words: Transcriptome sequencing, *Mythimna separate*, Noctuidae, Unigenes

* First author: 刘亚娟，研究生，工作于内蒙古大学生命科学学院

** Correspondence: Hada Wuriyanghan; E-mail: nmhadawu77@ imu. edu. cn